CAMBRIDGE MONOGRAPHS ON APPLIED AND COMPUTATIONAL MATHEMATICS

14 Simulating Hamiltonian Dynamics

The *Cambridge Monographs on Applied and Computational Mathematics* reflects the crucial role of mathematical and computational techniques in contemporary science. The series publishes expositions on all aspects of applicable and numerical mathematics, with an emphasis on new developments in this fast-moving area of research.

State-of-the-art methods and algorithms as well as modern mathematical descriptions of physical and mechanical ideas are presented in a manner suited to graduate research students and professionals alike. Sound pedagogical presentation is a prerequisite. It is intended that books in the series will serve to inform a new generation of researchers.

Also in this series:

1. A Practical Guide to Pseudospectral Methods, *Bengt Fornberg*
2. Dynamical Systems and Numerical Analysis, *A. M. Stuart and A. R. Humphries*
3. Level Set Methods and Fast Marching Methods, *J. A. Sethian*
4. The Numerical Solution of Integral Equations of the Second Kind, *Kendall E. Atkinson*
5. Orthogonal Rational Functions, *Adhemar Bultheel, Pablo González-Vera, Erik Hendiksen, and Olav Njåstad*
6. The Theory of Composites, *Graeme W. Milton*
7. Geometry and Topology for Mesh Generation *Herbert Edelsfrunner*
8. Schwarz–Christoffel Mapping *Tofin A. Driscoll and Lloyd N. Trefethen*
9. High-Order Methods for Incompressible Fluid Flow, *M. O. Deville, P. F. Fischer and E. H. Mund*
10. Practical Extrapolation Methods, *Avram Sidi*
11. Generalized Riemann Problems in Computational Fluid Dynamics, *Matania Ben-Artzi and Joseph Falcovitz*
12. Radial Basis Functions, *Martin D. Buhmann*
13. Iterative Krylov Methods for Large Linear Systems, *Henk A. van der Vorst.*
15. Collocation Methods for Volterra Integral and Related Functional Equations, *Hermann Brunner.*

Simulating Hamiltonian Dynamics

BENEDICT LEIMKUHLER
University of Leicester

SEBASTIAN REICH
Imperial College, London

CAMBRIDGE UNIVERSITY PRESS
Cambridge, New York, Melbourne, Madrid, Cape Town, Singapore,
São Paulo, Delhi, Dubai, Tokyo, Mexico City

Cambridge University Press
The Edinburgh Building, Cambridge CB2 8RU, UK

Published in the United States of America by Cambridge University Press, New York

www.cambridge.org
Information on this title: www.cambridge.org/9780521772907

First published 2004

A catalogue record for this publication is available from the British Library

Library of Congress Cataloguing in Publication data

Leimkuhler, B.
Simulating Hamiltonian dynamics / Benedict Leimkuhler, Sebastian Reich.
p. cm. – (Cambridge monographs on applied and computational mathematics ; 14)
Includes bibliographical references and index.
ISBN 0 521 77290 7
1. Hamiltonian systems. I. Reich, Sebastian. II. Title. III. Series.

QA614.83.L45 2004
515′.39 – dc22 2004045685

ISBN 978-0-521-77290-7 Hardback

Contents

Preface

About geometric integration

This book is about simulating dynamical systems, especially conservative systems such as arise in celestial mechanics and molecular models. We think of the *integrator* as the beating heart of any dynamical simulation, the scheme which replaces a differential equation in continuous time by a difference equation defining approximate snapshots of the solution at discrete timesteps. As computers grow in power, approximate solutions are computed over ever-longer time intervals, and the integrator may be iterated many millions or even billions of times; in such cases, the qualitative properties of the integrator itself can become critical to the success of a simulation. *Geometric integrators* are methods that exactly (i.e. up to rounding errors) conserve qualitative properties associated to the solutions of the dynamical system under study.

The increase in the use of simulation in applications has mirrored rising interest in the theory of dynamical systems. Many of the recent developments in mathematics have followed from the appreciation of the fundamentally chaotic nature of physical systems, a consequence of nonlinearities present in even the simplest useful models. In a chaotic system the individual trajectories are by definition inherently unpredictable in the exact sense: solutions *depend sensitively on the initial data*. In some ways, this observation has limited the scope and usefulness of results obtainable from mathematical theory. Most of the common techniques rely on local approximation and perturbation expansions, methods best suited for understanding problems which are "almost linear," while the new mathematics that would be needed to answer even the most basic questions regarding chaotic systems is still in its infancy. In the absence of a useful general theoretical method for analyzing complex nonlinear phenomena, simulation is increasingly pushed to the fore. It provides one of the few broadly applicable and practical means of shedding light on the behavior of complex nonlinear systems, and is now a standard tool in everything from materials modeling to bioengineering, from atomic theory to cosmology.

As models grow in complexity and dimension, and the demands placed on simulation have risen, the need for more sophisticated numerical methods and analytic techniques also grows. Longer time interval simulations require more stable methods. Larger problems call for more efficient schemes, tailored to a particular application or family of applications. And more intricate modeling of delicate behaviors or properties requires corresponding improvements in the resolution of those properties during simulation.

In writing this book for a broad audience of scientists, we have attempted to limit the introduction of technical detail, but in some places this cannot be avoided. The calculations are generally included for the benefit of students. We hope that appreciation of the general principles will not be lost in following the details of arguments. In the words of John Von Neumann, "One expects a mathematical theorem or a mathematical theory not only to describe and classify in a simple and elegant way numerous and a priori disparate special cases. One also expects elegance in its architectural, structural make-up. . . . If the deductions are lengthy or complicated there should be some simple general principle involved, which explains the complications and details, reduces the apparent arbitrariness to a few simple guiding motivations." If there is one such guiding principle underlying our work it is this: classical mechanics – on which all physical models are based – also provides the proper foundation for numerical simulation of those systems. We will attempt to show in this book that practical, efficient methods for simulating conservative systems can be realized by making judicious use of the methods of classical mechanics.

An emphasis on methods

In this book we address ourselves primarily to the following pair of questions:

Which properties should be fundamental to an integration method for a (conservative) model?

How can we design and implement schemes that respect physical principles regardless of timestep or traditional accuracy considerations?

Although our interest is always ultimately in the methods themselves and in quantifying the relative differences among them, we will find that in attempting to answer the above questions, we are drawn far afield from the usual domain of the numerical analyst. The first question will lead us into the field of mechanics so that we may appreciate something of the nature of those structures and symmetries that underlie physical models and contribute to their long-term evolution. The second question will take us outside even the areas that have traditionally been investigated by mathematicians, since the special forms of force functions, presence of constraints, relationships among the variables, or efficiency considerations dictate to a large extent the features of appropriate (i.e. practical) methods used in applications.

It is important to emphasize that our treatment is not comprehensive; we have made a selection from the literature which comprises – in our view – the most important material from the standpoint of practical application.

Beginning with the idea of *splitting* we will show how many simple but effective integrators can be generated by using a few building blocks. The same techniques can be used to derive more sophisticated schemes. For example, explicit higher-order methods have a very natural derivation in the case of canonical mechanical systems developed in terms of the "kinetic+potential" form of the energy.

We survey recent work on methods for constrained systems and consider various approaches to the simulation of rigid body systems, methods which offer an efficient and – in many cases – demonstrably superior geometric alternative to more widespread schemes. Variable stepsize geometric integrators will be introduced based on a rescaling of the time variable. Methods for mixed systems possessing both rapidly and slowly varying degrees of freedom – or weak and strong forces – also call for the construction of specialized schemes. In all cases, our aim will be to present the ideas in as general a form as is prudent, highlighting instances where a given technique might be of use in other applications. Molecular dynamics applications are an important source of challenging problems for geometric integration, so we devote some time to their particular characteristics. Conservative partial differential equations introduce many new issues for the development of geometric integrators, a topic we touch on in the final chapter of the book.

How to use this book

This book is intended, first, as a text for a course in computational mechanics or as a tool for self-instruction, and, second, as a basic reference for researchers and educators – regardless of discipline – interested in using and developing geometric integrators. The book should serve as a bridge from traditional training in the sciences to the recent research literature in the field of geometric integration. By emphasizing mathematical and computational issues and illustrating the various concepts and techniques with carefully developed model problems, it is hoped that the book can appeal to a wide audience, including mathematicians unfamiliar with modeling issues, and physicists, chemists, and engineers wishing to gain a better understanding of the mathematical underpinnings of existing methods or in developing effective methods for new applications.

The book assumes only that the reader has had undergraduate coursework in linear algebra and differential equations. At several points we introduce, but do not thoroughly develop, topics from dynamical systems. A good introductory text in dynamical systems is the book of VERHULST [197]. We develop – in Chapters 2 and 3 – most of the necessary preliminaries of numerical analysis and classical mechanics from the ground up, however the reader should be aware that the treatment provided here of the required background material is necessarily brief; only those elements that are essential to our later study of geometric integrators are given. For an introduction to numerical analysis, the reader is referred to the classic books of ATKINSON [11], BURDEN AND FAIRES [37], and DAHLQUIST AND BJÖRK [48]. The book of GEAR [70] can provide a useful introduction to the numerical solution of differential equations. The books of HAIRER, NØRSETT AND WANNER [82] and HAIRER AND WANNER [84] can serve as references for obtaining a more complete picture of the mathematical issues associated with construction of methods and error analysis for ordinary differential equations. ISERLES [91] has written an integrated text that introduces numerical methods for both ordinary and partial differential equations.

First published at the end of the 19th century, ROUTH's *Dynamics of a System of Rigid Bodies (Elementary Part)*[1] remains a marvelous introduction to classical mechanics and provides a wealth of examples and exercises for the student (many of which could now be revisited with the aid of the modern computational techniques developed in this book). For a more systematic treatment to Hamiltonian classical mechanics the reader is referred to the following texts: LANDAU AND LIFSHITZ [105], MARION [121], GOLDSTEIN [73], and ARNOLD [7], all of which are well-worn occupants of our bookshelves. These books are quite varied in their use of notation and even in the way in which they motivate and explain identical material, but we have found all of them to be helpful on various occasions. If only one book is to be consulted, the elegant book of LANCZOS [104] is remarkable both in terms of its readability and its breadth, owing partly to the absence of detailed proofs. A modern rigorous treatment of classical mechanics may be found in MARSDEN AND RATIU [124], a book which also contains a number of useful examples and notes on history and applications.

The book by SANZ-SERNA AND CALVO [172] was the first to cover symplectic integration methods and applications to classical mechanics and is still an excellent introduction to the subject. The more recent book by HAIRER, LUBICH, AND WANNER [80] covers a wide range of topics from geometric integration and should be very useful as an additional reference.

In a graduate course in applied mathematics or computational physics, it is probable that much of the material of Chapters 1–3 could be skipped or skimmed, depending on the backgrounds of the students and the interests of the teacher. Some caution should be exercised here. In particular, it is essential

[1]Reprinted in 1960 as a Dover Edition.

that the student understand the concepts of convergence and order of convergence for a numerical method, the definition of the flow map, first integrals, and at least the condition for a symplectic map in terms of the Jacobian of the flow map.

Molecular dynamics provides a rich source of problems for geometric integration, and we often draw on examples from this field for motivation and for evaluation of concepts and methods. Here again, it is likely that the reader may, on occasion, wish for a more detailed description of the problems or of typical approaches used by chemists and physicists. One reference stands out in this area for clarity of presentation and breadth: ALLEN AND TILDESLEY [4]. More recent books of FRANKEL AND SMIT [66] and SCHLICK [174] help to fill in the picture.

Exercises included at the end of each chapter are intended to be demanding but not overwhelming; some of the multi-part problems could be assigned as projects, especially those involving the use of computers.

Computer software

This book primarily emphasizes the mathematical properties of algorithms for solving differential equations. In later chapters, we will often see the methods introduced and analyzed as abstract maps of phase space. This approach, while essential to understanding and generalizing the methods, has the tendency to obscure both the intuitive basis for the theory and the ultimate importance of the subject. We would like emphasize that *the student must implement and test numerical methods in order to gain a full understanding of the subject.*

While any programming language and graphics package could, in principle, be employed, the need for flexibility in the coding and testing of methods and the need to be able to work easily with scientific functions of vectors and matrices makes a specialized, interpreted language system for mathematics the best environment for problem solving.

At the time of this writing, there are several widely distributed commercial software packages that support the simplified design and testing of algorithms, including the commercial packages MATLAB, MAPLE, and MATHEMATICA. Of these, the authors prefer the user interface and programming structure of MATLAB, but this is largely a matter of taste and any of the three mentioned systems would be suitable. These packages are all available to students at heavily discounted prices and run on a variety of computer platforms.

For a student on a severe budget, there are several widely available *free* alternatives to the commercial packages. These options include, notably, OCTAVE, which is distributed under the Free Software Foundation's GNU Public License. Our experience with this software is that it is adequate for most study purposes, although the commercial alternatives are generally superior with regard to ease-of-use, documentation, and reliability.

We will occasionally describe algorithms in this text, but we will attempt to avoid system-specific details, so the student is expected to supplement the mathematical study with study of the the user's guide for the software system they are using, in particular the appropriate sections on programming.

Notation

Let us summarize some basic notation used throughout the book. Dependent variables, such as positions $\boldsymbol{q}$ and velocities $\boldsymbol{v}$ are elements of a Euclidean space $\mathbb{R}^d$, where $d \geq 1$ is the appropriate dimension. More specifically, we will always identify dependent variables with *column vectors*. When two column vectors $\boldsymbol{u} \in \mathbb{R}^k$, $\boldsymbol{v} \in \mathbb{R}^l$ are given we may write $(\boldsymbol{u}, \boldsymbol{v})^T$ for the column vector in $\mathbb{R}^{k+l}$ obtained by concatenating the two vectors. The transpose is there to remind us that the result is again a column vector. We will often need to refer to a set indexed by a parameter for which we write $\{a_t\}_{t\in P}$, where P is the index set. As a short-hand, we will write $\{a_t\}$ if the index set is clear from the context.

The set of $k \times k$ matrices with real coefficients is $\mathbb{R}^{k\times k}$ and capital bold-face letter are used to denote matrices, e.g., $\boldsymbol{A}, \boldsymbol{B} \in \mathbb{R}^{k\times k}$. The k-dimensional identity is $\boldsymbol{I}_k$ or $\boldsymbol{I}$ as a short-hand if the dimension is clear from the context.

A vector-valued function $\boldsymbol{F} : \mathbb{R}^n \to \mathbb{R}^m$ will be assumed to map column vectors of dimension n to column vectors of dimension m. The vector of partial derivatives of a scalar-valued function $f(\boldsymbol{q})$ is identified with a *row vector* in $\mathbb{R}^d$ and is denoted by $f_{\boldsymbol{q}}(\boldsymbol{q})$ or, equivalently, by $\partial f/\partial \boldsymbol{q}\,(\boldsymbol{q})$. Hence the *Jacobian matrix* of a vector-valued function $\boldsymbol{F}(\boldsymbol{q})$ is identified with the $m \times n$ matrix $\boldsymbol{F}_{\boldsymbol{q}}(\boldsymbol{q})$.

The *scalar product*, *inner product*, or *dot product* of two column vectors $\boldsymbol{a}$ and $\boldsymbol{b}$ in $\mathbb{R}^d$ is denoted by $\langle \boldsymbol{a}, \boldsymbol{b}\rangle$ or $\boldsymbol{a}\cdot\boldsymbol{b}$, or, simply, $\boldsymbol{a}^T\boldsymbol{b}$, where $\boldsymbol{a}^T$ is the transpose of $\boldsymbol{a}$. The *cross product* of $\boldsymbol{a}$ and $\boldsymbol{b}$ is denoted $\boldsymbol{a} \times \boldsymbol{b}$. We will frequently use

$$\widehat{\boldsymbol{a}}\boldsymbol{b} = \boldsymbol{a} \times \boldsymbol{b},$$

where $\widehat{\boldsymbol{a}} \in \mathbb{R}^{3\times 3}$ is a skew-symmetric matrix related to the vector $\boldsymbol{a} = (a_1, a_2, a_3)^T$ by

$$\widehat{\boldsymbol{a}} = \begin{bmatrix} 0 & -a_3 & a_2 \\ a_3 & 0 & -a_1 \\ -a_2 & a_1 & 0 \end{bmatrix}.$$

The *norm* of a vector $\boldsymbol{a}$ is defined by

$$\|\boldsymbol{a}\| = \langle \boldsymbol{a}, \boldsymbol{a}\rangle^{1/2}.$$

A real-valued function f defined on an interval I of the real line is said to be square-integrable if $\int_I f(x)^2 dx$ is bounded. In that case we say that f lies in the function space L_2 on the interval I. We say that the L_2-norm of f is

$$\|f\|_2 = \left(\int_I f(x)^2 dx \right)^{\frac{1}{2}},$$

and we define the L_2-inner product of two square integrable functions f and g by

$$(f, g) = \int_I f(x) g(x) dx$$

The *gradient* $\nabla_{\boldsymbol{q}} V(\boldsymbol{q})$ of a scalar-valued differentiable function $V(\boldsymbol{q})$ is defined by

$$\langle \nabla_{\boldsymbol{q}} V(\boldsymbol{q}), \boldsymbol{u} \rangle = \lim_{\varepsilon \to 0} \frac{V(\boldsymbol{q} + \varepsilon \boldsymbol{u}) - V(\boldsymbol{q})}{\varepsilon},$$

where the equality is to hold for all vectors $\boldsymbol{u}$ of the same dimension as $\boldsymbol{q}$. This definition leads to the relation

$$\nabla_{\boldsymbol{q}} V(\boldsymbol{q}) = V_{\boldsymbol{q}}(\boldsymbol{q})^T,$$

and, hence, the gradient $\nabla_{\boldsymbol{q}} V(\boldsymbol{q})$ is a column vector.

The time derivative of a function $\boldsymbol{q}(t)$ will normally be denoted by $d\boldsymbol{q}/dt(t)$, but whenever it is more convenient we may instead use the short-hand $\dot{\boldsymbol{q}}(t)$. If clear from the context, we will also frequently drop the argument in functions and write, for example, $\dot{\boldsymbol{q}}$ instead of $\dot{\boldsymbol{q}}(t)$, or $\boldsymbol{F}$ instead of $\boldsymbol{F}(\boldsymbol{q})$. The same conventions apply to higher-order derivatives.

If a variable u depends on several independent variables such as time t and space x, then the partial derivatives are often denoted by u_t and u_x, respectively.

Given two maps $\boldsymbol{\Psi}_1 : \mathbb{R}^n \to \mathbb{R}^n$ and $\boldsymbol{\Psi}_2 : \mathbb{R}^n \to \mathbb{R}^n$ with compatible range and domain, we define their composition $\boldsymbol{\Psi}_2 \circ \boldsymbol{\Psi}_1$ by

$$[\boldsymbol{\Psi}_2 \circ \boldsymbol{\Psi}_1](\boldsymbol{z}) = \boldsymbol{\Psi}_2(\boldsymbol{\Psi}_1(\boldsymbol{z})),$$

for all $\boldsymbol{z} \in \mathbb{R}^n$. The inverse of a one-to-one map $\boldsymbol{\Psi}$ is denoted by $\boldsymbol{\Psi}^{-1}$. Hence

$$\boldsymbol{\Psi}^{-1} \circ \boldsymbol{\Psi} = \mathbf{id},$$

where $\mathbf{id}(\boldsymbol{z}) = \boldsymbol{z}$ the identity map.

Finally the Landau-order notation $\mathcal{O}(\Delta t)$ is used to denote a quantity that goes to zero as rapidly as Δt goes to zero. More generally, we will write $g(\Delta t) = \mathcal{O}(\Delta t^p)$ if $g(\Delta t)/\Delta t^p$ is bounded as $\Delta t \to 0$ but $g(\Delta t)/\Delta t^q$ is unbounded if $q > p$.

Acknowledgements

We have to admit that when we began this project, we had no appreciation for its ultimate complexity. The only thing that has kept the project going at some critical points was the broad support and encouragement we received from our families, our colleagues, and our students.

A key aspect of the book is the emphasis on applications. Brian Laird proposed the Lennard–Jones model discussed in Chapter 4 and made several other helpful comments. Chris Sweet contributed most of the numerical experiments in Chapter 6, including the long-term simulations of the solar system, and Eric Barth provided the alanine dipeptide simulation in Chapter 11. Jason Frank helped with many of the experiments in Chapter 12. Jason's tireless proofreading of the entire manuscript must also be remarked!

We hope that the book will be useful for teaching upper-division undergraduate and graduate courses. In 2002, Ph.D. students from various fields and various parts of Europe – Martin Dahlberg (chemistry), Thomas Sauvaget (mathematics/physics), Marko Vranicar (physics), and Fredrick Hedman (mathematics) – attended an informal summer school in Leicester taught from an early draft of the book. Part of the book formed the basis for a series of survey lectures given by one of us in Bari, Italy in 2002. Mitch Luskin, Claudia Wulff, and Steve Bond have subjected their classes to a preliminary draft of the book for testing in this setting, and provided useful comments.

We would like also to thank Teijo Arponen, Jeff Cash, Colin Cotter, Ernst Hairer, Zhidong Jia, Per-Christian Moan, Brian Moore, and Antonella Zanna for reading various parts of the manuscript and providing feedback. We have also relied on the extraordinary patience of Cambridge University Press, particularly our editors David Tranah and Joseph Bottrill.

1

Introduction

This book is about numerical methods for use in simulating dynamical phenomena governed by conservative processes. In this chapter, we review a few basic principles regarding conservative models. In general, we are concerned here with initial value problems for systems of ordinary differential equations (ODEs) of the form

$$\frac{d}{dt}\mathbf{z} = \mathbf{f}(\mathbf{z}), \qquad \mathbf{z}(t_0) = \mathbf{z}^0,$$

where $\mathbf{z} : \mathbb{R} \to \mathbb{R}^k$. The basic questions encountered early on in a first course on ODEs concern existence and uniqueness of solutions, a topic addressed, for example, by Picard's theorem. Discussion then turns to various techniques for analytically solving the differential equations when $\mathbf{f}$ has a prescribed form. In particular, the scalar case $k = 1$ is an instance of a *separable differential equation* and such models are in principle solvable in quadratures (i.e. by evaluating certain integrals and solving certain algebraic equations). Linear systems are exactly solvable after determination of the eigenvalues and eigenvectors (or generalized eigenvectors, in the degenerate case). Beyond these and a few other special cases, most models are not exactly integrable. In this book we are mostly interested in complex models that do not admit exact solutions.

The emphasis of this book is on the particular models which are formulated naturally as *conservative systems* of ODEs, most importantly *Hamiltonian systems.* As a general rule, mechanical systems resulting from physical principles are Hamiltonian until (usually for prudent modeling purposes) subjected to simplifying reductions or truncations. For example, in typical fluid dynamics applications, the incorporation of diffusive effects due to friction with a boundary plays an essential role in the modeling. However, in many situations, the conservative paradigm can be retained and remains the most appropriate foundation for the construction of models, since it is in no small measure due to properties such as conservation of energy and angular momentum that matter behaves as it does.

The existence of Kepler's laws which approximately describe the motion of the planets in the solar system are reflections of the conservative nature of gravitational dynamics. The celebrated Kolmogorov–Arnold–Moser theory which discusses the local stability of nonlinear dynamical systems in the vicinity of certain critical points applies only to conservative systems. Even dissipative systems typically retain certain conservation laws (for example conservation of mass in fluid dynamics), and many of the ideas developed in this book are still applicable to such problems.

1.1 N-body problems

Conservative dynamical systems most often originate through application of Newton's second law which describes the motion of a body in an applied force field. In a classical N-body system (Fig. 1.1), several point masses are involved and the forces acting on any one body arise from the presence of neighboring bodies or some external field.

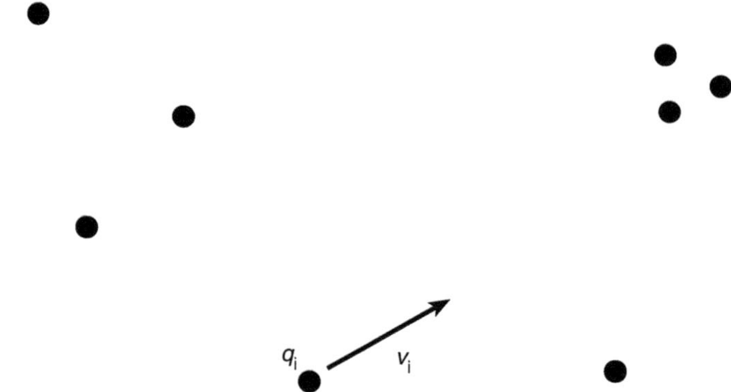

Figure 1.1 An N-body system.

Let the ith body be assigned a mass m_i, an instantaneous position $\boldsymbol{q}_i$ (with respect to some appropriate reference frame), and a velocity $\boldsymbol{v}_i$, $i = 1, \ldots, N$. Let $\boldsymbol{F}_i$ represent the force acting on body i (due, for example, to interactions with the other particles). We assume that the force can be obtained as the negative gradient of a potential energy function V with respect to the ith particle position $\boldsymbol{q}_i$, i.e.

$$\boldsymbol{F}_i = -\nabla_{\boldsymbol{q}_i} V(\boldsymbol{q}_1, \boldsymbol{q}_2, \ldots, \boldsymbol{q}_N).$$

The N-point particles then move according to Newton's equations of motion

$$\frac{d}{dt}\boldsymbol{q}_i = \boldsymbol{v}_i, \qquad (1.1)$$

$$m_i \frac{d}{dt}\boldsymbol{v}_i = \boldsymbol{F}_i, \qquad i = 1, 2, \ldots, N. \qquad (1.2)$$

N-body problems can be naturally formulated to describe motion in any Euclidean space $\mathbb{R}^\nu$, $\nu > 0$, i.e. with $\boldsymbol{q}_i$, $\boldsymbol{v}_i$, and $\boldsymbol{F}_i$ all in $\mathbb{R}^\nu$. Such a system is said to have νN *degrees of freedom*. We will say that the *phase space* of an N-body problem is the $2\nu N$-dimensional set consisting of all possible positions $\boldsymbol{q} = (\boldsymbol{q}_1, \boldsymbol{q}_2, \ldots, \boldsymbol{q}_N)^T$ and velocities $\boldsymbol{v} = (\boldsymbol{v}_1, \boldsymbol{v}_2, \ldots, \boldsymbol{v}_N)^T$ of the particles. Under mild smoothness assumptions on the potential energy function V, there exists, at least locally through any point $(\boldsymbol{q}^0, \boldsymbol{v}^0)$ of phase space, a unique *trajectory* of the mechanical system: a solution of the equations (1.1)–(1.2) subject to the initial conditions $\boldsymbol{q}(0) = \boldsymbol{q}^0$, $\boldsymbol{v}(0) = \boldsymbol{v}^0$. At a *critical point* $\boldsymbol{q} = \bar{\boldsymbol{q}}$, all of the forces acting on the particles in the system vanish; hence the trajectory through $(\bar{\boldsymbol{q}}, \boldsymbol{0})$ reduces to a single point.

The total energy associated to the mechanical system (1.1)–(1.2) is the sum of kinetic and potential terms

$$E(\boldsymbol{q}, \boldsymbol{v}) = \frac{1}{2}\sum_{i=1}^{N} m_i \|\boldsymbol{v}_i\|^2 + V(\boldsymbol{q}).$$

It is easy to see that the energy is constant along a trajectory, since

$$\begin{aligned}\frac{d}{dt}E &= \sum_{i=1}^{N} m_i \boldsymbol{v}_i \cdot \dot{\boldsymbol{v}}_i + \sum_{i=1}^{N} \nabla_{\boldsymbol{q}_i} V(\boldsymbol{q}) \cdot \dot{\boldsymbol{q}}_i \\ &= \sum_{i=1}^{N} m_i \boldsymbol{v}_i \cdot \left(\frac{1}{m_i}\boldsymbol{F}_i\right) - \sum_{i=1}^{N} \boldsymbol{F}_i \cdot \boldsymbol{v}_i = 0.\end{aligned}$$

(Refer to the preface for details on the notation used in this derivation and later in the book.) A system with an energy function constant along solutions is referred to as a *conservative* system.

1.2 Problems and applications

Let us briefly survey a few of the most important recurring N-body applications. Examples of these problems, along with a number of other types of models, are developed in more detail in various places in the book.

The historical origin of the N-body problem lies in *gravitational modeling*, and these problems remain of substantial current interest. Simulations are being conducted on a wide variety of astronomical systems, including planetary systems (for understanding both their formation and their long-term stability), systems of interacting stars or binaries, galaxies, and globular clusters. Closely related problems arise in semi-classical studies of atomic systems. As an example, the three-body gravitational problem involving bodies of unequal mass has the potential energy

$$V(\boldsymbol{q}_1, \boldsymbol{q}_2, \boldsymbol{q}_3) = -\frac{Gm_1m_2}{\|\boldsymbol{q}_1 - \boldsymbol{q}_2\|} - \frac{Gm_2m_3}{\|\boldsymbol{q}_2 - \boldsymbol{q}_3\|} - \frac{Gm_1m_3}{\|\boldsymbol{q}_1 - \boldsymbol{q}_3\|},$$

where G is the universal gravitational constant. Such a three-body problem has no general, analytical solution, so simulation is needed to enhance understanding, sometimes in conjunction with partial theoretical analysis, for example to determine the stability of certain configurations of the bodies. Chaotic solutions of the three-body problem may include arbitrarily close approaches of the bodies, in which case the singularity in the potential may cause significant difficulty for numerical simulation and some sort of regularizing transformations of time and/or coordinates are needed. We will return to consider some of these issues in one of the book's later chapters.

Classical mechanics is also the basis of many *molecular models* in chemistry, physics, and biology, including those commonly used for studying liquids and gases, materials, proteins, nucleic acids, and other polymers. In these applications, V is composed of a sum of several heterogeneous nonlinear contributions based on the distances between pairs of particles, varying both in functional form and in relative intensity. These terms may be "local" ("short-range") meaning that they effectively involve only contributions from nearby particles, or they may be "long-range." A commonly treated system with only local interactions is the simplified model of a gas or liquid, consisting of N identical atoms of a certain prescribed mass, interacting in a *Lennard–Jones* pair potential

$$\varphi_{\text{L.J.}}(r) = \epsilon\left[\left(\frac{\bar{r}}{r}\right)^{12} - 2\left(\frac{\bar{r}}{r}\right)^{6}\right]. \tag{1.3}$$

The total potential energy is

$$V = \sum_{1\le i<j\le N} \varphi_{\text{L.J.}}(\|\boldsymbol{q}_i - \boldsymbol{q}_j\|). \tag{1.4}$$

Note that such models are always simplifications of vastly more complex quantum-mechanical models. The parameters ϵ and $\bar{r}$ of the Lennard–Jones potential provide a fit to experiment, but would depend on the temperature and pressure at which the simulation is performed. Strictly speaking, the pair interactions between atoms would include all pairs, no matter how distant, but since the energy decays like r^{-6}, the forces are generally found to be so small outside of some critical radius that the potential can simply be cut off beyond this distance. In practice, this is usually done by introducing a smooth transition of the potential energy function to a constant value.

Regardless of what other potentials may be present, the presence of the Lennard–Jones potential ensures that the particle forces are ultimately strongly repulsive at short range. These repulsive forces are a very important aspect of molecular systems. The potential will be singular where particle positions overlap, but otherwise, the potential is smooth and the solution is *globally defined*: solutions started away from singularities can be extended without bound in t. Thus molecular systems do not undergo the extreme collisions encountered in strictly Coulombic problems such as gravitation.

Still other classes of conservative systems arise through discretization of *partial differential equations*. A semi-linear wave equation of the form

$$u_{tt} = u_{xx} - f(u), \qquad u = u(x,t),$$

is conservative under certain prescriptions of boundary and initial data. If we assume, for example, that solutions are defined on the interval $[0, L]$ and are periodic with period L, then the energy functional is

$$E[u] = \int_0^L \left[\frac{1}{2}u_t^2 + \frac{1}{2}u_x^2 + F(u)\right] dx,$$

where $F(u) = \int_0^u f(s)ds$. The equations of motion could be written in the "Newton-like" form

$$v_t = -\delta_u V[u], \qquad u_t = v,$$

where $V[u] = \int_0^L \left[\frac{1}{2}u_x^2 + F(u)\right] dx$ represents the potential energy, and δ_u, termed the *variational derivative*, is the analogue of the gradient appearing in the Newtonian equations of motion,

$$(\delta_u G[u], \delta u) = \lim_{\varepsilon \to 0} \frac{G[u + \varepsilon\delta u] - G[u]}{\varepsilon}, \tag{1.5}$$

where G is a functional, like the potential energy V, that assigns a real number to functions $u(x)$ and the equality holds for all sufficiently regular periodic functions $\delta u(x)$, i.e. $\delta u(x) = \delta u(x+L)$. This definition is formally equivalent to the defining relation for the gradient mentioned in the preface with $\langle .,. \rangle$ replaced by the L_2 inner product,

$$(u, v) = \int_0^L u(x)v(x)dx.$$

Note that $\delta_u G[u]$ is itself a function of x.

The simplest centered finite difference spatial discretization takes the form, for $i = 1, 2, \ldots, N$,

$$\frac{d}{dt}u_i = v_i,$$
$$\frac{d}{dt}v_i = \frac{u_{i-1} - 2u_i + u_{i+1}}{\Delta x^2} - f(u_i),$$

where $u_i \approx u(i\Delta x, t)$, $\Delta x = L/N$, and the periodic boundary condition leads to the definitions $u_0 \equiv u_N$, $u_{N+1} \equiv u_1$. This is in the form of a standard N-body system in one dimension with positions $\boldsymbol{u} = (u_1, u_2, \ldots, u_N)^T$, velocities $\boldsymbol{v} = (v_1, v_2, \ldots, v_N)^T$, and potential

$$V(\boldsymbol{u}) = \sum_{i=1}^{N} \left(\frac{u_{i+1} - u_i}{\Delta x} \right)^2 + \sum_{i=1}^{N} F(u_i).$$

1.3 Constrained dynamics

In the setting of modern applications, we will need to consider generalizations of the traditional N-body problem in which the basic modeling unit is not the point particle moving in Euclidean space but an object moving in some *constrained* space.

For instance, in molecular dynamics, the bond stretch between two atoms is typically modeled by a spring with rest length $L > 0$, say

$$V_{12}(\boldsymbol{q}_1, \boldsymbol{q}_2) = \frac{\alpha}{2} \left(\|\boldsymbol{q}_1 - \boldsymbol{q}_2\| - L \right)^2,$$

where $\boldsymbol{q}_1$ and $\boldsymbol{q}_2$ are the positions of the atoms and α is a positive parameter. When α is large, the vibrational frequency is also large, while the variation in the length of the stretch from L will typically be small. It is then common practice to replace one or more of these bonds by rigid length constraints, i.e. to introduce a constraint of the form

$$\|\boldsymbol{q}_1 - \boldsymbol{q}_2\|^2 = L^2.$$

If enough constraints among a set of particles are imposed simultaneously, the group becomes completely rigid. Such *rigid bodies* have very interesting dynamical properties in and of themselves. For example, in molecular dynamics, it is standard practice to replace small polyatomic molecules (for example, H_2O) by rigid bodies. As another illustration, while it may be appropriate to treat the bodies in the solar system as point masses for many purposes, in more delicate situations, the nonspherical rigid body structure of the planets may need to be taken into consideration.

Let us begin by extending Newton's equations for particle motion to the constrained case. Imagine a particle of mass m moving on a constraint surface defined as the zero set of some smooth function g. At any time, the particle is acted on by two types of forces: *applied forces* defined in the usual way by a potential energy function V, and *constraint* forces which act in such a way as to make the particle lie on the constraint surface. Although we do not in general know anything about the directionality of the applied force, we may take as our starting point the *principle of D'Alembert*: *the constraint force acts along the normal direction to the constraint surface*, i.e. along the direction of the gradient to the function g at the point of contact (Fig. 1.2).

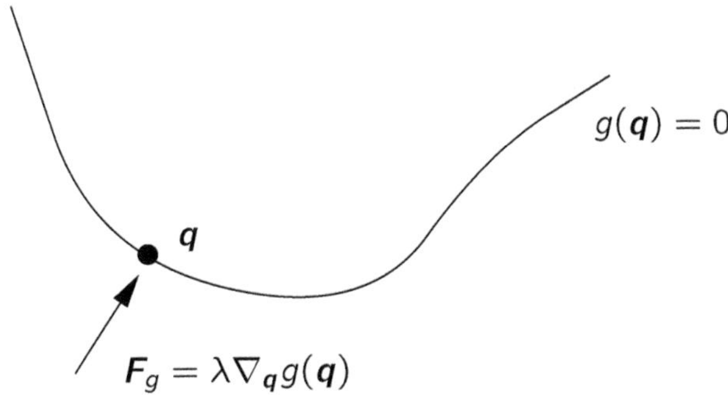

Figure 1.2 D'Alembert's Principle: the constraint force acts in the normal direction to the constraint surface at the point of contact.

Thus, if we denote the constraint forces by $\boldsymbol{F}_g$, we have

$$\boldsymbol{F}_g \quad || \quad \nabla_{\boldsymbol{q}} g(\boldsymbol{q}),$$

or

$$\boldsymbol{F}_g = \lambda \nabla_{\boldsymbol{q}} g(\boldsymbol{q}),$$

where λ is a scalar.

Using Newton's second law, the equations of motion then take the form

$$m\dot{\boldsymbol{v}} = -\nabla_{\boldsymbol{q}} V(\boldsymbol{q}) + \lambda \nabla_{\boldsymbol{q}} g(\boldsymbol{q}), \tag{1.6}$$

$$\dot{\boldsymbol{q}} = \boldsymbol{v}, \tag{1.7}$$

$$g(\boldsymbol{q}) = 0. \tag{1.8}$$

The parameter λ is an unknown which is uniquely determined by the condition that $\boldsymbol{q}(t)$ satisfy (1.8) at all points on the trajectory and that the trajectory be smooth. Specifically, if we differentiate the equation $g(\boldsymbol{q}(t)) = 0$ twice with respect to time, we find first

$$\frac{d}{dt} g(\boldsymbol{q}) = \nabla_{\boldsymbol{q}} g(\boldsymbol{q}) \cdot \dot{\boldsymbol{q}} = \nabla_{\boldsymbol{q}} g(\boldsymbol{q}) \cdot \boldsymbol{v} = 0, \tag{1.9}$$

and, then

$$\frac{d^2}{dt^2}g(\boldsymbol{q}) = \langle \boldsymbol{v}, g_{\boldsymbol{qq}}(\boldsymbol{q})\boldsymbol{v}\rangle + m^{-1}\nabla_{\boldsymbol{q}} g(\boldsymbol{q}) \cdot [-\nabla_{\boldsymbol{q}} V(\boldsymbol{q}) + \lambda \nabla_{\boldsymbol{q}} g(\boldsymbol{q})] = 0, \qquad (1.10)$$

where $g_{\boldsymbol{qq}}(\boldsymbol{q})$ represents the Hessian matrix of g. Provided that $\nabla_{\boldsymbol{q}} g(\boldsymbol{q}) \neq 0$, the equation (1.10) has a unique solution $\lambda = \Lambda(\boldsymbol{q}, \boldsymbol{v})$

$$\Lambda(\boldsymbol{q}, \boldsymbol{v}) = \frac{m}{\|\nabla_{\boldsymbol{q}} g(\boldsymbol{q})\|^2}\left(\frac{1}{m}\langle \nabla_{\boldsymbol{q}} g(\boldsymbol{q}), \nabla_{\boldsymbol{q}} V(\boldsymbol{q})\rangle - \langle \boldsymbol{v}, g_{\boldsymbol{qq}}(\boldsymbol{q})\boldsymbol{v}\rangle\right).$$

Equations (1.6)–(1.8) are a special case of the constrained Euler–Lagrange equations. As a simple illustration, we mention the example of a bead (of mass m) moving in gravity in two dimensions (coordinates (x, z)) along a wire described by the curve $\Gamma : z = f(x)$. The constraint is $g(x, z) := z - f(x)$, and the equations of motion take the form

$$m\ddot{x} = -\lambda f'(x), \qquad (1.11)$$
$$m\ddot{z} = -mg + \lambda, \qquad (1.12)$$
$$z = f(x). \qquad (1.13)$$

Here g represents the earth's gravitational constant.

As a second illustration, consider the spherical pendulum consisting of a bob of mass m suspended from a fixed point on a rigid massless rod of length $L > 0$. We formulate the problem in cartesian coordinates (x, y, z) with energy

$$E = \frac{1}{2m}(\dot{x}^2 + \dot{y}^2 + \dot{z}^2) + mgz,$$

and equations of motion

$$\begin{aligned} m\ddot{x} &= 2\lambda x, \\ m\ddot{y} &= 2\lambda y, \\ m\ddot{z} &= -mg + 2\lambda z, \\ 0 &= x^2 + y^2 + z^2 - L^2. \end{aligned}$$

1.4 Exercises

1. *Scalar nonlinear models.* Consider a single-degree-of-freedom problem of the form

$$\begin{aligned} \dot{q} &= v, \\ m\dot{v} &= -\varphi'(q). \end{aligned}$$

a. Write the energy function $E(q, v)$ for the above system and verify that it is conserved along trajectories of the system.

b. Set the energy function to a constant value, say E_0, and show that the resulting equation can be solved for v as a function of q, subject to a choice of the sign of v. Using this, together with the differential equations, show that the equations of motion reduce to a first-order differential equation for q of the form

$$\dot{q} = \pm\sqrt{\left(\frac{2}{m}\right)(E_0 - \varphi(q))}.$$

(Observe that this equation is separable, and hence the solution can in principle be recovered by integration.)

Discuss first the case $v \neq 0$ and continue with an investigation of the solution behavior in the vicinity of $v = 0$.

2. *Morse oscillator.* (See problem 1 above.) The Morse oscillator is a one-degree-of-freedom conservative system consisting of a single particle of unit mass moving in the potential $\varphi(q) = D(1 - e^{-\beta q})^2$. In the following set $D = 1$ and $\beta = 1$.

a. Sketch the graph of φ as a function of q.

b. Sketch several of the level curves ($E(q, v) = E_0$, E_0 fixed) of the energy function. In particular, observe that the system has bounded trajectories for $E < E_*$. What is E_*? What can be said about an orbit with energy $E = E_*$?

c. Sketch the graphs of several solution curves as functions of t. [Hint: the velocity field can be sketched by using the result of problem 1b.]

3. *Pendulum.* The planar version of the pendulum is described by the equations

$$\begin{aligned} m\ddot{x} &= 2\lambda x, \\ m\ddot{y} &= -mg + 2\lambda y, \\ 0 &= x^2 + y^2 - L^2. \end{aligned}$$

Introduce coordinates $x = L \sin\theta$, $y = -L\cos\theta$ and show that the equations of motion can be reduced to a single second-order differential equation for θ which is in the form of a nonlinear oscillator.

4. *Bead-on-wire.* Consider the "bead-on-wire" problem (1.11)–(1.13). Show that the equations of motion can be reduced to a second-order unconstrained differential equation for x of the form

$$\ddot{x} = -f'(x)\frac{g + f''(x)\dot{x}^2}{1 + f'(x)^2}.$$

5. *Variational derivative.* Using the definition (1.5) of the variational derivative and integration by parts, verify that

$$\delta_u V[u] = -u_{xx} + F'(u),$$

for

$$V[u] = \int_0^L \left[\frac{1}{2}u_x^2 + F(u)\right] dx.$$

2

Numerical methods

In this chapter, we introduce the concepts of accuracy and stability for a numerical method approximating the solution of an ordinary differential equation initial value problem

$$\frac{d}{dt}z = f(z), \qquad z(t_0) = z^0 \in \mathbb{R}^k. \tag{2.1}$$

Here $z(t)$ represents the solution at a particular time t; $z = z(t)$ thus defines a parameterized trajectory. We assume that trajectories are defined for all initial values $z^0 \in \mathbb{R}^n$ and for all times $t \geq t_0$. For simplicity, we typically take $t_0 = 0$. One also often uses the notation $z(t; z^0)$ to distinguish the trajectory to a given initial value z^0.

It is an important feature of the initial value problems we consider, that the solution value at any given point on the trajectory determines the solution at all later points on the trajectory, through integration of an appropriate initial value problem. In effect, the solutions $z(t; z^0)$ of the differential equation define a mapping, or rather a one-parametric family of mappings, $\{\Phi_t\}_{t\geq 0}$, which take initial data to later points along trajectories, i.e.

$$\Phi_t(z^0) = z(t; z^0), \qquad z^0 \in \mathbb{R}^k.$$

We term the map $\Phi_t : \mathbb{R}^k \to \mathbb{R}^k$ the *flow map* of the given system (see Fig. 2.1).

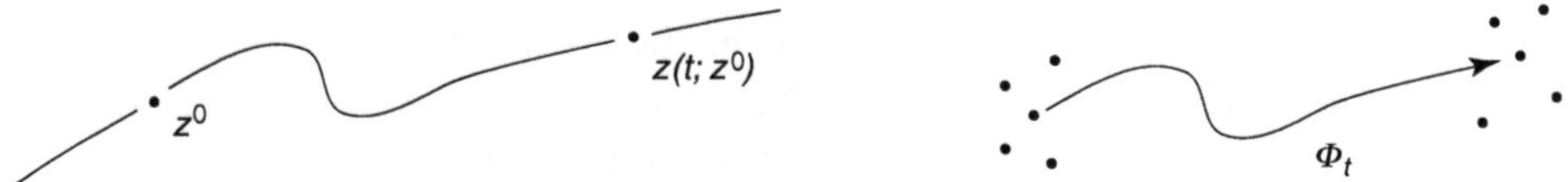

Figure 2.1 The flow map. The existence of a trajectory through each point of phase space (left) implies the existence of a flow map Φ_t taking points of phase space to their evolution through t units of time (right).

In general, the flow map has the following property: if we solve the differential equations from a given initial point z^0 up to a time t_1, then solve from the resulting point forward t_2 units of time, the effect is the same as solving the equations with initial value z^0 up to time $t_1 + t_2$. In terms of the mapping, $\boldsymbol{\Phi}_{t_1} \circ \boldsymbol{\Phi}_{t_2} = \boldsymbol{\Phi}_{t_1+t_2}$. Such a family of mappings is sometimes referred to as a one-parametric *semigroup*.[1]

Given a differential equation and an initial value, a discrete version of a trajectory of the system could be obtained by taking snapshots of the solution at equally spaced points in time t_0, $t_1 = t_0 + \Delta t$, $t_2 = t_1 + \Delta t, \ldots$. The idea of most numerical methods for solving the initial value problem is to provide a simple rule for computing approximations to a discrete trajectory. Assume that this is done by a sequential iterative procedure: starting at t_0, and given the initial value $z(t_0) = z^0$, we compute, via some equation or system of equations, an approximation z^1 to $z(t_1)$, then, by the same means, $z^2 \approx z(t_2)$, etc. In this way, as much of the solution as desired can be obtained. This computational paradigm was understood by Euler, and was used – long before the advent of computers – as a means of studying the theoretical properties of differential equations. The framework was refined and applied to track the motion of charged particles and planets in the first decades of this century, and molded into a practical, effective tool in various fields of science and engineering during the computer revolution of the 1950s and 1960s.

Even the simplest general timestepping methods can, in principle, be used for a wide variety of dynamics simulations, but shortcomings soon become evident in simulations of large, nonlinear dynamical systems. Standard error analysis can be used to demonstrate that a certain numerical method converges in the limit of a small timestep, but in any simulation the ability to take small timesteps is in direct conflict with the cost of a timestep and the need to perform integrations on time intervals long enough to elicit relevant macroscopic behavior. Moreover, even the seemingly elementary supposition that accuracy should be the foundation for analyzing methods is in question in many modern simulations, since, as we have noted in the introduction, accuracy of a particular trajectory may not be directly relevant to simulations of chaotic systems on long time intervals.

In this chapter, we will discuss some elementary properties of timestepping schemes for ordinary differential equations, including the notions of accuracy and stability, and we discuss the application of various schemes to some simple classes of differential equations, notably linear conservative models.

[1] In typical applications we consider in this book, the trajectories can be viewed as defined not only for all $t \geq 0$ but also for all negative times, and the flow maps form a one-parametric *group* (see problem 7 in the Exercises).

2.1 One-step methods

Given a discrete trajectory up to time t_n, there are, in general, few restrictions on the way in which the next approximation z^{n+1} is computed. For example, it might be based on a formula involving the previous computed points along the discrete trajectory, on the derivative of the solution at previous points, or on the values of higher derivatives of the solution at the previous timestep. A common choice is to base the next approximation only on the l previously computed points $z^{n-l+1}, z^{n-l+2}, \ldots, z^n$ and the corresponding values of the derivative at those points, $\dot{z}^{n-l+1}, \dot{z}^{n-l+2}, \ldots, \dot{z}^n$ (where $\dot{z}^i \equiv f(z^i)$). A *linear multistep method* is defined by a linear recurrence relation involving these data together with the associated values at the subsequent time point, z^{n+1} and $\dot{z}^{n+1}$.

We will be primarily concerned with the case $l = 1$, generalized *one-step methods*. By iterating the flow map, we know that we obtain a series of snapshots of the true trajectory

$$z^0, \boldsymbol{\Phi}_{\Delta t}(z^0), \boldsymbol{\Phi}_{\Delta t} \circ \boldsymbol{\Phi}_{\Delta t}(z^0), \ldots$$

or, compactly, $\{\boldsymbol{\Phi}^n_{\Delta t}(z^0)\}_{n=0}^{\infty}$, where the composition power $\boldsymbol{\Phi}^n_{\Delta t}$ is the identity if $n = 0$ and is otherwise the n-fold composition of $\boldsymbol{\Phi}_{\Delta t}$ with itself. For a one-step method, the approximating trajectory can be viewed as the iteration of another mapping $\boldsymbol{\Psi}_{\Delta t} : \mathbb{R}^k \to \mathbb{R}^k$ of the underlying space, so that

$$z^n = \boldsymbol{\Psi}^n_{\Delta t}(z^0). \tag{2.2}$$

The mapping $\boldsymbol{\Psi}_{\Delta t}$ is generally nonlinear, will depend on the function $\boldsymbol{f}$ and/or its derivatives, and may be quite complicated – perhaps even defined implicitly as the solution of some nonlinear algebraic system. Because one-step methods generate a mapping of the phase space, they can be studied with the same techniques and concepts that one employs for understanding the flow map itself. To benefit from this natural correspondence, we will primarily focus on one-step methods throughout this book.

2.1.1 Derivation of one-step methods

One-step methods can be derived in various ways. One way is to first integrate both sides of (2.1) on a small interval $[t, t + \Delta t]$, obtaining

$$z(t + \Delta t) - z(t) = \int_0^{\Delta t} f(z(t + \tau))d\tau.$$

The right-hand side can then be replaced by a suitable quadrature formula resulting in an approximation of the form

$$z(t + \Delta t) \approx z(t) + \sum_{i=1}^{s} b_i \boldsymbol{f}(z(t + \tau_i)),$$

for an appropriate set of weights $\{b_i\}$ and quadrature points $\{\tau_i\}$. In *Euler's method*, the quadrature rule used is just

$$\int_0^{\Delta t} \boldsymbol{f}(\boldsymbol{z}(t+\tau))d\tau = \Delta t \boldsymbol{f}(\boldsymbol{z}(t)) + \mathcal{O}(\Delta t^2).$$

The *implicit Euler method*, on the other hand, is obtained by replacing the value of $\boldsymbol{f}$ at the left endpoint of the interval with its value at the right endpoint. The term *implicit* here refers to the fact that a nonlinear system of equations must be solved to advance the step. The *trapezoidal rule* is based on

$$\int_0^{\Delta t} \boldsymbol{f}(\boldsymbol{z}(t+\tau))d\tau = \frac{1}{2}\Delta t\,[\boldsymbol{f}(\boldsymbol{z}(t)) + \boldsymbol{f}(\boldsymbol{z}(t+\Delta t))] + \mathcal{O}(\Delta t^3),$$

and the *implicit midpoint method* is defined by

$$\int_0^{\Delta t} \boldsymbol{f}(\boldsymbol{z}(t+\tau))d\tau = \Delta t \boldsymbol{f}\left(\frac{\boldsymbol{z}(t)+\boldsymbol{z}(t+\Delta t)}{2}\right) + \mathcal{O}(\Delta t^3).$$

In each case, the associated one-step methods are obtained by using the truncated approximation formula to relate subsequent points $\boldsymbol{z}^n$ and $\boldsymbol{z}^{n+1}$ along a discrete trajectory.

ELEMENTARY ONE-STEP METHODS FOR $\frac{d}{dt}\boldsymbol{z} = \boldsymbol{f}(\boldsymbol{z})$	
Euler's method	Implicit Euler
$\boldsymbol{z}^{n+1} = \boldsymbol{z}^n + \Delta t \boldsymbol{f}(\boldsymbol{z}^n)$	$\boldsymbol{z}^{n+1} = \boldsymbol{z}^n + \Delta t \boldsymbol{f}(\boldsymbol{z}^{n+1})$
Trapezoidal rule	Implicit midpoint
$\boldsymbol{z}^{n+1} = \boldsymbol{z}^n + \frac{\Delta t}{2}\left[\boldsymbol{f}(\boldsymbol{z}^n) + \boldsymbol{f}(\boldsymbol{z}^{n+1})\right]$	$\boldsymbol{z}^{n+1} = \boldsymbol{z}^n + \Delta t \boldsymbol{f}\left(\frac{\boldsymbol{z}^n+\boldsymbol{z}^{n+1}}{2}\right)$

This can of course be generalized to design more accurate approximations, as we shall shortly see.

Let us show in detail how we would apply such methods to treat a second-order system of differential equations

$$\ddot{\boldsymbol{q}} = \boldsymbol{g}(\boldsymbol{q}),$$

such as arises from Newton's second law. We begin by writing the differential equations in a first-order form

$$\dot{q} = v,$$
$$\dot{v} = g(q).$$

Next, we identify the pair of dependent variables (or vectors of dependent variables) with a new vector z, and define a vector field $f(z)$ by pairing the right-hand sides of each equation

$$z = \begin{bmatrix} q \\ v \end{bmatrix}, \qquad f(z) = f\left(\begin{bmatrix} q \\ v \end{bmatrix}\right) = \begin{bmatrix} v \\ g(q) \end{bmatrix}.$$

We can then apply any of the numerical methods given to the equation $\dot{z} = f(z)$. Thus Euler's method for the second-order system becomes

$$q^{n+1} = q^n + \Delta t v^n,$$
$$v^{n+1} = v^n + \Delta t g(q^n).$$

2.1.2 Error analysis

Error analysis for a one-step method such as those discussed above centers on two issues: (i) the accuracy of the one-step approximation, as determined by comparison of the Taylor series expansions of true solutions and the numerical approximation in a timestep, and (ii) the accumulation of the error during computation of a discrete trajectory. The errors due to round-off in the computations (to which all floating point calculations are subject) are typically assumed to be much smaller than the errors due to the introduction of the approximation scheme, and are simply ignored.

We will use the term *local error* for the difference between the exact and approximate solutions on a single timestep of size Δt, say starting from a point $\bar{z}$ along some particular trajectory. Let us denote the local error at this point by $\mathbf{le}(\Delta t; \bar{z})$ (see Fig. 2.2). Using Taylor series, we can obtain an expansion in powers of Δt of the form

$$\mathbf{le}(\Delta t; \bar{z}) = \mathbf{C}_{p+1}(\bar{z})\Delta t^{p+1} + \mathbf{C}_{p+2}(\bar{z})\Delta t^{p+2} + \dots.$$

The integer p is termed the *order* of the method. Below we give the leading terms of the local error for the elementary one-step methods introduced earlier. We use the notation $z^{(k)}$ to denote the kth derivative of $z(t)$ at $z(t) = \bar{z}$.

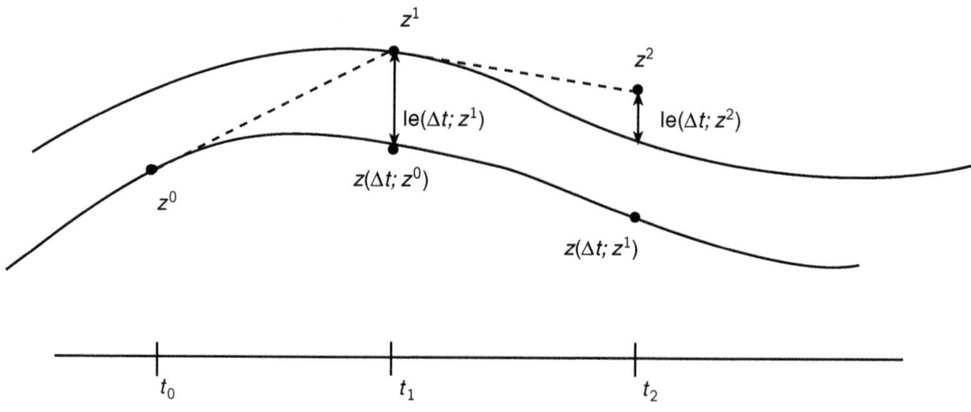

Figure 2.2 Local errors in numerical integration.

LOCAL ERROR FOR ELEMENTARY ONE-STEP METHODS

Euler's method	Implicit Euler
$\mathbf{le}(\Delta t; \bar{z}) = -\frac{1}{2}\ddot{z}\Delta t^2 + \mathcal{O}(\Delta t^3)$	$\mathbf{le}(\Delta t; \bar{z}) = \frac{1}{2}\ddot{z}\Delta t^2 + \mathcal{O}(\Delta t^3)$
Trapezoidal rule	**Implicit midpoint**
$\mathbf{le}(\Delta t; \bar{z}) = \frac{1}{12}z^{(3)}\Delta t^3 + \mathcal{O}(\Delta t^4)$	$\mathbf{le}(\Delta t; \bar{z}) = \frac{1}{12}z^{(3)}\Delta t^3 + \mathcal{O}(\Delta t^4)$

To make sense of these formulas, observe that any derivative of a trajectory $z(t)$ at the point where $z(t) = \bar{z}$ could be rewritten as a function just of $\bar{z}$ itself by using the differential equations. For example

$$\dot{z}(t) = \boldsymbol{f}(\bar{z}),$$

and

$$\ddot{z}(t) = \frac{d}{dt}\boldsymbol{f}(z(t))|_z(t) = \bar{z} = \boldsymbol{f}'(\bar{z})\boldsymbol{f}(\bar{z}),$$

where $\boldsymbol{f}'$ denotes the Jacobian matrix of $\boldsymbol{f}$. In this way, the leading terms in the local error given above can be viewed as products of a coefficient function of a point $\bar{z}$ with a power of Δt.

At the first integration step, the local error introduced is $\mathbf{le}(\Delta t; z^0)$. At the next step, the relevant term is $\mathbf{le}(\Delta t; z^1)$. A similar error is introduced at each step of the method. The issue then becomes how these errors accumulate over a certain time interval $[0, T]$. The form that the global error development takes depends on the type of method involved. A global error bound for Euler's method, for example, is easy to obtain under the assumption that $\boldsymbol{f}$ obeys a *Lipschitz condition* in an appropriate open set $D \subset \mathbb{R}^k$ containing the exact solution

$$\|\boldsymbol{f}(\boldsymbol{u}) - \boldsymbol{f}(\boldsymbol{v})\| \leq L\|\boldsymbol{u} - \boldsymbol{v}\|, \qquad \text{uniformly for all } \boldsymbol{u}, \boldsymbol{v} \in D, \quad (2.3)$$

where L is a positive constant. The following theorem then tells us that the largest error in the numerical solution is proportional to the stepsize.

Theorem 1 *Under the assumption (2.3), and provided the exact solution is twice continuously differentiable, then the error for Euler's method admits a bound of the form*

$$\|\boldsymbol{z}(t_n) - \boldsymbol{z}^n\| \leq K\left(e^{t_n L} - 1\right)\Delta t \leq K\left(e^{TL} - 1\right)\Delta t, \quad n = 1, 2, \ldots, N, \quad (2.4)$$

where K is independent of the integration interval T and the stepsize $\Delta t = T/N$. □

Proof. We first construct a recurrence relation for the numerical error defined by

$$\boldsymbol{e}^n = \boldsymbol{z}(t_n) - \boldsymbol{z}^n.$$

We make use of the Taylor expansion of the solution on the one hand and of the numerical method on the other, then regroup terms

$$\begin{aligned}
\boldsymbol{e}^{n+1} &= (\boldsymbol{z}(t_n) + \Delta t\dot{\boldsymbol{z}}(t_n) + \frac{1}{2}\Delta t^2\ddot{\boldsymbol{z}}(\tau)) - (\boldsymbol{z}^n + \Delta t\boldsymbol{f}(\boldsymbol{z}^n)) \\
&= (\boldsymbol{z}(t_n) + \Delta t\boldsymbol{f}(\boldsymbol{z}(t_n)) + \frac{1}{2}\Delta t^2\ddot{\boldsymbol{z}}(\tau)) - (\boldsymbol{z}^n + \Delta t\boldsymbol{f}(\boldsymbol{z}^n)) \\
&= (\boldsymbol{z}(t_n) - \boldsymbol{z}^n) + \Delta t(\boldsymbol{f}(\boldsymbol{z}(t_n)) - \boldsymbol{f}(\boldsymbol{z}^n)) + \frac{1}{2}\Delta t^2\ddot{\boldsymbol{z}}(\tau),
\end{aligned}$$

$\tau \in [t_n, t_{n+1}]$. Next we take norms, apply the triangle inequality, and use the Lipschitz condition (2.3) to obtain

$$\begin{aligned}
\|\boldsymbol{e}^{n+1}\| &\leq \|\boldsymbol{z}(t_n) - \boldsymbol{z}^n\| + \Delta t\|\boldsymbol{f}(\boldsymbol{z}(t_n)) - \boldsymbol{f}(\boldsymbol{z}^n)\| + \frac{1}{2}\Delta t^2\|\ddot{\boldsymbol{z}}(\tau)\| \\
&\leq (1 + \Delta t L)\|\boldsymbol{e}^n\| + \frac{1}{2}\Delta t^2\|\ddot{\boldsymbol{z}}(\tau)\|.
\end{aligned}$$

Since the solution is twice continuously differentiable, we may bound $\ddot{\boldsymbol{z}}(\tau)$ on $[0, T]$ by a constant M. Observe that a linear recurrence relation of the form

$$a_{n+1} \leq Ca_n + D$$

satisfies the bound $a_n \leq C^n a_0 + \frac{C^n-1}{C-1}D$. The bound (2.4) with $K = \frac{M}{2L}$ follows by applying this result to the case at hand. In particular, we set $a_n = \|e^n\|$, $C = 1 + \Delta t L$, $D = \Delta t^2 M/2$ and observe that $a^0 = 0$ as well as

$$(1 + \Delta t L)^n \leq e^{n\Delta t L}.$$

□

It is important to note several features of this bound, which is representative of the types of error bounds obtainable for numerical difference equations. First, the positive constant L may be quite large (it can be obtained as the norm of the Jacobian matrix $\boldsymbol{f}'$ of $\boldsymbol{f}$). The error bound thus grows exponentially in time and will quickly have little to tell us quantitatively about the numerical solution.

Fixing the length of the time interval, T, the error inequality does provide us with a precise upper bound on the error and suggests that the error is reduced in proportion to the reduction of the stepsize. In practical applications, such a linear scaling relationship between the timestep and error *is* usually observed, even if the scaling factor is generally found to be rather smaller (but still exponential in T) in comparison with that given by the bound (Ke^{TL}).

2.2 Numerical example: the Lennard–Jones oscillator

In many situations, the error will exhibit a nonuniform growth, due to abrupt changes in the dynamics. Let us illustrate this phenomenon by examining the growth of error when Euler's method is applied to a Lennard–Jones oscillator

$$\dot{q} = v,$$
$$\dot{v} = -\varphi'(q),$$

where $\varphi(q) = q^{-12} - 2q^{-6}$. It is clear that the potential will rise rapidly without bound for $q \to 0$, thus solutions of the system must be bounded away from $q = 0$ if the energy $E = v^2/2 + \varphi(q)$ is to remain constant along trajectories. Moreover, it can be shown that the solutions are bounded for $E \leq 0$ and unbounded for $E > 0$. A trajectory is shown on the left in Fig. 2.3. On the right in the same figure, the energy error is plotted as a function of time, for a sample trajectory with stepsize $\Delta t = 0.001$.

For studying examples such as this one, it is often useful to keep in mind a local propagation relation of the form

$$\|\boldsymbol{e}^{n+1}\| \leq (1 + \Delta t a(t_n))\|\boldsymbol{e}^n\| + \Delta t^2 b(t_n) + O(\Delta t^3). \tag{2.5}$$

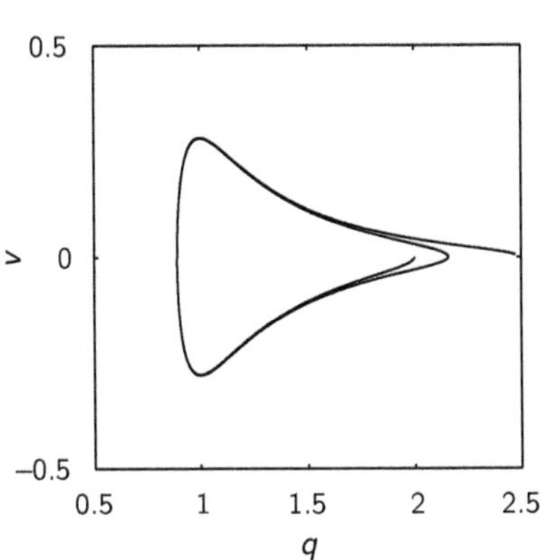

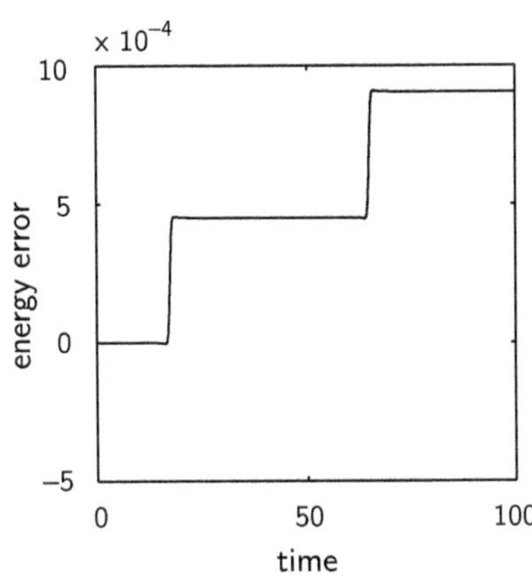

Figure 2.3 Solution curves in the qv-plane (left) and energy error (right) for Euler's method applied to the Lennard–Jones oscillator.

The functions $a = a(t)$ and $b = b(t)$ depend on a particular trajectory of the system. Such an estimate would hold for any first-order method. For Euler's method, we have

$$a(t_n) = \|\boldsymbol{f}'(\boldsymbol{z}(t_n))\|,$$
$$b(t_n) = \max_{\tau \in [t_n, t_{n+1}]} \frac{1}{2}\|\ddot{\boldsymbol{z}}(\tau)\|,$$

where $\boldsymbol{z} = \boldsymbol{z}(t)$ represents the particular trajectory of interest, and $\boldsymbol{f}'$ denotes the Jacobian matrix of $\boldsymbol{f}$. A proof is developed in the exercises of this chapter. While still only a bound, (2.5) indicates that increase in the error in Euler's method at a timestep arises from two sources: the sensitivity of the solution to perturbations in the initial data (measured by $a(t)$) and the local smoothness of the solution (measured by $b(t)$).

Let us examine the bound (2.5) in the case of a nonlinear oscillator. In terms of the variables q and $v = \dot{q}$, the vector field has the expression

$$\boldsymbol{f}(\boldsymbol{z}) = \begin{bmatrix} v \\ -\varphi'(q) \end{bmatrix}.$$

We can write the 2×2 Jacobian matrix of $\boldsymbol{f}$ as

$$\boldsymbol{f}'(\boldsymbol{z}) = \begin{bmatrix} 0 & 1 \\ -\varphi''(q) & 0 \end{bmatrix}.$$

The 2-norm of a matrix $\boldsymbol{A}$ is the square root of the magnitude of the largest eigenvalue of $\boldsymbol{A}^T\boldsymbol{A}$, which for our Jacobian matrix is just $\max\{1, |\varphi''(q)|\}$. On the other hand, the second derivative of a solution $z = z(t)$ can be evaluated in terms of the vector field, i.e.

$$\ddot{z} = \begin{bmatrix} \ddot{q} \\ \ddot{v} \end{bmatrix} = \begin{bmatrix} -\varphi' \\ -\varphi'' v \end{bmatrix},$$

hence

$$\|\ddot{z}(t)\| = \sqrt{(\varphi'(q(t)))^2 + (\varphi''(q(t)))^2 v(t)^2}.$$

We can use the energy relation $\frac{1}{2}v^2 + \varphi(q) = E$ to replace the latter expression by one involving q only. Summarizing, the two numbers $a(t)$ and $b(t)$ which govern the potential growth of errors in the bound (2.5) are

$$a(t) = \max\{1, |\varphi''(q(t))|\},$$
$$b(t) = \frac{1}{2}\sqrt{(\varphi'(q(t)))^2 + 2(\varphi''(q(t)))^2(E - \varphi(q(t))}.$$

If we graph these two numbers against time along a trajectory, we observe fluctuations in a, b and the error as shown in Fig. 2.4. Both a and b increase rapidly as q approaches the singularity at $q = 0$. The graphs indicate that the largest growth in the error is well correlated with the increase in a and b.

This example highlights an important issue for numerical simulation: the need to incorporate adaptivity in the timestep when solving problems for which the solution exhibits a varying degree of smoothness on the relevant time interval. We will return to this issue in Chapter 9.

2.3 Higher-order methods

The Euler method is a *first-order* integrator, meaning that it has global error proportional to the stepsize. Higher-order methods have global error which satisfies a higher power law in the stepsize.

Suppose that a given one-step method $\boldsymbol{\Psi}_{\Delta t}$ is such that, at some given point $\bar{z}$ of phase space, $\boldsymbol{\Psi}_{\Delta t}(\bar{z})$ approximates the solution $z(\Delta t, \bar{z})$ through $\bar{z}$ for small Δt. The quality of this local approximation can be measured by comparing the Taylor series expansions of $\boldsymbol{\Psi}_{\Delta t}(\bar{z})$ and $z(\Delta t, \bar{z})$ in terms of Δt. The order of the numerical one-step method is then defined as the largest integer $p \geq 1$ such that coefficients of the stepsize powers Δt^i, $i = 1, \ldots, p$, are identical in the Taylor series expansions or, in other words,

$$||\boldsymbol{\Psi}_{\Delta t}(\bar{z}) - z(\Delta t, \bar{z})|| \leq C\Delta t^{p+1},$$

for all $\bar{z}$ in the domain of interest, where $C > 0$ is an appropriate constant.

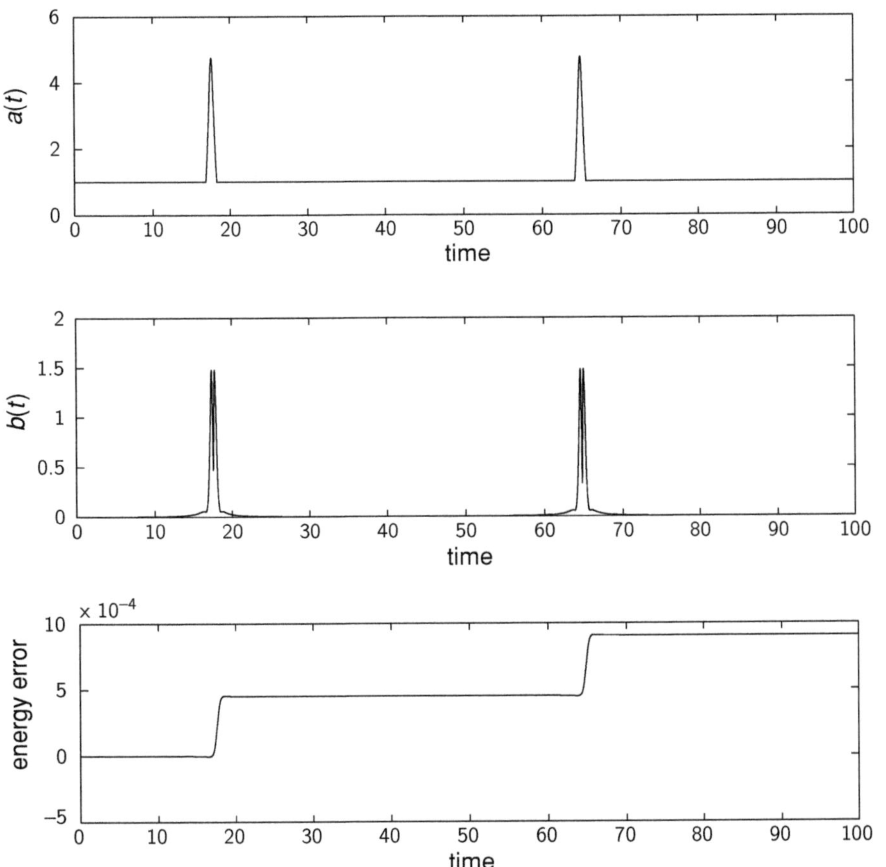

Figure 2.4 Comparison of the variation in coefficients a and b with the evolution of the error.

Let us assume that our vector field satisfies the Lipschitz condition (2.3). For many classes of one-step methods, including the Runge–Kutta methods considered in the next subsection, it is then possible to show that the map $\boldsymbol{\Psi}_{\Delta t}$ itself obeys a Lipschitz condition of the form:

$$||\boldsymbol{\Psi}_{\Delta t}(z_1) - \boldsymbol{\Psi}_{\Delta t}(z_2)|| \leq (1 + \bar{L}\Delta t)||z_1 - z_2||, \qquad \bar{L} \geq L.$$

Using an argument very similar to that used to prove the convergence of Euler's method (see, for example, [82] for details) one then finds that over a fixed time interval, the approximation of the flow by timesteps of size Δt will be accurate to within $\mathcal{O}(\Delta t^p)$, i.e.,

$$||\,\boldsymbol{\Psi}^n_{\Delta t}(z^0) - z(n\Delta t, z^0)\,|| \leq K\left(e^{\bar{L}T} - 1\right)\Delta t^p, \qquad 1 \leq n \leq N, \quad (2.6)$$

where $K > 0$ is an appropriate constant.

As mentioned above, the practical value of this type of estimate is severely limited by the presence of the exponential term involving the time interval, by the involvement of the Lipschitz constant which can be exceedingly large at certain points during the course of integration, and by the fact that the estimate does not take into account any structure of the map $\boldsymbol{\Psi}_{\Delta t}$, such as whether it is expanding or contracting. Nonetheless, the global error bound is useful on shorter time intervals.

Using relation (2.6) and the leading terms in the local error expansions for various methods worked out earlier, we can conclude that the Euler and implicit Euler methods are first-order accurate, while the trapezoidal rule and implicit midpoint methods are of second order.

2.4 Runge–Kutta methods

All of the methods discussed so far are special cases of Runge–Kutta methods. The class of general s-stage Runge–Kutta methods is given below.

RUNGE–KUTTA METHODS FOR $\frac{d}{dt}z = f(z)$

$$z^{n+1} = z^n + \Delta t \sum_{i=1}^{s} b_i \boldsymbol{f}(\boldsymbol{Z}_i),$$

where, for $i = 1, 2, \ldots, s$,

$$\boldsymbol{Z}_i = z^n + \Delta t \sum_{j=1}^{s} a_{ij} \boldsymbol{f}(\boldsymbol{Z}_j).$$

The number of stages s and the constant coefficients $\{b_i\}$, $\{a_{ij}\}$ completely characterize a Runge–Kutta method. In general, such a method is implicit and leads to a nonlinear system in the s internal stage variables $\boldsymbol{Z}_i$. In some cases the formulas are layered in such a way that the first stage variable $\boldsymbol{Z}_1$ is given explicitly in terms of z^n, then $\boldsymbol{Z}_2$ is determined from z^n and $\boldsymbol{Z}_1$, etc.

An example of a fourth-order explicit Runge–Kutta method is given next.

A FOURTH-ORDER RUNGE–KUTTA METHOD (RK-4)

$$Z_1 = z^n, \tag{2.7}$$

$$Z_2 = z^n + \frac{1}{2}\Delta t f(Z_1), \tag{2.8}$$

$$Z_3 = z^n + \frac{1}{2}\Delta t f(Z_2), \tag{2.9}$$

$$Z_4 = z^n + \Delta t f(Z_3), \tag{2.10}$$

$$z_{n+1} = z_n + \frac{\Delta t}{6}\left[f(Z_1) + 2f(Z_2) + 2f(Z_3) + f(Z_4)\right]. \tag{2.11}$$

To see that this method has fourth order, one views (2.7)–(2.11) as defining a function $z^{n+1} = \Psi_{\Delta t}(z^n)$. We take the difference of this function with the solution $z(\Delta t; z^n)$ through z^n, then expand this difference in a Taylor series about $\Delta t = 0$. After some simplification, it will be found that the constant through fourth-order terms of this expansion in powers of Δt vanish identically, and we are left with

$$z^{n+1} - z(\Delta t; z^n) = C_5(z^n)\Delta t^5 + \mathcal{O}(\Delta t^6),$$

where the coefficient function $C_5(z)$ depends on the vector field f and its derivatives. Since the method introduces a local error of size proportional to Δt^5 at each step, it will exhibit fourth-order global accuracy.

In some cases the increased order of accuracy associated with a particular higher-order explicit Runge–Kutta method may provide a better approximation of the solution, but this increased accuracy must be balanced against the increased work involved in computing the timestep. Moreover, in very long-term simulations or at large stepsizes, nonphysical effects generally become apparent, such as energy drift or artificial dissipation.

This is well illustrated by reconsidering the Lennard–Jones oscillator solved earlier with Euler's method. The graph of solutions obtained using the fourth-order Runge–Kutta method given above is shown in Fig. 2.5 along with the energy error. It is immediately clear that the error is much smaller in magnitude, by a factor of more than ten, despite a much increased timestep of $\Delta t = 0.1$. But note, too, that the energy error grows in a similar way, in jumps associated to the approach of q to the singularity at the origin. If the simulation is carried out on a much longer time interval (but still very short in relation to an actual molecular dynamics simulation), the error steadily accumulates (Fig. 2.6, left) until the numerical solution bears little relation to the true orbit. Another

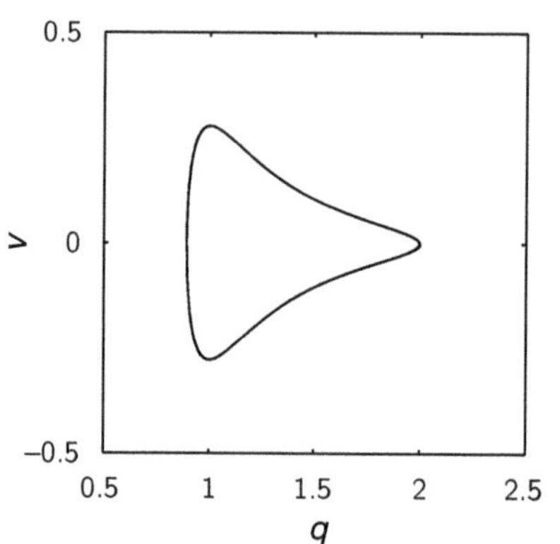

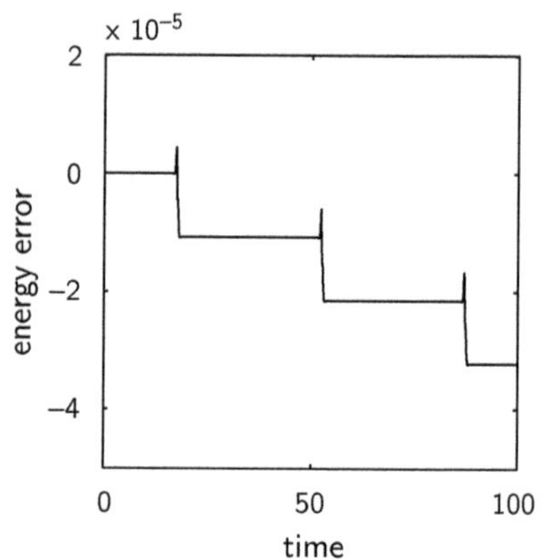

Figure 2.5 Solution curves in the qv-plane (left) and energy error (right) for the fourth-order Runge–Kutta method applied to the Lennard–Jones oscillator.

important observation to be made from (2.6) is that the drift in the energy error (right panel) can be very small compared with the error being introduced in the solution itself. Qualitatively, the picture for Euler's method and the picture for the fourth-order Runge–Kutta method are quite similar: steady drift is observed in the energy and in the solution trajectories, eventually rendering the results of numerical computation useless.

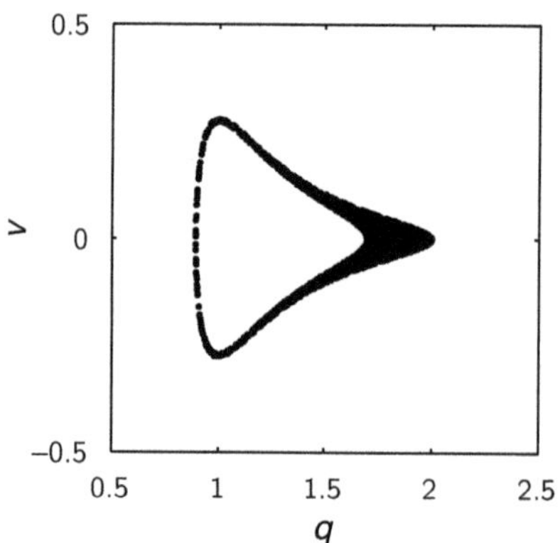

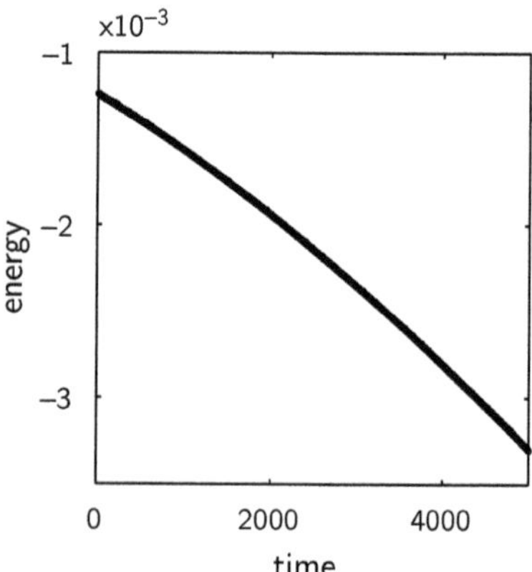

Figure 2.6 Solution curves in the qv-plane (left) and energy error (right) for the fourth-order Runge–Kutta method applied to the Lennard–Jones oscillator on a time interval $[0, 5000]$.

2.5 Partitioned Runge–Kutta methods

Yet another approach to approximating the solution trajectory is based on using different approximation formulas for different components of the solution. For example, we will often treat canonical Hamiltonian systems in this book which admit a natural dichotomy between positions and momenta. Partitioned Runge–Kutta (PRK) methods exploit this dichotomy by using different sets of quadrature rules for each subset of the variables. If the system can be written in the form $\frac{d}{dt}\boldsymbol{u} = \boldsymbol{g}(\boldsymbol{u}, \boldsymbol{v})$, $\frac{d}{dt}\boldsymbol{v} = \boldsymbol{h}(\boldsymbol{u}, \boldsymbol{v})$, then the associated Partitioned Runge–Kutta method of s stages uses two sets of coefficients $(\{\bar{b}_i\}, \{\bar{a}_{ij}\})$ and $(\{\tilde{b}_i\}, \{\tilde{a}_{ij}\})$, and computes the timestep as follows:

PARTITIONED RUNGE–KUTTA METHODS FOR $\frac{d}{dt}\boldsymbol{u} = \boldsymbol{g}(\boldsymbol{u}, \boldsymbol{v})$; $\frac{d}{dt}\boldsymbol{v} = \boldsymbol{h}(\boldsymbol{u}, \boldsymbol{v})$

$$\boldsymbol{u}^{n+1} = \boldsymbol{u}^n + \Delta t \sum_{i=1}^{s} \bar{b}_i \boldsymbol{g}(\boldsymbol{U}_i, \boldsymbol{V}_i),$$

$$\boldsymbol{v}^{n+1} = \boldsymbol{v}^n + \Delta t \sum_{i=1}^{s} \tilde{b}_i \boldsymbol{h}(\boldsymbol{U}_i, \boldsymbol{V}_i),$$

where, for $i = 1, 2, \ldots, s$,

$$\boldsymbol{U}_i = \boldsymbol{u}^n + \Delta t \sum_{j=1}^{s} \bar{a}_{ij} \boldsymbol{g}(\boldsymbol{U}_j, \boldsymbol{V}_j),$$

$$\boldsymbol{V}_i = \boldsymbol{v}^n + \Delta t \sum_{j=1}^{s} \tilde{a}_{ij} \boldsymbol{h}(\boldsymbol{U}_j, \boldsymbol{V}_j).$$

These methods are often used for treating mechanically derived problems of the form:

$$\boldsymbol{M}\frac{d^2}{dt^2}\boldsymbol{q} = -\nabla_{\boldsymbol{q}} V(\boldsymbol{q}),$$

with $\boldsymbol{u} = \boldsymbol{q}$ and $\boldsymbol{v} = \frac{d}{dt}\boldsymbol{q}$, in which setting they are typically referred to as Runge–Kutta–Nyström (RKN) methods.

One example of this type of Partitioned Runge–Kutta method is the method below

$$q^{n+1} = q^n + \Delta t v^n, \tag{2.12}$$
$$Mv^{n+1} = Mv^n - \Delta t \nabla_q V(q^{n+1}). \tag{2.13}$$

This method is very similar to Euler's method, except that the vector field is evaluated in one component at time level $n+1$ and in the other at time level n. It is immediately obvious that we could as well have proposed the alternative method

$$q^{n+1} = q^n + \Delta t v^{n+1}, \tag{2.14}$$
$$Mv^{n+1} = Mv^n - \Delta t \nabla_q V(q^n). \tag{2.15}$$

Both of these methods are explicit and first-order accurate. We call these methods collectively the Asymmetrical Euler methods. To distinguish these two schemes, we refer to the first (2.12)–(2.13) as *Asymmetrical Euler-A* and the second as *Asymmetrical Euler-B* (or, briefly, *Euler-A* and *Euler-B*).

Another Partitioned Runge–Kutta method is the *Störmer–Verlet* method for Newton's equations ($\dot{q} = v$, $M\dot{v} = -\nabla_q V(q)$), defined as follows:

STÖRMER–VERLET METHOD

$$q^{n+1} = q^n + \Delta t v^{n+\frac{1}{2}}, \tag{2.16}$$
$$Mv^{n+\frac{1}{2}} = Mv^n - \frac{\Delta t}{2}\nabla_q V(q^n), \tag{2.17}$$
$$Mv^{n+1} = Mv^{n+\frac{1}{2}} - \frac{\Delta t}{2}\nabla_q V(q^{n+1}). \tag{2.18}$$

Evidently, the scheme is fully explicit, since we can evaluate successively the formula (2.17), then (2.16) and finally (2.18) without solving any nonlinear system. Note further that the discretization can be solved in terms of the half-step velocites $v^{n-\frac{1}{2}}$, $v^{n+\frac{1}{2}}$, etc., resulting in

$$q^{n+1} = q^n + \Delta t v^{n+\frac{1}{2}},$$
$$Mv^{n+\frac{1}{2}} = Mv^{n-\frac{1}{2}} - \Delta t \nabla_q V(q^n).$$

One can altogether eliminate the velocities from the Störmer–Verlet method to obtain the following two-step method in the coordinates only

$$M\frac{q^{n+1} - 2q^n + q^{n+1}}{\Delta t^2} = -\nabla_q V(q^n). \tag{2.19}$$

This formulation is often called the *leapfrog* method. Both the leapfrog and the Störmer–Verlet formulation can be shown to be second-order accurate.

2.6 Stability and eigenvalues

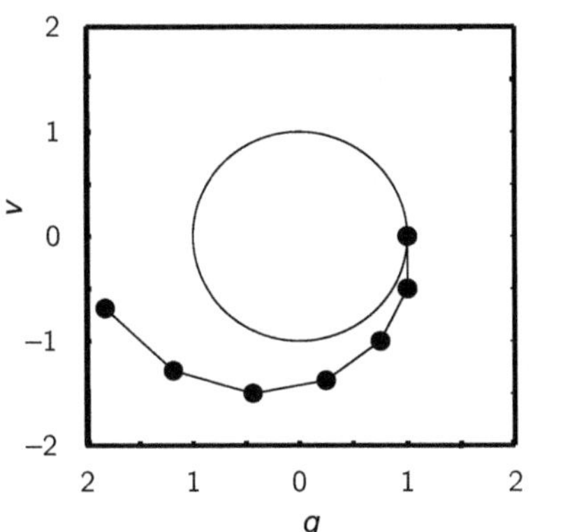

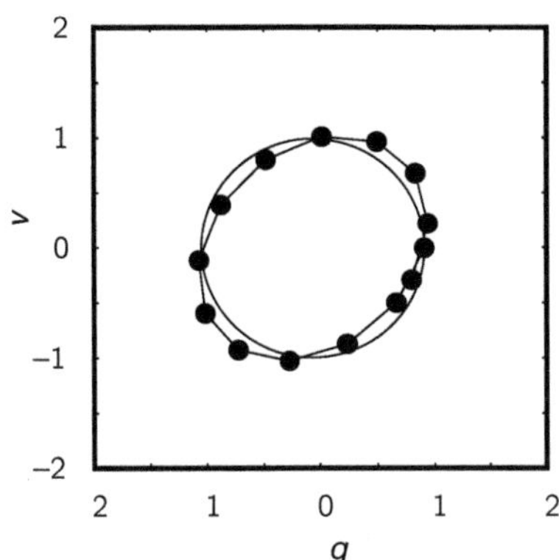

Figure 2.7 Numerical solution of the harmonic oscillator obtained using Euler's method (left) and the Euler-B method (right).

If we graph a solution of the harmonic oscillator $\dot{q} = v$; $\dot{v} = -q$ we should see a circle in the phase plane (the qv-plane). On the other hand, if we apply a numerical method and compute discrete points with the method, there is no reason to expect that these would lie on a circle. Numerical solutions obtained using two different methods are shown in Fig. 2.7, graphed along with the circle that would represent the exact solution for the chosen initial condition. Observe that the solutions spiral out in the case of the first method (left) and that they appear to lie on an ellipse in the case of the second method (right). The method used to produce the illustration on the left is Euler's method, while the scheme used on the right is the Euler-B method (2.14)–(2.15) we encountered in the previous section.

While neither method exactly replicates the circular orbits, it seems clear that there is an important qualitative difference between the two schemes. The terminology we use to discuss this distinction is *asymptotic stability*. In this section, we show that the long-term asymptotic dynamics of numerical methods can be thoroughly understood in the case of the harmonic oscillator, or, more generally, any linear mechanical system.

Write the equation for the harmonic oscillator in the form

$$\frac{d}{dt}z = \boldsymbol{A}z,$$

where

$$\boldsymbol{A} = \begin{bmatrix} 0 & 1 \\ -\omega^2 & 0 \end{bmatrix}.$$

The solution at any time can be defined by a matrix (the fundamental solution matrix), $\boldsymbol{R}(t)$

$$z(t; z^0) = \boldsymbol{R}(t)z^0, \qquad \boldsymbol{R}(t) = \begin{bmatrix} \cos\omega t & \frac{1}{\omega}\sin\omega t \\ -\omega\sin\omega t & \cos\omega t \end{bmatrix},$$

which has the eigenvalues $\mu_{1,2} = e^{\pm i\omega t}$, both of which lie on the unit circle. It is also easy to verify that the determinant of $\boldsymbol{R}(t)$ is equal to one.

The Euler's method approximation leads to the mapping

$$z^{n+1} = \hat{\boldsymbol{R}}(\Delta t)z^n, \qquad \hat{\boldsymbol{R}}(\Delta t) = \begin{bmatrix} 1 & \Delta t \\ -\Delta t\,\omega^2 & 1 \end{bmatrix},$$

where the propagation matrix $\hat{\boldsymbol{R}}(\Delta t)$ has the eigenvalues

$$\hat{\lambda}_{1,2} = 1 \pm i\Delta t\omega.$$

The growth of error through the powers $[\hat{\boldsymbol{R}}]^n$ is determined by the powers of the eigenvalues of that matrix. A numerical method is *asymptotically stable* if the growth of the solution for a linear model problem is asymptotically bounded. A sufficient condition for asymptotic stability is that the eigenvalues of the method are (i) in the unit disk in the complex plane, and (ii) simple (not repeated) if on the unit circle. Since the eigenvalues of Euler's method are both of modulus greater than one, their powers grow *exponentially fast* and the method is unstable.

Note that the asymptotic instability of Euler's method does not contradict the convergence of the method, since fixing any time interval $[0, T]$ and simultaneously driving the number of steps N to infinity as $\Delta t \to 0$ so that $N\Delta t = T$, we have

$$\lim_{N\to\infty} (\hat{\lambda}_{1,2})^N = \lim_{N\to\infty} (1 \pm i\Delta t\omega)^N = e^{\pm i\omega T} + O(\Delta t).$$

In some cases, it is possible to show that the eigenvalues of a numerical method applied to the harmonic oscillator also lie on the unit circle in the complex plane. Applying the Störmer–Verlet method to the harmonic oscillator results in the propagator

$$\hat{\boldsymbol{R}}(\Delta t) = \begin{bmatrix} 1 - \frac{\Delta t^2\omega^2}{2} & \Delta t \\ -\Delta t\omega^2\left(1 - \frac{\Delta t^2\omega^2}{4}\right) & 1 - \frac{\Delta t^2\omega^2}{2} \end{bmatrix}, \tag{2.20}$$

which has eigenvalues

$$\hat{\lambda}_{1,2} = 1 - \mu^2 \pm \sqrt{2\mu^2(\frac{1}{2}\mu^2 - 1)}, \qquad \mu^2 = \frac{1}{2}\Delta t^2\omega^2.$$

Note that $\hat{\lambda}_1\hat{\lambda}_2 = 1$ for all values of $\mu \geq 0$ implying that the determinant of $\hat{\boldsymbol{R}}(\Delta t)$ is equal to one. Furthermore, for $\mu^2 < 2$, the eigenvalues are complex, and

$$|\hat{\lambda}_{1,2}|^2 = (1-\mu^2)^2 + 2\mu^2(1-\frac{1}{2}\mu^2) = 1.$$

For $\mu^2 > 2$, the eigenvalues are real and one has modulus greater than unity. Thus the Störmer–Verlet rule has a stability condition of the form

$$\Delta t^2\omega^2 < 4, \tag{2.21}$$

when applied to the harmonic oscillator.

As we vary the stepsize Δt from zero, the eigenvalues of the Störmer–Verlet method move around the unit circle until $\Delta t = 2/\omega$, at which timestep both eigenvalues are at -1. For larger stepsizes, one eigenvalue heads toward the origin along the negative real axis, while the other goes off to infinity.

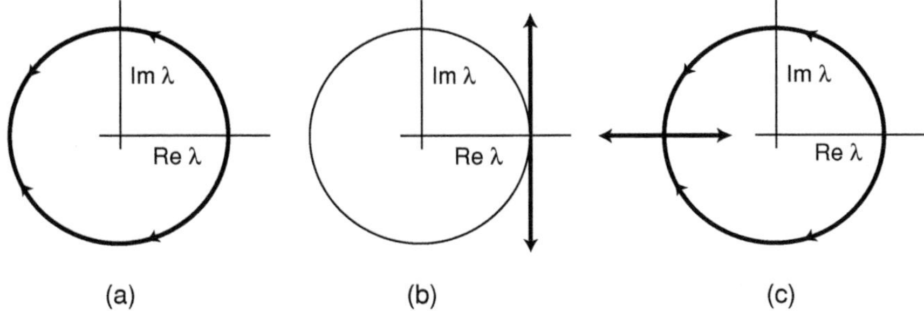

Figure 2.8 Comparison of eigenvalue curves (bold, marked by arrows) for (a) the true propagator, (b) Euler's method, and (c) the Störmer–Verlet method.

The eigenvalues for the propagators obtained by using Euler's method and the Störmer–Verlet method are diagrammed in Fig. 2.8(b) and (c). Compare these curves (marked by arrows) with the eigenvalues of the exact propagator (Fig. 2.8(a). It is more useful to visualize the stability properties of the different methods as curves in three-dimensional space, parameterized by adding a dimension of time (or timestep). For the true propagator, these eigenvalue curves are helices on the cylinder of radius one (see Fig. 2.9). Numerical methods like Euler's method (or the fourth-order Runge–Kutta method) have eigenvalue curves that leave the cylinder, even for small Δt, whereas for certain other methods such as Euler-B (see Exercises) and Störmer–Verlet, the eigenvalues remain on the surface of the cylinder. until $\Delta t = 2/\omega$. The curves for Euler and Störmer–Verlet are also shown in Fig. 2.9. In Chapter 5, we will see that this desirable linear stability property can be viewed as the elementary consequence of certain general geometric principles respected by the Störmer–Verlet and Euler-B methods. In the case considered here, these geometric principles reduce to the fact that the determinant of the matrices $\boldsymbol{R}(t)$ and $\hat{\boldsymbol{R}}(\Delta t)$, respectively, are equal to one.

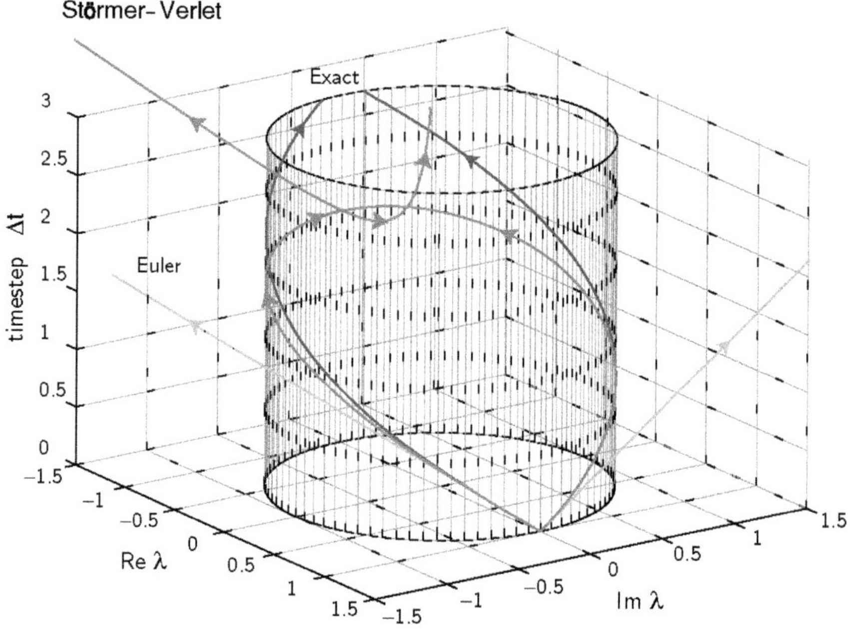

Figure 2.9 Eigenvalue curves as functions of time or timestep. The eigenvalues of the exact propagator for the harmonic oscillator are helices on the unit cylinder, whereas the eigenvalues of Euler's method immediately leave the cylinder. Those of Störmer–Verlet remain on the cylinder until $\Delta t = 2/\omega$.

We can easily extend this discussion to a $2N$-dimensional linear system of differential equations of the form

$$\frac{d}{dt}\boldsymbol{q} = \boldsymbol{v}, \tag{2.22}$$

$$\frac{d}{dt}\boldsymbol{v} = -\boldsymbol{K}\boldsymbol{q}. \tag{2.23}$$

We assume that $\boldsymbol{K}$ is an $N \times N$ constant symmetric matrix. (In general, if this system arose from Newton's laws of motion in linear forces, or through a process of linearization of a nonlinear mechanical system, we might expect to find mass coefficients multiplying the accelerations in the second equation, but we ignore these for simplicity.)

The behavior of solutions to a linear system such as this one over time is determined by the eigenvalue–eigenvector structure of the matrix $\boldsymbol{K}$. Recall that an eigenvalue–eigenvector pair for a symmetric $N \times N$ matrix $\boldsymbol{K}$ consists of a (real) scalar λ and a nonzero N-vector $\boldsymbol{v}$, related by the equation

$$\boldsymbol{K}\boldsymbol{v} = \lambda \boldsymbol{v}.$$

Furthermore the eigenvectors form a complete basis in $\mathbb{R}^N$. Denote the eigenvalue–eigenvector pairs by $(\lambda_i, \boldsymbol{v}_i)$, $i = 1, \ldots, N$. The eigenvectors can be viewed as the columns of a matrix $\boldsymbol{V}$ with the property

$$\boldsymbol{K}\boldsymbol{V} = [\boldsymbol{K}\boldsymbol{v}_1, \boldsymbol{K}\boldsymbol{v}_2, \ldots, \boldsymbol{K}\boldsymbol{v}_N] = [\lambda_1 \boldsymbol{v}_1, \lambda_2 \boldsymbol{v}_2, \ldots, \lambda_N \boldsymbol{v}_N] = \boldsymbol{V}\boldsymbol{\Lambda},$$

where $\boldsymbol{\Lambda}$ is a diagonal matrix with the eigenvalues on the diagonal. Since the columns of $\boldsymbol{V}$ are linearly independent, the matrix $\boldsymbol{V}$ is nonsingular, so we may premultiply both sides of the above equation by $\boldsymbol{V}^{-1}$ to obtain

$$\boldsymbol{V}^{-1}\boldsymbol{K}\boldsymbol{V} = \boldsymbol{\Lambda},$$

and we say that the matrix $\boldsymbol{K}$ is *diagonalizable.*

Returning to the linear mechanical system, the behavior of solutions is now easily analyzed. Starting from the linear differential equations (2.22–2.23), we make the nonsingular changes of variables

$$\boldsymbol{q}(t) = \boldsymbol{V}\boldsymbol{u}(t),$$

and

$$\boldsymbol{v}(t) = \boldsymbol{V}\boldsymbol{w}(t),$$

resulting in the equations

$$\frac{d}{dt}\boldsymbol{V}\boldsymbol{u} = \boldsymbol{V}\boldsymbol{w},$$
$$\frac{d}{dt}\boldsymbol{V}\boldsymbol{w} = -\boldsymbol{K}\boldsymbol{V}\boldsymbol{u}.$$

After premultiplying each equation by $\boldsymbol{V}^{-1}$, we arrive at

$$\frac{d}{dt}\boldsymbol{u} = \boldsymbol{w},$$
$$\frac{d}{dt}\boldsymbol{w} = -\boldsymbol{V}^{-1}\boldsymbol{K}\boldsymbol{V}\boldsymbol{u} = -\boldsymbol{\Lambda}\boldsymbol{u}.$$

Because $\boldsymbol{\Lambda}$ is diagonal, we see that the system of $2N$ equations reduces to N decoupled 2×2 systems of the form

$$\frac{d}{dt}u_i = w_i, \tag{2.24}$$
$$\frac{d}{dt}w_i = -\lambda_i u_i. \tag{2.25}$$

The eigenvalues λ_i are real, since $\boldsymbol{K}$ is symmetric. If the matrix $\boldsymbol{K}$ is, moreover, positive definite, then the λ_i are all positive, and the decoupled systems are harmonic oscillators.

The numerical analysis of this system proceeds along similar lines. We apply a discretization method to (2.22)–(2.23), then use the same set of linear transformations to decouple the variables. In the end, we find that the stability of the numerical method on the linear system is determined by the stability of the same method when applied to each of the oscillators (2.24)–(2.25).

Thus, for a linear mechanical system with potential $V = \frac{1}{2}\boldsymbol{q}^T\boldsymbol{K}\boldsymbol{q}$, with $\boldsymbol{K}$ a symmetric positive definite matrix, the harmonic stability condition such as that for the Störmer–Verlet method (2.21) must be applied with ω^2 replaced by each of the eigenvalues of the matrix $\boldsymbol{K}$. For large systems, the number of eigenvalues and different components or modes of the solution will be similarly large, but for the asymptotic stability of the numerical solution, *the restriction on the timestep for all of the components is determined by the timestep restriction for the fastest mode of the system.*

Much recent research in numerical methods for differential equations is aimed at developing new classes of methods which allow the slow modes to be propagated more efficiently in the presence of such high-frequency components, and we introduce some such schemes in Chapter 10.

2.7 Exercises

1. *Euler's method.* Assume that the vector field $\boldsymbol{f}$ is continuously differentiable and a given solution $z = z(t)$ is twice continuously differentiable. Assume the convergence of Euler's method, then use this to prove the following error growth relation for Euler's method

$$\|\boldsymbol{e}^{n+1}\| \leq (1 + \Delta t a(t_n))\|\boldsymbol{e}^n\| + \Delta t^2 b(t_n) + \mathcal{O}(\Delta t^3),$$

where

$$a(t_n) = \|\boldsymbol{f}'(z(t_n))\|,$$
$$b(t_n) = \max_{\tau\in[t_n,t_{n+1}]} \frac{1}{2}\|\ddot{z}(\tau)\|.$$

2. *Asymmetrical Euler: order of accuracy.* Show that the Euler-B method (2.14)–(2.15) is a Partitioned Runge–Kutta method by defining all of the relevant coefficients. Show that this method introduces an error proportional to Δt^2 in one step when applied to a general one-degree-of-freedom problem and find the leading term in the local error expansion for this method.

3. *Asymmetrical Euler: stability.* Determine the stability condition for the Euler-A method using the technique of Section 2.6, and graph the eigenvalues when applied to the harmonic oscillator. Compare with the corresponding stability conditions of the Störmer–Verlet and Euler methods.

4. *Stability: the pendulum.* Consider the planar pendulum described by

$$\ddot{\theta} = -\frac{g}{L}\sin\theta,$$

where g is the earth's gravitational constant and L the length of the pendulum. Using a linearization of this equation, determine the stepsize stability restriction for the Euler-B method applied to integrate the system near the hanging-down configuration ($\theta = 0$).

Note: In general, the stability of the discrete linearized problem at a critical point does *not* automatically imply the stability of the corresponding nonlinear equations. In order to say something about the latter, more powerful methods such as the KAM theory must typically be invoked (see, for example, [197]).

5. *Boundary of the stability region.* We have seen that for $\Delta t\omega < 2$, the eigenvalues of the Störmer–Verlet method applied to the Harmonic oscillator are both on the unit circle and are distinct. The special case $\Delta t\omega = 2$ is interesting, since it represents the boundary of the stability region. Is the Störmer–Verlet method stable when applied to the Harmonic oscillator (in the sense that the numerical solution is bounded asymptotically for all n) when $\Delta t\omega = 2$? Explain.

6. *Linear spring-mass system.* Consider a linear system in 1D consisting of N-point particles with nonuniform masses m_i, with the i, j particle pair joined by a zero rest-length spring (spring constant k_{ij}).

a. Show that the total potential energy can be written $V = \frac{1}{2}\boldsymbol{q}^T\boldsymbol{K}\boldsymbol{q}$, where $\boldsymbol{K}$ is a symmetric, positive semi-definite matrix, then show that the equations of motion are in the form of an N-body system

$$\boldsymbol{M}\frac{d^2}{dt^2}\boldsymbol{q} = -\boldsymbol{K}\boldsymbol{q},$$

where $\boldsymbol{M}$ is a diagonal matrix.

b. Introduce a scaling $\tilde{\boldsymbol{q}}_i = \sqrt{m_i}\boldsymbol{q}_i$ and show that in the new variables, the equations can be written

$$\frac{d^2}{dt^2}\tilde{\boldsymbol{q}} = -\tilde{\boldsymbol{K}}\tilde{\boldsymbol{q}},$$

where $\tilde{\boldsymbol{K}}$ is symmetric and positive semi-definite.

c. Consider the special case of a homogeneous nearest neighbor spring system with both boundaries fixed at the origin, N unit masses, and uniform spring constant κ. Write out the solution in this case. Hint: The $N \times N$ matrix $\boldsymbol{A} = (a_{ij})$ with

$$a_{ij} = \begin{cases} -2, & i = j \\ 1, & |i - j| = 1 \\ 0, & \text{else} \end{cases}$$

has eigenvalues

$$\lambda_k = -2(1 + \cos(\theta_k)),$$

and corresponding eigenvectors $\boldsymbol{u}_k$ with components

$$u_{kj} = \sin(j\theta_k), \qquad j = 1, \ldots, N,$$

where $\theta_k = \frac{\pi k}{N+1}$, $k = 1, 2, \ldots, N$.

d. Determine the numerical stability restriction for using Euler's method and the Störmer–Verlet method to solve this linear spring-mass system, as a relation involving N and Δt.

7. *Flow maps and numerical methods.* A set $\mathcal{G}$ together with an associative product relation $*$ is called a *group* if (i) $\mathcal{G}$ is closed under $*$, i.e. $g_1, g_2 \in \mathcal{G}$ implies $g_1 * g_2 \in \mathcal{G}$, (ii) $\mathcal{G}$ contains an identity element e such that $e * g = g * e = g$, and (iii) for any element $g \in \mathcal{G}$ there is an inverse element $g^{-1} \in \mathcal{G}$ such that $g * g^{-1} = g^{-1} * g = e$.

a. Suppose that a given differential equation

$$\frac{d}{dt}z = \boldsymbol{f}(z)$$

admits solutions defined for infinite time in both the positive and negative directions. Show that the flow maps $\{\boldsymbol{\Phi}_t\}_{t \in \mathbb{R}}$ defined as the mappings which take points $\bar{z}$ to points t units in time later along the solution passing through $\bar{z}$ constitute a one-parametric group, where the group operation $*$ is a composition of maps and

$$\boldsymbol{\Psi}_{t_1+t_2} = \boldsymbol{\Phi}_{t_2} \circ \boldsymbol{\Phi}_{t_1}.$$

b. Consider now the set $\hat{\mathcal{G}}$ which consists of all the mappings defined by applying Euler's method with timesteps $\Delta t > 0$. Does $\hat{\mathcal{G}}$ constitute a one-parametric group in the above sense with Δt taking the role of time t?

8. *Computer project with one-step methods.* In this exercise you will write a small computer program to test a numerical method for solving $\dot{q} = v$, $\dot{v} = -\varphi'(q)$, $q(0) = q^0$, $v(0) = v^0$. Refer to the preface for a discussion of computer software.

a. Write a module `eulerstep` that takes as input:

- an arbitrary function $\varphi : \mathbb{R} \to \mathbb{R}$,
- real scalars q^n, v^n, Δt,

and computes the result of taking a *single step* with Euler's method applied to the differential equation.

b. Write a computer program `stepper` which takes as inputs:

- an arbitrary function $\varphi : \mathbb{R} \to \mathbb{R}$,
- real scalars q^0, v^0, Δt,
- integer N,
- the name of a module (such as `eulerstep`) which implements a one-step method for the differential equation.

Then solve the system $\dot{q} = v$, $\dot{v} = -\varphi'(q)$ by taking N steps with Euler's method starting from q^0, v^0. The program should produce as output a pair of $(N + 1)$ one-dimensional arrays $\boldsymbol{Q}$, $\boldsymbol{V}$ consisting of the beginning and ending of positions and velocities and all intermediate steps.

c. Write modules `eulerAstep`, `eulerBstep`, `stormerstep` and `rk4step` with similar inputs and outputs to the module `eulerstep` but implementing a single timestep of the Euler-A, Euler-B, Störmer–Verlet, and fourth-order Runge–Kutta methods of the text.

d. Experiment with the various methods using the `stepper` routine. Examine the energy conservation of the various methods, when applied to a Morse oscillator with unit coefficients, $\varphi(q) = (1 - \exp(-q))^2$.

3

Hamiltonian mechanics

In this chapter we introduce the Hamiltonian formulation of classical mechanics. The elementary properties of Newton's equations such as the conservation of energy or momentum can be explained without much difficulty by the use of nothing more than the chain rule of calculus, but it turns out that there is another, deeper level of structure that relates to the properties of *bundles of trajectories* emanating from a set of initial conditions. The most useful tool for describing the solutions is the *flow map* already introduced in the previous chapter.

For any differential equation $\dot{z} = f(z)$, recall that we denote the solution through a given initial condition z^0 by $z(t; z^0)$. Assuming that this solution is globally defined, we may fix a value of t, and consider the trajectories starting from any arbitrary point z^0 in phase space. The flow map $\boldsymbol{\Phi}_t$ is the mapping from initial points z^0 to final points $z(t; z^0)$ along trajectories, thus

$$z(t; z^0) = \boldsymbol{\Phi}_t(z^0).$$

For conservative mechanical systems, it turns out that the flow map inherits certain global qualitative or *geometric* properties which are, in a very real sense, as fundamental as any of the physical laws which characterize our understanding of the properties of matter. An example of such a qualitative property is the conservation of the volume of a set of points of phase space under their simultaneous time evolution. Another example is found in the stringent restrictions placed on the asymptotic behavior of solutions of Hamiltonian systems near equilibrium points. In this chapter, we explore the geometric properties which are the consequences of Hamiltonian mechanics. This treatment is necessarily restricted; the reader is referred to the preface for suggestions for further reading.

3.1 Canonical and noncanonical Hamiltonian systems

Let us begin our discussion by recalling Newton's equations (1.1)–(1.2) for N particles moving in $\mathbb{R}^3$. To compactify the notation, we introduce the diagonal mass matrix $\boldsymbol{M} \in \mathbb{R}^{3N \times 3N}$ such that

$$\boldsymbol{M}\boldsymbol{v} = (m_1 \boldsymbol{v}_1, \ldots, m_N \boldsymbol{v}_N)^T.$$

We also eliminate the particle velocities and write (1.1)–(1.2) as a second-order differential equation

$$\boldsymbol{M}\ddot{\boldsymbol{q}} = \boldsymbol{F}(\boldsymbol{q}) \tag{3.1}$$

in the vector of particle coordinates

$$\boldsymbol{q} := (\boldsymbol{q}_1, \ldots, \boldsymbol{q}_N)^T.$$

The force $\boldsymbol{F}$ is given by the negative gradient of a potential energy function with respect to position

$$\boldsymbol{F}(\boldsymbol{q}) := -\nabla_{\boldsymbol{q}} V(\boldsymbol{q}).$$

The Hamiltonian formulation of Newtonian mechanics reduces (3.1) back to a system of first-order equations, but instead of including the particle velocities it relies on the vector of *linear momenta* $\boldsymbol{p} \in \mathbb{R}^{3N}$ defined by

$$\boldsymbol{p} := \boldsymbol{M}\dot{\boldsymbol{q}}.$$

There are obviously many ways to rewrite second-order equations as systems of first-order equations. A priori, there is no reason why one approach should have a significant advantage over another. The observation of Hamilton was that the differential equations defining both position and momentum are obtained by dual operations on the same total energy function.

For Newtonian mechanics, the differential equations take the following form:

HAMILTONIAN FORM OF NEWTONIAN MECHANICS

$$\frac{d}{dt}\boldsymbol{q} = \boldsymbol{M}^{-1}\boldsymbol{p}, \tag{3.2}$$

$$\frac{d}{dt}\boldsymbol{p} = -\nabla_{\boldsymbol{q}} V(\boldsymbol{q}). \tag{3.3}$$

The equations (3.2)–(3.3) are termed a *Hamiltonian system* with *Hamiltonian* (or energy)

$$H(\boldsymbol{q}, \boldsymbol{p}) := \frac{\boldsymbol{p}^T \boldsymbol{M}^{-1} \boldsymbol{p}}{2} + V(\boldsymbol{q}),$$

which is a *first integral* (*constant of motion*), i.e.

$$\frac{d}{dt}H(\boldsymbol{q}(t), \boldsymbol{p}(t)) = \boldsymbol{p}(t)^T \boldsymbol{M}^{-1}\dot{\boldsymbol{p}}(t) + \dot{\boldsymbol{q}}(t)^T \nabla_{\boldsymbol{q}} V(\boldsymbol{q}(t)) = 0,$$

along solutions $(\boldsymbol{q}(t), \boldsymbol{p}(t))$ of (3.2)–(3.3).

It is easy and useful to generalize equations (3.2)–(3.3). Given a phase space $\mathbb{R}^d \times \mathbb{R}^d$ of arbitrary (even) dimension $2d \geq 2$ and an arbitrary (smooth) Hamiltonian function $H : \mathbb{R}^d \times \mathbb{R}^d \to \mathbb{R}$, the corresponding *canonical* Hamiltonian equations of motion are

$$\frac{d}{dt}\boldsymbol{q} = +\nabla_{\boldsymbol{p}} H(\boldsymbol{q}, \boldsymbol{p}), \tag{3.4}$$

$$\frac{d}{dt}\boldsymbol{p} = -\nabla_{\boldsymbol{q}} H(\boldsymbol{q}, \boldsymbol{p}). \tag{3.5}$$

It now becomes evident that the change of variables to positions and momenta has uncovered a symmetry in their function (or, rather, an *antisymmetry*, since we also have a change of sign). We will see shortly that this antisymmetry, together with the smoothness of the solutions with respect to perturbation of the initial data, has important ramifications for the flow map of the system.

When discussing Hamiltonian systems, it is often convenient to use the notation

$$\boldsymbol{z} := (\boldsymbol{q}, \boldsymbol{p})^T,$$

with $\boldsymbol{q}, \boldsymbol{p} \in \mathbb{R}^d$, $\boldsymbol{z} \in \mathbb{R}^{2d}$, and to introduce the $2d \times 2d$ *canonical structure matrix* $\boldsymbol{J}$,

$$\boldsymbol{J} := \begin{bmatrix} \boldsymbol{0} & +\boldsymbol{I}_d \\ -\boldsymbol{I}_d & \boldsymbol{0} \end{bmatrix}. \tag{3.6}$$

Then the Hamiltonian system (3.4)–(3.5) can be rewritten in compact form:

<u>HAMILTONIAN SYSTEM</u>

$$\frac{d}{dt}\boldsymbol{z} = \boldsymbol{J}\nabla_{\boldsymbol{z}} H(\boldsymbol{z}). \tag{3.7}$$

The term "canonical" is reserved for Hamiltonian systems on an even dimensional Euclidean space with $\boldsymbol{J}$ as in (3.6), but Hamiltonian systems can be generalized in various ways without altering the discussion of their geometric properties in any essential way. For example, we may allow $\boldsymbol{J}$ to be an arbitrary invertible constant skew-symmetric matrix (with $\boldsymbol{J}^T = -\boldsymbol{J}$). Still more generally, we may allow, under suitable restrictions, the structure matrix itself to depend on the phase space variable $\mathbf{z}$, i.e. $\mathbf{J} = \mathbf{J}(\mathbf{z})$, or the phase space may be replaced by an appropriate smooth, even-dimensional manifold.[1]

The system (3.7) is an example of an *autonomous* differential equation, meaning that the vector field is formally independent of time. Time-dependent or *non-autonomous* Hamiltonian systems with $H = H(\mathbf{z}, t)$ also arise frequently in applications. The equations of motion are then given by

$$\frac{d}{dt}\mathbf{z} = \mathbf{J}\nabla_{\mathbf{z}} H(\mathbf{z}, t).$$

In most cases, the special structures associated to autonomous Hamiltonian systems are easily extended to non-autonomous systems.

3.2 Examples of Hamiltonian systems

In this section, we survey some of the many types of Hamiltonian systems that arise in the physical sciences. It is our experience that having a feel for the types of applications which may arise is important to understanding both the theoretical and numerical issues associated with their study.

3.2.1 Linear systems

A canonical *linear Hamiltonian system* is defined by a quadratic Hamiltonian $H = \frac{1}{2}\mathbf{z}^T \mathbf{L}\mathbf{z}$, where $\mathbf{L}$ is a symmetric $2d \times 2d$ matrix and $\mathbf{J}$ a structure matrix of type (3.6). The equations of motion are thus

$$\frac{d}{dt}\mathbf{z} = \mathbf{J}\mathbf{L}\mathbf{z}. \tag{3.8}$$

A matrix of the form $\mathbf{A} = \mathbf{J}\mathbf{L}$ with $\mathbf{L}$ symmetric is typically referred to as a *Hamiltonian matrix*. For example, the harmonic oscillator is a one-degree-of-freedom linear Hamiltonian system with

$$\mathbf{L} = \begin{bmatrix} \omega^2 & 0 \\ 0 & 1 \end{bmatrix},$$

and $\mathbf{z} = (q, p)^T$.

The solution of a linear Hamiltonian system is formally computable in terms of the eigenvalues and eigenvectors of the matrix $\mathbf{A} = \mathbf{J}\mathbf{L}$. (In practice, the computation of the eigenstructure may be quite involved, so the solution is often propagated instead by an appropriate numerical method.) It is interesting to

[1]We will consider such *constrained Hamiltonian systems* beginning in Chapter 7.

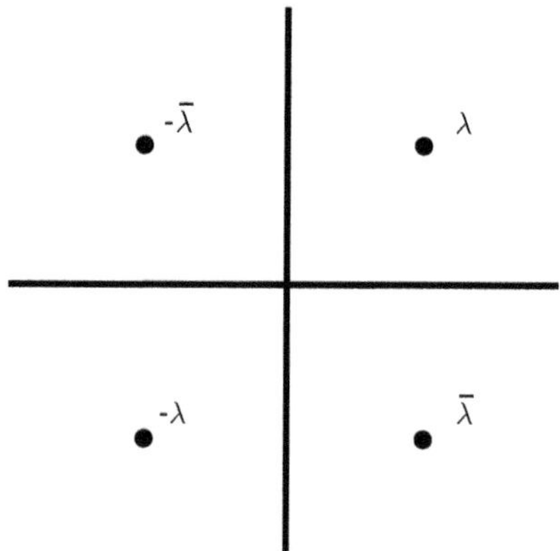

Figure 3.1 If λ is an eigenvalue of a Hamiltonian matrix, then so are $\bar{\lambda}$, $-\lambda$, and $-\bar{\lambda}$.

consider some of the properties of this eigenstructure. Since $\boldsymbol{A} = \boldsymbol{JL}$ is real, it follows that the eigenvalues occur in complex conjugate pairs. Moreover, if λ is an eigenvalue of $\boldsymbol{A}$, it follows that λ is also an eigenvalue of $\boldsymbol{A}^T = \boldsymbol{L}^T\boldsymbol{J}^T = -\boldsymbol{LJ}$, i.e.

$$-\boldsymbol{LJu} = \lambda \boldsymbol{u}$$

for some $\boldsymbol{u}$. Let $\boldsymbol{v} = \boldsymbol{Ju}$, then we have

$$-\boldsymbol{Lv} = \lambda \boldsymbol{J}^{-1}\boldsymbol{v},$$

and, after premultiplying by $\boldsymbol{J}$

$$-\boldsymbol{JLv} = \lambda \boldsymbol{v}.$$

This means that $-\lambda$ is also an eigenvalue of $\boldsymbol{A}$ (with eigenvector $\boldsymbol{Ju}$). Summarizing, if λ is an eigenvalue of a Hamiltonian matrix, then so are $\bar{\lambda}$, $-\lambda$, and, by extension, $-\bar{\lambda}$. A Hamiltonian matrix therefore has a symmetric spectrum with respect to both coordinate axes of the complex plane (Fig. 3.1).

3.2.2 Single-degree-of-freedom problems

Anharmonic one-degree-of-freedom oscillators with Hamiltonians of the form $H(q,p) = p^2 + \varphi(q)$ are also of interest. In the previous chapters we encountered the Lennard–Jones oscillator, with potential

$$\varphi_{\mathrm{L.J.}}(q) = \epsilon\left[\left(\frac{\bar{r}}{q}\right)^{12} - 2\left(\frac{\bar{r}}{q}\right)^{6}\right].$$

As another example, the *plane pendulum* can be described in terms of the angle q made with the vertical axis by the Hamiltonian

$$H = \frac{1}{2}p^2 - \frac{g}{L}\cos q,$$

where L is the length of the pendulum and g the gravitational constant.

The *phase portrait* of a planar system qualitatively summarizes the global dynamics by depicting a few representative orbits in the q, p domain. These orbits can obviously be identified with pieces of the curves defined by the equation $H(q, p) = \text{constant}$, thus we may view the orbits as level curves of the Hamiltonian function. The graph of the surface $E = H(q, p)$ in (q, p, E)-space characterizes the dynamics of a single-degree-of-freedom system. In particular, the equilibria are associated with critical points on this surface: local minima correspond to stable equilibria, while saddle points correspond to unstable equilibria. In Fig. 3.2, we show the graph of $E = H(q, p)$ together with the phase portrait (level curves of H) for the Lennard–Jones oscillator ($\epsilon = 0.25$, $\bar{r} = 1$) and the pendulum ($g/L = 1$).

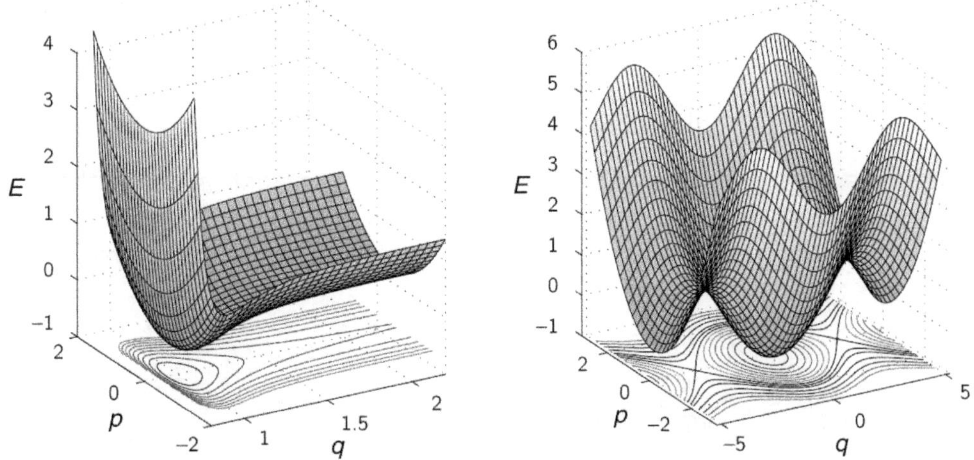

Figure 3.2 Energy surfaces for one-degree-of-freedom problems, and phase portraits. Left, Lennard–Jones oscillator. Right, plane pendulum.

3.2.3 Central forces

Next, consider the three-degree-of-freedom nonlinear system describing the motion of a body of unit mass in $\mathbb{R}^3$, i.e. $\boldsymbol{q} = (q_1, q_2, q_3)^T$, with a potential energy of interaction given by $\varphi = \varphi(r)$ where r represents the distance from the body to the origin. This is an instance of a *central-force* problem. If $\varphi(r) = -1/r$, then we have the *Kepler problem* with Hamiltonian

$$H = \frac{1}{2}(p_1^2 + p_2^2 + p_3^2) - \frac{1}{\sqrt{q_1^2 + q_2^2 + q_3^2}}. \tag{3.9}$$

The level sets of constant energy are five dimensional surfaces (!), hence nontrivial to visualize, but, as we shall see in Section 3.3, a complete description of the dynamics of the system is still easily obtained by using some additional properties of the equations.

3.2.4 Charged particle in a magnetic field

Up to normalization of certain constants, the Kepler problem can be viewed as a model of gravitational dynamics. It also represents, qualitatively, the classical model of an electron in the field of a positively charged fixed nucleus. It is interesting to consider in this context a generalization in which the electron moves not only under the Coulombic potential, but also in an applied (constant) magnetic field. If $\boldsymbol{b} = (b_1, b_2, b_3)^T$ is a vector representing such a field, the equations of motion for the particle are found to take the form

$$m\ddot{\boldsymbol{q}} = -\gamma \frac{\boldsymbol{q}}{\|\boldsymbol{q}\|^3} + \boldsymbol{b} \times \dot{\boldsymbol{q}}, \tag{3.10}$$

where m is the mass of the particle, γ is a positive constant, and $\times$ is the usual cross product of vectors.

Defining the momenta in the usual way, we have the first-order system

$$\frac{d}{dt}\boldsymbol{q} = \frac{1}{m}\boldsymbol{p}, \tag{3.11}$$

$$\frac{d}{dt}\boldsymbol{p} = -\gamma \frac{\boldsymbol{q}}{\|\boldsymbol{q}\|^3} - \frac{1}{m}\boldsymbol{p} \times \boldsymbol{b}. \tag{3.12}$$

Here $\boldsymbol{q}$, and $\boldsymbol{p}$ are assumed to be vectors in $\mathbb{R}^3$. It is easy to show that this is *not* a canonical Hamiltonian system, since, if it were, we would have to have

$$\nabla_p H = \frac{1}{m}\boldsymbol{p}, \qquad \nabla_q H = \gamma \frac{\boldsymbol{q}}{\|\boldsymbol{q}\|^3} + \frac{1}{m}\boldsymbol{p} \times \boldsymbol{b},$$

and equality of mixed partials would be violated. However, a Hamiltonian description (3.7) does follow if we define instead

$$\boldsymbol{J} = \begin{bmatrix} 0 & \boldsymbol{I} \\ -\boldsymbol{I} & \hat{\boldsymbol{b}} \end{bmatrix},$$

where $\widehat{\boldsymbol{b}}$ is the skew symmetric matrix defined by

$$\widehat{\boldsymbol{b}} = \begin{bmatrix} 0 & -b_3 & b_2 \\ b_3 & 0 & -b_1 \\ -b_2 & b_1 & 0 \end{bmatrix},$$

taking as Hamiltonian[2]

$$H(\boldsymbol{q}, \boldsymbol{p}) = \frac{1}{2m}\|\boldsymbol{p}\|^2 - \frac{\gamma}{\|\boldsymbol{q}\|}.$$

3.2.5 Lagrange's equation

Given a Lagrangian function $L(\boldsymbol{q}, \dot{\boldsymbol{q}})$, Lagrange's equation of motion is

$$\frac{d}{dt}\nabla_{\dot{q}} L(\boldsymbol{q}, \dot{\boldsymbol{q}}) - \nabla_{q} L(\boldsymbol{q}, \dot{\boldsymbol{q}}) = \boldsymbol{0}. \tag{3.13}$$

This equation is the Euler–Lagrange equation minimizing the action integral

$$\mathcal{L}[\boldsymbol{q}] = \int_{t_0}^{t_1} L(\boldsymbol{q}(t), \dot{\boldsymbol{q}}(t))\, dt.$$

For more details see [7, 8, 73]. Lagrange's equation (3.13) is second order in time. It is reduced to a system of first-order equations by introducing the *conjugate momenta*

$$\boldsymbol{p} = \nabla_{\dot{q}} L(\boldsymbol{q}, \dot{\boldsymbol{q}}). \tag{3.14}$$

We require that this relation defines a one-to-one map between $\boldsymbol{p}$ and $\dot{\boldsymbol{q}}$ for fixed $\boldsymbol{q}$. It can be shown (see problem 1 in the Exercises) that the equation (3.14) together with the reformulation

$$\dot{\boldsymbol{p}} = \nabla_{q} L(\boldsymbol{q}, \dot{\boldsymbol{q}}) \tag{3.15}$$

of Lagrange's equation (3.13) are canonical with Hamiltonian

$$H(\boldsymbol{q}, \boldsymbol{p}) = \boldsymbol{p} \cdot \dot{\boldsymbol{q}} - L(\boldsymbol{q}, \dot{\boldsymbol{q}}). \tag{3.16}$$

For example, take a particle of unit mass moving in a central field. The Lagrangian is

$$L = \frac{1}{2}(\dot{q}_1^2 + \dot{q}_2^2 + \dot{q}_3^2) + \frac{1}{\sqrt{q_1^2 + q_2^2 + q_3^2}}.$$

[2] The charged particle in a magnetic field is described here as a *noncanonical* Hamiltonian system. It is interesting to note that a canonical formulation of this system is also possible by modifying the definition of the momentum; see [122].

Note the change in sign compared with the Hamiltonian (3.9). The conjugate momenta is

$$p = \dot{q},$$

as expected.

The charged particle in a magnetic field has Lagrangian

$$L = \frac{m}{2}\|\dot{q}\|^2 + \frac{\gamma}{\|q\|} + \frac{1}{2}q^T\hat{b}\dot{q}. \tag{3.17}$$

According to (3.14), the conjugate momenta is

$$p = m\dot{q} - \frac{1}{2}\hat{b}q. \tag{3.18}$$

This definition of p is different from the one used in Section 3.2.4! In Problem 1 in the Exercises you will be asked to derive the associated Hamiltonian and to verify the canonical equations of motion (3.32)–(3.33). This example emphasizes the point that the Hamiltonian formulation of a problem need not be unique and that there is a certain freedom in the choice of the Hamiltonian H and the structure matrix J.

3.2.6 N-body problem

Finally, let us return to the homogeneous system of N bodies moving in $\mathbb{R}^3$ with masses m_i, $i = 1, \ldots, N$, interacting through a particle–particle interaction potential (*pair-potential*) $\varphi(r)$, with r the distance between two particles. The corresponding canonical equations of motion are

$$\frac{d}{dt}q_i = \frac{1}{m_i}p_i, \tag{3.19}$$

$$\frac{d}{dt}p_i = -\sum_{i\neq j}\frac{\varphi'(r_{ij})}{r_{ij}}(q_i - q_j), \qquad i = 1, 2, \ldots, N, \tag{3.20}$$

where $r_{ij} = \|q_i - q_j\|$; the Hamiltonian function is

$$H = \frac{1}{2}\sum_{i=1}^{N}\frac{\|p_i\|^2}{m_i} + \sum_{i=1}^{N-1}\sum_{j=i+1}^{N}\varphi(r_{ij}).$$

3.3 First integrals

A *first integral*, *constant of motion*, or *conserved quantity* of a general differential equation is a function $G : \mathbb{R}^{2d} \to \mathbb{R}$ which is constant along all solution curves $z(t; z^0)$ of the system, i.e.

$$G(z(t; z^0)) = G(z^0),$$

for all $z^0 \in \mathbb{R}^{2d}$ and all $t \in \mathbb{R}$. We say that having a first integral is a "geometric property" of a system of differential equations because its existence implies that the solutions are at least partly described by the geometry of the lower-dimensional manifolds $\{z \in \mathbb{R}^{2d} \mid G(z) = \text{constant}\}$.

What are the conditions that ensure that a particular function G is a first integral for a canonical Hamiltonian system (3.7)? Assume that $z(t)$ is a solution curve of (3.7), then

$$\begin{aligned}\frac{d}{dt}G(z(t)) &= \nabla_z G(z(t)) \cdot \dot{z}(t) \\ &= [\nabla_z G(z(t))]^T \dot{z}(t) \\ &= [\nabla_z G(z(t))]^T J \nabla_z H(z(t)).\end{aligned}$$

If this quantity is to vanish for every trajectory $z(t; z^0)$ with initial conditions chosen from some open set in phase space, we must have that the *Poisson bracket* of G and H, defined by

$$\{G, H\}(z) := [\nabla_z G(z)]^T J \nabla_z H(z),$$

vanishes identically, i.e.

$$\{G, H\}(z) \equiv 0.$$

This is a necessary and sufficient condition for G to be a first integral of a Hamiltonian system with Hamiltonian H.

More generally, the time evolution of an arbitrary, smooth function $G : \mathbb{R}^{2d} \to \mathbb{R}$ along a solution of a canonical system with Hamiltonian H is given by

$$\frac{d}{dt}G(z) = \{G, H\}(z).$$

For example, take $G(q, p) := q_i$, then

$$\frac{d}{dt}q_i = \{q_i, H\}(q, p) = \frac{\partial}{\partial p_i}H(q, p),$$

as we would expect.

The Poisson bracket is a bilinear, antisymmetric operation.[3] This means that, for functions $F, G, H : \mathbb{R}^{2d} \to \mathbb{R}$, and scalars α, β

$$\{F, \alpha G + \beta H\} = \alpha\{F, G\} + \beta\{F, H\},$$

and

$$\{F, G\} = -\{G, F\}.$$

[3] The Poisson bracket also satisfies the Jacobi identity $\{\{F, G, \}, H\} + \{\{H, F\}, G\} + \{\{G, H\}, F\} = 0$ [149, 124]. This identity is automatically satisfied for all constant structure matrices J.

Note that the latter identity implies linearity in the first argument as well. Moreover, $\{F, F\} = -\{F, F\}$, hence $\{F, F\} = 0$, i.e. the Poisson bracket of a function with itself vanishes identically.

We have already seen that the Hamiltonian (or energy) H of a Hamiltonian system is a first integral. Using the Poisson bracket notation, this statement reduces to the observation $\{H, H\} = 0$.

Example 1 The Kepler problem in $\mathbb{R}^3$ possesses an interesting and nontrivial integral invariant structure. First, the energy,

$$H = \frac{1}{2}\boldsymbol{p}^T\boldsymbol{p} - \frac{1}{||\boldsymbol{q}||} = \frac{1}{2}(p_1^2 + p_2^2 + p_3^2) - \frac{1}{\sqrt{q_1^2 + q_2^2 + q_3^2}},$$

is a constant of motion. Moreover, one easily verifies that the components of the *angular momentum* vector,

$$\boldsymbol{m} = \boldsymbol{q} \times \boldsymbol{p} = \begin{bmatrix} m_1 \\ m_2 \\ m_3 \end{bmatrix}$$

are conserved quantities. For example

$$\begin{aligned}\{\boldsymbol{m}_1, H\} &= \{q_2p_3 - q_3p_2, H\} \\ &= p_3\frac{\partial H}{\partial p_2} - p_2\frac{\partial H}{\partial p_3} + q_3\frac{\partial H}{\partial q_2} - q_2\frac{\partial H}{\partial q_3} \\ &= p_3p_2 - p_2p_3 - q_3q_2/||\boldsymbol{q}||^3 + q_2q_3/||\boldsymbol{q}||^3 \\ &= 0.\end{aligned}$$

Since, for any vectors $\boldsymbol{a}$ and $\boldsymbol{b}$, we have

$$\boldsymbol{a} \cdot (\boldsymbol{a} \times \boldsymbol{b}) = 0, \qquad \boldsymbol{b} \cdot (\boldsymbol{a} \times \boldsymbol{b}) = 0,$$

it follows that $\boldsymbol{q}$ and $\boldsymbol{p}$ lie permanently in the plane perpendicular to $\boldsymbol{m}$ (see Fig. 3.3). For simplicity, it is typically assumed that this plane of motion is oriented such that $\boldsymbol{m}$ points in the q_3 direction, in which case we have $p_3 \equiv 0$, implying q_3 is constant along trajectories, and the more traditional planar Kepler problem results with Hamiltonian reduced to

$$H = \frac{1}{2}(p_1^2 + p_2^2) - \frac{1}{\sqrt{q_1^2 + q_2^2}}.$$

This is a system with two degrees of freedom and two first integrals; energy H and the third component of the angular momentum vector $\boldsymbol{m}$. Hence any bounded motion is restricted to a two-dimensional surface which is diffeomorph to a torus.

Furthermore, the special form of the Kepler force field creates an additional vector, namely the *Runge–Lenz* vector

$$\boldsymbol{e} = \boldsymbol{p} \times \boldsymbol{m} - \frac{\boldsymbol{q}}{\|\boldsymbol{q}\|},$$

which is constant along solution curves (see problem 2 in the Exercises). This invariance implies that the bounded trajectories of the Kepler problem are closed in the (q_1, q_2)-plane [7]. □

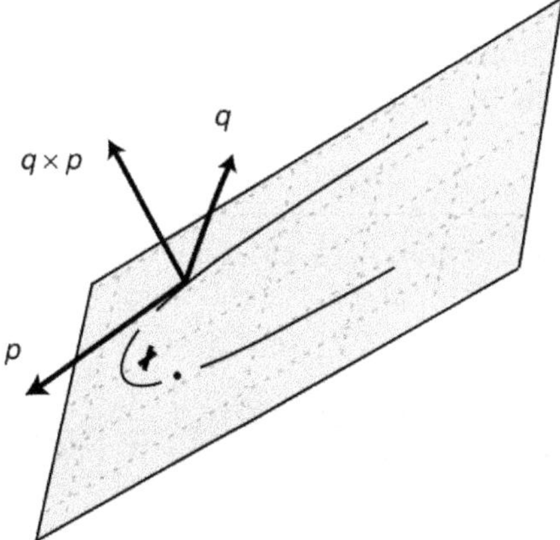

Figure 3.3 Significance of the angular momentum in the Kepler problem: orbits lie in the plane perpendicular to $\boldsymbol{m} = \boldsymbol{q} \times \boldsymbol{p}$.

Angular momentum $\boldsymbol{m} = \boldsymbol{q} \times \boldsymbol{p}$ is a first integral for any particle moving under a central force field (see problem 2 in the Exercises), i.e. with Hamiltonian of the form

$$H = \frac{1}{2}\boldsymbol{p}^T\boldsymbol{p} + \varphi(\|\boldsymbol{q}\|),$$

thus all such systems have planar motion.

In general, each first integral reduces the number of degrees of freedom by one as an intersection of invariant manifolds. The Kepler problem is an example of an *integrable* Hamiltonian system, meaning, intuitively, that the system possesses an independent first integral G_i for each of its degrees of freedom. The solutions of an integrable system of d degrees of freedom are then restricted to d-dimensional level sets

$$\mathcal{M}_{\mathrm{c}} = \{(\boldsymbol{q}, \boldsymbol{p}) \in \mathbb{R}^{2d} : G_i(\boldsymbol{q}, \boldsymbol{p}) = c_i, \ i = 1, \ldots, d\}.$$

If these level sets are compact sets (i.e. closed and bounded), then they can be viewed as d-dimensional tori, i.e. the level sets can be transformed by a smooth (and smoothly invertible) mapping to a set of the form $T = S^1 \times S^1 \times S^1 \ldots \times S^1$ (d times), where S^1 is the circle.[4] Upon introducing *angle variables* $\boldsymbol{\phi}$ to describe the tori and identifying each torus by its *action* $\mathbf{I} \in \mathbb{R}^d$ [7, 8], the integrable system can be transformed to the canonical system

$$\dot{\mathbf{I}} = \mathbf{0}, \qquad \dot{\boldsymbol{\phi}} = \nabla_{\mathbf{I}} S(\mathbf{I}),$$

with Hamiltonian $S(\mathbf{I})$. Thus integrable systems have regular quasi-periodic solutions with frequency $\boldsymbol{\omega} = \nabla_{\mathbf{I}} S(\mathbf{I})$.

Most systems do not possess a full complement of first integrals. The few first integrals that are typically present can be of substantial physical and practical importance. In complex nonlinear systems, they sometimes provide a simple way of distinguishing plausible trajectories from nonphysical ones or of assessing the quality of an approximation.

Example 2 The equations of motion (3.19)–(3.20) of N particles interacting through a distance-dependent pair-potential $\varphi(r)$ admit total linear momentum $\boldsymbol{p}_{\text{tot}} = \sum_i \boldsymbol{p}_i$ and total angular momentum $\boldsymbol{m}_{\text{tot}} = \sum_i \boldsymbol{q}_i \times \boldsymbol{p}_i$ as first integrals. For example:

$$\frac{d}{dt}\boldsymbol{p}_{tot} = \sum_i \frac{d}{dt}\boldsymbol{p}_i = \sum_i \sum_{i \neq j} -\frac{\varphi'(r_{ij})}{r_{ij}}(\boldsymbol{q}_i - \boldsymbol{q}_j) = \mathbf{0},$$

since $r_{ij} = r_{ji}$. □

In molecular applications, *periodic boundary conditions* are often introduced as a modeling device. In this case, the total linear momentum remains a conserved quantity, but the total angular momentum is sacrificed.

3.4 The flow map and variational equations

Each Hamiltonian system gives rise to a family of flow maps parameterized by time t. Contrary to the previously used notation $\boldsymbol{\Phi}_t$, we denote the flow map from now on by $\boldsymbol{\Phi}_{t,H}$ to indicate its dependence on the Hamiltonian H. As we have mentioned in the previous chapter, the family of flow maps is closed under the composition operation, i.e. for any $t = \tau_1$, $t = \tau_2$

$$\boldsymbol{\Phi}_{\tau_1,H} \circ \boldsymbol{\Phi}_{\tau_2,H} = \boldsymbol{\Phi}_{\tau_2,H} \circ \boldsymbol{\Phi}_{\tau_1,H} = \boldsymbol{\Phi}_{\tau_1+\tau_2,H}.$$

The flow map $\boldsymbol{\Phi}_{0,H}$ at $t = 0$ is the identity map (abbreviated **id**). Every flow map evidently has an inverse in the family, since

$$\boldsymbol{\Phi}_{-t,H} \circ \boldsymbol{\Phi}_{t,H} = \boldsymbol{\Phi}_{0,H} = \mathbf{id}.$$

[4] Some additional technical assumptions have to be made [7, 8].

The one-parameter family of time t flow maps of a Hamiltonian system thus defines a commutative group.

Example 3 Consider a particle of mass m that is moving in $\mathbb{R}^3$ without any force acting on it. The corresponding Hamiltonian is the kinetic energy of the particle and the flow map is given by

$$\boldsymbol{\Phi}_{t,H}(\boldsymbol{q}, \boldsymbol{p}) = \begin{bmatrix} \boldsymbol{q} + \frac{t}{m}\boldsymbol{p} \\ \boldsymbol{p} \end{bmatrix}.$$

□

Example 4 For the linear Hamiltonian system (3.8), the flow map $\boldsymbol{\Phi}_{t,H}$ is given by a matrix exponential

$$\boldsymbol{\Phi}_{t,H}(\boldsymbol{z}) = \exp(t\boldsymbol{J}\boldsymbol{L})\,\boldsymbol{z}.$$

Let us consider in detail the example of the harmonic oscillator

$$\frac{d}{dt}q = p,$$
$$\frac{d}{dt}p = -\omega^2 q.$$

The first step is to introduce new coordinates $\hat{\boldsymbol{z}} := \boldsymbol{S}^{-1}\boldsymbol{z}$, $\boldsymbol{z} = (q, p)^T$, with

$$\boldsymbol{S} = \begin{bmatrix} 1/\sqrt{\omega} & 0 \\ 0 & \sqrt{\omega} \end{bmatrix}.$$

In these new coordinates the matrix

$$\boldsymbol{J}\boldsymbol{L} = \begin{bmatrix} 0 & 1 \\ -\omega^2 & 0 \end{bmatrix}$$

gets transformed to

$$\boldsymbol{S}^{-1}\boldsymbol{J}\boldsymbol{L}\boldsymbol{S} = \begin{bmatrix} 0 & \omega \\ -\omega & 0 \end{bmatrix}.$$

The corresponding transformed linear system

$$\frac{d}{dt}\hat{q} = \omega\hat{p},$$
$$\frac{d}{dt}\hat{p} = -\omega\hat{q},$$

$\hat{\boldsymbol{z}} = (\hat{q}, \hat{p})^T$, generates the flow map

$$\hat{\boldsymbol{\Phi}}_{t,H} = \exp(t\boldsymbol{S}^{-1}\boldsymbol{J}\boldsymbol{L}\boldsymbol{S}),$$
$$= \begin{bmatrix} \cos\omega t & \sin\omega t \\ -\sin\omega t & \cos\omega t \end{bmatrix}.$$

We obtain the flow map $\boldsymbol{\Phi}_{t,H}$ by transforming $\hat{\boldsymbol{\Phi}}_{t,H}$ back to the variable $z = \boldsymbol{S}\hat{z}$. Thus

$$\begin{aligned}
\boldsymbol{\Phi}_{t,H}(z) &= \exp(t\boldsymbol{J}\boldsymbol{L})z \\
&= \boldsymbol{S}\hat{\boldsymbol{\Phi}}_{t,H}\boldsymbol{S}^{-1}z \\
&= \begin{bmatrix} 1/\sqrt{\omega} & 0 \\ 0 & \sqrt{\omega} \end{bmatrix} \begin{bmatrix} \cos\omega t & \sin\omega t \\ -\sin\omega t & \cos\omega t \end{bmatrix} \begin{bmatrix} \sqrt{\omega} & 0 \\ 0 & 1/\sqrt{\omega} \end{bmatrix} z \\
&= \begin{bmatrix} \cos\omega t & \omega^{-1}\sin\omega t \\ \omega\sin\omega t & \sin\omega t \end{bmatrix} z.
\end{aligned}$$

Hence the flow of the harmonic oscillator is described by a product of three matrices: a scaling, a rotation, and the inverse scaling. □

Hamiltonian systems with bounded smooth Hamiltonian function H admit smooth flow maps. The inverse map $\boldsymbol{\Phi}_{t,H}^{-1} \equiv \boldsymbol{\Phi}_{-t,H}$ is also smooth. A map which is smooth, invertible, and whose inverse map is also smooth is called a *diffeomorphism*.

Even when the potential energy is not smooth (for example, when it has singular points) we can still often define a flow map subject to a suitable domain restriction. For example, for an N-body system subject to a pair potential φ with a singularity at $r = 0$, ultimately repulsive at short range ($\lim_{r\to 0+}\varphi(r) = +\infty$), we can define a global flow map on the complement of the singular set $\{(\boldsymbol{q}_1, \boldsymbol{q}_2, \ldots, \boldsymbol{q}_N) | \boldsymbol{q}_i = \boldsymbol{q}_j, \text{ some } i, j,\ i \neq j\}$.

In general, the flow map cannot be written down explicitly but can only be approximated numerically. We will take up this task in the following chapters.

The flow map notation allows us to emphasize solution properties that are shared by bundles of trajectories. For example, instead of considering a single initial point z^0, we can look at a neighborhood $\mathcal{U} \subset \mathbb{R}^{2d}$ of initial points near the given z^0 and can ask how this whole set of initial conditions is transported by the flow map, i.e. we can discuss the transformed sets

$$\begin{aligned}
\mathcal{U}_t &:= \boldsymbol{\Phi}_{t,H}(\mathcal{U}), \\
&:= \{z \in \mathbb{R}^{2d} : z = \boldsymbol{\Phi}_{t,H}(\hat{z}) \text{ with } \hat{z} \in \mathcal{U}\},
\end{aligned} \tag{3.21}$$

$t > 0$. This point of view will allow us to better understand some of the qualitative properties of Hamiltonian equations of motion.

A first step in this direction is to investigate the solution behavior of a system near a given trajectory $z(t; z^0)$ via *linearization* of the system along $z(t; z^0)$. Let $z(t; \bar{z}^0)$ denote another solution with an almost identical initial condition, i.e. $\bar{z}^0 \approx z^0$. Then it is often of interest to know how this small difference

$$\delta z^0 := \bar{z}^0 - z^0$$

in the initial conditions is propagated in time. Of course, the answer is obtained by simply comparing the two solution curves, i.e.

$$\delta z(t) := z(t; \bar{z}^0) - z(t; z^0).$$

However, if only $z(t; z^0)$ is explicitly known, then an approximation for $\delta z(t)$ can be obtained via Taylor series expansion of the flow map in δz^0

$$\begin{aligned}\delta z(t) &= \boldsymbol{\Phi}_{t,H}(\bar{z}^0) - \boldsymbol{\Phi}_{t,H}(z^0)\\ &= \boldsymbol{\Phi}_{t,H}(z^0 + \delta z^0) - \boldsymbol{\Phi}_{t,H}(z^0)\\ &\approx \frac{\partial}{\partial z}\boldsymbol{\Phi}_{t,H}(z^0)\,\delta z^0.\end{aligned}$$

This motivates the definition of the time-dependent vector

$$\boldsymbol{\xi}(t) = \frac{\partial}{\partial z}\boldsymbol{\Phi}_{t,H}(z^0)\,\delta z^0. \tag{3.22}$$

We can differentiate with respect to time making use of

$$\frac{d}{dt}z(t) = \frac{\partial}{\partial t}\boldsymbol{\Phi}_{t,H}(z(0)) = J\nabla_z H(\boldsymbol{\Phi}_{t,H}(z(0))) = J\nabla_z H(z(t))$$

to obtain

$$\begin{aligned}\frac{d}{dt}\boldsymbol{\xi}(t) &= \frac{\partial}{\partial t}\frac{\partial}{\partial z}\boldsymbol{\Phi}_{t,H}(z^0)\,\delta z^0\\ &= \frac{\partial}{\partial z}\left[\frac{\partial}{\partial t}\boldsymbol{\Phi}_{t,H}(z^0)\right]\,\delta z^0\\ &= \frac{\partial}{\partial z}\left[J\nabla_z H(\boldsymbol{\Phi}_{t,H}(z^0))\right]\,\delta z^0\\ &= JH_{zz}(z(t))\left[\frac{\partial}{\partial z}\boldsymbol{\Phi}_{t,H}(z^0)\,\delta z^0\right]\\ &= JH_{zz}(z(t))\,\boldsymbol{\xi}(t).\end{aligned}$$

Hence we have derived a linear time-dependent differential equation in $\boldsymbol{\xi} \in \mathbb{R}^{2d}$, called the *variational equations* along a solution $z(t) = z(t; z^0)$:

VARIATIONAL EQUATIONS

$$\frac{d}{dt}\boldsymbol{\xi} = JH_{zz}(z(t))\,\boldsymbol{\xi}. \tag{3.23}$$

Note that the quality of the approximation $\delta z(t) \approx \boldsymbol{\xi}(t)$ depends on the smallness of δz^0 and will only be valid, in general, over relatively short time intervals. However, the variational equations are important in their own right as we will see in the following section.

In the particular situation that $\boldsymbol{z}(t;\boldsymbol{z}^0)$ is an equilibrium solution $\boldsymbol{z}(t;\boldsymbol{z}^0) \equiv \bar{\boldsymbol{z}} = \boldsymbol{z}^0$, the variational equations (3.23) reduce to the time-independent linear system

$$\frac{d}{dt}\boldsymbol{\xi} = J H_{zz}(\bar{\boldsymbol{z}})\,\boldsymbol{\xi}.$$

For the Newtonian mechanical system (3.2)–(3.3) with equilibrium $\bar{\boldsymbol{z}} = (\bar{\boldsymbol{q}}, \boldsymbol{0})$, the variational equations about $\boldsymbol{z} = \bar{\boldsymbol{z}}$ are

$$\frac{d}{dt}\boldsymbol{\xi} = \begin{bmatrix} \boldsymbol{0} & M^{-1} \\ -V_{qq}(\bar{\boldsymbol{q}}) & \boldsymbol{0} \end{bmatrix} \boldsymbol{\xi}.$$

The study of this linear system often allows one to understand the behavior of solutions of the nonlinear system in the vicinity of the equilibrium point. For example, if the matrix $V_{qq}(\bar{\boldsymbol{q}})$ is positive definite, the equilibrium point $\bar{\boldsymbol{q}}$ will be stable for both the linear system *and* the nonlinear system.[5]

Let us abbreviate the Jacobian of the flow map $\boldsymbol{\Phi}_{t,H}$ by $\boldsymbol{F}(t) \in \mathbb{R}^{2d \times 2d}$, i.e.

$$\boldsymbol{F}(t) = \frac{\partial}{\partial \boldsymbol{z}}\boldsymbol{\Phi}_{t,H}(\boldsymbol{z}^0).$$

Then formula (3.22) can be written as $\boldsymbol{\xi}(t) = \boldsymbol{F}(t)\delta\boldsymbol{z}^0$, which holds for any vector $\delta\boldsymbol{z}^0$. It is then easy to verify that the matrix $\boldsymbol{F}(t)$ itself is the solution of the matrix-valued variational equation

$$\frac{d}{dt}\boldsymbol{F} = J H_{zz}(\boldsymbol{z}(t))\,\boldsymbol{F}, \tag{3.24}$$

with initial condition $\boldsymbol{F}(0) = \boldsymbol{I}_{2d}$.

3.5 Symplectic maps and Hamiltonian flow maps

This and following sections are devoted to the definition of symplecticness of a smooth mapping $\boldsymbol{\Psi} : \mathbb{R}^{2d} \to \mathbb{R}^{2d}$ and its implications for the solution behavior of canonical Hamiltonian systems. The presentation in this section is divided into three parts. We start with an algebraic definition of symplecticness and show that the flow map $\boldsymbol{\Phi}_{T,H}$ of a canonical Hamiltonian system is symplectic. In the second part, we discuss the implications on the solution behavior for single-degree-of-freedom systems. Finally, in the third part, we give another derivation

[5] Another frequent application of the variational equations occurs in the context of *periodic solutions*. A solution $\{\boldsymbol{z}(t)\}_{t\in\mathbb{R}}$ is called *periodic* if there exists a constant $T > 0$ such that $\boldsymbol{z}(t) = \boldsymbol{z}(t+T)$. In this case the variational equations reduce to a T-periodic time-dependent linear differential equation and its properties can be analyzed using Floquet theory [76].

of symplecticness, making clear its geometric interpretation in arbitrary finite dimensions.

A smooth map $\boldsymbol{\Psi}$ on the phase space $\mathbb{R}^{2d}$ is called a *symplectic map* with respect to the (constant and invertible) structure matrix $\boldsymbol{J}$ if its Jacobian $\boldsymbol{\Psi}_z(\boldsymbol{z})$ satisfies:

SYMPLECTICNESS OF A MAP

$$[\boldsymbol{\Psi}_z(\boldsymbol{z})]^T \boldsymbol{J}^{-1} \boldsymbol{\Psi}_z(\boldsymbol{z}) = \boldsymbol{J}^{-1}, \tag{3.25}$$

for all $\boldsymbol{z}$ in the domain of definition of $\boldsymbol{\Psi}$. In case $\boldsymbol{J}$ in (3.25), given by (3.6), one sees the term "canonical map" used as a synonym for the term "symplectic map."

For a linear transformation $\boldsymbol{\Psi}(\boldsymbol{z}) = \boldsymbol{B}\boldsymbol{z}$, the condition of symplecticness reduces to $\boldsymbol{B}^T \boldsymbol{J}^{-1} \boldsymbol{B} = \boldsymbol{J}^{-1}$ and $\boldsymbol{B}$ is referred to as a *symplectic matrix*.

Example 5 Consider the Hénon map $\boldsymbol{\Psi}_{a,b} : \mathbb{R}^2 \to \mathbb{R}^2$ defined by

$$\boldsymbol{\Psi}_{a,b}(q, p) := \begin{bmatrix} p \\ 1 + bq + ap^2 \end{bmatrix},$$

$a, b \neq 0$ parameters. To check that this map is canonical, we compute the Jacobian matrix

$$\frac{\partial}{\partial \boldsymbol{z}} \boldsymbol{\Psi}_{a,b}(\boldsymbol{z}) = \begin{bmatrix} 0 & 1 \\ b & 2ap \end{bmatrix},$$

of $\boldsymbol{\Psi}_{a,b}$. Then a straightforward calculation yields

$$\left[\frac{\partial}{\partial \boldsymbol{z}} \boldsymbol{\Psi}_{a,b}\right]^T \boldsymbol{J}^{-1} \frac{\partial}{\partial \boldsymbol{z}} \boldsymbol{\Psi}_{a,b} = \begin{bmatrix} 0 & b \\ 1 & 2ap \end{bmatrix} \begin{bmatrix} 0 & -1 \\ 1 & 0 \end{bmatrix} \begin{bmatrix} 0 & 1 \\ b & 2ap \end{bmatrix} = -b\boldsymbol{J}^{-1}.$$

(Here $\boldsymbol{J}$ is the 2×2 canonical symplectic structure matrix.) Hence the Hénon map is symplectic for $b = -1$. □

Example 6 Often, a non-linear change of position in coordinates from $\boldsymbol{q}$ to $\hat{\boldsymbol{q}}$ is prescribed by

$$\hat{\boldsymbol{q}} = \boldsymbol{h}(\boldsymbol{q}),$$

where $\boldsymbol{h} : \mathcal{U} \subset \mathbb{R}^d \to \mathbb{R}^d$ is a diffeomorphism (i.e. a smooth map with a smooth inverse).

For example, consider the transformation

$$q_1 = r \sin\theta,$$
$$q_2 = r \cos\theta$$

from polar (r, θ) coordinates to Cartesian (q_1, q_2) coordinates.

One then requires a definition of the corresponding transformed conjugate momenta $\hat{\boldsymbol{p}}$ such that the overall transformation $\boldsymbol{\Psi}$ from $(\boldsymbol{q}, \boldsymbol{p})$ to $(\hat{\boldsymbol{q}}, \hat{\boldsymbol{p}})$ is symplectic. Such a definition is provided by

$$\hat{\boldsymbol{p}} := [\boldsymbol{h}_q(\boldsymbol{q})]^{-T}\boldsymbol{p}.$$

The overall coordinate transformation $\boldsymbol{\Psi}$ is hence given by

$$\boldsymbol{\Psi}(\boldsymbol{q}, \boldsymbol{p}) = \begin{bmatrix} \boldsymbol{h}(\boldsymbol{q}) \\ [\boldsymbol{h}_q(\boldsymbol{q})]^{-T}\,\boldsymbol{p} \end{bmatrix}. \tag{3.26}$$

This type of symplectic transformation is sometimes called a *canonical point transformation* [73, 124].

The Jacobian of the map $\boldsymbol{\Psi}$ is

$$\boldsymbol{\Psi}_z(\boldsymbol{z}) = \begin{bmatrix} \boldsymbol{h}_q(\boldsymbol{q}) & \boldsymbol{0} \\ \boldsymbol{G}(\boldsymbol{q}, \boldsymbol{p}) & [\boldsymbol{h}_q(\boldsymbol{q})]^{-T} \end{bmatrix},$$

where $\boldsymbol{G}(\boldsymbol{q}, \boldsymbol{p})$ is the matrix of partial derivative of the vector $[\boldsymbol{h}_q(\boldsymbol{q})]^{-T}\boldsymbol{p}$ with respect to $\boldsymbol{q}$. It can be verified by explicit computation that

$$\boldsymbol{\Psi}_z(\boldsymbol{z})^T \boldsymbol{J}^{-1} \boldsymbol{\Psi}_z(\boldsymbol{z}) = \boldsymbol{J}^{-1},$$

where $\boldsymbol{J}$ is the canonical structure matrix. A different proof for the symplecticness of canonical point transformations will be given in Section 3.6.

For the coordinate transformation from polar to cartesian coordinates, we obtain the expression

$$\begin{bmatrix} p_1 \\ p_2 \end{bmatrix} = \frac{1}{r}\begin{bmatrix} r\sin\theta & \cos\theta \\ r\cos\theta & -\sin\theta \end{bmatrix}\begin{bmatrix} p_r \\ p_\theta \end{bmatrix}$$

for the conjugate momenta p_r, p_θ. This implies that the kinetic energy of a particle in $\mathbb{R}^2$ with mass m transforms according to

$$\frac{1}{2m}\left(p_1^2 + p_2^2\right) = \frac{1}{2m}\left(p_r^2 + \frac{1}{r^2}p_\theta^2\right).$$

□

The main result of this section is contained in the following:

Theorem 1 *The flow map $\boldsymbol{\Phi}_{\tau,H}$ of a Hamiltonian system (3.7) is symplectic.* □

Proof. Recall that the Jacobian of the flow map

$$\boldsymbol{F}(t) = \frac{\partial}{\partial \boldsymbol{z}}\boldsymbol{\Phi}_{t,H}(\boldsymbol{z})$$

satisfies the variational equation (3.24) with initial condition $\boldsymbol{F}(0) = \boldsymbol{I}_{2d}$. To prove the proposition, we have to verify that

$$\boldsymbol{F}(t)^T \boldsymbol{J}^{-1} \boldsymbol{F}(t) = \boldsymbol{J}^{-1}.$$

Since $\boldsymbol{F}(0) = \boldsymbol{I}_{2d}$ the statement is true for $t = 0$. Hence we only have to show that

$$\frac{d}{dt}\boldsymbol{K} = 0, \qquad \boldsymbol{K}(t) := \boldsymbol{F}(t)^T \boldsymbol{J}^{-1} \boldsymbol{F}(t).$$

Indeed, we obtain

$$\begin{aligned}
\frac{d}{dt}\boldsymbol{K} &= \boldsymbol{F}^T \boldsymbol{J}^{-1} \dot{\boldsymbol{F}} + \left[\dot{\boldsymbol{F}}\right]^T \boldsymbol{J}^{-1} \boldsymbol{F} \\
&= \boldsymbol{F}^T \boldsymbol{J}^{-1} \left[\boldsymbol{J} H_{zz}(z(t)) \boldsymbol{F}\right] + \left[\boldsymbol{F}^T H_{zz}(z(t)) \boldsymbol{J}^T\right] \boldsymbol{J}^{-1} \boldsymbol{F} \\
&= \boldsymbol{F}^T H_{zz}(z(t)) \boldsymbol{F} - \boldsymbol{F}^T H_{zz}(z(t)) \boldsymbol{F} \\
&= \boldsymbol{0}.
\end{aligned}$$

□

The symplecticness of the flow map implies the existence of certain global conservation laws or *integral invariants* related to the evolution of subsets of phase space. In particular, we will show in the following subsection that symplecticness implies preservation of area for one-degree-of-freedom systems ($d = 1$). In higher dimensions this invariant is replaced by the preservation of volume ($d > 1$). Conservation of volume under the flow map of a canonical Hamiltonian system follows from Liouville's Theorem [7] (see problem 7 in the Exercises), but it can also be considered as a consequence of the symplecticness of the flow map.[6] In other words, the sets $\mathcal{U}_t$, $t > 0$, defined by (3.21) might change their shapes significantly but they will always occupy the same volume in phase space as time t increases. The existence of this integral invariant implies severe and important restrictions on the possible solution behavior of Hamiltonian dynamical systems. As we will see throughout this book, preservation of the symplectic structure under numerical discretization is often a very desirable property for long-term approximate integration of Hamiltonian systems.

3.5.1 One-degree-of-freedom systems

Let $\boldsymbol{\Psi} : \mathbb{R}^2 \to \mathbb{R}^2$ be a symplectic map in the plane. We write the Jacobian $\boldsymbol{\Psi}_z(z)$ as

$$\boldsymbol{\Psi}_z(z) = \begin{bmatrix} a & b \\ c & d \end{bmatrix}.$$

Using this in (3.25), we obtain the equivalent condition

$$ad - bc = 1.$$

[6]Symplecticness also implies the existence of other integral invariants for $d > 1$. Thus the concept of symplecticness is a *stronger* property than conservation of volume.

However, this is the same as saying

$$\det[\boldsymbol{\Psi}_z(z)] = 1.$$

Thus, for planar maps $\boldsymbol{\Psi}$, the condition of symplecticness (3.25) is that the determinant of the Jacobian $\boldsymbol{\Psi}_z(z)$ is everywhere equal to one.

What is the implication of this for the behavior of the map $\boldsymbol{\Psi}$? Let $\mathcal{R}$ be a bounded subset of phase space. The oriented area[7] $\alpha(\mathcal{R})$ is defined by the integral

$$\alpha(\mathcal{R}) = \int_{\mathcal{R}} dq dp.$$

Let $\hat{\mathcal{R}}$ denote the image of $\mathcal{R}$ under $\boldsymbol{\Psi}$, i.e. $\hat{\mathcal{R}} = \boldsymbol{\Psi}(\mathcal{R})$. Then, following the standard rules for changing variables in an integral

$$\begin{aligned}\alpha(\hat{\mathcal{R}}) &= \int_{\hat{\mathcal{R}}} d\hat{q} d\hat{p} \\ &= \int_{\mathcal{R}} \det[\boldsymbol{\Psi}_z(z)]\, dq dp \\ &= \int_{\mathcal{R}} dq dp \\ &= \alpha(\mathcal{R}).\end{aligned}$$

Thus a symplectic map of the plane is area preserving.

The implications of symplecticness for the behavior of maps in higher dimensions are taken up in the following section.

3.5.2 The symplectic structure of phase space

We now turn to the general situation and discuss the symplectic structure associated to $2d$-dimensional phase space. The book of LANDAU AND LIFSHITZ [105] gives a clear explanation of the symplectic structure. A more advanced and mathematically precise treatment will be found in the book of ARNOLD [7].

Let us start with some notation from differential geometry. A *two-form* on $\mathbb{R}^{2d}$ is a skew-symmetric bilinear function $\Omega(\boldsymbol{\xi}, \boldsymbol{\eta})$ with arguments $\boldsymbol{\xi}$ and $\boldsymbol{\eta}$ in $\mathbb{R}^{2d}$. *Bilinearity* means that the function is linear in each of its two arguments

[7] The area of $\mathcal{R}$ is given by the absolute value of $\alpha(\mathcal{R})$. The sign of $\alpha(\mathcal{R})$ depends on the orientation of $\mathcal{R}$.

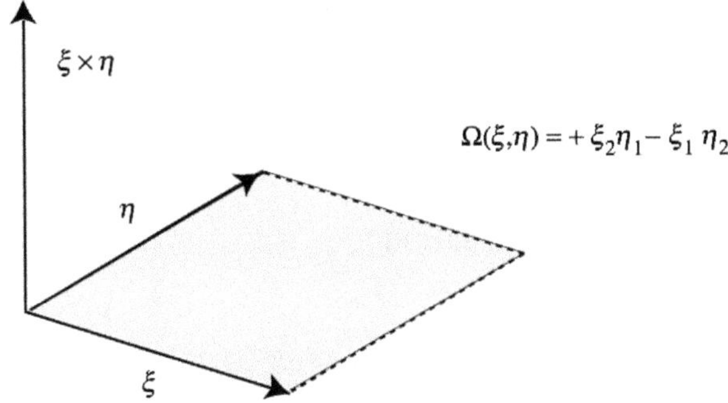

Figure 3.4 The two-form Ω for $d = 1$ is the oriented area of the parallelogram described by a pair of vectors in $\mathbb{R}^2$, i.e. it is the transverse coordinate of the cross product of the pair of vectors viewed as embedded in $\mathbb{R}^3$.

separately, and skew-symmetry means

$$\Omega(\boldsymbol{\xi}, \boldsymbol{\eta}) = -\Omega(\boldsymbol{\eta}, \boldsymbol{\xi}),$$

for any $\boldsymbol{\xi}$, $\boldsymbol{\eta}$.

The symplectic structure matrix

$$\boldsymbol{J} = \begin{bmatrix} \mathbf{0} & +\boldsymbol{I}_d \\ -\boldsymbol{I}_d & \mathbf{0} \end{bmatrix},$$

introduces the *symplectic two-form* Ω on the phase space $\mathbb{R}^{2d}$

$$\Omega(\boldsymbol{\xi}, \boldsymbol{\eta}) := \boldsymbol{\xi}^T \boldsymbol{J}^{-1} \boldsymbol{\eta}, \qquad \boldsymbol{\xi}, \boldsymbol{\eta} \in \mathbb{R}^{2d}. \tag{3.27}$$

Let us give a geometric interpretation of the two-form Ω. For $d = 1$, $\Omega(\boldsymbol{\xi}, \boldsymbol{\eta})$ is the oriented area of the parallelogram spanned by the two vectors $\boldsymbol{\xi}$ and $\boldsymbol{\eta}$ in $\mathbb{R}^2$. This is easily verified since, for $d = 1$

$$\boldsymbol{\xi}^T \boldsymbol{J}^{-1} \boldsymbol{\eta} = \xi_2 \eta_1 - \xi_1 \eta_2, \qquad \boldsymbol{\xi} = (\xi_1, \xi_2)^T, \boldsymbol{\eta} = (\eta_1, \eta_2)^T \in \mathbb{R}^2.$$

The orientation of area refers to the fact that $\Omega(\boldsymbol{\xi}, \boldsymbol{\eta})$ is either equal to the area spanned by $\boldsymbol{\xi}$ and $\boldsymbol{\eta}$ or is equal to the negative of this area. Viewing the two planar vectors as embedded in $\mathbb{R}^3$, the two-form Ω is nothing but the z-coordinate of the cross product of the two vectors (Fig. 3.4).

For $d > 1$, we project $\boldsymbol{\xi} \in \mathbb{R}^{2d}$ and $\boldsymbol{\eta} \in \mathbb{R}^{2d}$ down on to the (q_i, p_i)-coordinate planes, $i = 1, \ldots, d$. Denote the corresponding vectors in $\mathbb{R}^2$ by $\boldsymbol{\xi}^{(i)}$, $\boldsymbol{\eta}^{(i)}$ respectively, $i = 1, \ldots, d$. Then $\Omega(\boldsymbol{\xi}, \boldsymbol{\eta})$ is equivalent to the sum of the oriented areas of the parallelograms spanned by the pair of vectors $\boldsymbol{\xi}^{(i)}$ and $\boldsymbol{\eta}^{(i)}$. To be precise, if we now denote by Ω_0 the standard two-form of a pair of vectors in the plane, define

$$\Omega(\boldsymbol{\xi}, \boldsymbol{\eta}) = \sum_{i=1}^{d} \Omega_0(\boldsymbol{\xi}^{(i)}, \boldsymbol{\eta}^{(i)}).$$

In fact, this is synonymous with saying

$$\Omega(\boldsymbol{\xi}, \boldsymbol{\eta}) = \boldsymbol{\xi}^T \boldsymbol{J}^{-1} \boldsymbol{\eta},$$

where $\boldsymbol{J}$ refers to the symplectic structure matrix in $\mathbb{R}^{2d}$.

Next we consider the integral of Ω over a smooth two-dimensional surface $\mathcal{S} \subset \mathbb{R}^{2d}$ of phase space. We can view such a two-surface as being described[8] by a smooth mapping $\boldsymbol{\sigma}$ from an open subset $\mathcal{R}$ of the (u, v)-plane into phase space $\mathbb{R}^{2d}$ such that the partial derivatives $\boldsymbol{\sigma}_u(u, v)$ and $\boldsymbol{\sigma}_v(u, v)$ are linearly independent vectors for all $(u, v) \in \mathcal{R}$. We define the integral of the two-form Ω over the surface $\mathcal{S}$ by

$$\begin{aligned} \int_{\mathcal{S}} \Omega &:= \int_{\mathcal{R}} \Omega(\boldsymbol{\sigma}_u du, \boldsymbol{\sigma}_v dv) \\ &= \int_{\mathcal{R}} \boldsymbol{\sigma}_u^T \boldsymbol{J}^{-1} \boldsymbol{\sigma}_v \, du dv; \end{aligned}$$

the last integral is to be understood as the usual integral of the function $f := \boldsymbol{\sigma}_u^T \boldsymbol{J}^{-1} \boldsymbol{\sigma}_v$ over $\mathcal{R} \subset \mathbb{R}^2$. To give a geometric interpretation to the integral, we separate the mapping $\boldsymbol{\sigma}$ into d components $\boldsymbol{\sigma}_1, \boldsymbol{\sigma}_2, \ldots, \boldsymbol{\sigma}_d$, with $\boldsymbol{\sigma}_i$ taking values in the (p_i, q_i)-coordinate plane. Denote the image of $\mathcal{R}$ under $\boldsymbol{\sigma}_i$ by $\mathcal{S}_i \subset \mathbb{R}^2$. To interpret the integral of Ω over the surface $\mathcal{S}$, we first partition $\mathcal{R}$ into a mesh of small rectangles with dimensions $\Delta u \times \Delta v$. Such a partition results in corresponding partitions of each of the map images (the coordinate plane projections of the surface) into small curved regions (see Fig. 3.5). For a fine enough partition, the "rectangle images" are approximated by parallelograms (see Fig. 3.6) spanned by the vectors

$$\boldsymbol{\xi}^{(i)} = \boldsymbol{\sigma}_{i,u} \Delta u \in \mathbb{R}^2, \qquad \boldsymbol{\eta}^{(i)} = \boldsymbol{\sigma}_{i,v} \Delta v \in \mathbb{R}^2.$$

The integral of the two-form Ω is now approximated by the sum of the oriented areas of these parallelograms. The integral of Ω over $\mathcal{S}$ is then taken to be the limiting value of this sum as the diameter of the underlying mesh imposed on $\mathcal{R}$ tends uniformly to zero. In other words

$$\int_{\mathcal{S}} \Omega = \sum_i \alpha(\mathcal{S}_i),$$

with $\alpha(\mathcal{S}_i)$ being the oriented area of $\mathcal{S}_i \subset \mathbb{R}^2$.

[8] In general, the surface must be divided into a collection of overlapping regions called *charts*, in each of which a local parameterization is defined [7].

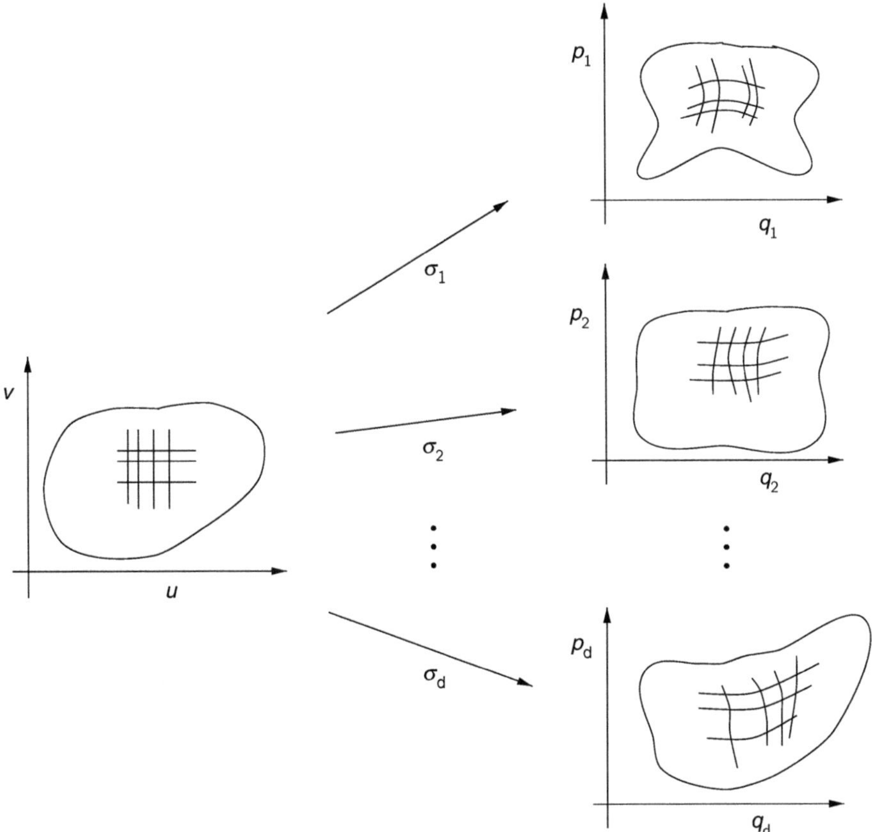

Figure 3.5 A surface is defined by a parameterization: a collection of d mappings from a parameter space in the uv-plane to the canonical (q_i, p_i)-coordinate planes. A rectangular mesh in parameter space induces a curved mesh in the coordinate planes.

Now consider how the symplectic two-form Ω changes under a coordinate transformation $\boldsymbol{\Psi} : \mathbb{R}^{2d} \to \mathbb{R}^{2d}$. We define the transformed differential form $\boldsymbol{\Psi}^*\Omega$ by

$$\boldsymbol{\Psi}^*\Omega\,(\boldsymbol{\xi}, \boldsymbol{\eta}) := \Omega\,(\boldsymbol{\Psi}_z(z)\boldsymbol{\xi}, \boldsymbol{\Psi}_z(z)\boldsymbol{\eta}).$$

In other words, $\boldsymbol{\Psi}^*\Omega(\boldsymbol{\xi}, \boldsymbol{\eta})$ is equal to $\Omega(\hat{\boldsymbol{\xi}}, \hat{\boldsymbol{\eta}})$ where $\hat{\boldsymbol{\xi}}$, $\hat{\boldsymbol{\eta}}$ are the transports of the vectors $\boldsymbol{\xi}$ and $\boldsymbol{\eta}$, respectively, under the linearization (Jacobian matrix) of $\boldsymbol{\Psi}$. A transformation $\boldsymbol{\Psi}$ is called *symplectic*, if

$$\boldsymbol{\Psi}^*\Omega = \Omega,$$

or, equivalently

$$\boldsymbol{\Psi}_z(z)^T \boldsymbol{J}^{-1} \boldsymbol{\Psi}_z(z) = \boldsymbol{J}^{-1}, \tag{3.28}$$

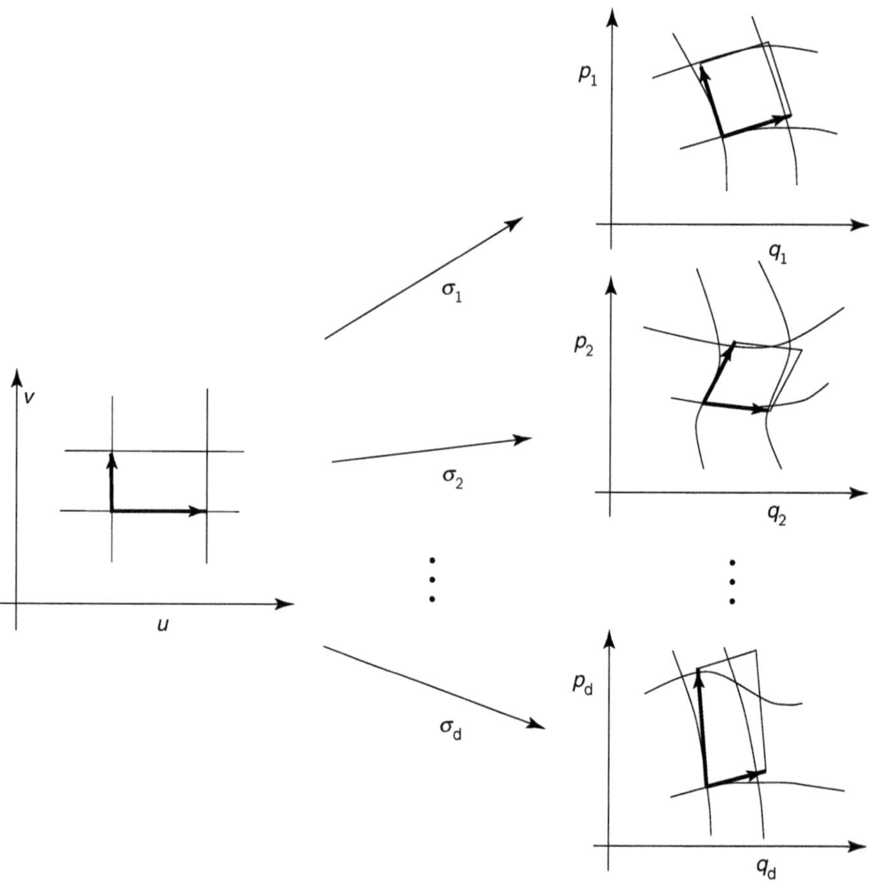

Figure 3.6 For a small rectangular patch in parameter space, the image under the parameterization is a small region that is approximated by a parallelogram whose oriented area is computed in the usual way. The integral of the two-form over a two-surface is then obtained as the limit, for decreasing meshwidth in parameter space, of the sum of the induced two-forms.

which is the definition we gave before. In other words, symplectic maps leave the symplectic two-form Ω invariant. Thus, for $d = 1$, a map $\boldsymbol{\Psi}$ is symplectic if the parallelogram spanned by $\boldsymbol{\xi}$ and $\boldsymbol{\eta}$ has the same area as the parallelogram spanned by the two vectors $\boldsymbol{\Psi}_z(z)\boldsymbol{\xi}$ and $\boldsymbol{\Psi}_z(z)\boldsymbol{\eta}$. This implies that, for $d = 1$, a symplectic map conserves area.

More generally, this important integral invariance can be written as

$$\begin{aligned} \int_{\mathcal{S}} \Omega &:= \int_{\mathcal{S}} \boldsymbol{\Psi}^* \Omega \\ &= \int_{\mathcal{R}} \Omega \left(\boldsymbol{\Psi}_z \boldsymbol{\sigma}_u du, \boldsymbol{\Psi}_z \boldsymbol{\sigma}_v dv \right) \end{aligned}$$

$$= \int_{\mathcal{R}} \Omega\left(\bar{\boldsymbol{\sigma}}_u du, \bar{\boldsymbol{\sigma}}_v dv\right)$$
$$= \int_{\bar{\mathcal{S}}} \Omega,$$

where $\bar{\mathcal{S}} := \boldsymbol{\Psi}(\mathcal{S})$ and $\bar{\boldsymbol{\sigma}} := \boldsymbol{\Psi} \circ \boldsymbol{\sigma} : \mathcal{R} \to \bar{\mathcal{S}}$ is a parameterization of $\bar{S}$. For $d > 1$, this formula implies the conservation of a sum of oriented areas of projections of $\mathcal{S}$ on to the (q_i, p_i)-coordinate planes, i.e.

$$\sum_i \alpha(\mathcal{S}_i) = \sum_i \alpha(\bar{\mathcal{S}}_i).$$

We refer to this quantity as the symplectic area.

We have already mentioned that the flows of Hamiltonian systems preserve volume in phase space. In fact, it can be shown that conservation of the symplectic two-form Ω alone implies that the sum over all distinct k-tuples $1 \leq i_1 < i_2 < \ldots < i_k \leq d$ of oriented volumes of projections of a $2k$-dimensional submanifold on to the $(p_{i_1}, p_{i_2}, \ldots, p_{i_k}, q_{i_1}, q_{i_2}, \ldots, q_{i_k})$-coordinate system is also conserved under the map [7]; as a special case ($k = d$), this implies the conservation of phase space volume.

We also mention that, although our geometric interpretation above relied on the fact that $\boldsymbol{J}$ was the canonical structure matrix, a corresponding geometric property would hold locally for any Hamiltonian system. This is a consequence of Darboux's theorem [7, 8].

3.6 Differential forms and the wedge product

The definition (3.28) is not always the most convenient approach to check the symplecticness of a given map $\boldsymbol{\Psi}$. This is true in particular if the map is given implicitly or the definition involves additional variables that could, in principle, be eliminated. In those cases, it is best to use implicit differentiation combined with a definition of symplecticness in terms of differential one- and two-forms. We develop the necessary mathematical material in this section.

Recall from calculus that if $f : \mathbb{R}^m \to \mathbb{R}$ is a smooth function, then its *directional derivative* along a vector $\boldsymbol{\xi} \in \mathbb{R}^m$, denoted here by $df(\boldsymbol{\xi})$, is given by

$$df(\boldsymbol{\xi}) = \frac{\partial f}{\partial z_1}\xi_1 + \frac{\partial f}{\partial z_2}\xi_2 + \cdots + \frac{\partial f}{\partial z_m}\xi_m,$$

where the partial derivatives of f are computed at a fixed location $\boldsymbol{z} \in \mathbb{R}^m$. It is clear from the definition that $df(\boldsymbol{\xi})$ is linear in $\boldsymbol{\xi}$. The linear functional $df(.)$ is called the *differential* of f at $\boldsymbol{z}$ and is an example of a differential one-form.

A particularly simple class of differentials is provided by the coordinate functions $f(z) = z_i$ and

$$dz_i(\boldsymbol{\xi}) = \xi_i,$$

$i = 1, \ldots, m$. Indeed, the differentials dz_i can be used as a basis for representing any other differential df since

$$df(\boldsymbol{\xi}) = \frac{\partial f}{\partial z_1} dz_1(\boldsymbol{\xi}) + \frac{\partial f}{\partial z_2} dz_2(\boldsymbol{\xi}) + \cdots + \frac{\partial f}{\partial z_m} dz_m(\boldsymbol{\xi}),$$

and, consequently

$$df = \frac{\partial f}{\partial z_1} dz_1 + \frac{\partial f}{\partial z_2} dz_2 + \cdots + \frac{\partial f}{\partial z_m} dz_m.$$

To summarize, the set of all differentials at a point $\boldsymbol{z} \in \mathbb{R}^m$ forms a linear space with the differentials dz_i as a basis.

We now apply a coordinate transformation $\boldsymbol{\Psi} : \mathbb{R}^m \to \mathbb{R}^m$ and define

$$\hat{\boldsymbol{z}} = \boldsymbol{\Psi}(\boldsymbol{z}),$$

as well as the transformed function

$$\hat{f}(\boldsymbol{z}) = f(\hat{\boldsymbol{z}}) = f(\boldsymbol{\Psi}(\boldsymbol{z})).$$

Denoting the ith component of the transformation $\boldsymbol{\Psi}$ by Ψ^i, i.e. $\hat{z}_i = \Psi^i(\boldsymbol{z})$, the standard chain rule implies

$$\begin{aligned} d\hat{f} &= \sum_j \frac{\partial f}{\partial \hat{z}_j} \frac{\partial \Psi^j}{\partial z_1} dz_1 + \sum_j \frac{\partial f}{\partial \hat{z}_j} \frac{\partial \Psi^j}{\partial z_2} dz_2 + \cdots + \sum_j \frac{\partial f}{\partial \hat{z}_j} \frac{\partial \Psi^j}{\partial z_m} dz_m, \\ &= \frac{\partial f}{\partial \hat{z}_1} \sum_i \frac{\partial \Psi^1}{\partial z_i} dz_i + \frac{\partial f}{\partial \hat{z}_2} \sum_i \frac{\partial \Psi^2}{\partial z_i} dz_i + \cdots + \frac{\partial f}{\partial \hat{z}_m} \sum_i \frac{\partial \Psi^m}{\partial z_i} dz_i, \end{aligned}$$

which we rewrite as

$$d\hat{f} = \frac{\partial f}{\partial \hat{z}_1} d\hat{z}_1 + \frac{\partial f}{\partial \hat{z}_2} d\hat{z}_2 + \cdots + \frac{\partial f}{\partial \hat{z}_m} d\hat{z}_m,$$

in the new basis

$$d\hat{z}_j = \sum_i \frac{\partial \Psi^j}{\partial z_i} dz_i. \tag{3.29}$$

To compactify the notation, we introduce the column vector of differential one-forms

$$d\boldsymbol{z} = (dz_1, dz_2, \ldots, dz_m)^T,$$

with the property that $dz(\boldsymbol{\xi}) = \boldsymbol{\xi}$ and hence

$$df = f_z \cdot dz.$$

The transformation rule (3.29) becomes

$$d\hat{z} = \boldsymbol{\Psi}_z(z)\, dz,$$

and, consequently

$$d\hat{z}(\boldsymbol{\xi}) = \boldsymbol{\Psi}_z(z)\, dz(\boldsymbol{\xi}) = \boldsymbol{\Psi}_z(z)\, \boldsymbol{\xi}.$$

Let us now come back to symplectic transformations and a splitting of variables $\boldsymbol{z} = (\boldsymbol{q}, \boldsymbol{p}) \in \mathbb{R}^{2d}$ with associated one-forms dq_i, dp_i, $i = 1, \ldots, d$. Given any two vectors $\boldsymbol{\xi}, \boldsymbol{\eta} \in \mathbb{R}^{2d}$, the symplectic two-form Ω is defined by

$$\Omega(\boldsymbol{\xi}, \boldsymbol{\eta}) = \boldsymbol{\xi}^T J^{-1} \boldsymbol{\eta}.$$

Using the ordering of the two vectors $\boldsymbol{\xi}$ and $\boldsymbol{\eta}$ into a sequence of projections $\boldsymbol{\xi}^{(i)}, \boldsymbol{\eta}^{(i)} \in \mathbb{R}^2$ down on to the (q_i, p_i)-coordinate planes, as introduced in the previous section, it is straightforward to verify that

$$\Omega(\boldsymbol{\xi}, \boldsymbol{\eta}) = \sum_i \Omega_0(\boldsymbol{\xi}^{(i)}, \boldsymbol{\eta}^{(i)}) = \sum_i \left[dp_i(\boldsymbol{\xi}) dq_i(\boldsymbol{\eta}) - dq_i(\boldsymbol{\xi}) dp_i(\boldsymbol{\eta})\right],$$

where, by definition, $dq_i(\boldsymbol{\xi}) = \xi_1^{(i)}$, $dp_i(\boldsymbol{\xi}) = \xi_2^{(i)}$ etc. This suggests to introduce the wedge product of two differentials df and dg via

$$(df \wedge dg)(\boldsymbol{\xi}, \boldsymbol{\eta}) := dg(\boldsymbol{\xi}) df(\boldsymbol{\eta}) - df(\boldsymbol{\xi}) dg(\boldsymbol{\eta}),$$

and, in particular, to write

$$\Omega = \sum_i dq_i \wedge dp_i.$$

Using vector notation, we shorten this to

$$\Omega = d\boldsymbol{q} \wedge d\boldsymbol{p}.$$

Conservation of symplecticness, under a transformation

$$\hat{\boldsymbol{q}} = \boldsymbol{\Psi}^1(\boldsymbol{q}, \boldsymbol{p}),$$
$$\hat{\boldsymbol{p}} = \boldsymbol{\Psi}^2(\boldsymbol{q}, \boldsymbol{p})$$

reduces now to the statement that

$$d\hat{\boldsymbol{q}} \wedge d\hat{\boldsymbol{p}} = d\boldsymbol{q} \wedge d\boldsymbol{p}, \tag{3.30}$$

where

$$d\hat{\boldsymbol{q}} = \boldsymbol{\Psi}_q^1(\boldsymbol{q}, \boldsymbol{p})d\boldsymbol{q} + \boldsymbol{\Psi}_p^1(\boldsymbol{q}, \boldsymbol{p})d\boldsymbol{p},$$
$$d\hat{\boldsymbol{q}} = \boldsymbol{\Psi}_q^2(\boldsymbol{q}, \boldsymbol{p})d\boldsymbol{q} + \boldsymbol{\Psi}_p^2(\boldsymbol{q}, \boldsymbol{p})d\boldsymbol{p}.$$

We will prove this result below.

One can immediately verify several algebraic properties of the wedge product which prove useful in calculations:

PROPERTIES OF THE WEDGE PRODUCT

Let $d\boldsymbol{a}$, $d\boldsymbol{b}$, $d\boldsymbol{c}$, be k-vectors of differential one-forms on $\mathbb{R}^m$, then the following properties hold.

1. *Skew-symmetry*

$$d\boldsymbol{a} \wedge d\boldsymbol{b} = -d\boldsymbol{b} \wedge d\boldsymbol{a}.$$

2. *Bilinearity:* for any $\alpha, \beta \in \mathbb{R}$

$$d\boldsymbol{a} \wedge (\alpha d\boldsymbol{b} + \beta d\boldsymbol{c}) = \alpha d\boldsymbol{a} \wedge d\boldsymbol{b} + \beta d\boldsymbol{a} \wedge d\boldsymbol{c}.$$

3. *Rule of matrix multiplication* (as a consequence of Property 2 and the definition)

$$d\boldsymbol{a} \wedge (\boldsymbol{A}d\boldsymbol{b}) = (\boldsymbol{A}^T d\boldsymbol{a}) \wedge d\boldsymbol{b},$$

for any $k \times k$ matrix $\boldsymbol{A}$. (See Problem 9 in the Exercises.)

Note that Properties 1 and 3 of the wedge product imply that, if $\boldsymbol{A}$ is a symmetric matrix, then

$$d\boldsymbol{a} \wedge (\boldsymbol{A}d\boldsymbol{a}) = 0.$$

We now prove (3.30). We first verify that

$$d\boldsymbol{q} \wedge d\boldsymbol{p} = \frac{1}{2}(\boldsymbol{J}^{-1}dz) \wedge dz.$$

Indeed, we have

$$
\begin{aligned}
(\boldsymbol{J}^{-1} d\boldsymbol{z}) \wedge d\boldsymbol{z} &= \sum_{i=1}^{d} [dz_i \wedge dz_{d+i} - dz_{d+i} \wedge dz_i] \\
&= \sum_{i=1}^{d} [dq_i \wedge dp_i - dp_i \wedge dq_i] \\
&= 2d\boldsymbol{q} \wedge d\boldsymbol{p}.
\end{aligned}
$$

Hence (3.30) is equivalent to

$$
(\boldsymbol{J}^{-1} d\hat{\boldsymbol{z}}) \wedge d\hat{\boldsymbol{z}} = (\boldsymbol{J}^{-1} d\boldsymbol{z}) \wedge d\boldsymbol{z}, \tag{3.31}
$$

where

$$
d\hat{\boldsymbol{z}} = \boldsymbol{\Psi}_z(\boldsymbol{z}) d\boldsymbol{z}.
$$

Applying the rule of matrix multiplication, we obtain

$$
\begin{aligned}
(\boldsymbol{J}^{-1} d\hat{\boldsymbol{z}}) \wedge d\hat{\boldsymbol{z}} &= \left(\boldsymbol{J}^{-1} \boldsymbol{\Psi}_z(\boldsymbol{z}) d\boldsymbol{z}\right) \wedge (\boldsymbol{\Psi}_z(\boldsymbol{z}) d\boldsymbol{z}), \\
&= \left(\boldsymbol{\Psi}_z(\boldsymbol{z})^T \boldsymbol{J}^{-1} \boldsymbol{\Psi}_z(\boldsymbol{z}) d\boldsymbol{z}\right) \wedge d\boldsymbol{z},
\end{aligned}
$$

and symplecticness of the map $\boldsymbol{\Psi}$, i.e.

$$
\boldsymbol{\Psi}_z(\boldsymbol{z})^T \boldsymbol{J}^{-1} \boldsymbol{\Psi}_z(\boldsymbol{z}) = \boldsymbol{J}^{-1}
$$

implies (3.31).

As indicated in the introduction, the wedge product notation can be combined with implicit differentiation which makes it a powerful tool to verify symplecticness of an implicitly given transformation $\boldsymbol{\Psi}$. As an illustration, let us come back to the canonical point transformation

$$
\begin{aligned}
\hat{\boldsymbol{q}} &= \boldsymbol{h}(\boldsymbol{q}), \\
\boldsymbol{h}_q(\boldsymbol{q})^T \hat{\boldsymbol{p}} &= \boldsymbol{p},
\end{aligned}
$$

as already discussed in Example 6. Implicit differentiation yields

$$
\begin{aligned}
d\hat{\boldsymbol{q}} &= \boldsymbol{h}_q(\boldsymbol{q}) d\boldsymbol{q}, \\
\boldsymbol{h}_q(\boldsymbol{q})^T d\hat{\boldsymbol{p}} + \sum_i \hat{p}_i h^i_{\boldsymbol{qq}} d\boldsymbol{q} &= d\boldsymbol{p},
\end{aligned}
$$

where $h^i(\boldsymbol{q})$ denotes the ith component of the transformation $\boldsymbol{h}(\boldsymbol{q})$. Taking wedge products with $d\hat{\boldsymbol{p}}$ and $d\boldsymbol{q}$, respectively, leads to

$$
d\hat{\boldsymbol{q}} \wedge d\hat{\boldsymbol{p}} = (\boldsymbol{h}_q(\boldsymbol{q}) d\boldsymbol{q}) \wedge d\hat{\boldsymbol{p}} = d\boldsymbol{q} \wedge \left(\boldsymbol{h}_q(\boldsymbol{q})^T d\hat{\boldsymbol{p}}\right),
$$

and, since

$$\sum_i d\boldsymbol{q} \wedge (\hat{p}_i h^i_{\boldsymbol{qq}}(\boldsymbol{q}) d\boldsymbol{q}) = \sum_i \hat{p}_i \left[d\boldsymbol{q} \wedge (h^i_{\boldsymbol{qq}}(\boldsymbol{q}) d\boldsymbol{q}) \right] = 0,$$

to

$$d\boldsymbol{q} \wedge d\boldsymbol{p} = d\boldsymbol{q} \wedge \left(\boldsymbol{h}_{\boldsymbol{q}}(\boldsymbol{q})^T d\hat{\boldsymbol{p}} + \sum_i \hat{p}_i h^i_{\boldsymbol{qq}} d\boldsymbol{q} \right) = d\boldsymbol{q} \wedge \left(\boldsymbol{h}_{\boldsymbol{q}}(\boldsymbol{q})^T d\hat{\boldsymbol{p}} \right),$$

respectively. This implies

$$d\hat{\boldsymbol{q}} \wedge d\hat{\boldsymbol{p}} = d\boldsymbol{q} \wedge \left(\boldsymbol{h}_{\boldsymbol{q}}(\boldsymbol{q})^T d\hat{\boldsymbol{p}} \right) = d\boldsymbol{q} \wedge d\boldsymbol{p},$$

as desired.

We will encounter more applications of this basic procedure in the following chapter.

3.7 Exercises

1. *Lagrange's equation.*

a. Assume that equation (3.14) defines, for each fixed $\boldsymbol{q}$, a one-to-one map between the conjugate momenta $\boldsymbol{p}$ and the velocities $\dot{\boldsymbol{q}}$. Let us write the inverse relation as $\dot{\boldsymbol{q}} = \boldsymbol{v}(\boldsymbol{p}, \boldsymbol{q})$. Use this map in the Hamiltonian (3.16) to evaluate $\nabla_{\boldsymbol{q}} H$ and $\nabla_{\boldsymbol{p}} H$. The result shows that the Lagrange equation (3.13) is equivalent to a canonical Hamiltonian system (3.4)–(3.5).

b. Given the Lagrangian (3.17) verify the definition of the conjugate momenta (3.18) and find an explicit expression for the Hamiltonian H in terms of $\boldsymbol{q}$ and $\boldsymbol{p}$ only. Show that

$$\frac{d}{dt}\boldsymbol{q} = \frac{1}{m}\boldsymbol{p} + \frac{1}{2m}\widehat{\boldsymbol{b}}\boldsymbol{q}, \tag{3.32}$$

$$\frac{d}{dt}\boldsymbol{p} = -\frac{\gamma}{\|\boldsymbol{q}\|^3}\boldsymbol{q} + \frac{1}{2m}\widehat{\boldsymbol{b}}\boldsymbol{p} - \frac{1}{4m}\widehat{\boldsymbol{b}}^T\widehat{\boldsymbol{b}}\boldsymbol{q} \tag{3.33}$$

are the associated canonical equations of motion and that these equations are equivalent to (3.10).

2. *Angular momentum.*

a. Show that, angular momentum $\boldsymbol{m} = \boldsymbol{q} \times \boldsymbol{p}$ is conserved for canonical equations with Hamiltonian

$$H(\boldsymbol{q}, \boldsymbol{p}) = \frac{1}{2}\|\boldsymbol{p}\|^2 + \varphi(\|\boldsymbol{q}\|).$$

If in addition $\varphi(r) = -1/r$ (Kepler problem), verify that the Runge–Lenz vector

$$\boldsymbol{e} = \boldsymbol{p} \times \boldsymbol{m} - \frac{\boldsymbol{q}}{\|\boldsymbol{q}\|}$$

is also a first integral. You may use the vector identity

$$\boldsymbol{a} \times (\boldsymbol{b} \times \boldsymbol{c}) = \boldsymbol{b}(\boldsymbol{a} \cdot \boldsymbol{c}) - \boldsymbol{c}(\boldsymbol{a} \cdot \boldsymbol{b}).$$

b. Show that the canonical Equations (3.32)–(3.33) for a charged particle in a constant magnetic field $\boldsymbol{b}$ possess the scalar quantity

$$\mu = \boldsymbol{b} \cdot (\boldsymbol{q} \times \boldsymbol{p})$$

as a first integral. The preserved quantity is also called the *magnetic momentum.* In terms of the noncanonical Hamiltonian formulation (3.11)–(3.12), the magnetic momentum is given by

$$\mu = \boldsymbol{b} \cdot \left(\boldsymbol{q} \times \left(\boldsymbol{p} - \frac{1}{2} \boldsymbol{b} \times \boldsymbol{q} \right) \right).$$

3. *Integrals of the N-body problem.* Show that the total linear and angular momenta

$$\boldsymbol{p}_{\text{tot}} = \sum_{i=1}^{N} \boldsymbol{p}_i, \quad \boldsymbol{m}_{\text{tot}} = \sum_{i=1}^{N} \boldsymbol{q}_i \times \boldsymbol{p}_i$$

are conserved quantities for an N-body system with distance-based pair potentials (the system does not need to be homogeneous).

4. *Variational equations.* Consider a planar central-force problem

$$H = \frac{1}{2}\|\boldsymbol{p}\|^2 + \varphi(\|\boldsymbol{q}\|), \qquad \boldsymbol{q}, \boldsymbol{p} \in \mathbb{R}^2.$$

a. Find the variational equations for this problem.

b. Let $\boldsymbol{q}^*$ be an equilibrium point ($\varphi'(\|\boldsymbol{q}^*\|) = 0$). Determine the eigenvalues and eigenvectors of the variational equations. [Hint: consider separately perturbations in (i) the radial ($\boldsymbol{q}$) direction and (ii) the tangential directions.] What does the eigenstructure suggest about the behavior of nearby solutions?

5. *Canonical transformations.* Consider the planar central-force problem of problem 4 and the canonical change to polar coordinates described in Example 6.

a. Show that the momentum corresponding to the angle variable, p_θ, is just the third component $m_3 = q_1 p_2 - q_2 p_1$ of the angular momentum vector m, and, hence, is constant along any solution.

b. Using the result of part [a], show how to reduce the planar central-force problem to a one-degree-of-freedom problem. In particular, show that the Kepler problem can be reduced to a one-degree-of-freedom problem with potential energy function

$$\varphi_{\text{reduced}}(r) = \frac{m_3^2}{2r^2} - \frac{1}{r}.$$

Graph this potential and discuss the behavior of solutions.

6. *Canonical transformations.* Show that the map $\boldsymbol{\Psi}_\epsilon$, defined by

$$\begin{aligned} \boldsymbol{Q} &= \boldsymbol{q} + \varepsilon \boldsymbol{p}, \\ \boldsymbol{P} &= \boldsymbol{p} + \varepsilon\, \boldsymbol{F}(\boldsymbol{Q}), \end{aligned}$$

is symplectic for $\varepsilon > 0$, provided the Jacobian matrix of the map $\boldsymbol{F} : \mathbb{R}^d \to \mathbb{R}^d$ at $\boldsymbol{Q}$ is symmetric.

7. *Volume preservation.* The divergence of a vector field $\boldsymbol{f} : \mathbb{R}^k \to \mathbb{R}^k$ is defined by

$$\operatorname{div} \boldsymbol{f} := \sum_{i=1}^{k} \frac{\partial}{\partial z_i} f_i .$$

Liouville's theorem [7] states that the phase flow of a divergence-free vector field preserves volume. Show that Hamiltonian vector fields are divergence free, i.e., $\operatorname{div}[\boldsymbol{J}\nabla_{\boldsymbol{z}} H]\,(\boldsymbol{z}) \equiv 0$, which implies conservation of volume for Hamiltonian systems.

8. *Volume preservation.* The linear differential equation

$$\frac{d}{dt} z = \boldsymbol{A} z,$$

$z \in \mathbb{R}^4$ and

$$\boldsymbol{A} = \begin{bmatrix} 3 & 0 & 0 & 0 \\ 0 & -1 & 0 & 0 \\ 0 & 0 & -1 & 0 \\ 0 & 0 & 0 & -1 \end{bmatrix},$$

generates a flow map that is volume preserving. Show that the flow map does not preserve the symplectic two-form (3.27) ($d = 2$) for any permutation of the variables.

9. *Wedge product.* Let $d\boldsymbol{a}$, $d\boldsymbol{b}$ be two k-vectors of differential one-forms on $\mathbb{R}^m$. Verify the rule of matrix multiplication, i.e., show that

$$d\boldsymbol{a} \wedge (\boldsymbol{M} d\boldsymbol{b}) = (\boldsymbol{M}^T d\boldsymbol{a}) \wedge d\boldsymbol{b},$$

for any $k \times k$ matrix $\boldsymbol{M}$. Hint: Start from

$$d\boldsymbol{a} \wedge (\boldsymbol{M} d\boldsymbol{b}) = \sum_i da_i \wedge \left(\sum_j m_{ij} db_j \right).$$

4

Geometric integrators

In Chapter 2, we introduced the concept of a numerical integrator as a mapping which approximates the flow-map of a given system of differential equations. We have also seen a few instances of how such integrators behave, demonstrating concepts such as convergence and order of accuracy. We observed that the typical picture is a locally accurate approximation that gradually drifts further from the true trajectory (see Fig. 2.3, Fig. 2.5 and the left panel of Fig. 2.7); the rate of drift can be reduced by reducing the stepsize (and thereby also increasing the amount of computational work), but the qualitative picture does not change in any significant way.

What stands out as remarkable, therefore, is the behavior, illustrated in the right panel of Fig. 2.7, of the Euler-B method, which retains bounded trajectories when applied to the harmonic oscillator. In Chapter 2, we provided an explanation for this in the form of a linear stability analysis showing that certain methods, including Störmer–Verlet and Euler-B, have eigenvalues on the unit circle when applied to the harmonic oscillator (or any other oscillatory linear system), if the stepsize is below some threshold value. The Euler-B and Störmer–Verlet methods (among others) possess a strong asymptotic stability property for linear systems.

It is interesting to note that a related long-term stability property extends to nonlinear models. If we apply, for example, the Störmer–Verlet methods to the Lennard–Jones oscillator, we obtain the results illustrated in Fig. 4.1 (compare with Fig. 2.3 and Fig. 2.5). As in the case of the same method applied to the harmonic oscillator, the numerical method appears to be finding solution points on a periodic orbit not much different than the exact one. Note that it would be quite extraordinary to imagine that the actual period itself would be exactly resolved, so that we would expect the true and approximate solutions to drift slowly apart with time. On the other hand, the energy returns, in each period, to its initial value, and there is no noticeable steady accumulation of error in the energy. This is reflected in a qualitative agreement between the numerical and exact solutions. There is clearly some important feature of the Störmer–Verlet method, one that cannot be explained with the techniques (error analysis or linear stability analysis) mentioned in Chapter 2.

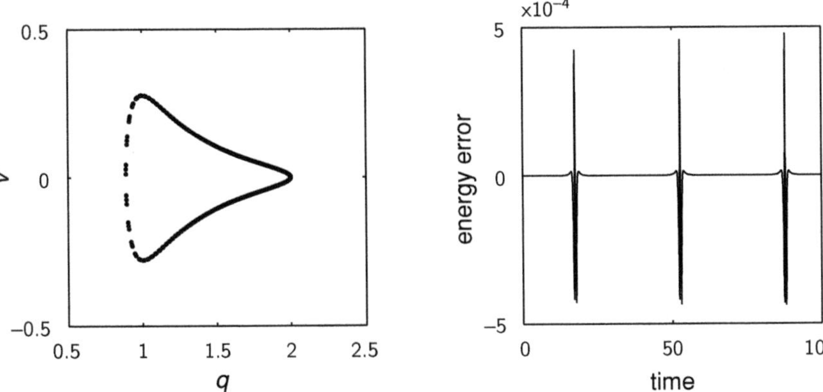

Figure 4.1 Solution curves in the qv-plane (left) and energy error (right) for the Störmer-Verlet method applied to the Lennard–Jones oscillator.

If these differences were observed solely in one-degree-of-freedom systems such as nonlinear oscillators, they would only be of academic interest, but this is not the case. The situation becomes still more dramatic when we look at a more complicated example. For example, we will consider in this section a simple molecular model consisting of a system of N-point masses interacting in pair potentials. We will see that the Störmer–Verlet method facilitates integration on very long time intervals, when applied to this complicated multidimensional nonlinear model problem.

In Chapter 3, we have seen some hints of a deeper theory when we discussed the structure of the flow map itself, focusing on those special properties of Hamiltonian systems

$$\frac{d}{dt}z = J\nabla_z H(z), \tag{4.1}$$

such as the conservation of first integrals and the symplectic structure. We have seen that the symplectic property carries with it geometric implications regarding the way in which the flow map acts on sets of initial conditions. For example, the volume of a set of points in phase space is conserved by a Hamiltonian flow. The flows of conservative systems have a *nonlinear structural stability property*: solutions through points of a region of phase space cannot be squeezed together over time into a smaller region of phase space. Observations like this one led DE VOGELAERE in the 1950s [49] and later RUTH [166], CHANNEL [43] and FENG KANG [57] to wonder if the structural properties of the flow map of a Hamiltonian system, if imposed on the numerical integrator itself, might lead to improved behavior in simulations.

Recall that a map $\boldsymbol{\Psi}$ from $\mathbb{R}^{2d}$ to $\mathbb{R}^{2d}$ is said to be symplectic if

$$\left[\frac{\partial}{\partial \boldsymbol{z}}\boldsymbol{\Psi}(\boldsymbol{z})\right]^T \boldsymbol{J}^{-1} \left[\frac{\partial}{\partial \boldsymbol{z}}\boldsymbol{\Psi}(\boldsymbol{z})\right] = \boldsymbol{J}^{-1}. \tag{4.2}$$

Equivalently, if we write

$$\begin{bmatrix} \boldsymbol{Q} \\ \boldsymbol{P} \end{bmatrix} = \boldsymbol{\Psi}\left(\begin{bmatrix} \boldsymbol{q} \\ \boldsymbol{p} \end{bmatrix}\right),$$

then the symplecticness of the map is summarized by the condition

$$d\boldsymbol{Q} \wedge d\boldsymbol{P} = d\boldsymbol{q} \wedge d\boldsymbol{p},$$

in terms of the wedge product of differential one-forms.

As in Chapter 2, we view a numerical method $\boldsymbol{\Psi}_{\Delta t}$ as a mapping from one time level to another which, via iteration, generates discrete approximations $(\boldsymbol{q}^n, \boldsymbol{p}^n)$. Any reasonable numerical integrator applied to (4.1) will preserve the symplecticness relation up to a certain error which is proportional to a power of the stepsize Δt. We will term a numerical method a *symplectic integrator* if the symplecticness condition

$$d\boldsymbol{q}^{n+1} \wedge d\boldsymbol{p}^{n+1} = d\boldsymbol{q}^n \wedge d\boldsymbol{p}^n \tag{4.3}$$

is preserved *exactly*. The problem now is to derive such symplectic integrators and to enquire in detail about the benefits of using such methods for the simulation of nonlinear systems.

To illustrate, let us walk through the simple one-degree-of-freedom example of the nonlinear oscillator discretized using the Euler-B method

$$q^{n+1} = q^n + \Delta t p^{n+1}, \tag{4.4}$$
$$p^{n+1} = p^n - \Delta t \varphi'(q^n). \tag{4.5}$$

Taking differentials of both sides of equations (4.4)–(4.5), and using the bilinearity of the wedge product, we have

$$\begin{aligned} dq^{n+1} \wedge dp^{n+1} &= d(q^n + \Delta t p^{n+1}) \wedge dp^{n+1} \\ &= dq^n \wedge dp^{n+1} + \Delta t dp^{n+1} \wedge dp^{n+1}. \end{aligned}$$

Now the wedge product of anything with itself is zero, so we have

$$\begin{aligned} dq^{n+1} \wedge dp^{n+1} &= dq^n \wedge dp^{n+1} \\ &= dq^n \wedge d(p^n - \Delta t \varphi'(q^n)) \\ &= dq^n \wedge dp^n - \Delta t dq^n \wedge d\varphi'(q^n). \end{aligned}$$

Since $d\varphi'(q^n) = \varphi''(q^n)dq^n$, the right-hand side reduces to

$$dq^n \wedge dp^n - \Delta t dq^n \wedge (\varphi''(q^n)dq^n) = dq^n \wedge dp^n,$$

and this implies that

$$dq^{n+1} \wedge dp^{n+1} = dq^n \wedge dp^n,$$

hence $\boldsymbol{\Psi}_{\Delta t}$ is a two-dimensional symplectic map, i.e. an area-preserving map of the plane. In this and later chapters, we will introduce more general classes of symplectic methods for arbitrary Hamiltonian functions and show how their symplecticness enhances their numerical performance.

There are many challenges that might cause this program to fail. First, it is not a priori clear that such symplectic integrators would, in typical cases, be more computable than the flow map itself. Fortunately, as we shall see in this and later chapters, for many types of Hamiltonian systems we can even develop *explicit* symplectic integrators whose per timestep cost is similar to that of the most popular non-symplectic methods. Another possible problem might arise from the following fact: the symplecticness condition is more general than the condition for having a Hamiltonian flow map. That is, there are symplectic maps that are not the flow map of any Hamiltonian system. Could it be that there are additional structures in a Hamiltonian flow that give rise to important aspects of some interesting physical phenomena or which contribute to the long-term stability of the dynamics? This topic will be visited in Chapter 5.

Before proceeding with the development of symplectic integrators it is natural to ask first if there are other – perhaps more elementary – properties of a Hamiltonian flow whose preservation would improve the outlook for simulation.

In particular, what of the energy integral? Isn't it reasonable to demand that a numerical integrator for a conservative problem should retain the total energy as a first integral? In fact, it is possible to design such schemes. Unfortunately, these methods often exhibit disappointing numerical properties. They are generally implicit, requiring the solution of nonlinear equations at each timestep. Moreover in typical cases, they do not demonstrate the same level of agreement with the qualitative solution behavior that we observe using simpler and more easily implemented symplectic methods. Nonetheless there are often good reasons to maintain some integrals, such as angular momentum, during numerical simulation and we explain how this can be done.

Another geometric property of Hamiltonian systems arising in typical applications such as the N-body problem is the so-called time-reversal symmetry. We believe that this property can have very important ramifications for the behavior of integrators, in many cases comparable to the effects of preserving symplectic structure. The reversing symmetry is therefore also introduced and discussed in this chapter.

4.1 Symplectic maps and methods

The Euler-B method can be applied, in a modified form, to the general Hamiltonian $H(\boldsymbol{q},\boldsymbol{p})$ with $d \geq 1$ degrees of freedom. The equations are

$$\boldsymbol{q}^{n+1} = \boldsymbol{q}^n + \Delta t \nabla_{\boldsymbol{p}} H(\boldsymbol{q}^n, \boldsymbol{p}^{n+1}), \tag{4.6}$$
$$\boldsymbol{p}^{n+1} = \boldsymbol{p}^n - \Delta t \nabla_{\boldsymbol{q}} H(\boldsymbol{q}^n, \boldsymbol{p}^{n+1}). \tag{4.7}$$

By arguments similar to those mentioned in Chapter 2, this method is easily seen to be first-order accurate. Observe that it is not possible, as it was for the nonlinear oscillator, to solve for the variables at one time level in terms of the previous one: the propagation requires in general the solution of a nonlinear system for $\boldsymbol{p}^{n+1}$. This method is implicit. For sufficiently small Δt, the solution of this system is guaranteed by the implicit function theorem. Let us examine this argument briefly, since it is typical of arguments used to justify the application of implicit methods. Define a function $\boldsymbol{F}$ by

$$\boldsymbol{F}(\boldsymbol{p}^n, \boldsymbol{q}^n, \Delta t, \boldsymbol{u}) = \boldsymbol{u} - \boldsymbol{p}^n - \Delta t \nabla_{\boldsymbol{q}} H(\boldsymbol{q}^n, \boldsymbol{u}).$$

We assume here that H is twice differentiable, so that $\boldsymbol{F}$ is differentiable in $\boldsymbol{u}$. Observe that $\boldsymbol{F}(\boldsymbol{p}^n, \boldsymbol{q}^n, 0, \boldsymbol{p}^n) = \boldsymbol{0}$, and that $\frac{\partial \boldsymbol{F}}{\partial \boldsymbol{u}}(\boldsymbol{p}^n, \boldsymbol{q}^n, 0, \boldsymbol{p}^n) = \boldsymbol{I}_d$ is nonsingular. It follows from the implicit function theorem (see BUCK [35]) that we can solve the equation

$$\boldsymbol{F}(\boldsymbol{p}^n, \boldsymbol{q}^n, \Delta t, \boldsymbol{u}) = \boldsymbol{0},$$

for $\boldsymbol{u}$ in terms of the other arguments of $\boldsymbol{F}$ whenever Δt is sufficiently small. Observe that the need to solve these nonlinear equations places a local restriction on the size of the timestep which could, in the event of rapid solution variation, be quite stringent.

Once $\boldsymbol{p}^{n+1}$ is known, calculation of $\boldsymbol{q}^{n+1}$ from (4.6) is then completely explicit.

The corresponding *Euler-A* method is given by

$$\boldsymbol{q}^{n+1} = \boldsymbol{q}^n + \Delta t \nabla_{\boldsymbol{p}} H(\boldsymbol{q}^{n+1}, \boldsymbol{p}^n), \tag{4.8}$$
$$\boldsymbol{p}^{n+1} = \boldsymbol{p}^n - \Delta t \nabla_{\boldsymbol{q}} H(\boldsymbol{q}^{n+1}, \boldsymbol{p}^n). \tag{4.9}$$

It can be shown that this method is also first-order accurate.

Euler-A and Euler-B methods are canonically symplectic. Let us show this for the Euler-B method using the wedge product notation. Taking differentials of the numerical method yields the implicit system of linear equations

$$d\boldsymbol{q}^{n+1} = d\boldsymbol{q}^n + \Delta t \left[H_{\boldsymbol{pq}} d\boldsymbol{q}^n + H_{\boldsymbol{pp}} d\boldsymbol{p}^{n+1} \right], \tag{4.10}$$

$$d\boldsymbol{p}^{n+1} = d\boldsymbol{p}^n - \Delta t \left[H_{\boldsymbol{qq}} d\boldsymbol{q}^n + H_{\boldsymbol{qp}} d\boldsymbol{p}^{n+1} \right], \tag{4.11}$$

which can be thought of as the discrete variational equations associated with (4.6)–(4.7). For notational convenience, we have dropped the arguments in the second-order partial Jacobian matrices $H_{\boldsymbol{qq}}$, $H_{\boldsymbol{qp}} = H_{\boldsymbol{pq}}^T$, $H_{\boldsymbol{pp}}$, i.e.

$$H_{\boldsymbol{qq}} = \left(\frac{\partial^2 H}{\partial q_i \partial q_j}\right), \quad H_{\boldsymbol{qp}} = \left(\frac{\partial^2 H}{\partial q_i \partial p_j}\right), \quad H_{\boldsymbol{pp}} = \left(\frac{\partial^2 H}{\partial p_i \partial p_j}\right).$$

Observe that

$$d\boldsymbol{q}^n \wedge H_{\boldsymbol{qq}} d\boldsymbol{q}^n = 0,$$

and

$$d\boldsymbol{p}^{n+1} \wedge H_{\boldsymbol{pp}} d\boldsymbol{p}^{n+1} = 0,$$

because $H_{\boldsymbol{qq}}$ and $H_{\boldsymbol{pp}}$ are symmetric matrices. If we take the wedge product of (4.10) with $d\boldsymbol{p}^{n+1}$ from the right, we find

$$d\boldsymbol{q}^{n+1} \wedge d\boldsymbol{p}^{n+1} = d\boldsymbol{q}^n \wedge d\boldsymbol{p}^{n+1} + \Delta t H_{\boldsymbol{pq}} d\boldsymbol{q}^n \wedge d\boldsymbol{p}^{n+1}. \tag{4.12}$$

Similarly

$$\begin{aligned} d\boldsymbol{q}^n \wedge d\boldsymbol{p}^{n+1} &= d\boldsymbol{q}^n \wedge d\boldsymbol{p}^n - \Delta t d\boldsymbol{q}^n \wedge H_{\boldsymbol{qp}} d\boldsymbol{p}^{n+1}, \\ &= d\boldsymbol{q}^n \wedge d\boldsymbol{p}^n - \Delta t H_{\boldsymbol{qp}}^T d\boldsymbol{q}^n \wedge d\boldsymbol{p}^{n+1}, \end{aligned} \tag{4.13}$$

by taking the wedge product of (4.11) with $d\boldsymbol{q}^n$ from the left. Upon plugging (4.13) into (4.12), we obtain the desired equality (4.3). A similar proof goes through for the Euler-A method (4.8)–(4.9). Therefore, *the general forms of the Euler-A and Euler-B methods are canonically symplectic.* Euler-A and Euler-B are sometimes referred to collectively as the *symplectic Euler* methods.

We can also find second-order symplectic methods without much difficulty. The *implicit midpoint method* is defined as follows

$$\boldsymbol{z}^{n+1} = \boldsymbol{z}^n + \Delta t \boldsymbol{J} \nabla H(\boldsymbol{z}^{n+1/2}), \qquad \boldsymbol{z}^{n+1/2} = (\boldsymbol{z}^{n+1} + \boldsymbol{z}^n)/2.$$

This method is evidently implicit. For implementation purposes, it is useful to write it as a composition of two "half-steps": an *implicit Euler* step

$$\boldsymbol{z}^{n+1/2} = \boldsymbol{z}^n + \frac{1}{2}\Delta t \boldsymbol{J} \nabla H(\boldsymbol{z}^{n+1/2}),$$

followed by an *explicit Euler* step

$$\boldsymbol{z}^{n+1} = \boldsymbol{z}^{n+1/2} + \frac{1}{2}\Delta t \boldsymbol{J} \nabla H(\boldsymbol{z}^{n+1/2}).$$

For implicit midpoint, the equation for $d\boldsymbol{z}^n$ is equivalent to

$$d\boldsymbol{z}^{n+1} = d\boldsymbol{z}^n + \Delta t \boldsymbol{J} H_{\boldsymbol{zz}} \frac{d\boldsymbol{z}^n + d\boldsymbol{z}^{n+1}}{2}.$$

Taking the wedge product with $J^{-1}dz^n$ and $J^{-1}dz^{n+1}$, respectively, from the left, we obtain

$$\begin{aligned}J^{-1}dz^{n+1} \wedge dz^n &= J^{-1}dz^n \wedge dz^n + \Delta t J^{-1} J H_{zz} \frac{dz^n + dz^{n+1}}{2} \wedge dz^n \\ &= J^{-1}dz^n \wedge dz^n + \frac{\Delta t}{2} H_{zz} dz^{n+1} \wedge dz^n,\end{aligned}$$

and

$$\begin{aligned}J^{-1}dz^{n+1} \wedge dz^{n+1} &= J^{-1}dz^n \wedge dz^{n+1} + \Delta t J^{-1} J H_{zz} \frac{dz^n + dz^{n+1}}{2} \wedge dz^{n+1} \\ &= J^{-1}dz^{n+1} \wedge dz^n - \frac{\Delta t}{2} H_{zz} dz^{n+1} \wedge dz^n.\end{aligned}$$

Here we have made use of Property 3 of the wedge product as stated in Section 3.6 and, in particular, used that

$$dz^n \wedge \boldsymbol{A} dz^n = dz^{n+1} \wedge \boldsymbol{A} dz^{n+1} = 0,$$

for any symmetric matrix $\boldsymbol{A} \in \mathbb{R}^{2d \times 2d}$. Summing up, we arrive at the equality

$$J^{-1}dz^{n+1} \wedge dz^{n+1} = J^{-1}dz^n \wedge dz^n,$$

implying symplecticness of the scheme with respect to the structure matrix $\boldsymbol{J}$. Observe that this argument required at no point that $\boldsymbol{J}$ be the canonical structure matrix, only that it be constant, skew-symmetric, and invertible, thus we have shown: *the implicit midpoint method preserves any constant symplectic structure.*

4.2 Construction of symplectic methods by Hamiltonian splitting

In the above discussion, we have shown that several integrators are symplectic when applied to integrate Hamiltonian systems. In this section, we show that there is a simple technique that can often be used to produce good symplectic methods.

Suppose that we can split the Hamiltonian H into the sum of $k \geq 2$ Hamiltonians H_i, $i = 1, \ldots, k$, i.e.

$$H(z) = \sum_{i=1}^{k} H_i(z),$$

with each Hamiltonian vector field

$$\frac{d}{dt} z = J \nabla_z H_i(z)$$

explicitly solvable. A symplectic integrator is then derived as an appropriate composition of the corresponding flow maps. Since each flow map is obviously symplectic and any composition of symplectic maps yields a symplectic map, the resulting numerical method is symplectic. See problem 2 in the Exercises.

As a simple example, consider a nonlinear oscillator $H(q,p) = \frac{1}{2}p^2 + \varphi(q)$. The energy can be decomposed into kinetic and potential terms

$$H = H_1 + H_2, \qquad H_1 = \frac{1}{2}p^2, \qquad H_2 = \varphi(q).$$

Now each term is exactly integrable. The equations of motion for H_1 are

$$\dot{q} = p,$$
$$\dot{p} = 0,$$

which has the flow map

$$\boldsymbol{\Phi}_{t,H_1}\left(\begin{bmatrix} q \\ p \end{bmatrix}\right) = \begin{bmatrix} q + tp \\ p \end{bmatrix}.$$

Similarly, H_2 has flow map

$$\boldsymbol{\Phi}_{t,H_2}\left(\begin{bmatrix} q \\ p \end{bmatrix}\right) = \begin{bmatrix} q \\ p - t\varphi'(q) \end{bmatrix}.$$

Each of these maps is symplectic (each is the flow map of a Hamiltonian system), hence the map defined by

$$\boldsymbol{\Psi}_{\Delta t} := \boldsymbol{\Phi}_{\Delta t,H_1} \circ \boldsymbol{\Phi}_{\Delta t,H_2}$$

is also symplectic.

We still need to show that such a composition method approximates the flow map with at least first order.

FIRST-ORDER SPLITTING

If $H = H_1 + H_2 + \ldots + H_k$ is any splitting into twice differentiable terms, then the composition method

$$\boldsymbol{\Psi}_{\Delta t} = \boldsymbol{\Phi}_{\Delta t,H_1} \circ \boldsymbol{\Phi}_{\Delta t,H_2} \circ \cdots \circ \boldsymbol{\Phi}_{\Delta t,H_k}$$

is (at least) a first-order symplectic integrator.

Let us walk through a proof for the case $k = 2$. In Chapter 5 a more detailed result will be given. We wish to compare the flow map $\boldsymbol{\Phi}_{\Delta t,H}$ with $\boldsymbol{\Psi}_{\Delta t}$. It is enough to compare the images of an arbitrary point z^0 under the two maps. Using a Taylor series expansion in powers of Δt and the definition of the flow map, it is easy to show that

$$\begin{aligned}\boldsymbol{\Phi}_{\Delta t,H}(z^0) &= z^0 + \Delta t\dot{z}(0) + \mathcal{O}(\Delta t^2)\\ &= z^0 + \Delta t \boldsymbol{J}\nabla H(z^0) + \mathcal{O}(\Delta t^2).\end{aligned}$$

We can argue in a similar way that

$$\boldsymbol{\Phi}_{\Delta t,H_i}(z^0) = z^0 + \Delta t \boldsymbol{J}\nabla H_i(z^0) + \mathcal{O}(\Delta t^2),$$

$i = 1, 2$. And then, performing similar expansions in Δt, we arrive at

$$\begin{aligned}\boldsymbol{\Psi}_{\Delta t}(z^0) &= \boldsymbol{\Phi}_{\Delta t,H_1}\left(\boldsymbol{\Phi}_{\Delta t,H_2}(z^0)\right)\\ &= \boldsymbol{\Phi}_{\Delta t,H_2}(z^0) + \Delta t \boldsymbol{J}\nabla H_1\left(\boldsymbol{\Phi}_{\Delta t,H_2}(z^0)\right) + \mathcal{O}(\Delta t^2)\\ &= z^0 + \Delta t \boldsymbol{J}\nabla H_2(z^0) + \Delta t \boldsymbol{J}\nabla H_1(z^0) + \mathcal{O}(\Delta t^2)\\ &= z^0 + \Delta t \boldsymbol{J}\nabla H(z^0) + \mathcal{O}(\Delta t^2)\\ &= \boldsymbol{\Phi}_{\Delta t,H}(z^0) + \mathcal{O}(\Delta t^2).\end{aligned}$$

This estimate of the local error, together with the evident smoothness of the flow map, proves that the composition method is at least first order.

The only apparent drawback of this approach is that it requires the splitting of the given Hamiltonian into explicitly solvable subproblems. This may not always be possible or desirable. In many cases, the system may admit a partitioning, but without the individual terms being exactly integrable. In these cases, one may be able to construct effective schemes by substituting another symplectic integrator for the exact flow map at some stage. The splitting technique may in this way simplify the development of an effective method by breaking down a complicated problem into a series of lesser challenges.

A curious special case arises when H_1 and H_2 are first integrals of each other, i.e., $\{H_1, H_2\}$. Then the two flow maps commute (see problem 4 in the Exercises) and the composition method is exact

$$\boldsymbol{\Phi}_{\Delta t,H_1+H_2} = \boldsymbol{\Phi}_{\Delta t,H_1} \circ \boldsymbol{\Phi}_{\Delta t,H_2} = \boldsymbol{\Phi}_{\Delta t,H_1} \circ \boldsymbol{\Phi}_{\Delta t,H_2}.$$

4.2.1 Separable Hamiltonian systems

The splitting described earlier for the special case of the oscillator is applicable to any separable Hamiltonian of the form

$$H(\boldsymbol{q}, \boldsymbol{p}) = T(\boldsymbol{p}) + V(\boldsymbol{q}).$$

As we have seen in earlier chapters, such systems are ubiquitous in chemical and physical modeling, being the standard form for N-body simulations with a flat (i.e. Euclidean) kinetic energy metric.

As before, the form of the energy function suggests a natural splitting into kinetic energy

$$H_1(\boldsymbol{p}) = T(\boldsymbol{p}),$$

and potential energy

$$H_2(\boldsymbol{q}) := V(\boldsymbol{q}).$$

The differential equations corresponding to H_2 can be written

$$\frac{d}{dt}\boldsymbol{q} = \boldsymbol{0},$$
$$\frac{d}{dt}\boldsymbol{p} = -\nabla_{\boldsymbol{q}} V(\boldsymbol{q}).$$

These equations are completely integrable, since $\boldsymbol{q}$ is constant along solutions and $\boldsymbol{p}$ therefore varies linearly with time. The flow map is

$$\boldsymbol{\Phi}_{\tau,V}(\boldsymbol{q},\boldsymbol{p}) = \begin{bmatrix} \boldsymbol{q} \\ \boldsymbol{p} - \tau\nabla_{\boldsymbol{q}} V(\boldsymbol{q}) \end{bmatrix}.$$

Similarly, we can derive the flow map for the kinetic term (H_1)

$$\boldsymbol{\Phi}_{\tau,T}(\boldsymbol{q},\boldsymbol{p}) = \begin{bmatrix} \boldsymbol{q} + \tau\nabla_{\boldsymbol{p}} T(\boldsymbol{p}) \\ \boldsymbol{p} \end{bmatrix}.$$

Now consider the composition of these two maps for $\tau = \Delta t$,

$$\boldsymbol{\Psi}_{\Delta t} := \boldsymbol{\Phi}_{\Delta t,T} \circ \boldsymbol{\Phi}_{\Delta t,V}.$$

Applying this map to a point of phase space $(\boldsymbol{q}^n,\boldsymbol{p}^n)$, we first compute a point $(\bar{\boldsymbol{q}},\bar{\boldsymbol{p}})$

$$\bar{\boldsymbol{q}} = \boldsymbol{q}^n,$$
$$\bar{\boldsymbol{p}} = \boldsymbol{p}^n - \Delta t\nabla_{\boldsymbol{q}} V(\boldsymbol{q}^n).$$

Next, apply $\boldsymbol{\Phi}_{\Delta t,T}$ to this point, i.e.

$$\boldsymbol{q}^{n+1} = \bar{\boldsymbol{q}} + \Delta t\nabla_{\boldsymbol{p}} T(\bar{\boldsymbol{p}}),$$
$$\boldsymbol{p}^{n+1} = \bar{\boldsymbol{p}}.$$

These equations can be simplified by the elimination of the intermediate values, to yield

$$\boldsymbol{q}^{n+1} = \boldsymbol{q}^n + \Delta t\nabla_{\boldsymbol{p}} T(\boldsymbol{p}^{n+1}),$$
$$\boldsymbol{p}^{n+1} = \boldsymbol{p}^n - \Delta t\nabla_{\boldsymbol{q}} V(\boldsymbol{q}^n).$$

This is evidently the Euler-B method introduced in the previous section specialized to the case of a mechanical Hamiltonian. While the Euler-B method is normally implicit, it becomes explicit when applied to this special class of Hamiltonians due to the separation of variable dependencies.

By reversing the order in which the two maps are applied, we obtain another composition method, $\boldsymbol{\Phi}_{\Delta t,V} \circ \boldsymbol{\Phi}_{\Delta t,T}$, which reduces after a similar calculation to the Euler-A method.

An interesting point that should be mentioned is that not all symplectic maps are given by a splitting. For the general Hamiltonian, it is clear that the Euler-A and Euler-B methods are not obtained from any splitting. It is only for the special case of a separable Hamiltonian that these methods can be viewed in this special way.

4.2.2 A second-order splitting method

Higher-order splitting methods are also easily constructed. We will take up this theme in more detail in Chapter 6. For now, as an illustration, consider again the mechanical Hamiltonian and the splitting

$$H = H_1 + H_2 + H_3,$$

with

$$H_1 = \frac{1}{2}V(\boldsymbol{q}), \qquad H_2 = T(\boldsymbol{p}), \qquad H_3 = \frac{1}{2}V(\boldsymbol{q}).$$

The associated composition method is equivalent to

$$\boldsymbol{\Psi}_{\Delta t} = \boldsymbol{\Phi}_{\Delta t/2,V} \circ \boldsymbol{\Phi}_{\Delta t,T} \circ \boldsymbol{\Phi}_{\Delta t/2,V}.$$

After simplification, it becomes clear that this integrator is nothing other than the second-order Störmer–Verlet method (2.16)–(2.18) of Chapter 2 written in terms of the canonical coordinates

$$\boldsymbol{p}^{n+1/2} = \boldsymbol{p}^n - \frac{1}{2}\Delta t \nabla_{\boldsymbol{q}} V(\boldsymbol{q}^n), \tag{4.14}$$

$$\boldsymbol{q}^{n+1} = \boldsymbol{q}^n + \Delta t \nabla_{\boldsymbol{p}} T(\boldsymbol{p}^{n+1/2}), \tag{4.15}$$

$$\boldsymbol{p}^{n+1} = \boldsymbol{p}^{n+1/2} - \frac{1}{2}\Delta t \nabla_{\boldsymbol{q}} V(\boldsymbol{q}^{n+1}). \tag{4.16}$$

Since it is a splitting method we may infer: *the Störmer–Verlet method (4.14)–(4.16) is canonically symplectic.*

Just as the implicit midpoint method turned out to be the composition of implicit and explicit Euler steps, notice that the Störmer–Verlet method can be expressed as a composition of half-steps using the Euler-A and Euler-B methods, i.e.

$$\begin{aligned}\boldsymbol{\Phi}_{\Delta t,\frac{1}{2}V} \circ \boldsymbol{\Phi}_{\Delta t,T} \circ \boldsymbol{\Phi}_{\Delta t,\frac{1}{2}V} &= \boldsymbol{\Phi}_{\Delta t,\frac{1}{2}V} \circ \boldsymbol{\Phi}_{\Delta t,\frac{1}{2}T} \circ \boldsymbol{\Phi}_{\Delta t,\frac{1}{2}T} \circ \boldsymbol{\Phi}_{\Delta t,\frac{1}{2}V} \\ &= \boldsymbol{\Phi}_{\frac{1}{2}\Delta t,V} \circ \boldsymbol{\Phi}_{\frac{1}{2}\Delta t,T} \circ \boldsymbol{\Phi}_{\frac{1}{2}\Delta t,T} \circ \boldsymbol{\Phi}_{\frac{1}{2}\Delta t,V}.\end{aligned}$$

4.3 Time-reversal symmetry and reversible discretizations

An important geometric property of Newton's equations of motion is related to the invariance of a Hamiltonian $H = H(\boldsymbol{q}, \boldsymbol{p})$ under the *reflection symmetry*

$$\boldsymbol{p} \mapsto -\boldsymbol{p},$$

i.e. the Hamiltonian is an even function in the momentum $\boldsymbol{p}$. We consider the consequences of this property for the solution behavior of the corresponding Hamiltonian system

$$\begin{aligned}\frac{d}{dt}\boldsymbol{q} &= \nabla_{\boldsymbol{p}} H(\boldsymbol{q}, \boldsymbol{p}), \\ \frac{d}{dt}\boldsymbol{p} &= -\nabla_{\boldsymbol{q}} H(\boldsymbol{q}, \boldsymbol{p}).\end{aligned}$$

The key fact is this: if $(\boldsymbol{q}(t), \boldsymbol{p}(t))$ is a solution then also $(\hat{\boldsymbol{q}}(t), \hat{\boldsymbol{p}}(t)) := (\boldsymbol{q}(-t), -\boldsymbol{p}(-t))$ is.

To prove this observe that

$$\frac{d}{dt}\hat{\boldsymbol{q}}(t) = -\dot{\boldsymbol{q}}(-t) = -\nabla_{\boldsymbol{p}} H(\boldsymbol{q}(-t), \boldsymbol{p}(-t)) = \nabla_{\boldsymbol{p}} H(\hat{\boldsymbol{q}}(t), \hat{\boldsymbol{p}}(t)),$$

and

$$\frac{d}{dt}\hat{\boldsymbol{p}}(t) = \dot{\boldsymbol{p}}(-t) = -\nabla_{\boldsymbol{q}} H(\boldsymbol{q}(-t), \boldsymbol{p}(-t)) = -\nabla_{\boldsymbol{q}} H(\hat{\boldsymbol{q}}(t), \hat{\boldsymbol{p}}(t)).$$

Here we used the general fact that if $h(x, y)$ is an even function of y, then h_x is also even in y, and h_y is odd in y, meaning that $h_x(x, -y) = h_x(x, y)$, and $h_y(x, -y) = -h_y(x, y)$.

The invariance of the Hamiltonian with respect to $\boldsymbol{p} \mapsto -\boldsymbol{p}$ evidently implies that for every solution of the Hamiltonian system, there is another solution which traverses the same positional curve but in the opposite direction, with a negated momentum.

Another way of saying this is that if we evolve the solution τ units in time forward from a given point $\boldsymbol{z}^0 = (\boldsymbol{q}^0, \boldsymbol{p}^0)$, then negate the momentum, evolve

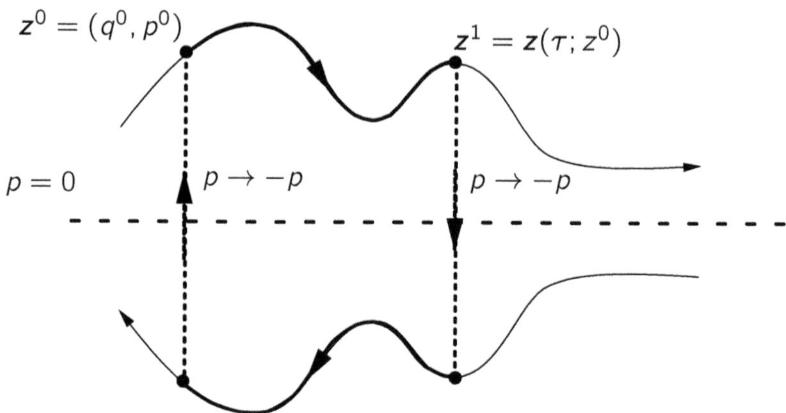

Figure 4.2 The time-reversal symmetry in a one-degree-of-freedom system. A point z^0 of the phase plane is mapped τ units forward along the trajectory, reflected about the line $p = 0$, again evolved forward in time τ units, and, finally, reflected again about the p-axis, arriving back at the initial point z^0.

the solution τ units further, then negate the momentum again, we end up back at our starting point z^0. A system that has this property is said to be *time reversible*.

In the planar case (one-degree-of-freedom case), time reversibility implies that the phase portrait is symmetric with respect to the line $p = 0$, with directionality of trajectories reversed about this axis (Fig. 4.2).

4.3.1 Time-reversible maps

Let us develop the concept of time-reversible systems in a somewhat more general setting. Upon introducing the matrix

$$\boldsymbol{S} = \begin{bmatrix} +\boldsymbol{I}_d & \boldsymbol{0} \\ \boldsymbol{0} & -\boldsymbol{I}_d \end{bmatrix},$$

the time-reversal symmetry can be rewritten as

$$H(\boldsymbol{z}) = H(\boldsymbol{S}\boldsymbol{z}), \qquad \boldsymbol{z} = (\boldsymbol{q}, \boldsymbol{p})^T.$$

This property of a Hamiltonian H implies an associated transformation property for the vector field

$$\begin{aligned}\nabla_{\boldsymbol{z}} H(\boldsymbol{z}) &= \nabla_{\boldsymbol{z}} H(\boldsymbol{S}\boldsymbol{z}) \\ &= \boldsymbol{S}^T \nabla_{\hat{\boldsymbol{z}}} H(\boldsymbol{S}\boldsymbol{z}),\end{aligned}$$

$\hat{z} = \boldsymbol{S}z$, implying

$$\begin{aligned} \boldsymbol{J}\nabla_z H(z) &= \boldsymbol{J}\boldsymbol{S}^T \nabla_{\hat{z}} H(\boldsymbol{S}z) \\ &= -\boldsymbol{S}\boldsymbol{J}\nabla_{\hat{z}} H(\boldsymbol{S}z), \end{aligned}$$

or, with $\dot{z} = \boldsymbol{f}(z) = \boldsymbol{J}\nabla_z H(z)$,

$$\boldsymbol{f}(z) = -\boldsymbol{S}\boldsymbol{f}(\boldsymbol{S}z). \tag{4.17}$$

We call a differential equation with this property *time reversible* with respect to the *involution* $\boldsymbol{S}$.[1]

Now solving $dz/dt = -\boldsymbol{S}\boldsymbol{f}(\boldsymbol{S}z)$ is precisely equivalent to applying the combination of a coordinate transformation $z \to \boldsymbol{S}z$ with a time transformation $dt/d\tau = -1$ to the differential equation $dz/dt = \boldsymbol{f}(z)$. In other words, applying the symmetry and simultaneously negating time does not effect the system. The flow map therefore satisfies

$$\boldsymbol{\Phi}_{\tau,H}(z) = \boldsymbol{S}\boldsymbol{\Phi}_{-\tau,H}(\boldsymbol{S}z).$$

Let $[\boldsymbol{\Phi}_{\tau,H}]^{-1}$ denote the inverse of the mapping $\boldsymbol{\Phi}_{\tau,H}$. Then we know that $\boldsymbol{\Phi}_{-\tau,H} = [\boldsymbol{\Phi}_{\tau,H}]^{-1}$ for any flow map, we therefore obtain

$$\boldsymbol{\Phi}_{\tau,H}(z) = \boldsymbol{S}\,[\boldsymbol{\Phi}_{\tau,H}]^{-1}(\boldsymbol{S}z).$$

We call any invertible mapping $\boldsymbol{\Psi}$ that satisfies

$$\boldsymbol{\Psi}(z) = \boldsymbol{S}\boldsymbol{\Psi}^{-1}(\boldsymbol{S}z) \tag{4.18}$$

time reversible under the involution $\boldsymbol{S}$.

4.3.2 Linear-reversible maps

Now let us discuss the case where the underlying map is a linear function. We say that

$$\boldsymbol{\Psi}(z) = \boldsymbol{P}z$$

is time reversible if the matrix $\boldsymbol{P}$ satisfies the identity

$$\boldsymbol{P} = \boldsymbol{S}\boldsymbol{P}^{-1}\boldsymbol{S}.$$

Since $\boldsymbol{S} = \boldsymbol{S}^{-1}$, the matrix $\boldsymbol{P}^{-1}$ is a similarity transformation of $\boldsymbol{P}$; i.e. $\boldsymbol{P}$ and $\boldsymbol{P}^{-1}$ have identical eigenvalues. Thus, as for symplectic matrices, real matrices satisfying time-reversal symmetry have an eigenvalue λ if and only if also $1/\lambda$, $\bar{\lambda}$, and $1/\bar{\lambda}$ lie in the spectrum.

[1] A matrix $\boldsymbol{S} \in \mathbb{R}^{d\times d}$ is called an involution if $\boldsymbol{S}\boldsymbol{S} = \boldsymbol{I}_d$.

4.3.3 Time-reversible methods by symmetric composition

The method $\boldsymbol{\Psi}^*_{\Delta t}$ defined by

$$\boldsymbol{\Psi}^*_{\Delta t} = [\boldsymbol{\Psi}_{-\Delta t}]^{-1}$$

is called the *adjoint method* of $\boldsymbol{\Psi}_{\Delta t}$. In other words, given a method $\boldsymbol{\Psi}_{\Delta t}$ its adjoint method $\boldsymbol{\Psi}^*_{\Delta t}$ is (implicitly) defined by

$$z^n = \boldsymbol{\Psi}_{-\Delta t}(z^{n+1}),$$

and

$$\boldsymbol{\Psi}^*_{\Delta t}(z^n) := z^{n+1}.$$

From this definition we readily obtain that the adjoint of the adjoint method is the original method, i.e. $\boldsymbol{\Psi}_{\Delta t} = [\boldsymbol{\Psi}^*_{\Delta t}]^*$. It is also readily verified that the local error of $\boldsymbol{\Psi}_{\Delta t}$ and its adjoint method are of the same order with respect to the stepsize Δt. A method $\boldsymbol{\Psi}_{\Delta t}$ is called symmetric if $\boldsymbol{\Psi}_{\Delta t} = \boldsymbol{\Psi}^*_{\Delta t}$, i.e., $\boldsymbol{\Psi}_{-\Delta t} = \boldsymbol{\Psi}^{-1}_{\Delta t}$.

As an illustration, the explicit Euler method

$$z^{n+1} = z^n + \Delta t \boldsymbol{J} \nabla_z H(z^n),$$

is not symmetric. The inverse $[\boldsymbol{\Psi}_{\Delta t}]^{-1}$ is obtained by exchanging z^n and z^{n+1}, i.e.

$$z^n = z^{n+1} + \Delta t \boldsymbol{J} \nabla_z H(z^{n+1}).$$

Upon replacing Δt by $-\Delta t$ and rearranging terms, we get the implicit Euler method

$$z^{n+1} = z^n + \Delta t \boldsymbol{J} \nabla_z H(z^{n+1}).$$

Thus the implicit Euler method is the adjoint of the explicit Euler method and vice versa. The symplectic Euler-B method

$$\begin{aligned} \boldsymbol{q}^{n+1} &= \boldsymbol{q}^n + \Delta t \nabla_p H(\boldsymbol{q}^n, \boldsymbol{p}^{n+1}), \\ \boldsymbol{p}^{n+1} &= \boldsymbol{p}^n - \Delta t \nabla_q H(\boldsymbol{q}^n, \boldsymbol{p}^{n+1}) \end{aligned}$$

is also not symmetric. Its adjoint is the symplectic Euler-A method

$$\begin{aligned} \boldsymbol{q}^{n+1} &= \boldsymbol{q}^n + \Delta t \nabla_p H(\boldsymbol{q}^{n+1}, \boldsymbol{p}^n), \\ \boldsymbol{p}^{n+1} &= \boldsymbol{p}^n - \Delta t \nabla_q H(\boldsymbol{q}^{n+1}, \boldsymbol{p}^n). \end{aligned}$$

Knowing a numerical method $\boldsymbol{\Psi}_{\Delta t}$ and its adjoint method, a symmetric method $\hat{\boldsymbol{\Psi}}_{\Delta t}$ is immediately available through the following *concatenation* (composition) of the two methods

$$\hat{\boldsymbol{\Psi}}_{\Delta t} := \boldsymbol{\Psi}^*_{\Delta t/2} \circ \boldsymbol{\Psi}_{\Delta t/2}.$$

Let us verify this statement

$$\begin{aligned}\hat{\boldsymbol{\Psi}}_{-\Delta t} &= \boldsymbol{\Psi}^*_{-\Delta t/2} \circ \boldsymbol{\Psi}_{-\Delta t/2} \\ &= \left[\boldsymbol{\Psi}_{\Delta t/2}\right]^{-1} \circ \left[\boldsymbol{\Psi}^*_{\Delta t/2}\right]^{-1} \\ &= \left[\boldsymbol{\Psi}^*_{\Delta t/2} \circ \boldsymbol{\Psi}_{\Delta t/2}\right]^{-1} \\ &= \left[\hat{\boldsymbol{\Psi}}_{\Delta t}\right]^{-1}.\end{aligned}$$

Take the implicit Euler method as $\boldsymbol{\Psi}_{\Delta t}$. Then $\hat{\boldsymbol{\Psi}}_{\Delta t}$ is the concatenation of the implicit with the explicit Euler method, i.e.

$$\begin{aligned}z^{n+1/2} &= z^n + \frac{\Delta t}{2} J\nabla_z H(z^{n+1/2}), \\ z^{n+1} &= z^{n+1/2} + \frac{\Delta t}{2} J\nabla_z H(z^{n+1/2}),\end{aligned}$$

which, since

$$z^{n+1/2} = \frac{z^n + z^{n+1}}{2},$$

is equivalent to the implicit midpoint method. Choosing $\boldsymbol{\Psi}_{\Delta t}$ to be the map corresponding to the explicit Euler method, we obtain the trapezoidal rule for $\hat{\boldsymbol{\Psi}}_{\Delta t}$.

Similarly, concatenation of half steps with the Euler-B method with the Euler-A method yields the series of steps

$$\begin{aligned}\boldsymbol{q}^{n+1/2} &= \boldsymbol{q}^n + \frac{1}{2}\Delta t \nabla_{\boldsymbol{p}} H(\boldsymbol{q}^n, \boldsymbol{p}^{n+1/2}), \\ \boldsymbol{p}^{n+1/2} &= \boldsymbol{p}^n - \frac{1}{2}\Delta t \nabla_{\boldsymbol{q}} H(\boldsymbol{q}^n, \boldsymbol{p}^{n+1/2}), \\ \boldsymbol{q}^{n+1} &= \boldsymbol{q}^{n+1/2} + \frac{1}{2}\Delta t \nabla_{\boldsymbol{p}} H(\boldsymbol{q}^{n+1}, \boldsymbol{p}^{n+1/2}), \\ \boldsymbol{p}^{n+1} &= \boldsymbol{p}^n - \frac{1}{2}\Delta t \nabla_{\boldsymbol{q}} H(\boldsymbol{q}^{n+1}, \boldsymbol{p}^{n+1/2}).\end{aligned}$$

After elimination of redundancy, we arrive at the generalized Störmer–Verlet/leapfrog method

$$\begin{aligned}\boldsymbol{q}^{n+1} &= \boldsymbol{q}^n + \frac{1}{2}\Delta t (\nabla_{\boldsymbol{p}} H(\boldsymbol{q}^n, \boldsymbol{p}^{n+1/2}) + \nabla_{\boldsymbol{p}} H(\boldsymbol{q}^{n+1}, \boldsymbol{p}^{n+1/2})), \\ \boldsymbol{p}^{n+1/2} &= \boldsymbol{p}^n - \frac{1}{2}\Delta t \nabla_{\boldsymbol{q}} H(\boldsymbol{q}^n, \boldsymbol{p}^{n+1/2}), \\ \boldsymbol{p}^{n+1} &= \boldsymbol{p}^n - \frac{1}{2}\Delta t \nabla_{\boldsymbol{q}} H(\boldsymbol{q}^{n+1}, \boldsymbol{p}^{n+1/2}).\end{aligned}$$

This is an implicit second-order symplectic method which specializes to the Störmer–Verlet method in the case of N-body systems.

It is worthwhile to note that, in both of the above examples, the concatenation of non-symmetric first-order methods yields a symmetric method of second order. More generally, one can prove the following:

Theorem 1 *The order of a symmetric method is necessarily even.* □

Proof. Suppose by way of contradiction that the order of a given symmetric method is odd, so that the leading-order term in the local error expansion is even. We have

$$\boldsymbol{\Psi}_{\Delta t}(\boldsymbol{z}) = \boldsymbol{\Phi}_{\Delta t}(\boldsymbol{z}) + \boldsymbol{C}_{2k}(\boldsymbol{z})\Delta t^{2k} + \mathcal{O}(\Delta t^{2k+1}),$$

and

$$\boldsymbol{\Psi}_{-\Delta t}(\boldsymbol{z}) = \boldsymbol{\Phi}_{-\Delta t}(\boldsymbol{z}) + \boldsymbol{C}_{2k}(\boldsymbol{z})\Delta t^{2k} + \mathcal{O}(\Delta t^{2k+1}).$$

Because of $\boldsymbol{\Psi}_{\Delta t}(\boldsymbol{z}) = \boldsymbol{z} + \mathcal{O}(\Delta t)$ and a similar property of $\boldsymbol{\Phi}_{\Delta t}$, the above two expansions immediately lead to

$$\boldsymbol{\Phi}_{-\Delta t} \circ \boldsymbol{\Psi}_{\Delta t}(\boldsymbol{z}) = \boldsymbol{z} + \boldsymbol{C}_{2k}(\boldsymbol{z})\Delta t^{2k} + \mathcal{O}(\Delta t^{2k+1}), \tag{4.19}$$

and

$$\boldsymbol{\Psi}_{-\Delta t} \circ \boldsymbol{\Phi}_{\Delta t}(\boldsymbol{z}) = \boldsymbol{z} + \boldsymbol{C}_{2k}(\boldsymbol{z})\Delta t^{2k} + \mathcal{O}(\Delta t^{2k+1}).$$

We can easily show the following: a smooth map $\boldsymbol{F}$ defined by

$$\boldsymbol{F}(\boldsymbol{z}) = \boldsymbol{z} + \epsilon \boldsymbol{G}(\boldsymbol{z})$$

is invertible for sufficiently small ϵ, and

$$\boldsymbol{F}^{-1}(\boldsymbol{z}) = \boldsymbol{z} - \epsilon \boldsymbol{G}(\boldsymbol{z}) + O(\epsilon^2).$$

Applying this with $\boldsymbol{F} = \boldsymbol{\Psi}_{-\Delta t} \circ \boldsymbol{\Phi}_{\Delta t}$, $\boldsymbol{G}(\boldsymbol{z}) = \boldsymbol{C}_{2k}(\boldsymbol{z})$ and $\epsilon = \Delta t^{2k}$, we find

$$\boldsymbol{\Phi}_{\Delta t}^{-1} \circ \boldsymbol{\Psi}_{-\Delta t}^{-1}(\boldsymbol{z}) = \boldsymbol{z} - \boldsymbol{C}_{2k}(\boldsymbol{z})\Delta t^{2k} + \mathcal{O}(\Delta t^{2k+1}).$$

On the other hand, since $\boldsymbol{\Phi}_{\Delta t}$ is self-adjoint and we have assumed the same property for $\boldsymbol{\Psi}_{\Delta t}$, it follows that

$$\boldsymbol{\Phi}_{-\Delta t} \circ \boldsymbol{\Psi}_{\Delta t}(\boldsymbol{z}) = \boldsymbol{z} - \boldsymbol{C}_{2k}(\boldsymbol{z})\Delta t^{2k} + \mathcal{O}(\Delta t^{2k+1}).$$

Comparing this formula with (4.19) we conclude that $\boldsymbol{C}_{2k} = \boldsymbol{0}$, contradicting our assumption that the leading term in the local error was of even order. □

Given a time-reversible Hamiltonian system or a general differential equation

$$\frac{d}{dt}z = f(z),$$

satisfying (4.17), then a numerical method is called *time reversible* if

$$\Psi_{\Delta t}(z^n) = S\Psi_{\Delta t}^{-1}(Sz^n).$$

This definition is in accordance with (4.18). Provided the numerical method is symmetric this definition is equivalent to

$$\Psi_{\Delta t}(z^n) = S\Psi_{-\Delta t}(Sz^n).$$

However this equality is satisfied by all numerical methods we have encountered so far when applied to a differential equation satisfying (4.17). Hence we can conclude that symmetric methods are time-reversible. See problem 11 in the Exercises.

4.4 First integrals

First integrals play an important role in Hamiltonian mechanics [149, 124]. We have already seen several instances of first integrals in Chapter 3 and noted that first integrals lead essentially to a reduction in the number of degrees of freedom. Can this property be reproduced by a numerical method? The answer is no in general if one looks at the class of symplectic integration methods. However, there are a number of important exceptions and we will discuss the most important ones in the following two subsections.

4.4.1 Preservation of first integrals by splitting methods

Splitting methods can sometimes be shown to exactly conserve certain first integrals. We begin by considering the important example of an N-body system with pairwise distance-dependent interactions

$$H = \sum_{i=1}^{N} \frac{\|p_i\|^2}{2m_i} + \sum_{i=1}^{N-1} \sum_{j=i+1}^{N} \varphi_{ij}(\|q_i - q_j\|). \tag{4.20}$$

Such a system is in the form $H(q, p) = T(p) + V(q)$, hence the splitting technique discussed in Section 4.2.1 is directly applicable. In particular the Störmer–Verlet method provides an elementary time-reversible and symplectic discretization for this problem.

On the other hand, as shown in Chapter 3, (4.20) has other properties and it is interesting to ask if these are also conserved under discretization. Specifically, both the total linear momentum $\boldsymbol{p}_{\rm tot} = \sum_{j=1}^{N} \boldsymbol{p}_j$ and the total angular momentum $\boldsymbol{m}_{\rm tot} = \sum_{j=1}^{N} \boldsymbol{q}_j \times \boldsymbol{p}_j$ are conserved by the flow. Let $\boldsymbol{p}^n_{\rm tot}$ and $\boldsymbol{m}^n_{\rm tot}$ represent the discrete forms of the linear and angular momenta; it is desirable that these be preserved under iteration, i.e. that

$$\boldsymbol{p}^{n+1}_{\rm tot} = \boldsymbol{p}^n_{\rm tot}, \qquad \boldsymbol{m}^{n+1}_{\rm tot} = \boldsymbol{m}^n_{\rm tot}.$$

To examine these directly, we can begin by writing out the discrete equations for each particle

$$\boldsymbol{q}_j^{n+1} = \boldsymbol{q}_j^n + \frac{1}{m_j}\boldsymbol{p}_j^{n+1/2}, \tag{4.21}$$

$$\boldsymbol{p}_j^{n+1/2} = \boldsymbol{p}_j^n + \sum_{i\neq j} \frac{\varphi'_{ij}(r_{ij}^n)}{r_{ij}^n}(\boldsymbol{q}_j^n - \boldsymbol{q}_i^n), \tag{4.22}$$

$$\boldsymbol{p}_j^{n+1} = \boldsymbol{p}_j^{n+1/2} + \sum_{i\neq j} \frac{\varphi'_{ij}(r_{ij}^{n+1})}{r_{ij}^{n+1}}(\boldsymbol{q}_j^{n+1} - \boldsymbol{q}_i^{n+1}). \tag{4.23}$$

Now the angular momentum can be computed directly as follows

$$\begin{aligned}
\boldsymbol{m}^{n+1}_{\rm tot} &= \sum_{j=1}^{N} \boldsymbol{q}_j^{n+1} \times \boldsymbol{p}_j^{n+1} \\
&= \sum_{j=1}^{N} \boldsymbol{q}_j^{n+1} \times (\boldsymbol{p}_j^{n+1/2} + \sum_{i\neq j} \frac{\varphi'_{ij}(r_{ij}^{n+1})}{r_{ij}^{n+1}}(\boldsymbol{q}_j^{n+1} - \boldsymbol{q}_i^{n+1})) \\
&= \sum_{j=1}^{N} \boldsymbol{q}_j^{n+1} \times \boldsymbol{p}_j^{n+1/2} + \sum_{j=1}^{N} \boldsymbol{q}_j^{n+1} \times \sum_{i\neq j} \tau_{ij}(\boldsymbol{q}_j^{n+1} - \boldsymbol{q}_i^{n+1})),
\end{aligned}$$

where we have written

$$\tau_{ij} := \frac{\varphi'_{ij}(r_{ij}^{n+1})}{r_{ij}^{n+1}}.$$

Continuing

$$\begin{aligned}
\boldsymbol{m}^{n+1}_{tot} &= \sum_{j=1}^{N} \boldsymbol{q}_j^{n+1} \times \boldsymbol{p}_j^{n+1/2} + \sum_{j=1}^{N}\sum_{i\neq j} \boldsymbol{q}_j^{n+1} \times \tau_{ij}(\boldsymbol{q}_j^{n+1} - \boldsymbol{q}_i^{n+1}) \\
&= \sum_{j=1}^{N} \boldsymbol{q}_j^{n+1} \times \boldsymbol{p}_j^{n+1/2} + \sum_{j=1}^{N}\sum_{i\neq j} \tau_{ij}(\boldsymbol{q}_j^{n+1} \times \boldsymbol{q}_j^{n+1} - \boldsymbol{q}_j^{n+1} \times \boldsymbol{q}_i^{n+1}) \\
&= \sum_{j=1}^{N} \boldsymbol{q}_j^{n+1} \times \boldsymbol{p}_j^{n+1/2} + \sum_{j=1}^{N}\sum_{i\neq j} \tau_{ij}\boldsymbol{q}_j^{n+1} \times \boldsymbol{q}_i^{n+1}.
\end{aligned}$$

Note that the second term on the right-hand side can be broken down into a sum of pairs of terms ($i < j$)

$$\tau_{ij}\boldsymbol{q}_j^{n+1} \times \boldsymbol{q}_i^{n+1} + \tau_{ji}\boldsymbol{q}_i^{n+1} \times \boldsymbol{q}_j^{n+1}.$$

It is easy to see that $\tau_{ij} = \tau_{ji}$, since these coefficients depend only on the distances between particles, thus, taking into account the antisymmetry of the cross product, we must have

$$\tau_{ij}\boldsymbol{q}_j^{n+1} \times \boldsymbol{q}_i^{n+1} + \tau_{ji}\boldsymbol{q}_i^{n+1} \times \boldsymbol{q}_j^{n+1} = \boldsymbol{0},$$

and

$$\boldsymbol{m}_{\text{tot}}^{n+1} = \sum_{j=1}^{N} \boldsymbol{q}_j^{n+1} \times \boldsymbol{p}_j^{n+1/2}.$$

Using (4.21) and simplifying, we find

$$\begin{aligned}\boldsymbol{m}_{\text{tot}}^{n+1} &= \sum_{j=1}^{N} (\boldsymbol{q}_j^n + \frac{1}{m_j}\boldsymbol{p}_j^{n+1/2}) \times \boldsymbol{p}_j^{n+1/2} \\ &= \sum_{j=1}^{N} \boldsymbol{q}_j^n \times \boldsymbol{p}_j^{n+1/2}.\end{aligned} \tag{4.24}$$

Upon substituting (4.22) into (4.24), a line of argumentation analogous to that we have used above leads to

$$\boldsymbol{m}_{\text{tot}}^{n+1} = \sum_{j=1}^{N} \boldsymbol{q}_j^n \times \boldsymbol{p}_j^n = \boldsymbol{m}_{\text{tot}}^n.$$

An even easier argument establishes the conservation of total linear momentum.

More generally, consider a splitting method with $H = H_1 + H_2 + \ldots + H_k$, and suppose that a given first integral F is also a first integral for each of the Hamiltonians H_i, i.e.

$$\{F, H_i\} = 0, \qquad i = 1, \ldots, k.$$

In this case, a numerical method constructed based on a concatenation of the flows of each of the Hamiltonians H_i will also automatically preserve F. Thus we might as well have shown the conservation of angular momentum above by showing that the Poisson brackets of each component of $\boldsymbol{m}_{\text{tot}}$ with T and with V all vanish.

Example 1 Let us consider the Kepler problem in $\mathbb{R}^3$ with Hamiltonian

$$H = \frac{1}{2}\boldsymbol{p}\cdot\boldsymbol{p} - \frac{1}{\|\boldsymbol{q}\|}.$$

A splitting method based on the splitting into kinetic and potential energy will certainly preserve linear and angular momentum. As a result the numerical solution will stay in a plane as determined by the initial data. However the numerically computed positions will not stay on a closed orbit. This is because the Runge–Lenz vector

$$\boldsymbol{e} = \boldsymbol{p}\times\boldsymbol{m} - \frac{\boldsymbol{q}}{\|\boldsymbol{q}\|}$$

is a first integral of H but not of T and V individually. □

4.4.2 Implicit midpoint preserves quadratic first integrals

There are other instances of exact conservation of first integrals under symplectic discretization. For example, the implicit midpoint method exactly preserves any quadratic first integral of the form

$$F = \frac{\boldsymbol{z}^T\boldsymbol{A}\boldsymbol{z}}{2} + \boldsymbol{b}^T\boldsymbol{z},$$

where $\boldsymbol{A}$ is a symmetric matrix. Let us prove this for an arbitrary differential equation

$$\frac{d}{dt}\boldsymbol{z} = \boldsymbol{f}(\boldsymbol{z})$$

that satisfies

$$\frac{d}{dt}F = (\boldsymbol{A}\boldsymbol{z}+\boldsymbol{b})^T\,\boldsymbol{f}(\boldsymbol{z}) = 0.$$

The implicit midpoint method yields

$$\boldsymbol{z}^{n+1} = \boldsymbol{z}^n + \Delta t\boldsymbol{f}(\boldsymbol{z}^{n+1/2}), \qquad \boldsymbol{z}^{n+1/2} = \frac{\boldsymbol{z}^{n+1}+\boldsymbol{z}^n}{2}, \tag{4.25}$$

and we have to show that

$$\frac{(\boldsymbol{z}^{n+1})^T\boldsymbol{A}\boldsymbol{z}^{n+1}}{2} + \boldsymbol{b}^T\boldsymbol{z}^{n+1} = \frac{(\boldsymbol{z}^n)^T\boldsymbol{A}\boldsymbol{z}^n}{2} + \boldsymbol{b}^T\boldsymbol{z}^n.$$

We multiply the first equation in (4.25) by $(\boldsymbol{A}\boldsymbol{z}^{n+1/2}+\boldsymbol{b})^T$ from the left. We obtain

$$\begin{aligned}\left(\boldsymbol{A}\boldsymbol{z}^{n+1/2}+\boldsymbol{b}\right)^T\boldsymbol{z}^{n+1} &= \left(\boldsymbol{A}\boldsymbol{z}^{n+1/2}+\boldsymbol{b}\right)^T\boldsymbol{z}^n + \Delta t\left(\boldsymbol{A}\boldsymbol{z}^{n+1/2}+\boldsymbol{b}\right)^T\boldsymbol{f}\left(\boldsymbol{z}^{n+1/2}\right)\\ &= \left(\boldsymbol{A}\boldsymbol{z}^{n+1/2}+\boldsymbol{b}\right)^T\boldsymbol{z}^n,\end{aligned}$$

and the desired result follows since

$$\frac{1}{2}(Az^n)^T z^{n+1} = \frac{1}{2}\left(Az^{n+1}\right)^T z^n.$$

This implies, for example, that the implicit midpoint method exactly conserves linear and angular momentum and the total energy of a linear Hamiltonian system.

Similarly, one can show that the symplectic Euler methods and the generalized leapfrog method preserve any first integral of the form

$$F = q^T A p + b^T z.$$

(See problem 9 in the Exercises.)

4.5 Case studies

The splitting technique is a remarkably versatile tool for designing symplectic methods. It will be encountered frequently in the later chapters of this book. In the remainder of this chapter, we give several detailed examples of how splitting methods can be easily constructed for some special applications.

4.5.1 Application to N-body systems: a molecular dynamics model problem

We apply the Störmer–Verlet integrator to simulate a simplified molecular system consisting of N equal mass particles interacting pairwise via a Lennard–Jones potential $\varphi_{\mathrm{LJ}}(r) = \epsilon\left[(\bar{r}/r)^12 - 2(\bar{r}/r)^6\right]$. We chose $\epsilon = 0.4$ and $\bar{r} = 1$. In this experiment we study the natural rearrangement of the particles over time.

In our numerical simulation we took $N = 400$ and placed all particles on to a regular cartesian lattice at the points $(i,j) \in \mathbb{R}^2$, $i,j = 0, 1, \ldots, 19$. Next we perturbed each of the particles from the exact lattice positions by a small uniformly distributed random offset (magnitude no more than 0.01) so that the resulting particle positions were near – but not exactly at – the lattice sites. We then simulated the system using the Störmer–Verlet method. If the cartesian lattice were a stable equilibrium structure for the Lennard–Jones system, we would expect the particles to oscillate (chaotically) within small domains near the lattice sites, but the cartesian lattice is *not* stable: the system instead rearranges itself into a more favorable configuration which is near to a *triangular* lattice. This behavior is illustrated in Fig. 4.3, where snapshots of the dynamics of a system of 400 particles at various times are shown.

At this energy level, the particles are mostly held together by the attractive effect of the Lennard–Jones potential, although it is possible for particles to be ejected occasionally from the larger cluster. The triangular lattice is the local

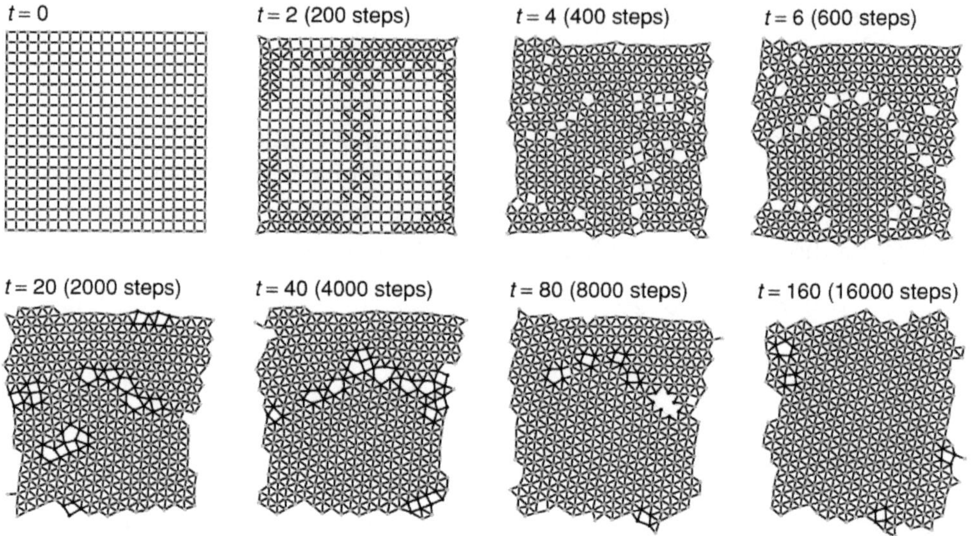

Figure 4.3 Snapshots of the dynamics of a 400 atom planar Lennard–Jones system at various times between $t = 0$ and $t = 160$.

structure that minimizes the energy. The evolving cluster of particles tends to spend a substantially greater proportion of time in the vicinity of this type of lattice configuration. One typically observes that the evolution is not entirely uniform: particles arrange themselves into ordered regions (sometimes referred to as *domains*) surrounded by so-called *lines of dislocation* or *grain boundaries*. The shape of each domain and of the entire system is irregular and may continue to evolve with time, but the lattice structure will generally remain.

The structure of the lattice can be characterized by means of a so-called *order parameter*. We define the order parameter as follows. For each particle, we first define the *neighbor set* $\mathcal{I}_i$ by

$$\mathcal{I}_i = \{j \neq i : \|\boldsymbol{q}_j - \boldsymbol{q}_i\| < \Delta\}$$

typically fixing Δ at a small multiple of the displacement $\bar{r}$ used in the definition of the Lennard–Jones potential, say $\Delta = 1.3\bar{r}$. For each pair of particles in the neighbor set, with indices j and k, say, we define the angle θ_{jik} with vertex at $\boldsymbol{q}_i$ and sides defined by $\boldsymbol{u} = \boldsymbol{q}_j - \boldsymbol{q}_i$ and $\boldsymbol{v} = \boldsymbol{q}_k - \boldsymbol{q}_i$. The so-called C_m, $m \geq 1$, order parameter is defined in terms of these angles by

$$\mu_m = \sum_{i=1}^{N} \sum_{j,k\in\mathcal{I}_i} \cos(m\,\theta_{jik}).$$

The order parameter μ_6 is maximized when the atoms are at the vertices of a hexagonal lattice. If we graph these order parameters against time we see the curves shown in Fig. 4.4. Observe that, following a relaxation phase, there is transition in both order parameters, illustrating the transition from rectangular to hexagonal lattice configurations.

At these energies, the defects are generally not stable structures; instead they can be observed to evolve very slowly with time, eventually disappearing from the lattice. After 16000 steps we have arrived at very near a hexagonal lattice. The elimination of the final defects is typically associated to ejection of a particle, as we see in the final frame of Fig. 4.3. The μ_6 order parameter (Fig. 4.4) continues to rise until about $t = 200$, where it stabilizes around $\mu_6 = 4000$.

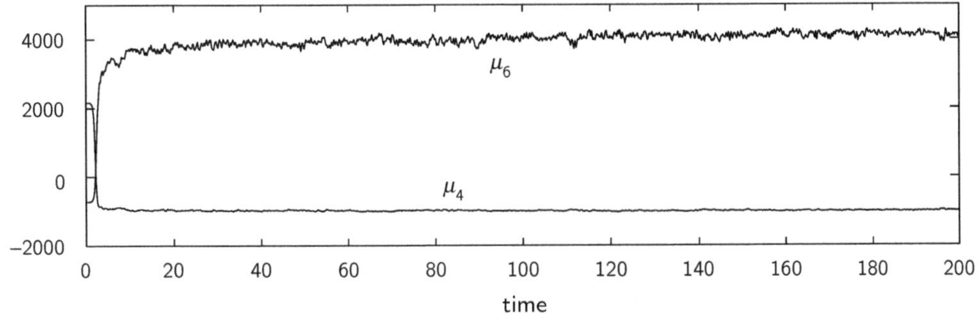

Figure 4.4 Long-term evolution of parameters μ_4 and μ_6 for the 400-atom Lennard–Jones system.

It is now natural to ask what meaning can be assigned to the simulation results. We will take up this issue in greater detail in Chapter 5. For now, we employ a crude measure of the accuracy of the simulation, the energy, which we have plotted against time in Fig. 4.5 for a relatively long simulation of 100000 timesteps (i.e. to $t = 1000$). Despite the presence of a large number of degrees of freedom and a complicated (chaotic) motion, the energy fluctuates within a narrow band around the true value.

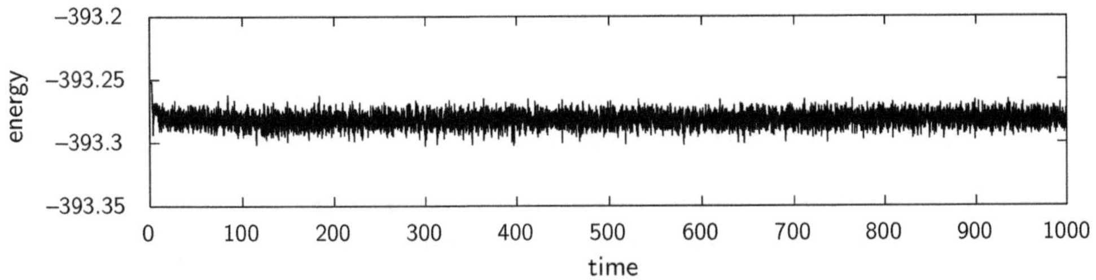

Figure 4.5 Energy vs. time for the 400-atom planar Lennard–Jones system solved on the time interval $[0, 1000]$.

Whereas for the Lennard–Jones oscillator (Fig. 4.1) the behavior might be explained as a consequence of the numerical method inheriting periodic solutions (a desirable property, no doubt, but one that might be restricted to very low-dimensional systems), the results here seem to indicate a more generic stability property.

The challenges evident in this planar molecular model become much more acute when we turn to three-dimensional problems, where the local minima proliferate. For such applications, integrators exhibiting good long-term stability are very important indeed.

4.5.2 Particle in a magnetic field

We next present a method to simulate the motion of a particle moving in a magnetic field. This scheme demonstrates the versatility of the splitting formalism. As discussed in Chapter 3, a classical particle of mass m moving in a potential field V and subject to a constant magnetic field $\boldsymbol{b} = (b_1, b_2, b_3)^T$ can be modeled as a Hamiltonian system with

$$H = \frac{1}{2m}\|\boldsymbol{p}\|^2 + V(\boldsymbol{q}),$$

and symplectic structure matrix

$$J = \begin{bmatrix} \boldsymbol{0} & +\boldsymbol{I} \\ -\boldsymbol{I} & \widehat{\boldsymbol{b}} \end{bmatrix}, \qquad \text{where} \qquad \widehat{\boldsymbol{b}} = \begin{bmatrix} 0 & -b_3 & b_2 \\ b_3 & 0 & -b_1 \\ -b_2 & b_1 & 0 \end{bmatrix}.$$

A splitting method can be based on the same splitting which gave the Euler-B method in the last section. The only difference lies in the modification of the structure matrix to incorporate the magnetic field.

Consider first the potential term $H_2 = V(\boldsymbol{q})$. Upon writing out the differential equations $\dot{\boldsymbol{z}} = J\nabla_{\boldsymbol{z}} H_2(\boldsymbol{z})$ in terms of $\boldsymbol{q}$ and $\boldsymbol{p}$ we find

$$\frac{d}{dt}\boldsymbol{q} = \boldsymbol{0},$$

$$\frac{d}{dt}\boldsymbol{p} = -\nabla_{\boldsymbol{q}} V(\boldsymbol{q}),$$

which has the solution after one timestep $\bar{\boldsymbol{q}} = \boldsymbol{q}^n$, $\bar{\boldsymbol{p}} = \boldsymbol{p}^n - \Delta t \nabla_{\boldsymbol{q}} V(\boldsymbol{q}^n)$. On the other hand, the differential equations for the kinetic term $H_1 = \|\boldsymbol{p}\|^2/2m$ are

$$\frac{d}{dt}\boldsymbol{q} = \boldsymbol{p}/m, \tag{4.26}$$

$$\frac{d}{dt}\boldsymbol{p} = \frac{1}{m}\boldsymbol{b} \times \boldsymbol{p} = \frac{1}{m}\widehat{\boldsymbol{b}}\boldsymbol{p}. \tag{4.27}$$

This is a linear system and hence is integrable. In fact, the second equation can be written

$$\frac{d}{dt}\boldsymbol{p} = \boldsymbol{\Omega}\boldsymbol{p} = \begin{bmatrix} 0 & -\omega_3 & \omega_2 \\ \omega_3 & 0 & -\omega_1 \\ -\omega_2 & \omega_1 & 0 \end{bmatrix} \boldsymbol{p},$$

$\boldsymbol{\Omega} := \widehat{\boldsymbol{b}}/m$, which has solution $\boldsymbol{p}(t) = \exp(\boldsymbol{\Omega}t)\,\boldsymbol{p}(0)$. This exponential can be explicitly computed using "Rodrigues' Formula"

$$\exp(\boldsymbol{\Omega}t) = \boldsymbol{I} + \frac{\sin(\omega t)}{\omega}\boldsymbol{\Omega} + 2\left[\frac{\sin(\omega t/2)}{\omega}\right]^2 \boldsymbol{\Omega}^2,$$

where $\omega = \sqrt{\omega_1^2 + \omega_2^2 + \omega_3^2}$. The solution for $\boldsymbol{q}$ is then obtained by integration, i.e.

$$\begin{aligned} \boldsymbol{q}(t) &= \boldsymbol{q}(0) + \frac{1}{m}\int_0^t \exp(\boldsymbol{\Omega}s)\boldsymbol{p}(0)\,ds \\ &= \boldsymbol{q}(0) + \frac{1}{m}\left[t\boldsymbol{I} + \frac{1-\cos(\omega t)}{\omega^2}\boldsymbol{\Omega} - \frac{\sin(\omega t) - \omega t}{\omega^3}\boldsymbol{\Omega}^2\right]\boldsymbol{p}(0) \\ &= \boldsymbol{q}(0) + \frac{1}{m}\boldsymbol{F}(t)\boldsymbol{p}(0), \end{aligned}$$

where the last equation is to be understood as the defining equation for the 3×3 matrix $\boldsymbol{F}(t)$.

As in the canonical case, the symmetric "leapfrog splitting" method

$$\boldsymbol{\Psi}_{\Delta t} = \boldsymbol{\Phi}_{\frac{1}{2}\Delta t, V} \circ \boldsymbol{\Phi}_{\Delta t, T} \circ \boldsymbol{\Phi}_{\frac{1}{2}\Delta t, V} \tag{4.28}$$

again yields a second-order method. The idea of using a splitting in this way to propagate a particle in a magnetic field was suggested to one of us by J.C. SCOVEL, therefore we feel that it is appropriate to name the scheme accordingly.

SCOVEL'S METHOD

$$\boldsymbol{p}^{n+1/2} = \boldsymbol{p}^n - \frac{\Delta t}{2}\nabla_{\boldsymbol{q}} V(\boldsymbol{q}^n), \tag{4.29}$$

$$\boldsymbol{q}^{n+1} = \boldsymbol{q}^n + \frac{1}{m}\boldsymbol{F}(\Delta t)\boldsymbol{p}^{n+1/2}, \tag{4.30}$$

$$\boldsymbol{p}^{n+1} = \exp(\Delta t \boldsymbol{\Omega})\boldsymbol{p}^{n+1/2} - \frac{\Delta t}{2}\nabla_{\boldsymbol{q}} V(\boldsymbol{q}^{n+1}). \tag{4.31}$$

Numerical experiment

Let us compare with numerical experiments Scovel's method and the symplectic implicit midpoint method, treating the case of a particle of mass $m = 1$ moving in a magnetic field $\boldsymbol{b} = (0, 0, 1)^T$ under the influence of an attractive Coloumb potential $V = -1/||\boldsymbol{q}||$.

To evaluate the quality of a numerical solution to this problem, we look to the first integrals, namely the energy H and the *magnetic momentum*

$$\mu = \boldsymbol{b} \cdot \left(\boldsymbol{q} \times \left(\boldsymbol{p} - \frac{1}{2} \boldsymbol{b} \times \boldsymbol{q} \right) \right).$$

(See the Exercises in Chapter 3.)

The initial conditions we used were $\boldsymbol{q} = (1, 1, 1)^T$ and $\boldsymbol{p} = (0, 0, 0)^T$. In Fig. 4.6, we plot the relative error in energy and the projection of the solution trajectory on to the (q_1, q_2)-plane for Scovel's method and $\Delta t = 0.1$.

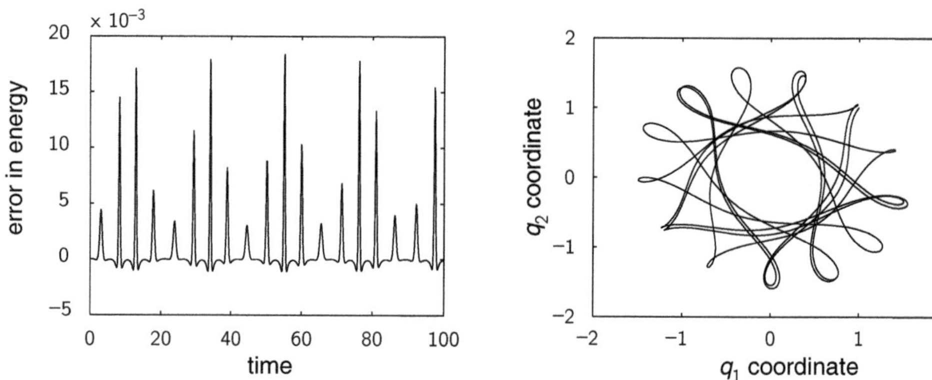

Figure 4.6 Energy and projected trajectory using Scovel's method, $\Delta t = 0.1$.

While Scovel's method is explicit, the implicit midpoint method requires the solution of a nonlinear system of equations at each timestep. For this purpose, we use the following procedure. The nonlinear equations defining the half-step in the implicit midpoint scheme are solved by a fixed-point iteration in $(\bar{\boldsymbol{q}}_k, \bar{\boldsymbol{p}}_k)$

$$\bar{\boldsymbol{q}}_{k+1} = \boldsymbol{q}^n + \frac{\Delta t}{2} \bar{\boldsymbol{p}}_k,$$
$$\bar{\boldsymbol{p}}_{k+1} = \boldsymbol{p}^n + \frac{\Delta t}{2} \boldsymbol{b} \times \bar{\boldsymbol{p}}_k - \frac{\Delta t}{2} \nabla_q V(\bar{\boldsymbol{q}}_k),$$

for $k = 0, 1, 2, \ldots$ with $\bar{\boldsymbol{p}}^0 = \boldsymbol{p}^n$. The iteration is stopped once

$$||\bar{\boldsymbol{q}}_k - \bar{\boldsymbol{q}}_{k-1}||_\infty + ||\bar{\boldsymbol{p}}_k - \bar{\boldsymbol{p}}_{k-1}||_\infty \leq \epsilon,$$

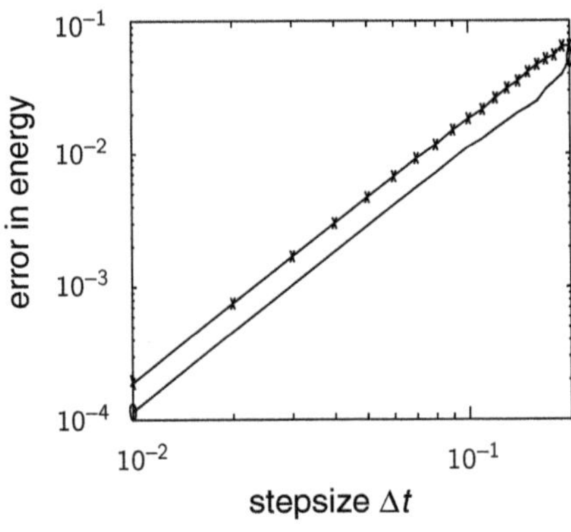

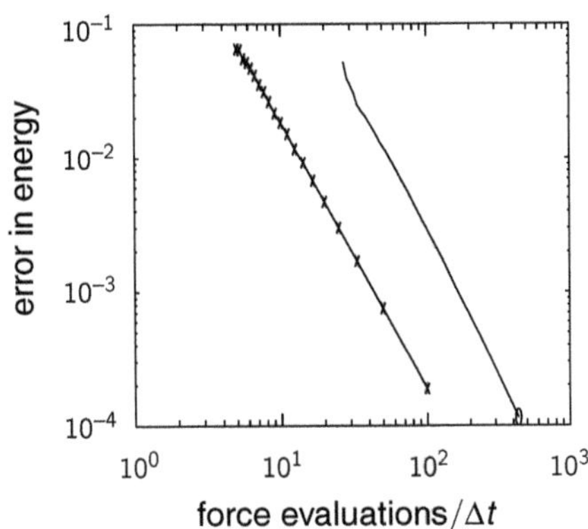

Figure 4.7 (Left) Maximum error in the energy versus stepsize for the implicit midpoint rule (solid) and Scovel's method (x). (Right) Precision versus work (number of force evaluations per unit time interval) diagram for the implicit midpoint method (solid) and Scovel's method (x).

with $\epsilon = 10^{-9}$. We then set

$$\boldsymbol{q}^{n+1/2} = \bar{\boldsymbol{q}}_k, \qquad \boldsymbol{p}^{n+1/2} = \bar{\boldsymbol{p}}_k,$$

and proceed with half a timestep of the explicit Euler method. This completes one timestep with the implicit midpoint method.

In Fig. 4.7(a), we plot the maximum relative error in energy, i.e.

$$E_{\max} = \max_{n=1,2,\ldots} \frac{|H(\boldsymbol{z}^0) - H(\boldsymbol{z}^n)|}{|H(\boldsymbol{z}^0)|},$$

as a function of the stepsize Δt. We see that the implicit midpoint rule is slightly more accurate than Scovel's method. In Fig. 4.7(b), we graph the numerical error against the number of force evaluations per unit time interval. Scovel's method method is the clear winner. The efficiency differences between the explicit and implicit methods generally become more pronounced in systems with higher dimension, or with more costly force calculations. This rule of thumb might not apply in cases where highly accurate solutions are required and higher-order methods need to be used.

4.5.3 Weakly coupled systems

Suppose that our Hamiltonian H can be written as a sum of k decoupled terms H_i together with a coupling term H_c, i.e.

$$H = H_1 + H_2 + \ldots + H_k + H_c, \tag{4.32}$$

and

$$\{H_i, H_j\} = 0, \tag{4.33}$$

for all $i, j = 1, \ldots, k$. The condition (4.33) implies that the flow map associated with the Hamiltonian

$$\tilde{H} = H_1 + H_2 + \cdots + H_k$$

satisfies

$$\boldsymbol{\Phi}_{t,\tilde{H}} = \boldsymbol{\Phi}_{t,H_1} \circ \boldsymbol{\Phi}_{t,H_2} \circ \cdots \circ \boldsymbol{\Phi}_{t,H_k}.$$

If, in addition, all the Hamiltonians H_i and H_c are explicitly integrable, then we can easily obtain a splitting method from the composition of the flow maps $\boldsymbol{\Phi}_{t,\tilde{H}}$ and $\boldsymbol{\Phi}_{t,H_c}$.

As an illustration, consider the gravitational model for a pair of planets orbiting a star. In the star-centered frame, the Hamiltonian is

$$H = \frac{1}{2m_1}\|\boldsymbol{p}_1\|^2 + \frac{1}{2m_2}\|\boldsymbol{p}_2\|^2 - \frac{Gm_1 m_\odot}{\|\boldsymbol{q}_1\|} - \frac{Gm_2 m_\odot}{\|\boldsymbol{q}_2\|} - \frac{Gm_1 m_2}{\|\boldsymbol{q}_1 - \boldsymbol{q}_2\|},$$

where G is the universal gravitational constant, and m_1, m_2, and $m_\odot$ represent the masses of the two planets and the star, respectively. We divide H into three parts

$$H_1 = \frac{1}{2m_1}\|\boldsymbol{p}_1\|^2 - \frac{Gm_1 m_\odot}{\|\boldsymbol{q}_1\|},$$

$$H_2 = \frac{1}{2m_2}\|\boldsymbol{p}_2\|^2 - \frac{Gm_2 m_\odot}{\|\boldsymbol{q}_2\|},$$

and

$$H_c = -\frac{Gm_1 m_2}{\|\boldsymbol{q}_1 - \boldsymbol{q}_2\|}.$$

A second-order splitting method is then given by

$$\boldsymbol{\Psi}_{\Delta t} = \boldsymbol{\Phi}_{\frac{1}{2}\Delta t,H_c} \circ \boldsymbol{\Phi}_{\Delta t,H_1+H_2} \circ \boldsymbol{\Phi}_{\frac{1}{2}\Delta t,H_c}.$$

The Hamiltonians H_1 and H_2 describe completely decoupled Kepler problems, hence they can be solved in principle. The coupling term H_c depends only on position variables, so it, too, can be integrated exactly. A method of this type was used to advantage by WISDOM AND HOLMAN [203] in a celebrated study of the long-term dynamics of the Solar System.

There are some subtleties here: while Kepler problems are exactly solvable using the available integrals, not all methods are suitable for use in a timestepping setting, where many such computations may be needed. Another aspect that will govern the overall efficiency of the method is the extent to which the perturbation due to the coupling term H_{12} can be viewed as small compared with the Kepler terms. If the coupling is strong, we can expect large errors to crop up at each step of computation. More discussion of gravitational N-body problems will be found in Chapter 9.

4.5.4 Linear/nonlinear splitting

There are many phenomena which are described, in zeroth-order approximation, by a quadratic Hamiltonian. For example, we might have

$$H = \frac{1}{2}\boldsymbol{p}^T \boldsymbol{M}^{-1}\boldsymbol{p} + \frac{1}{2}\boldsymbol{q}^T \boldsymbol{K}\boldsymbol{q} + \epsilon H_{\mathrm{N}}(\boldsymbol{q}, \boldsymbol{p}),$$

where ϵ can be viewed as a small perturbation parameter. For these problems it is often useful to employ a splitting technique in which the quadratic part is evolved separately from H_{N}.

In most cases, H_{N} may be regarded as position dependent only and, hence, is integrable. If H_{N} is not exactly integrable, an alternate symplectic method can often be used for its evolution. An example for linear/nonlinear splitting is given by the Fermi–Pasta–Ulam problem (see problem 5 in the Exercises) [60, 192]. Some discussion of linear/nonlinear splittings in the context of Hamiltonian partial differential equations may be found in [129].

4.6 Exercises

1. *Canonical methods.* Show that the Euler-A method is canonical.

2. *Composition of symplectic maps.* Show that the composition of two symplectic maps $\boldsymbol{\Psi}_1$, $\boldsymbol{\Psi}_2$ yields a symplectic map.

3. *Kepler problem.* Discretize the planar Kepler problem with Hamiltonian

$$H(\boldsymbol{q}, \boldsymbol{p}) = \frac{1}{2}\boldsymbol{p}^T \boldsymbol{p} - \frac{1}{\|\boldsymbol{q}\|},$$

and initial conditions $\boldsymbol{q} = (1, 0)^T$ and $\boldsymbol{p} = (0, 1)^T$, by the explicit Störmer–Verlet method and the implicit midpoint rule. Use functional iteration to solve the nonlinear equations resulting from the implicit midpoint rule (see Section 4.5.2). Compare the two methods based on the conservation of energy versus stepsize and the number of force field evalations per timestep. Take the stepsize Δt from the interval $[0.01, 0.0001]$ and integrate over a time interval $[0, 10]$.

4. *Commuting flow maps.* Show that if the Poisson bracket of two functions F and G, $\{F, G\}$, vanishes identically, then the corresponding flow maps $\boldsymbol{\Phi}_{\tau,F}$ and $\boldsymbol{\Phi}_{\tau,G}$ satisfy

$$\phi \circ \boldsymbol{\Phi}_{\tau,F} \circ \boldsymbol{\Phi}_{\tau,G} - \phi \circ \boldsymbol{\Phi}_{\tau,G} \circ \boldsymbol{\Phi}_{\tau,F} = \mathcal{O}(\tau^2),$$

for any smooth function $\phi : \mathbb{R}^{2d} \to \mathbb{R}$. (This result is a first step to show that the two flow maps commute if and only if $\{F, G\} = 0$; see Chapter 5.) Hint: differentiate the given formula with respect to τ and make use of the Poisson bracket notation.

5. *Fermi–Pasta–Ulam problem.* The "Fermi–Pasta–Ulam problem" [60, 192] has Hamiltonian

$$H = \frac{1}{2m}\sum_{i=1}^{N-1} p_i^2 + \frac{\kappa}{2}\sum_{i=1}^{N-1} r_i^2 + \frac{\lambda}{s}\sum_{i=1}^{N-1} \epsilon r_i^s, \qquad r_i = q_{i+1} - q_i, \qquad q_0 = q_N = 0,$$

where κ is the linear elastic constant (Hooke's spring), λ is a small parameter, and s is a small positive integer (usually $s = 3$ or $s = 4$). Choosing $m = \kappa = 1$, $N = 32$, $s = 3$, and $\lambda = 1/4$,

a. Implement the Euler-B method for this problem.

b. Design and implement a first-order "linear/nonlinear" splitting method for this problem.

c. Apply each of the two methods with the following initial condition

$$q_i = \left(\frac{2}{N}\right)^{1/2} \sin\frac{i\pi}{N}, \qquad p_i = 0,$$

$i = 1, \ldots, N-1$, over a time interval of 200 periods of length $T = 2\pi/\omega$, $\omega = 2\sin(\pi/(2N))$ [192]. Monitor the total energy H and the harmonic energy

$$E_{\text{harmonic}} = \frac{1}{2m}\sum_{i=1}^{N-1} p_i^2 + \frac{\kappa}{2}\sum_{i=1}^{N-1} r_i^2.$$

Compare the methods in terms of accuracy and efficiency. For further comparison, also implement the fourth-order explicit Runge–Kutta method of Section 2.4. Changing the value of λ, what behavior in E_{harmonic} do you observe for larger and smaller values of λ?

6. *Discretized Schrödinger equation.* Differential equations involving complex variables arise in quantum mechanics following discretization in the spatial variables. In a simplified case such systems would take the form of a linear system

$$\dot{z} = -\mathrm{i}\boldsymbol{H}z, \tag{4.34}$$

where the dimension of the complex vector $\boldsymbol{z} \in \mathbb{C}^d$ is typically very large, and $\boldsymbol{H}$ is a real symmetric matrix. This can be viewed as a canonical Hamiltonian system over a symplectic structure on $\mathbb{C}^d$. We introduce the complex-valued Hermitian inner product

$$\langle \boldsymbol{u}, \boldsymbol{v} \rangle_C = \sum_{i=1}^{d} \bar{u}_i v_i,$$

where $\bar{w}$ denotes the complex conjugate of a complex number $w \in \mathbb{C}$. The symplectic two-form Ω is now defined in terms of the imaginary part of the Hermitian inner product [124]

$$\Omega(\boldsymbol{\xi}, \boldsymbol{\eta}) = \operatorname{Im} \langle \boldsymbol{\xi}, \boldsymbol{\eta} \rangle_C, \qquad \boldsymbol{\xi}, \boldsymbol{\eta} \in \mathbb{C}^d.$$

The real inner product needed to define the gradient is given by

$$\langle \boldsymbol{u}, \boldsymbol{v} \rangle = \operatorname{Re} \langle \boldsymbol{u}, \boldsymbol{v} \rangle_C.$$

a. Using the standard definition

$$\langle \nabla_z H(\boldsymbol{z}), \boldsymbol{u} \rangle = \lim_{\varepsilon \to 0} \frac{H(\boldsymbol{z} + \varepsilon \boldsymbol{u}) - H(\boldsymbol{z})}{\varepsilon}$$

of a gradient, show that

$$\nabla_z H(\boldsymbol{z}) = \boldsymbol{H}\boldsymbol{z},$$

for the quadratic Hamiltonian

$$H(\boldsymbol{z}) = \frac{1}{2} \langle \boldsymbol{z}, \boldsymbol{H}\boldsymbol{z} \rangle.$$

Also verify that the structure matrix associated with the symplectic two-form Ω is

$$\boldsymbol{J} = -\mathrm{i}\boldsymbol{I}_d,$$

i.e.

$$\Omega(\boldsymbol{\xi}, \boldsymbol{\eta}) = \langle \boldsymbol{\xi}, \boldsymbol{J}^{-1} \boldsymbol{\eta} \rangle.$$

b. Show that the two-norm $\|\boldsymbol{z}\|_2$ is a conserved quantity of the differential equation (4.34). Discretize the Schrödinger equation (4.34) by the implicit midpoint method. Is the resulting discretization symplectic and will it preserve the two-norm $\|\boldsymbol{z}\|_2$? Note: recall that the two-norm of a vector $\boldsymbol{z} = (z_1, z_2, \dots, z_d)^T$ in $\mathbb{C}^d$ is defined by

$$\|\boldsymbol{z}\| = \langle \boldsymbol{z}, \boldsymbol{z} \rangle^{1/2} = (|z_1|^2 + |z_2|^2 + \dots + |z_d|^2)^{1/2},$$

where $|z_i|$ represents the complex modulus of z_i, i.e. the square root of the sum of squares of the real and imaginary parts.

c. By separating $\boldsymbol{z} = \boldsymbol{x} + \mathrm{i}\boldsymbol{y}$ into its real and imaginary parts, show that (4.34) can be rewritten as a canonical and separable Hamiltonian system with d degrees of freedom. If the Störmer–Verlet method is applied to this system, is the two-norm of the solution still conserved?

7. *Time-dependent Hamiltonian systems.* A time-dependent or *non-autonomous* Hamiltonian system in $\boldsymbol{R}^{2d}$ has an energy function

$$H = H(\boldsymbol{q}, \boldsymbol{p}, t).$$

The differential equations associated to such a system are taken to be

$$\dot{\boldsymbol{q}} = +\nabla_{\boldsymbol{p}} H(\boldsymbol{q}, \boldsymbol{p}, t),$$
$$\dot{\boldsymbol{p}} = -\nabla_{\boldsymbol{q}} H(\boldsymbol{q}, \boldsymbol{p}, t).$$

The key difference between an autonomous (time-independent) and non-autonomous system is that, in the latter case, the energy is no longer a conserved quantity of the motion. On the other hand, it is possible to construct an *extended* system for any non-autonomous system by identifying t with an additional variable Q, corresponding momentum P, and Hamiltonian

$$\tilde{H} = H(\boldsymbol{q}, \boldsymbol{p}, Q) + P. \tag{4.35}$$

a. Write out the differential equations for the extended Hamiltonian $\tilde{H}$. Show that if the extended system is solved with initial conditions $\boldsymbol{q}(t_0) = \boldsymbol{q}^0$, $\boldsymbol{p}(t_0) = \boldsymbol{p}^0$, $Q(t_0) = t_0$, $P(t_0) = 0$, then the solution obtained is the same as that of the original Hamiltonian (H) for the initial conditions $\boldsymbol{q}(t_0) = \boldsymbol{q}^0$, $\boldsymbol{p}(t_0) = \boldsymbol{p}^0$. Thus a non-autonomous Hamiltonian system is equivalent to an autonomous system with an additional degree of freedom.

b. A symplectic integrator applied to the extended Hamiltonian system can typically be reduced to an integrator for the original system. By applying Euler-B to (4.35) and simplifying the resulting equations, show that, in terms of the original variables, the method is equivalent to

$$\boldsymbol{q}^{n+1} = \boldsymbol{q}^n + \Delta t \nabla_{\boldsymbol{p}} H(\boldsymbol{q}^n, \boldsymbol{p}^{n+1}, t_n),$$
$$\boldsymbol{p}^{n+1} = \boldsymbol{p}^n - \Delta t \nabla_{\boldsymbol{p}} H(\boldsymbol{q}^n, \boldsymbol{p}^{n+1}, t_n).$$

c. Find the appropriate generalization of the Störmer–Verlet method to non-autonomous Hamiltonian systems of the form $H(\boldsymbol{q}, \boldsymbol{p}) = T(\boldsymbol{p}) + V(\boldsymbol{q}, t)$.

d. *(Refers to problem 6.)* In many applications arising in quantum mechanics, the matrix $\boldsymbol{H}$ in (4.34) must be regarded as time dependent. Often, the system takes the form

$$\boldsymbol{H} = \boldsymbol{H}_0 + \varepsilon \boldsymbol{H}_1(t),$$

where $\boldsymbol{H}_0$ is independent of t and easily diagonalized (i.e. has eigenvalues that can easily be computed), while $\boldsymbol{H}_1(t)$ is very sparse (has many zero elements) or is possibly diagonal, and $\varepsilon > 0$ is a small coupling parameter. Based on this additive decomposition, and an extension of the Hamiltonian as outlined above, develop an appropriate symplectic splitting method for the non-autonomous, Hamiltonian system (4.34).

8. *General reversing symmetry and magnetic fields.* Recall that a general differential equation $dz/dt = \boldsymbol{f}(z)$ is reversible with respect to an involution $\boldsymbol{S}$ if $-\boldsymbol{f}(z) = \boldsymbol{S}\boldsymbol{f}(\boldsymbol{S}z)$.

a. With this definition, show that the equations of motion for a particle in a constant magnetic field are not time reversible with respect to the involution $\boldsymbol{p} \to -\boldsymbol{p}$.

b. Consider an augmented system obtained by treating the constant vector $\boldsymbol{b}$ as a *variable* of the system

$$\frac{d}{dt}\boldsymbol{q} = \boldsymbol{p}/m,$$
$$\frac{d}{dt}\boldsymbol{p} = -\nabla_{\boldsymbol{q}} V(\boldsymbol{q}) + \frac{1}{m}\boldsymbol{b} \times \boldsymbol{p},$$
$$\frac{d}{dt}\boldsymbol{b} = \boldsymbol{0}.$$

Show that this system is time reversible with respect to the involution defined by $\boldsymbol{p} \to -\boldsymbol{p}$, $\boldsymbol{b} \to -\boldsymbol{b}$.

c. Consider the implication of part (b) for the flow map of the system. What property should a method have to be considered time reversible for the magnetic field problem? In what sense is Scovel's method a time-reversible scheme?

9. *Preservation of first integrals.* Show for any of the two symplectic Euler methods that first integrals of the general form

$$F = \boldsymbol{q}^T \boldsymbol{A} \boldsymbol{p}$$

are exactly preserved. What can you conclude for the generalized leapfrog method?

10. *Canonical Hamiltonian with a magnetic field.* The equations of motion for a particle in a constant magnetic field can be recast in an alternative, canonical Hamiltonian form. For a constant magnetic field $\boldsymbol{b}$, this was discussed in Chapter 3. Here we generalize the canonical approach to nonconstant magnetic fields $\boldsymbol{b}(\boldsymbol{q})$, $\boldsymbol{q} \in \mathbb{R}^3$.

a. We assume that

$$\sum_{i=1}^{3} \frac{\partial}{\partial q_i} b_i(\boldsymbol{q}) = 0.$$

Hence, one can introduce a (non-unique) vector potential $\boldsymbol{A}$ such that $\nabla_{\boldsymbol{q}} \times \boldsymbol{A}(\boldsymbol{q}) = \boldsymbol{b}(\boldsymbol{q})$. For a constant vector $\boldsymbol{b}$, what is $\boldsymbol{A}(\boldsymbol{q})$?

b. Write out the differential equations for the canonical Hamiltonian

$$H_c(\boldsymbol{p}, \boldsymbol{q}) = \frac{1}{2m} \|\boldsymbol{p} + \boldsymbol{A}(\boldsymbol{q})\|^2 - \frac{\gamma}{\|\boldsymbol{q}\|}$$

and compare with (3.32)–(3.33) of Chapter 3.

c. Devise symplectic methods for the canonical equations of motion. Discuss their respective advantages and disadvantages.

11. *Time-reversible methods.* Show that the generalized Störmer–Verlet method and the implicit midpoint rule are both time reversible when applied to a time-reversible Hamiltonian system. Use the symmetry property of the two methods and show that

$$\boldsymbol{\Psi}_{\Delta t}(\boldsymbol{z}^n) = \boldsymbol{S}\boldsymbol{\Psi}_{-\Delta t}(\boldsymbol{S}\boldsymbol{z}^n).$$

On the contrary, both symplectic Euler methods also satisfy the above identity but, since they are not symmetric, they are not time reversible. Verify this statement for the symplectic Euler-B method.

5

The modified equations

We have seen in the previous chapter that integrators preserving symplectic structure and/or first integrals can often be constructed in a straightforward way. In this chapter, we consider the properties of those methods and the implications for long-term simulations.

The traditional approach of numerical analysis generally assumes that the purpose of simulation is the faithful reproduction of a particular solution or trajectory, but individual trajectories typically are not of primary interest in most modern, *scientific research*;[1] rather, the scientist typically treats the trajectory as a particular realization of a *fundamentally stochastic* evolution modelling in some way the myriad undetermined perturbations present in a "real-world" environment. It was the important discovery of LORENZ [119] that differential equations can exhibit a chaotic solution behavior that includes an essentially stochastic or "random" component. The scientist views the model being analyzed as representative of a *class of nearby models* based on parameters which are typically only empirically (and approximately) determined. Furthermore, exact initial conditions are also typically not available. Some classical examples of such a scenario are *molecular dynamics* and *numerical weather prediction*.

It is now apparent that most modern large-scale simulations are conducted with timesteps and time intervals such that the numerical solution cannot be thought of as close to any particular model solution. The purpose of wedding the development of integrators to the standard axiomatic principle of timestepping – that one is attempting to approximate a particular trajectory – is thus called into question. Although high accuracy often is not needed in nonlinear dynamics computations, we must recognize certain important constraints imposed by the laws of nature. For example, there is widespread agreement that a conservative system should be sampled on or near the surface of constant energy, although it should be evident by now that this alone is not enough of a restriction (any

[1] There are some exceptions, for example, the determination of satellite orbits, where highly accurate trajectories are sometimes needed, but these situations are relatively rare in the experience of the authors.

arbitrary sequence of random data points in the phase space could be easily projected to the energy surface, but would not be expected to provide a reasonable approximation of a trajectory). The presence of *first integrals* such as energy and weaker *adiabatic invariants* (which exhibit only a mild variation over long time intervals) can provide helpful criteria for assessing the validity of a simulation, even in a nonlinear setting.

In this chapter, we begin our investigation of the implications of geometric properties for the behavior of a numerical method. The cornerstone of this theory is the principle of *backward error analysis*. In the context of numerical linear algebra, backward error analysis was promulgated by WILKINSON [202] and others and has been used as a means of evaluating the propagation of rounding errors in various matrix algorithms. Let us illustrate with the problem of solving a system of N linear equations in N unknowns

$$\boldsymbol{A}\boldsymbol{x} = \boldsymbol{b},$$

where the $N \times N$ nonsingular matrix $\boldsymbol{A}$ and right-hand side vector $\boldsymbol{b} \in \mathbb{R}^N$ are known, and a vector $\boldsymbol{x} \in \mathbb{R}^N$ is to be computed by the standard technique of Gaussian elimination, i.e. reduction of the matrix $\boldsymbol{A}$ to an upper triangle matrix via successive elementary row operations (replacing a row of the matrix by the sum of the row and a scalar multiple of another), and corresponding modification of the right-hand side. At each stage of the calculation in finite precision, rounding errors are introduced. These rounding errors are compounded by successive operations, and this may result in a serious growth in error. In a *forward error analysis*, the potential magnitude of this error growth is evaluated, resulting in a bound for the total error as a function of the dimension of the system and the magnitude of the rounding errors introduced at each step.

For Gaussian elimination, it has been found that an alternative *backward error analysis* is far more meaningful. The philosophy of backward error is based on recognition that the problem being solved is, itself, typically only an approximation to the actual problem of interest, since the elements of the matrix $\boldsymbol{A}$ and vector $\boldsymbol{b}$ are, themselves, generally subject to small errors of measurement, prior computation, and/or finite representation. It can be shown that the solution obtained by a numerical finite-precision arithmetic implementation of Gaussian elimination is the *exact* solution of a *nearby* linear system, i.e. the approximate solution $\widehat{\boldsymbol{x}}$ obtained by the numerical scheme satisfies a perturbed linear system of the form

$$(\boldsymbol{A} + \Delta\boldsymbol{A})\widehat{\boldsymbol{x}} = \boldsymbol{b} + \Delta\boldsymbol{b}, \tag{5.1}$$

where the magnitudes of the matrix $\Delta\boldsymbol{A}$ and vector $\Delta\boldsymbol{b}$ can be bounded by the product of a constant, the *growth factor*, and the magnitude of the rounding errors.

It is this idea which we will now develop for the purposes of understanding geometric integrators. In this analogy, the linear system is replaced by a system of differential equations satisfying some geometric properties (for example, Hamiltonian, symplecticness, integral preserving). Truncation errors introduced by the integrator play the same role as rounding errors do in the linear algebra context. It is shown that, in a practical sense, the numerical solution obtained from an appropriate geometric integrator is the exact solution to a perturbed system of differential equations satisfying the same geometric property (or properties). The existence of such a backward error interpretation has direct implications for the qualitative behavior of the numerical solution and, ultimately, in the effectiveness of the method compared with others.

5.1 Forward v. backward error analysis

The discretization of a sufficiently smooth differential equation,

$$\frac{d}{dt}z = \boldsymbol{f}(z), \tag{5.2}$$

by a one-step method

$$z^{n+1} = \boldsymbol{\Psi}_{\Delta t}(z^n), \qquad t_{n+1} = t_n + \Delta t,$$

of order $p \geq 1$ implies that there exists a constant $M > 0$ such that

$$\|\,\boldsymbol{\Psi}_{\Delta t}(z) - \boldsymbol{\Phi}_{\Delta t,f}(z)\,\| \leq M\,\Delta t^{p+1}, \tag{5.3}$$

for all $\boldsymbol{z}$ in an appropriate subset of phase space $\mathbb{R}^k$ and for all Δt sufficiently small. Here $\boldsymbol{\Phi}_{\Delta t,f}$ denotes the exact time-Δt-flow of the vector field $\boldsymbol{f}$. As already mentioned in Section 2.1.2, the difference between the exact solution $\boldsymbol{z}(t_n)$ and the numerically computed approximation $\boldsymbol{z}^n$ at $t = t_n$ satisfies an upper bound

$$\|\,z(t_n) - z^n\,\| \leq K\left(e^{t_n L} - 1\right)\Delta t^p,$$

$L > 0$ the Lipschitz constant of $\boldsymbol{f}$ and $K > 0$ a constant independent of t_n and Δt. It is obvious that this estimate becomes useless whenever $t_n L \gg 1$ unless a very small timestep Δt is taken. This requirement is certainly not satisfied in simulations of large nonlinear systems where low-order methods are typically used to integrate to very long times t_n.

Backward error (or modified equations) analysis in the context of (partial) differential equations can be traced back to the work of Warming and Hyett [199]. The idea is simple: we derive a modified differential equation (the modified differential equation depends on the stepsize Δt)

$$\frac{d}{dt}z = \tilde{\boldsymbol{f}}(z;\Delta t), \tag{5.4}$$

and choose the modified vector field $\tilde{\boldsymbol{f}}(\Delta t) : \mathbb{R}^k \to \mathbb{R}^k$ such that the numerical method $\boldsymbol{\Psi}_{\Delta t}$ is a more accurate solution to this modified problem.

Let us demonstrate the basic idea by going through a simple example. Assume we discretize the differential equation (5.2) by the explicit Euler method

$$z^{n+1} = z^n + \Delta t\boldsymbol{f}(z^n). \tag{5.5}$$

This approximation has to be compared with the flow map $\boldsymbol{\Phi}_{t,\boldsymbol{f}}$ at $t = \Delta t$. The first two terms in the Taylor series expansion of $\boldsymbol{\Phi}_{t=\Delta t,\boldsymbol{f}}$ are given by

$$\begin{aligned} z(t_{n+1}) &= \boldsymbol{\Phi}_{\Delta t,\boldsymbol{f}}(z^n) \\ &= z^n + \Delta t\boldsymbol{f}(z^n) + \frac{\Delta t^2}{2}\boldsymbol{f}'(z^n)\boldsymbol{f}(z^n) + \mathcal{O}(\Delta t^3), \end{aligned}$$

where $\boldsymbol{f}'(z)$ denotes the Jacobian of $\boldsymbol{f}$ at z. Taking the difference of $\boldsymbol{\Psi}_{\Delta t}(z^n)$ and $\boldsymbol{\Phi}_{\Delta t,\boldsymbol{f}}(z^n)$, we obtain, as expected,

$$\boldsymbol{\Phi}_{\Delta t,\boldsymbol{f}}(z^n) - \boldsymbol{\Psi}_{\Delta t}(z^n) = \frac{\Delta t^2}{2}\boldsymbol{f}'(z^n)\boldsymbol{f}(z^n) + \mathcal{O}(\Delta t^3).$$

Let us now consider the modified differential equation

$$\frac{d}{dt}z = \tilde{\boldsymbol{f}}_1(z;\Delta t) := \boldsymbol{f}(z) - \frac{\Delta t}{2}\boldsymbol{f}'(z)\boldsymbol{f}(z), \tag{5.6}$$

with associated flow map $\boldsymbol{\Phi}_{t,\tilde{\boldsymbol{f}}_1}$. At $t = \Delta t$, we obtain (neglecting terms of order Δt^3 and higher)

$$\begin{aligned} \boldsymbol{\Phi}_{\Delta t,\tilde{\boldsymbol{f}}_1}(z^n) &\approx \left[z^n + t\tilde{\boldsymbol{f}}_1(z^n;\Delta t) + \frac{t^2}{2}\tilde{\boldsymbol{f}}_1'(z^n;\Delta t)\tilde{\boldsymbol{f}}_1(z^n;\Delta t) \right]_{t=\Delta t} \\ &\approx z^n + \Delta t\left[\boldsymbol{f}(z^n) - \frac{\Delta t}{2}\boldsymbol{f}'(z^n)\boldsymbol{f}(z^n) \right] + \frac{\Delta t^2}{2}\boldsymbol{f}'(z^n)\boldsymbol{f}(z^n). \end{aligned}$$

Thus

$$\boldsymbol{\Phi}_{\Delta t,\tilde{\boldsymbol{f}}_1}(z^n) = z^n + \Delta t\boldsymbol{f}(z^n) + \mathcal{O}(\Delta t^3),$$

and comparison with (5.5) shows that the Euler method is a second-order discretization of the modified differential equation (5.6). This procedure can be continued. We make the *ansatz*

$$\tilde{\boldsymbol{f}}_i(z;\Delta t) = \boldsymbol{f}(z) + \Delta t\boldsymbol{\delta f}_1(z) + \Delta t^2\boldsymbol{\delta f}_2(z) + \ldots + \Delta t^i\boldsymbol{\delta f}_i(z), \tag{5.7}$$

with $\boldsymbol{\delta f}_1(z) = -(1/2)\boldsymbol{f}'(z)\boldsymbol{f}(z)$. As for the modification $\boldsymbol{\delta f}_1$, the functions $\boldsymbol{\delta f}_j$, $j = 2, \ldots, i$, are obtained by computing the Taylor series expansion of the flow map $\boldsymbol{\Phi}_{t,\tilde{f}_i}$ and choosing the modified differential equation[2] such that

$$\boldsymbol{\Phi}_{t=\Delta t,\tilde{f}_i}(z^n) = z^n + \Delta t \boldsymbol{f}(z^n) + \mathcal{O}(\Delta t^{i+2}).$$

Hence Euler's method can be viewed as a discretization of order $p = i+1$ for the modified differential equation (5.7). If we assume for the moment that we can increase the index i to infinity and that the limit is well defined, then we may take $\tilde{\boldsymbol{f}}(\Delta t) := \tilde{f}_{i=\infty}(\Delta t)$ in (5.4) and the Euler method would be the exact solution to the modified differential equation (5.4) at $t = \Delta t$, i.e.

$$\boldsymbol{\Psi}_{\Delta t}(z) = \boldsymbol{\Phi}_{\Delta t,\tilde{f}}(z). \tag{5.8}$$

This equation is the equivalent to the backward error equation (5.1) for Gaussian elimination.

It is important to keep in mind that backward error analysis requires that the stepsize Δt is kept constant. Otherwise different modified differential equations are obtained at each integration step and the numerical method cannot be considered as the "exact" solution of a single modified differential equation.

Let us return to general numerical one-step methods of order $p \geq 1$. Ideally, we would like to find a modified differential equation (5.4) such that

(i) the modified vector field $\tilde{\boldsymbol{f}}(\Delta t)$ is close to the given vector field $\boldsymbol{f}$, i.e.,

$$||\, \boldsymbol{f}(z) - \tilde{\boldsymbol{f}}(z; \Delta t)\, || = \mathcal{O}(\Delta t^p),$$

$p \geq 1$ the order of the numerical method $\boldsymbol{\Psi}_{\Delta t}$, and

(ii) the numerical computed solutions can be considered as the exact solutions of the modified vector field $\tilde{\boldsymbol{f}}(\Delta t)$, i.e., the equality (5.8) holds.[3]

Let us suppose for the moment that (i) and (ii) can indeed be achieved. We may compare the solution behavior of the modified equation (5.4) to the solutions of the given problem (5.2). In general, not much has been gained yet: the solutions of these two differential equations will, in general, still diverge exponentially fast as the time interval t_n is increased. Backward error analysis becomes useful as soon as one can show that the modified equation shares some qualitative features with the given problem. We will come back to this point in Section 5.1.2.

[2] A general framework for computing modified equations for splitting methods will be given in Section 5.4.

[3] The equality (5.8) can be achieved for linear problems. For general nonlinear (analytic) problems, the equality in (5.8) holds except for an exponentially small term in the stepsize Δt. See Section 5.2. for more details

5.1.1 Linear systems

In this section, we focus on linear differential equations

$$\frac{d}{dt}z = \boldsymbol{A}z, \tag{5.9}$$

$z \in \mathbb{R}^k$, $\boldsymbol{A} \in \mathbb{R}^{k\times k}$.

While linear differential equations are certainly too special to be of practical relevance, restriction to this class of problems allows us to satisfy condition (5.8) exactly and to present the basic ideas of backward error analysis in a simple context. We have already encountered the basic ingredients in Section 2.6 when we discussed numerical methods for the harmonic oscillator using eigenvalues. Nonlinear problems have to be treated with different techniques and we will come back to them in Section 5.2.

The general solution of a linear equation (5.9) can be obtained in the following manner. We make the *ansatz*

$$z(t) = \mathrm{Re}\,\{\boldsymbol{v}\,\mathrm{e}^{\lambda t}\},$$

where $\boldsymbol{v} \in \mathbb{C}^k$ is a complex-valued vector, $\lambda \in \mathbb{C}$ is a complex number, and $\mathrm{Re}\,(z)$ denotes the real part of a complex number z. Upon substituting this trial solution into the equation (5.9), we obtain the algebraic condition

$$\mathrm{Re}\,\left\{[\boldsymbol{A} - \lambda \boldsymbol{I}_k]\,\boldsymbol{v}\mathrm{e}^{\lambda t}\right\} = \boldsymbol{0}.$$

Since this equation has to hold for all t, we have to have

$$[\boldsymbol{A} - \lambda \boldsymbol{I}_k]\,\boldsymbol{v} = \boldsymbol{0},$$

and, hence, λ is an eigenvalue of $\boldsymbol{A}$ and $\boldsymbol{v}$ an associated eigenvector. Denote the set of eigenvalues of $\boldsymbol{A}$ by λ_i, $i = 1, \ldots, k$, with associated eigenvectors $\boldsymbol{v}_i$, then the general solution of (5.9) can be represented as[4]

$$z(t) = \sum_i \mathrm{Re}\,\{c_i\,\boldsymbol{v}_i\,\mathrm{e}^{\lambda_i t}\},$$

where c_i, $i = 1, \ldots, k$, are complex numbers determined by the initial condition $z(0) = z_0$, i.e.,

$$z_0 = \sum_i \mathrm{Re}\,\{c_i\,\boldsymbol{v}_i\}.$$

[4]Here we have assumed, for simplicity, that all eigenvalues are distinct from each other and the eigenvectors form a complete basis in $\mathbb{R}^k$.

Example 1 Let us discuss the harmonic oscillator

$$\frac{d}{dt}q = \omega p, \tag{5.10}$$

$$\frac{d}{dt}p = -\omega q, \tag{5.11}$$

with frequency $\omega > 0$. The associated matrix

$$\boldsymbol{A} = \begin{bmatrix} 0 & \omega \\ -\omega & 0 \end{bmatrix}$$

has eigenvalues $\lambda_{1/2} = \pm i\omega$ and (normalized) eigenvectors

$$\boldsymbol{v}_1 = \frac{1}{\sqrt{2}}\begin{bmatrix} -i \\ 1 \end{bmatrix}, \qquad \boldsymbol{v}_2 = \frac{1}{\sqrt{2}}\begin{bmatrix} 1 \\ -i \end{bmatrix}.$$

The general solution can be written as

$$q(t) = c_1 \mathrm{Re}\,\{-i\mathrm{e}^{i\omega t}\} + c_2 \mathrm{Re}\,\{\mathrm{e}^{-i\omega t}\} = c_1 \sin(\omega t) + c_2 \cos(\omega t),$$

and

$$p(t) = c_1 \mathrm{Re}\,\{\mathrm{e}^{i\omega t}\} + c_2 \mathrm{Re}\,\{-i\mathrm{e}^{-i\omega t}\} = c_1 \cos(\omega t) - c_2 \sin(\omega t),$$

where $c_1 = p(0)$ and $c_2 = q(0)$. □

Let us return to numerical approximations of (5.9). Any one-step method discussed in Chapters 2 and 4 will lead to a discretization of type (see Section 2.6)

$$\begin{aligned} \boldsymbol{z}^{n+1} &= \boldsymbol{\Psi}_{\Delta t}(\boldsymbol{z}^n), \\ &= \widehat{\boldsymbol{R}}(\Delta t)\boldsymbol{z}^n. \end{aligned} \tag{5.12}$$

Here $\widehat{\boldsymbol{R}}(\Delta t)$ is a $k \times k$ matrix depending on the stepsize Δt. To be able to relate the numerical solutions $\boldsymbol{z}^n$ back to the analytic solution, we write

$$\boldsymbol{z}^n = \mathrm{Re}\,\{\boldsymbol{u}\,\mathrm{e}^{\tilde{\lambda} t_n}\}, \qquad t_n = n \cdot \Delta t,$$

and obtain the algebraic condition

$$\left[\widehat{\boldsymbol{R}}(\Delta t) - \mu \boldsymbol{I}_k\right] \boldsymbol{u} = \boldsymbol{0}$$

for $\boldsymbol{u}$ and

$$\mu = \mathrm{e}^{\tilde{\lambda}\Delta t}.$$

Hence, for any fixed value of Δt, we have to find all eigenvectors $\boldsymbol{u}_i$ and eigenvalues μ_i, $i = 1, \ldots, k$, of the matrix $\widehat{\boldsymbol{R}}(\Delta t)$. Then the numerical approximation

z^n is given by

$$z^n = \sum_i \mathrm{Re}\,\{d_i\, \boldsymbol{u}_i \mu_i^n\} = \sum_i \mathrm{Re}\,\{d_i\, \boldsymbol{u}_i e^{\tilde{\lambda}_i t_n}\}, \tag{5.13}$$

with

$$\tilde{\lambda}_i = \frac{1}{\Delta t} \ln \mu_i,$$

and the complex numbers d_i are determined by the initial $\boldsymbol{z}^0 = \boldsymbol{z}(0)$.

Example 2 The explicit Euler method, applied to (5.9), leads to

$$\widehat{\boldsymbol{R}}(\Delta t) = \boldsymbol{I}_k + \Delta t \boldsymbol{A}.$$

It is easy to verify that an eigenvector $\boldsymbol{v}$ of the matrix $\boldsymbol{A}$ is also an eigenvector of $\widehat{\boldsymbol{R}}(\Delta t)$, i.e. $\boldsymbol{v} = \boldsymbol{u}$. This observation leads immediately to

$$\mu_i = 1 + \Delta t \lambda_i,$$

where λ_i is an eigenvalue of $\boldsymbol{A}$. More generally, any (non-partitioned) Runge–Kutta (RK) method (see Section 2.4) leads to a matrix $\widehat{\boldsymbol{R}}(\Delta t)$ which has a set of eigenvectors identical to that of $\boldsymbol{A}$. Hence μ is equal to a rational polynomial in λ. For example, the implicit midpoint rule leads to

$$\mu_i = \frac{1 + (\Delta t/2)\lambda_i}{1 - (\Delta t/2)\lambda_i}.$$

Let us apply these formulas to the harmonic oscillator (5.10)–(5.11). The explicit Euler method yields

$$\mu_{1/2} = 1 \pm i\omega\Delta t = r e^{\pm i\tilde{\omega}\Delta t},$$

where

$$r = \sqrt{1 + (\omega\Delta t)^2}, \qquad \tilde{\omega} = \frac{\tan^{-1}(\omega\Delta t)}{\Delta t}.$$

This expression has to be compared with $e^{\pm i\omega\Delta t}$. We see that the numerical solutions grow in magnitude since $r > 1$ and that there is a shift in phase since $\tilde{\omega} < \omega$. On the other hand, the implicit midpoint rule leads to $r = 1$ and

$$\tilde{\omega} = \frac{2}{\Delta t} \tan^{-1}(\omega\Delta t/2).$$

Hence, the numerical solutions have constant magnitude but there is still a shift in phase, i.e., $\tilde{\omega} < \omega$. □

We now give (5.13) a time-continuous interpretation, i.e.,

$$\boldsymbol{z}(t) = \sum_i \mathrm{Re}\,\{d_i\, \boldsymbol{u}_i e^{\tilde{\lambda}_i t}\}, \tag{5.14}$$

and develop the modified equation corresponding to the scheme (5.12). For a fixed value of Δt, denote the matrix of eigenvectors $\boldsymbol{u}_i$ of $\widehat{\boldsymbol{R}}(\Delta t)$ by $\boldsymbol{U}(\Delta t)$ and the diagonal matrix of eigenvalues μ_i by $\boldsymbol{M}(\Delta t)$. Then, upon dropping the stepsize dependence of all matrices involved, the identity

$$\widehat{\boldsymbol{R}} = \boldsymbol{U}\,\boldsymbol{M}\,\boldsymbol{U}^{-1}$$

is obtained, which leads to the matrix

$$\tilde{\boldsymbol{A}} := \boldsymbol{U}\,\boldsymbol{\Lambda}\boldsymbol{U}^{-1},$$

where $\boldsymbol{\Lambda}$ is a diagonal matrix with its diagonal entries equal to $\tilde{\lambda}_i$. Indeed, since

$$\mathrm{e}^{t\,\tilde{\boldsymbol{A}}}{}_{|t=n\Delta t} = \boldsymbol{U}\,\mathrm{e}^{n\Delta t\,\boldsymbol{\Lambda}}\boldsymbol{U}^{-1} = \boldsymbol{U}\,\left[\mathrm{e}^{\Delta t\,\boldsymbol{\Lambda}}\right]^n \boldsymbol{U}^{-1} = \boldsymbol{U}\,[\boldsymbol{M}]^n\,\boldsymbol{U}^{-1} = \left[\widehat{\boldsymbol{R}}\right]^n,$$

we can conclude that (5.12) can be viewed as the exact solution of the modified linear differential equation

$$\frac{d}{dt}\boldsymbol{z} = \tilde{\boldsymbol{A}}(\Delta t)\,\boldsymbol{z} \tag{5.15}$$

sampled at time intervals Δt.

Example 3 Let us apply a general Runge–Kutta method to the harmonic oscillator (5.10)–(5.11). We write

$$\mu_{1/2} = r\mathrm{e}^{\pm i\tilde{\omega}\Delta t},$$

with both $\tilde{\omega}$ and r depending on Δt. Then the modified differential equation is of the form

$$\frac{d}{dt}\boldsymbol{z} = \tilde{\omega}(\Delta t)\boldsymbol{J}\boldsymbol{z} + \frac{\log r(\Delta t)}{\Delta t}\boldsymbol{z},$$

and, hence

$$\tilde{\boldsymbol{A}}(\Delta t) := \tilde{\omega}(\Delta t)\boldsymbol{J} + \frac{\log r(\Delta t)}{\Delta t}\boldsymbol{I}_2.$$

The harmonic oscillator has energy

$$H = \frac{\omega}{2}\left[q^2 + p^2\right].$$

If we evaluate the energy along solutions of the modified equations, then we obtain

$$\frac{d}{dt}H = \frac{2\log r(\Delta t)}{\Delta t}H.$$

Hence a RK method conserves energy if and only if $r(\Delta t) = 1$. This is the case for all Gauss–Legendre RK methods, which also happen to be symplectic. In fact, a Gauss–Legendre RK method, applied to the harmonic oscillator (5.10)–(5.11), can be thought of as the exact solution to a modified harmonic oscillator (Hamiltonian system) with modified Hamiltonian

$$\tilde{H} = \frac{\tilde{\omega}(\Delta t)}{2}\left[q^2 + p^2\right],$$

i.e. a harmonic oscillator with frequency $\tilde{\omega}(\Delta t)$. □

This last example showed that the modified equation (5.15) may inherit the Hamiltonian structure of the given equations of motion provided a proper discretization is used. In particular, for the harmonic oscillator, the symplectic Gauss–Legendre RK methods conserve energy exactly and can be viewed as exact solutions to a "modified" harmonic oscillator. In the following section we will explore the relation of symplectic methods and conservation of energy in the context of general nonlinear Hamiltonian systems before presenting more detailed results in Section 5.2.

5.1.2 The nearby Hamiltonian

Let us suppose that (5.2) is a Hamiltonian differential equation, then it would be desirable that the modified vector field $\tilde{\boldsymbol{f}}(\Delta t)$ is also Hamiltonian with modified Hamiltonian function $\tilde{H}(\Delta t)$. As we will discuss in more detail in Section 5.2, symplectic integration methods indeed give rise to modified differential equations that are Hamiltonian. In particular:

(i) the modified Hamiltonian $\tilde{H}(\Delta t)$ is close to the originally given Hamiltonian H, i.e.

$$|\,\tilde{H}(\boldsymbol{z};\Delta t) - H(\boldsymbol{z})\,| = \mathcal{O}(\Delta t^p),$$

$p \geq 1$ the order of the method, and

(ii)

$$\tilde{\boldsymbol{f}}(\boldsymbol{z};\Delta t) = \boldsymbol{J}\nabla_{\boldsymbol{z}}\tilde{H}(\boldsymbol{z};\Delta t).$$

The connection between symplectic integration methods and Hamiltonian modified equations has, for example, been explored in the work of AUERBACH AND FRIEDMAN [12], FENG [59], SANZ-SERNA [171], and YOSHIDA [206]. A more systematic treatment has, for example, been given by HAIRER [77], REICH [153], BENETTIN AND GIORGILLI [16], TANG [191], and CALVO, MURUA, AND SANZ-SERNA [40].

Thus a symplectic integrator solves a slightly perturbed Hamiltonian problem "exactly." If the qualitative solution behavior of the given problem is "stable" under small perturbations of the Hamiltonian, then, roughly speaking, a symplectic method will reproduce this qualitative solution behavior. Examples of such instances will be discussed in Section 5.2.3. On the other hand, non-symplectic methods, for which the modified vector field is not Hamiltonian, will change the qualitative solution behavior of a Hamiltonian problem in a significant way. In terms of the dynamics at equilibria for example, a non-Hamiltonian perturbation may change stable centers into sources or sinks.

Let us have a closer look at the explicit Euler method and the symplectic Euler-A method (4.8)–(4.9) from Section 4.1 when applied to Hamiltonian differential equations

$$\frac{d}{dt}\boldsymbol{q} = \boldsymbol{M}^{-1}\boldsymbol{p},$$
$$\frac{d}{dt}\boldsymbol{p} = -\nabla_q V(\boldsymbol{q}).$$

According to the general formula (5.6), the explicit Euler method gives rise to a modified system (5.7) with the first-order modification given by

$$\begin{aligned}\boldsymbol{\delta f}_1(\boldsymbol{q},\boldsymbol{p}) &= -\frac{1}{2}\begin{bmatrix} \boldsymbol{0} & \boldsymbol{M}^{-1} \\ -V_{\boldsymbol{qq}}(\boldsymbol{q}) & \boldsymbol{0}\end{bmatrix}\begin{bmatrix}\boldsymbol{M}^{-1}\boldsymbol{p} \\ -\nabla_q V(\boldsymbol{q})\end{bmatrix}, \\ &= \frac{1}{2}\begin{bmatrix}\boldsymbol{M}^{-1}\nabla_q V(\boldsymbol{q}) \\ V_{\boldsymbol{qq}}(\boldsymbol{q})\boldsymbol{M}^{-1}\boldsymbol{p}\end{bmatrix}.\end{aligned}$$

This vector field is not conservative. Next we consider the symplectic Euler-A method

$$\boldsymbol{q}^{n+1} = \boldsymbol{q}^n + \Delta t \boldsymbol{M}^{-1}\boldsymbol{p}^n, \tag{5.16}$$
$$\boldsymbol{p}^{n+1} = \boldsymbol{p}^n - \Delta t \nabla_q V(\boldsymbol{q}^{n+1}). \tag{5.17}$$

Taylor expansion of the second equation yields

$$\boldsymbol{p}^{n+1} = \boldsymbol{p}^n - \Delta t \nabla_q V(\boldsymbol{q}^n) - \Delta t^2 V_{\boldsymbol{qq}}(\boldsymbol{q}^n)\boldsymbol{M}^{-1}\boldsymbol{p}^n + \mathcal{O}(\Delta t^3).$$

If we compare this with the Taylor series expansion of the exact time-Δt-flow map, then we obtain the first-order modification

$$\boldsymbol{\delta f}_1(\boldsymbol{q},\boldsymbol{p}) = \frac{1}{2}\begin{bmatrix}\boldsymbol{M}^{-1}\nabla_q V(\boldsymbol{q}) \\ -V_{\boldsymbol{qq}}(\boldsymbol{q})\boldsymbol{M}^{-1}\boldsymbol{p}\end{bmatrix}$$

in the expansion (5.7). This term is conservative and can be written as

$$\boldsymbol{\delta f}_1(\boldsymbol{q},\boldsymbol{p}) = \begin{bmatrix}+\nabla_p \delta H_1(\boldsymbol{q},\boldsymbol{p}) \\ -\nabla_q \delta H_1(\boldsymbol{q},\boldsymbol{p})\end{bmatrix}, \quad \text{with} \quad \delta H_1 := \frac{1}{2}\boldsymbol{p}^T\boldsymbol{M}^{-1}\nabla_q V(\boldsymbol{q}).$$

Thus the symplectic Euler-A method is a second-order integrator for a modified Hamiltonian system with Hamiltonian

$$\tilde{H}_1(\Delta t) = \frac{\boldsymbol{p}^T \boldsymbol{M}^{-1}\boldsymbol{p}}{2} + V(\boldsymbol{q}) + \frac{\Delta t}{2}\boldsymbol{p}^T \boldsymbol{M}^{-1}\nabla_{\boldsymbol{q}} V(\boldsymbol{q}).$$

Example 4 Let us consider the reduced Kepler problem

$$\frac{d}{dt} r = p_r, \qquad \frac{d}{dt} p_r = \frac{l^2}{r^3} - \frac{m_3}{r^2}$$

with Hamiltonian

$$H(r, p_r) = \frac{1}{2}p_r^2 + \frac{l^2}{2r^2} - \frac{m_3}{r}.$$

Here $r > 0$ is the distance of the planet to the origin ("sun") and m_3 is the (constant) angular momentum of the planet. We discretize the equations of motion by the Euler-A method and obtain

$$r^{n+1} = r^n + \Delta t p_r^n, \quad p_r^{n+1} = p_r^n - \Delta t \left[\frac{m_3}{(r^{n+1})^2} - \frac{l^2}{(r^{n+1})^3} \right].$$

To first order in Δt, the associated modified Hamiltonian is

$$\tilde{H}_1(r, p_r; \Delta t) = H(r, p_r) + \frac{\Delta t}{2}\left[\frac{m_3}{r^2} - \frac{l^2}{r^3} \right] p_r.$$

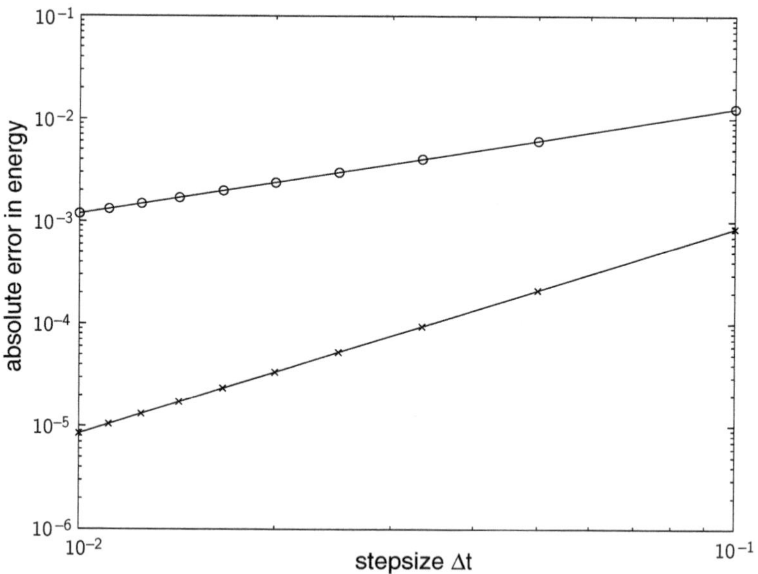

Figure 5.1 Maximum error in the energy with respect to the Hamiltonian H (o) and the modified Hamiltonian $\tilde{H}_1(\Delta t)$ (x) as a function of the stepsize Δt.

In Fig. 5.1, we plot the error in energy along numerically computed trajectories w.r.t. the Hamiltonian H and the modified Hamiltonian $\tilde{H}_1(\Delta t)$. The initial conditions $(r, p_r) = (1, 0.5)$ were used while the stepsize was chosen from $\Delta t \in [0.01, 0.1]$. It can be seen that the numerical solution converges with first order in Δt to $H = const.$ and with second order to $\tilde{H}_1(\Delta t) = const.$, as predicted. An important point is that both H and $\tilde{H}_1(\Delta t)$ possess the same qualitative solution behavior, i.e. the solutions are periodic. Its shape and period are, of course, different for the exact solution and its numerical approximation. This can be seen from Fig. 5.2. Note that the modified

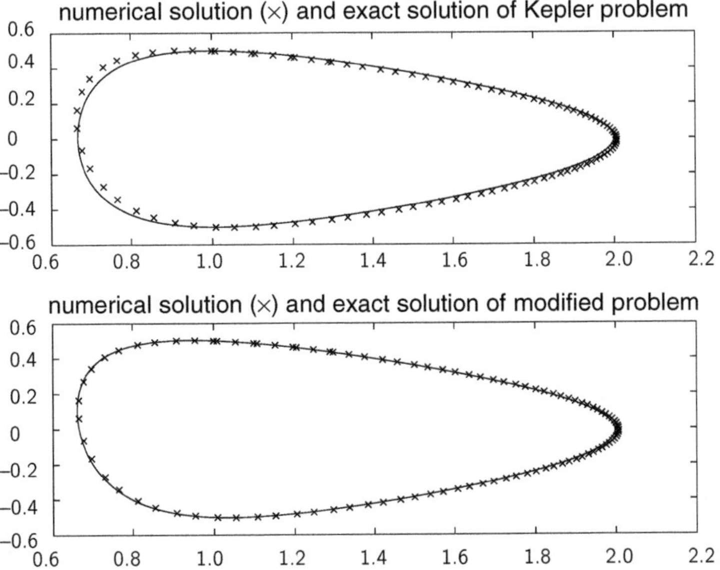

Figure 5.2 The numerical solution for $\Delta t = 0.2$ (x) as compared with the exact solution of the Kepler problem and the exact solution of the modified problem with Hamiltonian $\tilde{H}_1(\Delta t = 0.2)$.

Hamiltonian equations are

$$\frac{d}{dt}r = p_r + \frac{\Delta t}{2}\left[\frac{m_3}{r^2} - \frac{l^2}{r^3}\right], \quad \frac{d}{dt}p_r = \frac{c^2}{q^3} - \frac{m_3}{r^2} + p_r\frac{\Delta t}{2}\left[\frac{2}{r^3} - \frac{3l^2}{r^4}\right]. \qquad \square$$

5.2 The modified equations

In this section, we examine in more detail the development of the modified equations and discuss their geometric properties in the context of Hamiltonian dynamics and symplectic integration.

5.2.1 Asymptotic expansion of the modified equations

Let us return to the formal expansion (5.7) of the modified vector field $\tilde{f}_i(\Delta t)$. The correcting vector fields $\boldsymbol{\delta f}_j$, $j = 1, \ldots, i$, are determined by Taylor series expansion of the flow map $\boldsymbol{\Phi}_{t=\Delta t, \tilde{f}_i}$ and the numerical method $\boldsymbol{\Psi}_{\Delta t}$ in terms of Δt and by matching the first $i+1$ terms in the two expansions. Here we describe a somewhat more abstract recursive approach [158]. Let us assume that a modified vector field $\tilde{f}_i(\Delta t)$ has been found such that the numerical method $\boldsymbol{\Psi}_{\Delta t}$ is an integrator of order $p = i+1$ with respect to this modified differential equation, i.e., the flow map $\boldsymbol{\Phi}_{t,\tilde{f}_i}(z)$ satisfies

$$\boldsymbol{\Phi}_{\Delta t,\tilde{f}_i}(z) - \boldsymbol{\Psi}_{\Delta t}(z) = \mathcal{O}(\Delta t^{i+2}).$$

Then we define

$$\boldsymbol{\delta f}_{i+1}(z) = \lim_{\tau \to 0} \frac{\boldsymbol{\Psi}_\tau - \boldsymbol{\Phi}_{\tau,\tilde{f}_i(\tau)}(z)}{\tau^{i+2}}, \tag{5.18}$$

and introduce a new modified vector field $\tilde{f}_{i+1}(\Delta t)$ by

$$\tilde{f}_{i+1}(\Delta t) := \tilde{f}_i(\Delta t) + \Delta t^{i+1} \boldsymbol{\delta f}_{i+1}. \tag{5.19}$$

This numerical method $\boldsymbol{\Psi}_{\Delta t}$ is now a method of order $p = i+2$ with respect to the new modified vector field $\tilde{f}_{i+1}(\Delta t)$ as can be seen from

$$\begin{aligned}
\boldsymbol{\Phi}_{\Delta t,\tilde{f}_{i+1}}(z) - \boldsymbol{\Psi}_{\Delta t}(z) &= \boldsymbol{\Phi}_{\Delta t,\tilde{f}_i}(z) + \Delta t^{i+2} \boldsymbol{\delta f}_{i+1}(z) + \mathcal{O}(\Delta t^{i+3}) - \boldsymbol{\Psi}_{\Delta t}(z) \\
&= \Delta t^{i+2} \boldsymbol{\delta f}_{i+1}(z) + \boldsymbol{\Phi}_{\Delta t,\tilde{f}_i}(z) - \boldsymbol{\Psi}_{\Delta t}(z) + \mathcal{O}(\Delta t^{i+3}) \\
&= \mathcal{O}(\Delta t^{i+3}).
\end{aligned}$$

The recursion is started with $\tilde{f}_0 := \boldsymbol{f}$. For a method of order $p \geq 2$, the first $p-1$ vector fields $\boldsymbol{\delta f}_i$ are identical zero.

Unfortunately, this series does not, in general, converge as $i \to \infty$, i.e., the recursion (5.18)–(5.19) yields only an asymptotic expansion. Estimates for the difference between the flow maps of the modified equations and the numerical method are available if the involved maps are real analytic and bounded on an open (complex) neighborhood of a compact subset $\mathcal{K} \subset \mathbb{R}^k$ of phase space. In particular, BENETTIN AND GIORGILLI [16], HAIRER AND LUBICH [79], and REICH [158], using different techniques, derive an estimate of type

$$\| \boldsymbol{\Psi}_{\Delta t}(z) - \boldsymbol{\Phi}_{\Delta t,\tilde{f}_i}(z) \| \leq c_1 \Delta t \, (c_2 (i+1) \, \Delta t)^{i+1}, \tag{5.20}$$

for all $z \in \mathcal{K}$, where $c_1, c_2 > 0$ are appropriate constants independent of the iteration index i and the stepsize Δt. This formula indicates that the sequence of modified vector fields $\tilde{f}_i(\Delta t)$ converges before it starts to diverge for larger values

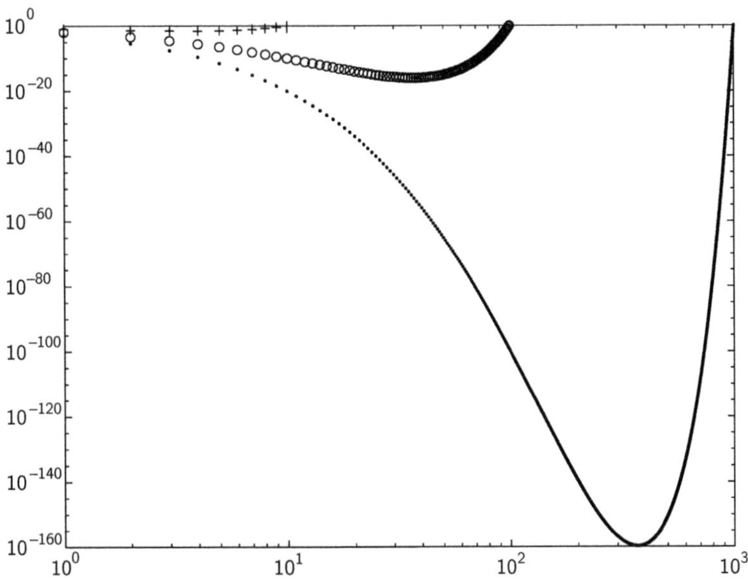

Figure 5.3 The function $f(i) = (i\Delta t)^i$ plotted for $\Delta t = 0.1$ (+), $\Delta t = 0.01$ (o), and $\Delta t = 0.001$ ($\cdot$) in the range $f(i) \leq 1$.

of i (Δt fixed). The point of divergence is shifted to larger and larger values of i as $\Delta t \to 0$. See Fig. 5.3 for an illustration.

Set i_* equal the integer part of

$$s(\Delta t) := \frac{1}{c_2\, e\, \Delta t} - 1.$$

This choice implies

$$c_2(i_* + 1)\Delta t < c_2(s+1)\Delta t = e^{-1}, \qquad \text{as well as} \qquad i_* + 1 > s,$$

and, hence, makes the expression on the right-hand side of (5.20) exponentially small in the stepsize Δt:

$$\begin{aligned} \|\boldsymbol{\Psi}_{\Delta t}(z) - \boldsymbol{\Phi}_{\Delta t, \tilde{f}_{i_*}}(z)\| &\leq c_1 \Delta t e^{-i_* - 1} \\ &\leq c_1 \Delta t e^{-s} \\ &\leq 3 c_1 \Delta t e^{-\gamma/\Delta t}, \end{aligned} \tag{5.21}$$

with $\gamma = 1/(c_2 e)$. The modified differential equation is now defined by

$$\frac{d}{dt} z = \tilde{\boldsymbol{f}}(z; \Delta t) := \tilde{\boldsymbol{f}}_{i_*}(z; \Delta t). \tag{5.22}$$

Following standard forward error analysis (see Sections 2.1.2 and 2.3), the difference of the exact solution $z(t_n)$ of the modified differential equation (5.22) and the numerical computed z^n at $t = t_n$ is bounded by

$$\| z(t_n) - z^n \| \leq K \left(e^{t_n \bar{L}} - 1 \right) e^{-\gamma/\Delta t} ,$$

$K > 0$ an appropriate constant and $\bar{L} > 0$ the Lipschitz constant of the numerical method as introduced in Section 2.3. In contrast to forward error analysis with respect to the given differential equation, this term remains (exponentially) small over periods of time

$$t_n \ll \frac{\gamma}{\Delta t \, \bar{L}}$$

[79]. Often this time interval is still not long enough, and more sophisticated error concepts, such as *shadowing* [76, 173] have to be used in addition to backward error analysis.

5.2.2 Conservation of energy for symplectic methods

If a Hamiltonian differential equation

$$\frac{d}{dt} z = J \nabla_z H(z) \tag{5.23}$$

is discretized by a symplectic method, then the modified vector fields $\tilde{f}_i(\Delta t)$ are Hamiltonian,[5] i.e., there exists a Hamiltonian $\tilde{H}_i(\Delta t)$ such that

$$\frac{d}{dt} z = \tilde{f}_i(z; \Delta t) = J \nabla_z \tilde{H}_i(z; \Delta t). \tag{5.24}$$

A proof of this result will be given in Section 5.3.

In case all the involved functions and maps are real analytic and bounded, the difference between the numerical method $\boldsymbol{\Psi}_{\Delta t}$ and the flow map $\boldsymbol{\Phi}_{\Delta t, \tilde{f}}$ of an optimally truncated modified equation (5.22) can be made exponentially small, i.e., an estimate (5.21) holds. Let us denote the corresponding Hamiltonian by $\tilde{H}(\Delta t)$. For a symplectic method of order $p \geq 1$, we have

$$H(z) - \tilde{H}(z; \Delta t) = \mathcal{O}(\Delta t^p). \tag{5.25}$$

[5] To be more precise: In general, the vector fields are only locally Hamiltonian [16]. But the modified Hamiltonian is global for symplectic splitting methods (compare Section 5.4) and for all symplectic Runge-Kutta methods [77, 80].

Let us now investigate the conservation of the modified Hamiltonian $\tilde{H}(\Delta t)$ along numerically computed solutions. After n steps, we obtain:

$$\begin{aligned} |\tilde{H}(z^n;\Delta t) - \tilde{H}(z^0;\Delta t)| &\leq \sum_{i=1}^{n} |\tilde{H}(z^i;\Delta t) - \tilde{H}(z^{i-1};\Delta t)| \\ &\leq \sum_{i=1}^{n} |\tilde{H}(\boldsymbol{\Psi}_{\Delta t}(z^{i-1});\Delta t) - \tilde{H}(\boldsymbol{\Phi}_{\Delta t,\tilde{f}}(z^{i-1});\Delta t)| \\ &\leq \sum_{i=1}^{n} \lambda\, \|\boldsymbol{\Psi}_{\Delta t}(z^{i-1}) - \boldsymbol{\Phi}_{\Delta t,\tilde{f}}(z^{i-1})\| \\ &\leq 3\,\lambda\, n\, \Delta t\, c_1\, e^{-\gamma/\Delta t}, \end{aligned}$$

$\lambda > 0$ the Lipschitz constant of $\tilde{H}(\Delta t)$. Here we have used the estimate (5.21) and the fact that $\tilde{H}(\Delta t)$ is a first integral of the modified vector field $\tilde{f}(\Delta t)$ which implies that

$$\tilde{H}(z^{i-1};\Delta t) = \tilde{H}(\boldsymbol{\Phi}_{\Delta t,\tilde{f}}(z^{i-1});\Delta t).$$

Thus the drift in the energy $\tilde{H}(\Delta t)$ remains exponentially small over exponentially long time intervals

$$t_N = n \cdot \Delta t \leq e^{\gamma/(2\Delta t)}.$$

This estimate and (5.25) imply the conservation of the given Hamiltonian H over an exponentially long period of time up to terms of order Δt^p. This result was first mentioned by NEISHTADT [143]. Explicit proofs can be found in the papers by BENETTIN AND GIORGILLI [16], HAIRER AND LUBICH [79], and REICH [158]

Example 5 The following experiment to illustrate the superior conservation of energy by a symplectic method follows an idea first used by BENETTIN AND GIORGILLI [16]. Consider the following one degree-of-freedom system

$$\dot{q} = p, \qquad \dot{p} = -V'(q), \qquad V(q) = e^{-q^2/2}.$$

For $|q|$ sufficiently large, e.g., $|q| > 20$, the force $F(q) = -V'(q)$ will be below machine precision in a standard double precision computation. Hence any numerical method will be essentially exact in that region of phase space and the modified equations can be identified with $\dot{q} = p$, $\dot{p} = 0$. Let us now perform a sequence of experiments with initial values $q(0) = -20$, $p(0) = 1$. We compute the solution up to the point where $|q(t)| > 20$ again. Now recall that we may assume that $\tilde{H} = H = p^2/2$ for $|q| > 20$. Hence we measure the change in energy from its initial to its final value, which gives us precisely the drift in energy due to the non-exact nature of backward error analysis. The numerically computed drift in the energy compared to the "fitted" exponential function $150e^{-11.8/\Delta t}$ can be found in Fig. 5.4. □

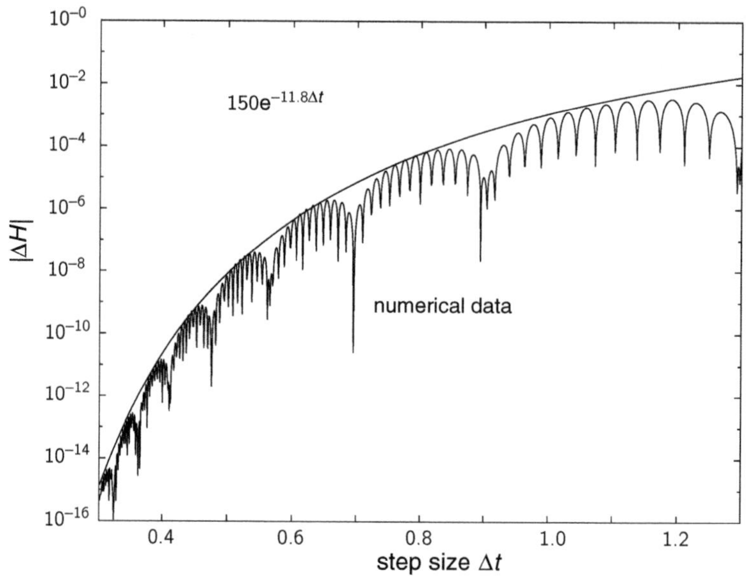

Figure 5.4 Demonstration of the exponential decay in the energy drift as $\Delta t \to 0$ for a simple one-dimensional collision problem.

The nearly exact conservation of a modified Hamiltonian $\tilde{H}$ has an interesting consequence in relation to exact conservation of energy. Following essentially the argument of GE AND MARSDEN [69], we assume that (5.23) does not possess any first integrals other than functions of the Hamiltonian H. By backward error analysis, there is a modified Hamiltonian $\tilde{H}(\Delta t)$ such that a symplectic method is the "exact" time-Δt-flow map corresponding to (5.24). If one would insist that the symplectic method also exactly conserves the Hamiltonian H, then one would obtain that the Poisson bracket of H and $\tilde{H}(\Delta t)$ is identically zero, i.e.

$$\{H, \tilde{H}(\Delta t)\} = 0.$$

This relation together with the assumption that H has no first integral except functions of H implies that the modified Hamiltonian has to be of the form

$$\tilde{H}(\boldsymbol{z}; \Delta t) = \rho(H(\boldsymbol{z}); \Delta t),$$

where ρ is some function of H and possibly Δt. However, this has the implication that, on level sets of constant energy $E = H$, the modified vector field is equivalent to the given vector field up to a multiplication by the constant factor $\rho'(E; \Delta t)$. Hence,

$$\boldsymbol{\Psi}_{\Delta t} = \boldsymbol{\Phi}_{\rho' \Delta t, H},$$

up to terms exponentially small in Δt and our symplectic and energy conserving method would solve the given Hamiltonian problem "exactly" up to a rescaling of time. This will be impossible, in general, and therefore exact conservation of energy and symplecticness are conflicting issues.

5.2.3 Applications

Let us now give a very brief outline of how backward error analysis and the existence of a modified Hamiltonian problem can guide us in understanding the numerical behavior of symplectic integration methods. We first consider the two extreme ends of possible solution behavior: (i) completely integrable and (ii) hyperbolic, i.e. fully chaotic. Finally, we briefly discuss the behavior of an adiabatic invariant under symplectic discretization.

Integrable systems

Let us assume that our given Hamiltonian system (5.23) can be, at least formally, transformed to a new set of canonical variables $(\mathbf{I}, \boldsymbol{\phi})$ and that the transformed Hamiltonian takes the simple form $H_0(\mathbf{I})$. The associated equations of motion

$$\frac{d}{dt}\mathbf{I} = \mathbf{0}, \qquad \frac{d}{dt}\boldsymbol{\phi} = \nabla_{\mathrm{I}} H_0(\mathbf{I})$$

are then solvable and the given problem is called integrable. The variables $(\mathbf{I}, \boldsymbol{\phi})$ are called action-angle variables [7, 8].

A classical problem in mechanics is the behavior of integrable systems under small perturbations. One is typically led to consider Hamiltonian functions of the form

$$H(\mathbf{I}, \boldsymbol{\phi}, \varepsilon) = H_0(\mathbf{I}) + \varepsilon H_1(\mathbf{I}, \boldsymbol{\phi}, \varepsilon), \tag{5.26}$$

where $\varepsilon > 0$ is small parameter and H_1 is 2π-periodic with respect to all the components in the angle variable $\boldsymbol{\phi}$. Let us assume that H_0 is convex in $\mathbf{I}$ and that both H_0 and H_1 are real-analytic functions. Then the NEKHOROSHEV theorem [144, 116] states that the action variable $\mathbf{I}(t)$ drifts by no more than terms of order $\mathcal{O}(\varepsilon^{1/2n})$ over an exponentially long time interval $|t| < e^{c/\varepsilon^{1/2n}}$. Here $c > 0$ is some constant and n is the number of degrees of freedom. See [116] for a precise statement of this result and its proof.

Let us now apply a symplectic integrator to an integrable Hamiltonian system of the form (5.23). Note that we do not make use of action-angle variables at this point. Under appropriate conditions, backward error analysis will lead to a modified Hamiltonian

$$\tilde{H}(z) = H(z) + \Delta t^p \delta H(z, \Delta t),$$

plus a remainder term exponentially small in Δt. The modified Hamiltonian $\tilde{H}$ can be, at least formally, transformed to action-angle variables $(\mathbf{I}, \boldsymbol{\phi})$ which yields a Hamiltonian of the form

$$\tilde{H}(\mathbf{I}, \boldsymbol{\phi}; \Delta t) = H_0(\mathbf{I}) + \Delta t^p \tilde{H}_1(\mathbf{I}, \boldsymbol{\phi}; \Delta t).$$

This perturbed Hamiltonian can be investigated by the NEKKHOROSHEV theorem with $\varepsilon = \Delta t^p$. In particular, the key ingredients of the elegant proof by LOSCHAK AND NEISHTADT [116] are conservation of total energy $\tilde{H}$, the conservation of an adiabatic invariant $J = \boldsymbol{\omega}_*^T \mathbf{I}$, where $\boldsymbol{\omega}_* = \nabla_{\mathbf{I}} H_0(\mathbf{I}_*)$ is a completely resonant frequency vector close to $\boldsymbol{\omega} = \nabla_{\mathbf{I}} H_0(\mathbf{I})$, and convexity of $H_0(\mathbf{I})$. These three conditions also hold for the modified equations and the long-time conservation of the action variables also applies to the numerical integration scheme. This result extends to the symplectic integration of perturbed integrable systems of the form (5.26). An elegant proof has been given by MOAN [135] using results of KUKSIN AND PÖSCHEL [100] on the embedding of symplectic maps into the flow of non-autonomous Hamiltonian systems.

An extensive discussion of integrable and near-integrable systems and their behavior under symplectic integration can also be found in the monograph [80].

Hyperbolic systems

Completely contrary to integrable systems, solutions of *hyperbolic systems* [76] diverge exponentially everywhere in phase space. This makes the solution behavior unpredictable over long time intervals and leads to "chaotic" dynamics. Strictly hyperbolic systems are difficult to find, but the notion of hyperbolicity is very fruitful for mathematical studies. Firstly, Hamiltonian hyperbolic systems remain hyperbolic under small changes in the Hamiltonian (structural stability), secondly the statistical mechanics of hyperbolic systems is quite well understood. Let us hence assume that a hyperbolic Hamiltonian system is integrated numerically by a symplectic method. We can assume that the associated modified Hamiltonian system is also hyperbolic. We also know that a symplectic method will approximately conserve energy over exponentially long time intervals. However, because of the exponential divergence of solutions, the numerical computed trajectory will *not* stay close to the exact solution of the modified problem over time periods larger than $\mathcal{O}(\Delta t^{-1})$. On the other hand, hyperbolic systems possess a *shadowing property* [76, 173]. Applied to our situation, one can conclude that any numerical trajectory can be shadowed by some exact solution of the modified problem over exponentially long periods of time (before a significant drift in energy is observed). In what sense is that information useful? Often one is not interested in an individual solution but in the time average of some observable $\mathcal{A}$ along trajectories. It turns out that for a hyperbolic Hamiltonian system the limit

$$\langle \mathcal{A} \rangle = \lim_{\mathcal{T}\to\infty} \frac{1}{\mathcal{T}} \int_0^{\mathcal{T}} \mathcal{A}(\boldsymbol{q}(t), \boldsymbol{p}(t))\, dt$$

is independent of the chosen trajectory for almost all initial values $(\boldsymbol{q}_0, \boldsymbol{p}_0)$ provided that the initial conditions are from one and the same energy E hypersurface, i.e.

$$H(\boldsymbol{q}_0, \boldsymbol{p}_0) = E.$$

In other words, the infinite time average $\langle \mathcal{A} \rangle$ depends only on the energy of the trajectory.[6] Computationally, we can only perform finite-time averages

$$\bar{\mathcal{A}}(\tau) = \frac{1}{\tau} \int_0^{\mathcal{T}} \mathcal{A}(\boldsymbol{q}(t), \boldsymbol{p}(t))\, dt.$$

But the existence of a *large deviation theorem* [207] implies that the set of all initial points for which

$$|\bar{\mathcal{A}}(\tau) - \langle \mathcal{A} \rangle| \geq \delta$$

has a measure that goes to zero exponentially fast in τ for a given fixed tolerance $\delta > 0$. Since, roughly speaking, numerical trajectories obtained from a symplectic integration method shadow some exact solution of a slightly perturbed hyperbolic Hamiltonian system, the same statement is true for numerically computed time averages. A precise formulation of such a result has been given by REICH in [158].

Numerical evidence for such a behavior can indeed be found even if the system is not provably hyperbolic. As a demonstration, we simulate a molecular N-body problem similar to what has been used in Section 4.5. A total of $N = 49$ particles move in the (x, y)-plane under the influence of a pair-wise repulsive potential and periodic boundary conditions are applied in the x and y directions. The equations of motion conserve energy and total linear momentum. We take the mean kinetic energy (temperature) as our observable; i.e.,

$$\mathcal{A} = \frac{1}{2N} \sum_{i=1}^{N} ||\boldsymbol{p}_i||^2.$$

We perform four simulations with random initial conditions from the constant energy level $E = H(\boldsymbol{q}_0, \boldsymbol{p}_0) = 2000$ with zero total linear momentum. The computed finite-time averages $\bar{\mathcal{A}}(\tau)$, $\tau \in [0, 10]$, can be found in Fig. 5.5. The averages at $\tau = 10$ are all within a $\delta = 0.01$ distance of each other.

This property of a symplectic integration method helps to explain the great success of the Störmer–Verlet method for molecular dynamics simulations [198, 4]. See Chapter 11 for further details.

[6] If further first integrals exists, such as total linear momentum and angular momentum, then $\langle \mathcal{A} \rangle$ also depends on those first integrals.

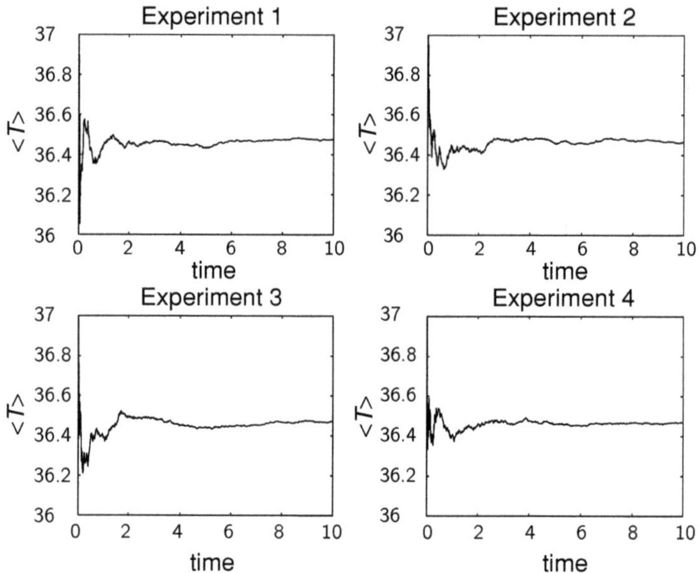

Figure 5.5 Time-averaged temperature computed along four different trajectories with one and the same initial energy $E = 2000$ and zero total linear momentum.

Adiabatic invariants

We have extensively discussed and used the harmonic oscillator

$$\ddot{q} = -\omega^2 q,$$

as an example throughout the book because of its simplicity. Let us now complicate the matter slightly by considering a harmonic oscillator with slowly varying frequency; for example

$$\omega(\varepsilon t) = \frac{1}{\sqrt{1 + 0.25\sin(2\pi\varepsilon t)}},$$

with $\varepsilon \ll 1$ a small parameter [7]. The behavior of such a time-dependent Hamiltonian system with Hamiltonian

$$H(p, q, \varepsilon t) = \frac{1}{2}\left(p^2 + \omega(\varepsilon t)^2 q^2\right),$$

becomes more transparent when going to action-angle variables (J, ϕ). Upon using the generating function [7, 8, 73] (compare also Section 6.4)

$$S(q, \phi, t) = \frac{1}{2}\omega(\varepsilon t)\, q^2 \cot\phi$$

the transformation is defined by

$$p = \frac{\partial S}{\partial q} = \omega(\varepsilon t)\, q \cot\phi,$$

and

$$J = -\frac{\partial S}{\partial \phi} = \frac{1}{2}\omega(\varepsilon t)\, q^2 \frac{1}{\sin^2\phi}.$$

We can solve this system for (q, p) to obtain

$$q = \sqrt{\frac{2J}{\omega}} \sin\phi,$$
$$p = \sqrt{2\omega J} \cos\phi.$$

The corresponding transformed Hamiltonian is

$$\begin{aligned}\bar{H}(J, \phi, \varepsilon t) &= H(p, q, \varepsilon t) + \frac{\partial S}{\partial t} \\ &= \omega(\varepsilon t) J + \frac{\varepsilon}{2} \frac{\omega'(\varepsilon t)}{\omega(\varepsilon t)} J \sin 2\phi \\ &= \omega(\varepsilon t) J + \varepsilon f(J, \phi, \varepsilon t),\end{aligned}$$

with $f(J, \phi, \varepsilon t) = \frac{\varepsilon}{2}\frac{\omega'(\varepsilon t)}{\omega(\varepsilon t)} J \sin 2\phi$.

Following the work of NEISHTADT [143], it is known that there exists another symplectic change of coordinates $(J, \phi) \to (\tilde{J}, \tilde{\phi})$ which is ε close to the identity such that the transformed Hamiltonian is of the form

$$\hat{H}(\tilde{J}, \tilde{\phi}, t) = \omega(\varepsilon t)\tilde{J} + \varepsilon^2 \tilde{g}(\tilde{J}, \varepsilon t; \varepsilon) + e^{-c/\varepsilon} \tilde{f}(\tilde{J}, \tilde{\phi}, \varepsilon t; \varepsilon),$$

$c > 0$ some constant and $\tilde{f}$ and $\tilde{g}$ are bounded functions. Hence we have

$$|J(0) - J(t)| = \mathcal{O}(\varepsilon), \qquad \text{for} \qquad |t| \le e^{c/(2\varepsilon)}, \tag{5.27}$$

and the action variable J is called an *adiabatic invariant* [7, 8, 73]. The action variable J can also be given a geometric interpretation. Over short time intervals, the motion is essentially periodic with a practically constant frequency $\omega_0 \approx \omega(\varepsilon t)$. Denote the area enclosed by the periodic orbits of

$$\dot{q} = p, \qquad \dot{p} = -\omega_0^2 q,$$

by A, then

$$J = \frac{1}{2\pi} \oint p\, dq = \frac{A}{2\pi}.$$

Note that the energy $H = \bar{H}$ will show a systematic drift since J is nearly constant and the product $\omega(\varepsilon t)J \approx H$ will follow changes in the frequency ω which can be of order one.

If the time-dependent equation of motion

$$\ddot{q} = -\omega(\varepsilon t)^2 q$$

is solved by a symplectic second-order integration method (see the Exercises of Chapter 4 as well as Example 6 below), then there exists a modified time-dependent Hamiltonian $\tilde{H}(\Delta t)$ which, written in action-angle variables, takes the form

$$\tilde{H}(J, \phi, \varepsilon t, \Delta t) = \omega(\varepsilon t)J + \varepsilon f(J, \phi, \varepsilon t) + \Delta t^2 g(J, \phi, \varepsilon t; \varepsilon, \Delta t),$$

g an appropriate function. Now assume that Δt^2 is bounded by ε: $\Delta t^2 < K\varepsilon$. Then it can be shown that an estimate of type (5.27) also holds for the numerical method [159]. See also [175] for further numerical experiments.

Example 6 Any time-dependent Hamiltonian $H(\mathbf{q}, \mathbf{p}, t)$ can be treated within the framework of autonomous Hamiltonian systems by enlarging the phase space by two additional variables (Q, P). This has been discussed in the Exercises of Chapter 4. Let us apply the idea to the Hamiltonian

$$H(q, p, t) = \frac{1}{2}\left[p + \omega(\varepsilon t)^2 q^2\right], \qquad \omega(\varepsilon t) = \frac{1}{\sqrt{1 + 0.25\sin(2\pi\varepsilon t)}}.$$

We define the extended Hamiltonian $\bar{H}(q, p, Q, P)$ as

$$\bar{H} = \frac{1}{2}\left[p + \omega(Q)^2 q^2\right] + \varepsilon P.$$

The associated equations of motion are

$$\begin{aligned}
\frac{d}{dt}q &= p, \\
\frac{d}{dt}p &= -\omega(Q)^2 q, \\
\frac{d}{dt}Q &= \varepsilon, \\
\frac{d}{dt}P &= -\omega(Q)\omega'(Q)q^2.
\end{aligned}$$

The variable Q is equal to slow time εt, i.e. $Q = \varepsilon t$. The equations can be integrated by a second-order splitting method using the split Hamiltonian

$$\bar{H}_1 = \frac{1}{4}\omega(Q)^2 q^2, \qquad \bar{H}_2 = \frac{1}{2}p^2 + \varepsilon P, \qquad \bar{H}_3 = \frac{1}{4}\omega(Q)^2 q^2.$$

One can iminate the variable P and obtains a modified Störmer–Verlet method.

We perform a series of experiments with constant Δt and increasing values of ε in the range $\varepsilon \in [0.001, 0.05]$. The initial conditions are $q(0) = 1$ and $p(0) = 0$. We monitor the time evolution of the adiabatic invariant

$$J = \frac{1}{2\omega}\left[p^2 + \omega^2 q^2\right],$$

and plot

$$\Delta J_{max} = \max_{t\in[0100]} |J(t) - J(0)|$$

as a function of ε for several values of Δt. See Fig. 5.6. As expected, we obtain a behavior of type

$$\Delta J_{max}(\Delta t, \varepsilon) \approx c_1 \Delta t^2 + c_2 \varepsilon.$$

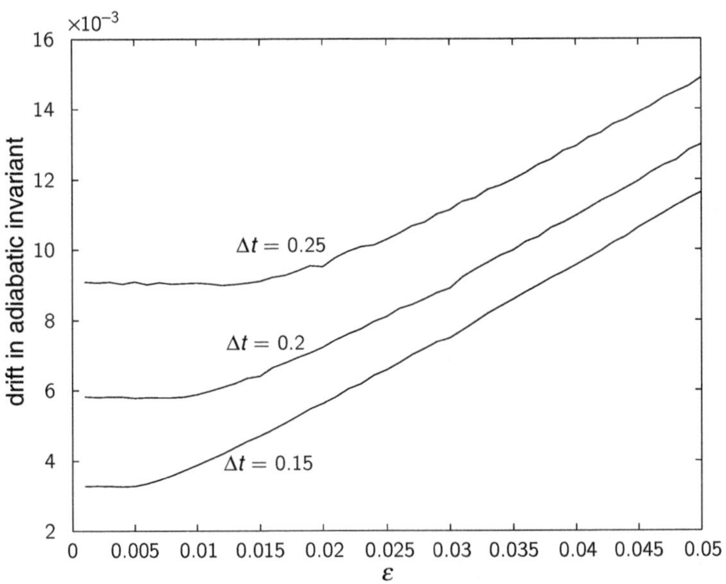

Figure 5.6 Variation in the adiabatic invariant J as a function of ε for different values of the stepsize Δt. □

5.3 Geometric integration and modified equations

An important aspect of backward error analysis is to show that the modified equations possess the same qualitative solution behavior as the given problem. A rigorous proof of such a statement is, in most cases, difficult. But it can be shown relatively easily that certain geometric aspects are preserved if the numerical method is chosen appropriately. We have discussed this already for

Hamiltonian systems and symplectic integration methods. The following more general result can be formulated:

Theorem 1 *Whenever the flow map of a given differential equation posesses some geometric properties such as existence of first integrals, time reversiblity, preservation of volume, symplecticness, and the numerical discretization preserves these properties exactly, then the flow map of the modified differential equation will also satisfy these geometric properties.* □

Proof. A proof of the theorem can be found in [153, 158]. See also [77, 16, 83, 74, 80].

Because of its importance, we discuss the case of symplectic methods in detail. Recall that the modified vector fields are recursively defined by

$$\boldsymbol{\delta f}_{i+1} := \lim_{\tau\to 0} \frac{\boldsymbol{\Psi}_\tau - \boldsymbol{\Phi}_{\tau,\tilde{\boldsymbol{f}}_i(\tau)}}{\tau^{i+2}}.$$

We introduce the short-hand notation

$$\boldsymbol{\Phi}_\tau := \boldsymbol{\Phi}_{\tau,\tilde{\boldsymbol{f}}_i(\tau)},$$

and obtain the relation

$$\partial_z \boldsymbol{\Psi}_\tau = \partial_z \boldsymbol{\Phi}_\tau + \tau^{i+2}\partial_z[\boldsymbol{\delta f}_{i+1}] + \mathcal{O}(\tau^{i+3}). \tag{5.28}$$

Let us now investigate the expression

$$\boldsymbol{F}_\tau := \frac{1}{\tau^{i+2}}\left[(\partial_z\boldsymbol{\Psi}_\tau)^T \boldsymbol{J}^{-1}\partial_z\boldsymbol{\Psi}_\tau - (\partial_z\boldsymbol{\Phi}_\tau)^T \boldsymbol{J}^{-1}\partial_z\boldsymbol{\Phi}_\tau\right].$$

Since both $\boldsymbol{\Psi}_\tau$ and $\boldsymbol{\Phi}_\tau$ are symplectic maps, we have $\boldsymbol{F}_\tau = \boldsymbol{0}$.

The next step is to make use of (5.28). This yields

$$\begin{aligned}\boldsymbol{F}_\tau &= -\left(\partial_z[\boldsymbol{\delta f}_{i+1}]\right)^T \boldsymbol{J}^{-1}\partial_z\boldsymbol{\Phi}_\tau - (\partial_z\boldsymbol{\Phi}_\tau)^T \boldsymbol{J}^{-1}\partial_z[\boldsymbol{\delta f}_{i+1}] + \mathcal{O}(\tau)\\ &= -\left(\partial_z[\boldsymbol{\delta f}_{i+1}]\right)^T \boldsymbol{J}^{-1} - \boldsymbol{J}^{-1}\partial_z[\boldsymbol{\delta f}_{i+1}] + \mathcal{O}(\tau).\end{aligned}$$

Hence, upon taking the limit $\tau \to 0$, one derives the condition

$$\left(\partial_z[\boldsymbol{\delta f}_{i+1}]\right)^T \boldsymbol{J}^{-1} + \boldsymbol{J}^{-1}\partial_z[\boldsymbol{\delta f}_{i+1}] = \boldsymbol{0},$$

which is equivalent to $\boldsymbol{\delta f}_{i+1}$ being locally Hamiltonian [7]. The existence of a global modified Hamiltonian can be deduced either from the fact that the domain $\boldsymbol{z} \in \Omega \subset \mathbb{R}^{2d}$ is simply connected or by explicit construction of a global Hamiltonian δH_{i+1}. For composition methods this is possible using the results of Section 5.4. Another explicit construction is provided by the *generating function* approach to the design of symplectic methods [16, 80]. We will discuss this approach briefly in Chapter 6.

Other applications of the theorem can be proven in a similar manner. See the exercises at the end of this chapter. □

Let us now focus on another application of Theorem 1. If a Hamiltonian H is in involution with a function F, i.e. $\{H, F\} = 0$, then F is a first integral of the equations of motion with Hamiltonian H. Let $\boldsymbol{\Phi}_{\tau,F}$ denote the flow map of

$$\frac{d}{d\tau}\boldsymbol{z} = J\nabla_z F(\boldsymbol{z}).$$

Since the Poisson bracket of F and H is zero, the two associated flow maps commute, i.e.

$$\boldsymbol{\Phi}_{t,H} \circ \boldsymbol{\Phi}_{\tau,F} = \boldsymbol{\Phi}_{\tau,F} \circ \boldsymbol{\Phi}_{t,H},$$

which is equivalent to

$$\boldsymbol{\Phi}_{t,H} = \boldsymbol{\Phi}_{-\tau,F} \circ \boldsymbol{\Phi}_{t,H} \circ \boldsymbol{\Phi}_{\tau,F} = \boldsymbol{\Phi}_{\tau,F}^{-1} \circ \boldsymbol{\Phi}_{t,H} \circ \boldsymbol{\Phi}_{\tau,F}.$$

Hence the flow map $\boldsymbol{\Phi}_{t,H}$ is invariant under the transformation $\boldsymbol{\psi}_\tau := \boldsymbol{\Phi}_{\tau,F}$. In fact, the reverse statement, called NOETHER's theorem, that any one-parameter family of symmetries gives rise to a first integral, is also true. See OLVER [149], ARNOLD [7], and MARSDEN AND RATIU [124] for an extensive discussion of symmetries in classical mechanics.

For example, consider the differential equation

$$\frac{d}{d\tau}\boldsymbol{q} = \boldsymbol{f}(\boldsymbol{q}),$$

and denote the associated flow map by $\boldsymbol{\phi}_\tau$. Then we can consider the one-parameter family of canonical point transformations $\boldsymbol{\Psi}_\tau$ defined by (compare Section 3.5)

$$\bar{\boldsymbol{q}}(\tau) = \boldsymbol{\phi}_\tau(\boldsymbol{q}), \tag{5.29}$$

$$\bar{\boldsymbol{p}}(\tau) = [\nabla_q \boldsymbol{\phi}_\tau(\boldsymbol{q})]^{-1}\,\boldsymbol{p}. \tag{5.30}$$

This transformation is equivalent to the flow map of a Hamiltonian system with Hamiltonian $F = \boldsymbol{p}^T\boldsymbol{f}(\boldsymbol{q})$, i.e.,

$$\boldsymbol{\Psi}_\tau = \boldsymbol{\Phi}_{\tau,F}.$$

Hence, if F is a first integral of H, i.e. $\{H, F\} = 0$, then $\boldsymbol{\Phi}_{t,H}$ is invariant under $\boldsymbol{\Psi}_\tau$ and vice versa.

Example 7 Let us look at the following particular class of linear transformations. Given a linear matrix group[7] $\mathcal{G} \subset \mathbb{R}^{d\times d}$, we can apply elements (matrices) of $\mathcal{G}$ to $\boldsymbol{q}$, hence defining new coordinates

$$\bar{q} = \boldsymbol{Q}\boldsymbol{q}, \qquad \boldsymbol{Q} \in \mathcal{G} \subset \mathbb{R}^{d\times d}.$$

Any matrix group $\mathcal{G}$ has an associated matrix algebra $\mathbf{g} \subset \mathbb{R}^{d\times d}$. The important property for our purposes is that the matrix exponential of any matrix $\boldsymbol{A} \in \mathbf{g}$ yields a matrix that is in the associated group $\mathcal{G}$. In fact any matrix $\boldsymbol{Q} \in \mathcal{G}$ can be represented by the matrix exponential of some matrix $\boldsymbol{A}$ in $\mathbf{g}$ [149]. Hence we consider the linear differential equation

$$\dot{q} = \boldsymbol{A}\boldsymbol{q}, \qquad \boldsymbol{A} \in \mathbf{g},$$

and obtain

$$\bar{q}(\tau) = \boldsymbol{Q}(\tau)\boldsymbol{q}, \qquad \boldsymbol{Q}(\tau) = e^{\tau \boldsymbol{A}}.$$

To derive a symplectic transformation from $(\boldsymbol{q}, \boldsymbol{p})$ to $(\bar{\boldsymbol{q}}, \bar{\boldsymbol{p}})$, we follow (5.30) and introduce canonical momenta

$$\bar{p}(\tau) = \boldsymbol{Q}(\tau)^{-T}\boldsymbol{p}.$$

The Hamiltonian associated with this one-parametric family of symplectic transformations is $F_{\boldsymbol{A}} = \boldsymbol{p}^T\boldsymbol{A}\boldsymbol{q}$.

The function $F_{\boldsymbol{A}}$ is a first integral if the given equations of motion are invariant under the symmetry group generated by $\mathcal{G}$. Furthermore, any matrix $\boldsymbol{A} \in \mathbf{g}$ will lead to a first integral. But only a finite number, equal to the dimension of the matrix algebra $\mathbf{g}$, of these integrals will be independent. For example, take the algebra of 3×3 skew-symmetric matrices which is the algebra associated to the group $SO(3)$ of orthogonal matrices. This algebra is three dimensional and any matrix in the algebra can be expressed as a linear combination of

$$\boldsymbol{A}_1 = \begin{bmatrix} 0 & 0 & 0 \\ 0 & 0 & -1 \\ 0 & 1 & 0 \end{bmatrix}, \quad \boldsymbol{A}_2 = \begin{bmatrix} 0 & 0 & 1 \\ 0 & 0 & 0 \\ -1 & 0 & 0 \end{bmatrix}, \quad \boldsymbol{A}_3 = \begin{bmatrix} 0 & -1 & 0 \\ 1 & 0 & 0 \\ 0 & 0 & 0 \end{bmatrix}.$$

The associated three first integrals are

$$F_{\boldsymbol{A}_1} = p_3q_2 - p_2q_3, \quad F_{\boldsymbol{A}_2} = p_1q_3 - p_3q_1, \quad F_{\boldsymbol{A}_3} = p_2q_1 - p_1q_2,$$

which are the three components of the angular momentum vector $\boldsymbol{m} = \boldsymbol{q} \times \boldsymbol{p}$. (We had already seen in Section 4.4 that the Störmer–Verlet method conserves angular momentum.) □

[7] A set $\mathcal{G}$ of invertible $d \times d$ matrices forms a matrix group in $\mathbb{R}^{d\times d}$, if the product of any two matrices $\boldsymbol{A}, \boldsymbol{B} \in \mathcal{G}$ is in $\mathcal{G}$; i.e., $\boldsymbol{AB} \in \mathcal{G}$ [149].

Let us now assume that a symplectic method $\boldsymbol{\Psi}_{\Delta t}$ is invariant under a family of symplectic transformations $\boldsymbol{\Phi}_{\tau,F}$ generated by a Hamiltonian F, i.e.

$$\boldsymbol{\Psi}_{\Delta t} = \boldsymbol{\Phi}_{\tau,F}^{-1} \circ \boldsymbol{\Psi}_{\Delta t} \circ \boldsymbol{\Phi}_{\tau,F},$$

for all $\Delta t \geq 0$. A result by GE [68] states that F is a first integral of $\boldsymbol{\Psi}_{\Delta t}$ up to a constant $c(\Delta t)$. Often this constant can be shown to be equal to zero for all stepsizes Δt. But even in the case that this is not possible one can still apply backward error analysis and Theorem 1 to obtain

$$\boldsymbol{\Phi}_{t,\tilde{H}} = \boldsymbol{\Phi}_{\tau,F}^{-1} \circ \boldsymbol{\Phi}_{t,\tilde{H}} \circ \boldsymbol{\Phi}_{\tau,F} = \boldsymbol{\Phi}_{-\tau,F} \circ \boldsymbol{\Phi}_{t,\tilde{H}} \circ \boldsymbol{\Phi}_{\tau,F},$$

where we neglected terms exponentially small in Δt. In other words, the two flow maps $\boldsymbol{\Phi}_{t,\tilde{H}}$ and $\boldsymbol{\Phi}_{\tau,F}$ commute and, hence, we must have $\{F, \tilde{H}\} = 0$. As a consequence we can conclude that the numerical method has F as a first integral (possibly up to a constant $c(\Delta t)$ exponentially small in Δt).

On the other hand, following the above proof of Theorem 1 for symplectic maps, it is also relatively easy to show that any numerical method that preserves a first integral F exactly has a modified equation that conserves F. (See the Exercises at the end of this chapter.)

5.4 Modified equations for composition methods

The flow map of a linear differential equation

$$\frac{d}{dt}z = \boldsymbol{A}z$$

can be written

$$\boldsymbol{\Phi}_{t,\boldsymbol{A}}(z) = \mathrm{e}^{t\boldsymbol{A}}z.$$

Here the matrix exponential is defined by the convergent series expansion

$$\mathrm{e}^{t\boldsymbol{A}} = \boldsymbol{I}_k + t\boldsymbol{A} + \frac{t^2}{2!}\boldsymbol{A}^2 + \frac{t^3}{3!}\boldsymbol{A}^3 + \cdots. \tag{5.31}$$

It is important to note that, contrary to

$$\mathrm{e}^{ta}\mathrm{e}^{tb} = \mathrm{e}^{t(a+b)},$$

we have

$$\mathrm{e}^{t\boldsymbol{A}}\mathrm{e}^{t\boldsymbol{B}} \neq \mathrm{e}^{t(\boldsymbol{A}+\boldsymbol{B})},$$

unless the two matrices $\boldsymbol{A}$ and $\boldsymbol{B}$ commute, i.e.

$$[\boldsymbol{A}, \boldsymbol{B}] := \boldsymbol{AB} - \boldsymbol{BA} = \boldsymbol{0}.$$

Instead there is a matrix $\boldsymbol{D}(t)$ such that

$$e^{t\boldsymbol{A}}e^{t\boldsymbol{B}} = e^{t\boldsymbol{D}(t)},$$

and the matrix $\boldsymbol{D}(t)$ is defined by a convergent expansion

$$\boldsymbol{D}(t) := \boldsymbol{D}_0 + t\boldsymbol{D}_1 + t^2\boldsymbol{D}_2 + t^3\boldsymbol{D}_3 + \cdots .$$

The matrices $\boldsymbol{D}_i$, $i = 0, 1, \ldots, \infty$, are given by the Baker–Campbell–Hausdorff (BCH) formula [196] which can be obtained from Taylor series expansion of the matrix exponential, i.e.

$$\begin{aligned}
e^{t\boldsymbol{A}}e^{t\boldsymbol{B}} &= \left(\boldsymbol{I}_k + t\boldsymbol{A} + \frac{t^2}{2}\boldsymbol{A}^2 + \cdots\right)\left(\boldsymbol{I}_k + t\boldsymbol{B} + \frac{t^2}{2}\boldsymbol{B}^2 + \cdots\right) \\
&= \boldsymbol{I}_k + t(\boldsymbol{A}+\boldsymbol{B}) + \frac{t^2}{2}\left(\boldsymbol{A}^2 + \boldsymbol{B}^2 + 2\boldsymbol{A}\boldsymbol{B}\right) + \cdots \\
&= \boldsymbol{I}_k + t(\boldsymbol{A}+\boldsymbol{B}) + \frac{t^2}{2}(\boldsymbol{A}+\boldsymbol{B})^2 + \frac{t^2}{2}(\boldsymbol{A}\boldsymbol{B} - \boldsymbol{B}\boldsymbol{A}) + \cdots \\
&= \boldsymbol{I}_k + t(\boldsymbol{A}+\boldsymbol{B}) + \frac{t^2}{2}(\boldsymbol{A}+\boldsymbol{B}) + \frac{t^2}{2}[\boldsymbol{A}, \boldsymbol{B}] + \cdots \\
&= e^{t(\boldsymbol{A}+\boldsymbol{B}) + (t^2/2)[\boldsymbol{A},\boldsymbol{B}] + \cdots}.
\end{aligned}$$

Hence

$$\begin{aligned}
\boldsymbol{D}_0 &= \boldsymbol{A} + \boldsymbol{B}, \\
\boldsymbol{D}_1 &= \frac{1}{2}[\boldsymbol{A}, \boldsymbol{B}],
\end{aligned}$$

and further expansion yields the next two terms

$$\begin{aligned}
\boldsymbol{D}_2 &= \frac{1}{12}\left([\boldsymbol{A}, [\boldsymbol{A}, \boldsymbol{B}]] + [\boldsymbol{B}, [\boldsymbol{B}, \boldsymbol{A}]]\right), \\
\boldsymbol{D}_3 &= \frac{1}{24}[\boldsymbol{A}, [\boldsymbol{B}, [\boldsymbol{B}, \boldsymbol{A}]]].
\end{aligned}$$

The BCH formula allows for a relatively simple backward error analysis for splitting methods applied to linear differential equation. For example, consider the linear differential equation

$$\frac{d}{dt}\boldsymbol{z} = (\boldsymbol{A} + \boldsymbol{B})\boldsymbol{z},$$

and assume that the matrix exponentials corresponding to the matrices $\boldsymbol{A}$ and $\boldsymbol{B}$ are easily computable. Then the composition method[8] $\boldsymbol{z}^{n+1} = \hat{\boldsymbol{R}}(\Delta t)\boldsymbol{z}^n$ with

$$\hat{\boldsymbol{R}}(\Delta t) = e^{\Delta t\boldsymbol{A}}e^{\Delta t\boldsymbol{B}}$$

[8] This composition method is often called the Trotter formula [194].

is a first-order method. According to the BCH formula, the matrix $\hat{\boldsymbol{R}}(\Delta t)$ is the exact exponential corresponding to the linear differential equation

$$\begin{aligned}\frac{d}{dt}z &= \boldsymbol{D}(\Delta t)z \\ &= \left(\boldsymbol{A}+\boldsymbol{B}+\frac{\Delta t}{2}[\boldsymbol{A},\boldsymbol{B}]+\frac{\Delta t^2}{12}([\boldsymbol{A},[\boldsymbol{A},\boldsymbol{B}]]+[\boldsymbol{B},[\boldsymbol{B},\boldsymbol{A}]])+\cdots\right)z,\end{aligned}$$

at time $t = \Delta t$. The second-order composition method[9] with

$$\hat{\boldsymbol{R}}(\Delta t) := \mathrm{e}^{(\Delta t/2)\boldsymbol{A}}\mathrm{e}^{\Delta t\boldsymbol{B}}\mathrm{e}^{(\Delta t/2)\boldsymbol{A}}$$

can be analyzed by multiple application of the BCH formula. We obtain the modified equation

$$\begin{aligned}\frac{d}{dt}z &= \boldsymbol{D}(\Delta t)z \\ &= \left(\boldsymbol{A}+\boldsymbol{B}+\frac{\Delta t^2}{24}(2[\boldsymbol{B},[\boldsymbol{B},\boldsymbol{A}]]-[\boldsymbol{A},[\boldsymbol{A},\boldsymbol{B}]])+\cdots\right)z. \qquad (5.32)\end{aligned}$$

The important point is that the modified differential equation is based on the matrices $\boldsymbol{A}$ and $\boldsymbol{B}$, the commutator $[\boldsymbol{A},\boldsymbol{B}]$, and repeated applications thereof.

We wish to extend this analysis to nonlinear differential equations

$$\frac{d}{dt}z = \boldsymbol{f}(z). \qquad (5.33)$$

Instead of the linear operator (matrix) $\boldsymbol{A}$, we now introduce the linear differential operator $L_{\boldsymbol{f}}$, called the *Lie derivative*, which acts on smooth scalar-valued functions g defined on phase space [7]. This operation, denoted by $L_{\boldsymbol{f}}g$, is defined by

$$L_{\boldsymbol{f}}g\,(z) := \nabla_z g(z)\cdot \boldsymbol{f}(z).$$

It naturally extends to vector-valued functions $\boldsymbol{g}$ by applying $L_{\boldsymbol{f}}$ to each component of $\boldsymbol{g}$. We denote this operation by $L_{\boldsymbol{f}}\boldsymbol{g}$. We can apply $L_{\boldsymbol{f}}$ repeatedly to itself, for example

$$L_{\boldsymbol{f}}(L_{\boldsymbol{f}}g) = L_{\boldsymbol{f}}^2 g,$$

and, as for the linear operator $\boldsymbol{A}$, we define the exponential of the operator $L_{\boldsymbol{f}}$ by the expansion (Lie series)

$$\mathrm{e}^{tL_{\boldsymbol{f}}} := \mathbf{id} + tL_{\boldsymbol{f}} + \frac{t^2}{2!}L_{\boldsymbol{f}}^2 + \frac{t^3}{3!}L_{\boldsymbol{f}}^3 + \cdots.$$

[9]This splitting is often called the Strang splitting [180].

This series can, formally, be obtained from (5.31) by replacing $\boldsymbol{A}$ by L_f and $\boldsymbol{I}$ by the identity map $\mathbf{id}$; $\mathbf{id}(z) = z$.

Let us now return to the flow map $\boldsymbol{\Phi}_{t,f}$ of (5.33). Repeated differentiation of (5.33) with respect to time yields the remarkable identity

$$\frac{d^n}{dt^n} z(t)_{|t=0} = L_f^{n-1} \boldsymbol{f}(z), \qquad z(t) = \boldsymbol{\Phi}_{t,f}(z).$$

Using this result, the Taylor series expansion of $\boldsymbol{\Phi}_{t,f}$ about $t = 0$ can be written as

$$\begin{aligned} \boldsymbol{\Phi}_{t,f}(z) &= z + t\frac{d}{dt}z(0) + \frac{t^2}{2!}\frac{d^2}{dt^2}z(0) + \frac{t^3}{3!}\frac{d^3}{dt^3}z(0) + \cdots \\ &= z + t\boldsymbol{f}(z) + \frac{t^2}{2!}L_f\boldsymbol{f}(z) + \frac{t^3}{3!}L_f^2\boldsymbol{f}(z) + \cdots . \end{aligned}$$

Let $L_f^i\mathbf{id}$, $i \geq 1$, denote the operation of L_f^i on the identity map $\mathbf{id}(z) = z$. It is easily checked that $L_f\mathbf{id} = \boldsymbol{f}$ and $L_f^2\mathbf{id} = L_f\boldsymbol{f}$. More generally

$$L_f^i\mathbf{id} = L_f^{i-1}\boldsymbol{f},$$

for any $i \geq 1$. Thus the flow map can be written as

$$\begin{aligned} \boldsymbol{\Phi}_{t,f} &= \mathbf{id} + t\boldsymbol{f} + \frac{t^2}{2!}L_f\boldsymbol{f} + \cdots \\ &= \left(\mathbf{id} + tL_f + \frac{t^2}{2!}L_f^2 + \cdots\right)\mathbf{id} \\ &= \mathrm{e}^{tL_f}\,\mathbf{id}. \end{aligned}$$

Furthermore, the action of the exponential on a function h amounts to

$$\mathrm{e}^{tL_f}\,h = h \circ \boldsymbol{\Phi}_{t,f}.$$

This can be seen from the expansion

$$\mathrm{e}^{tL_f}\,h = h + tL_f h + \frac{t^2}{2!}L_f^2 h + \cdots = h + t\frac{d}{dt}h + \frac{t^2}{2!}\frac{d^2}{dt^2}h + \cdots = h \circ \boldsymbol{\Phi}_{t,f}.$$

If we substitute h by the flow map of a second vector field $\boldsymbol{g}$, then we obtain the composition formula

$$\boldsymbol{\Phi}_{t,f} \circ \boldsymbol{\Phi}_{t,g} = \mathrm{e}^{tL_g} \circ \boldsymbol{\Phi}_{t,f} = \mathrm{e}^{tL_g}\mathrm{e}^{tL_f}\mathbf{id}.$$

Note that the order of the exponentials is opposite to what one might have expected.

Using exponential notation, a first-order composition method, corresponding to the differential equation

$$\frac{d}{dt}z = \boldsymbol{f}(z) + \boldsymbol{g}(z),$$

can now be written as

$$\boldsymbol{\Psi}_{\Delta t} = \boldsymbol{\Phi}_{\Delta t,f} \circ \boldsymbol{\Phi}_{\Delta t,g} = \mathrm{e}^{\Delta t L_g} \mathrm{e}^{\Delta t L_f} \mathbf{id}. \tag{5.34}$$

For

$$\boldsymbol{f} = \begin{bmatrix} \mathbf{0} \\ -\nabla_q V(\boldsymbol{q}) \end{bmatrix} \quad \text{and} \quad \boldsymbol{g} = \begin{bmatrix} M^{-1}\boldsymbol{p} \\ \mathbf{0} \end{bmatrix},$$

this composition method corresponds to the symplectic Euler-A method

$$\boldsymbol{q}^{n+1} = \boldsymbol{q}^n + \Delta t \boldsymbol{M}^{-1}\boldsymbol{p}^n,$$
$$\boldsymbol{p}^{n+1} = \boldsymbol{p}^n - \Delta t \nabla_{\boldsymbol{q}} V(\boldsymbol{q}^{n+1}).$$

The BCH formula can be generalized to interpret the composition method (5.34) as the "exact" solution of a modified problem. The only formal modification necessary is to replace the matrix commutator by the Lie commutator[10]

$$[L_f, L_g] := L_g L_f - L_f L_g. \tag{5.35}$$

Thus we, formally, obtain the modified differential equation

$$\frac{d}{dt} z = \tilde{f}(z; \Delta t),$$

with

$$\begin{aligned} \mathrm{e}^{\Delta t L_{\tilde{f}}} \mathbf{id} &= \left(\mathbf{id} + \Delta t L_g + \frac{\Delta t^2}{2!} L_g^2 + \cdots \right) \left(\mathbf{id} + \Delta t L_f + \frac{\Delta t^2}{2!} L_f^2 + \cdots \right) \mathbf{id} \\ &= \left(\mathbf{id} + \Delta t (L_f + L_g) + \frac{\Delta t^2}{2} \left((L_f + L_g)^2 + [L_f, L_g] \right) + \cdots \right) \mathbf{id} \end{aligned}$$

and, consequently

$$L_{\tilde{f}} = L_f + L_g + \frac{\Delta t}{2} [L_f, L_g] + \mathcal{O}(\Delta t^2). \tag{5.36}$$

Unlike the linear case, this series does *not*, in general, converge. A similar statement follows for the second-order splitting

$$\boldsymbol{\Psi}_{\Delta t} = \boldsymbol{\Phi}_{\Delta t/2,f} \circ \boldsymbol{\Phi}_{\Delta t,g} \circ \boldsymbol{\Phi}_{\Delta t/2,f} = \mathrm{e}^{(\Delta t/2) L_f} \circ \mathrm{e}^{\Delta t L_g} \circ \mathrm{e}^{(\Delta t/2) L_f} \mathbf{id}, \tag{5.37}$$

which is, formally, obtained from (5.32) by substituting $\boldsymbol{A}$ by L_f and $\boldsymbol{B}$ by L_g, respectively.

[10] The definition (5.35) is sometimes replaced by

$$[L_f, L_g] := L_f L_g - L_g L_f.$$

We prefer (5.35) because of its natural link to the Poisson bracket. See (5.38).

Let us now assume that the two vector fields $\boldsymbol{f}$ and $\boldsymbol{g}$ are Hamiltonian. We introduce the notation

$$L_H := L_{\{\mathbf{id},H\}}$$

to denote the differential operator $L_{\boldsymbol{f}}$ corresponding to the Hamiltonian vector field $\boldsymbol{f} = \{\mathbf{id}, H\}$ with Hamiltonian H and Lie–Poisson bracket $\{.,.\}$. Note that $L_H f = \{f, H\}$ [7]. For example, the Euler-A method can be written as

$$\boldsymbol{\Psi}_{\Delta t} = \boldsymbol{\Phi}_{\Delta t,V} \circ \boldsymbol{\Phi}_{\Delta t,T} = \mathrm{e}^{\Delta t L_T} \mathrm{e}^{\Delta t L_V} \mathbf{id}.$$

This method is symplectic and gives rise to a modified Hamiltonian differential equation

$$\frac{d}{dt} z = \tilde{\boldsymbol{f}}(z; \Delta t) = \boldsymbol{J} \nabla_z \tilde{H}(z; \Delta t),$$

with modified Hamiltonian $\tilde{H}(\Delta t)$. Again the BCH formula can be used. We just have to note that the commutator of two Hamiltonian vector fields H_1 and H_2 is related to the Poisson bracket of the two corresponding Hamiltonian functions via the formula

$$L_{\{H_1,H_2\}} = [L_{H_1}, L_{H_2}]. \tag{5.38}$$

Thus we obtain for the Euler-A method the modified Hamiltonian

$$\tilde{H} = T + V + \frac{\Delta t}{2}\{V, T\} + \frac{\Delta t}{12}\left(\{V, \{V, T\}\} + \{T, \{T, V\}\}\right) + \cdots.$$

Note that $\{V, T\}(\boldsymbol{q}, \boldsymbol{p}) = \boldsymbol{p}^T \boldsymbol{M}^{-1} \nabla_{\boldsymbol{q}} V(\boldsymbol{q})$ and we recover the modified Hamiltonian $\tilde{H}_1(\Delta t)$ from Section 5.1.2. The second-order splitting (5.37) becomes

$$\boldsymbol{\Psi}_{\Delta t} = \mathrm{e}^{(\Delta t/2)L_{H_1}} \mathrm{e}^{\Delta t L_{H_2}} \mathrm{e}^{(\Delta t/2)L_{H_1}} \mathbf{id}, \tag{5.39}$$

with modified Hamiltonian

$$\tilde{H} = H_1 + H_2 + \frac{\Delta t^2}{24}\left(2\{H_2, \{H_2, H_1\}\} - \{H_1, \{H_1, H_2\}\}\right) + \cdots. \tag{5.40}$$

In particular, the Störmer–Verlet method corresponds to the second-order splitting (5.39) with $H_1 = V$ and $H_2 = T$ where $T = \boldsymbol{p}^T \boldsymbol{M}^{-1} \boldsymbol{p}/2$ and $V = V(\boldsymbol{q})$. In other words, using exponential notation, the Störmer–Verlet method can be written more compactly as

$$\boldsymbol{\Psi}_{\Delta t} = \mathrm{e}^{(\Delta t/2)L_V} \mathrm{e}^{\Delta t L_T} \mathrm{e}^{(\Delta t/2)L_V} \mathbf{id},$$

and the first terms of the modified Hamiltonian $\tilde{H}(\Delta t)$ can be obtained from (5.40).

In summary, the Baker–Campbell–Hausdorff formula, originally derived for the composition of matrix exponentials, and a Lie algebraic treatment of composition methods are very useful for the derivation of higher-order methods and for the development of a backward error analysis. In this context, we like to mention, in particular, the work of DRAGT AND FINN [51], NERI [145], DRAGT *et al.* [52], SUZUKI [188, 189], and YOSHIDA [205].

5.5 Exercises

1. *Modified equations.* Find the exact modified equations for the symplectic scheme

$$q^{n+1} = q^n + \Delta t p^{n+1}, \qquad p^{n+1} = p^n - \Delta t \omega^2 q^n.$$

Recall the discussion in Section 2.6 of Chapter 2.

2. *Conservation of first integrals.* Show that the modified equations exactly conserve a first integral if the method does so. Hint: follow the proof of Theorem 1, denote the first integral by $G(\boldsymbol{z})$, and consider the behavior of the function

$$F_\tau = \frac{1}{\tau^{i+2}} \left[G(\boldsymbol{\Phi}_\tau) - G(\boldsymbol{\Psi}_\tau) \right],$$

where $\boldsymbol{\Phi}_\tau = \boldsymbol{\Phi}_{\tau, \tilde{f}_i(\tau)}$.

3. *Time-reversible methods and modified equations.* Show that the modified equations are time reversible if the method is time reversible. Hint: note that time reversibility of a method is equivalent to (compare Chapter 4)

$$\boldsymbol{\Psi}_{\Delta t} \circ (\boldsymbol{S}\boldsymbol{\Psi}_{\Delta t} \circ \boldsymbol{S}) - \mathbf{id} = \mathbf{0}.$$

Use this fact to find an appropriate function F_τ as used in the proof of Theorem 1.

4. *Symmetric methods and modified equations.* Show that the modified vector field expansion (5.7) contains only terms of even power in Δt if the numerical method $\boldsymbol{\Psi}_{\Delta t}$ is symmetric, i.e.

$$\boldsymbol{\Psi}_{-\Delta t} = \left[\boldsymbol{\Psi}_{\Delta t} \right]^{-1}. \tag{5.41}$$

This result implies that the order of a symmetric method is necessarily even (see Theorem 1 in Section 4.3.3. Hint: the flow map of the modified vector field

$$\tilde{\boldsymbol{f}}(\Delta t) = \sum_i \Delta t^i \boldsymbol{\delta f}_i$$

satisfies

$$\boldsymbol{\Phi}_{-\Delta t, \tilde{\boldsymbol{f}}(\Delta t)} = \left[\boldsymbol{\Phi}_{\Delta t, \tilde{\boldsymbol{f}}(\Delta t)}\right]^{-1} = \boldsymbol{\Phi}_{\Delta t, -\tilde{\boldsymbol{f}}(\Delta t)},$$

which gives the modified vector field for the map $[\boldsymbol{\Psi}_{\Delta t}]^{-1}$. Show also that $\boldsymbol{\Psi}_{-\Delta t}$ has the modified vector field

$$\hat{\boldsymbol{f}}(\Delta t) = \sum_i (-\Delta t)^i \boldsymbol{\delta f}_i.$$

Finally, use (5.41) and compare the corresponding modified vector fields.

5. *Canonical point transformation.* Prove that the transformation (5.29)–(5.30) is the flow map of the Hamiltonian vector field with Hamiltonian $F = \boldsymbol{p}^T \boldsymbol{f}(\boldsymbol{q})$. Hint: Rewrite (5.30) as

$$[\nabla_q \boldsymbol{\phi}_\tau(\boldsymbol{q})]\, \bar{\boldsymbol{p}}(\tau) = \boldsymbol{p},$$

differentiate with respect to τ, and set $\tau = 0$.

6. *Kepler problem and modified equations.* Apply the Störmer–Verlet method to the planar Kepler problem

$$H(\boldsymbol{q}, \boldsymbol{p}) = \frac{1}{2}\boldsymbol{p}^T\boldsymbol{p} - \frac{1}{||\boldsymbol{q}||}, \qquad \boldsymbol{q}, \boldsymbol{p} \in \mathbb{R}^2.$$

Use the BCH formula (5.40) to compute the second-order corrections of the modified Hamiltonian $\tilde{H}$ for this particular problem. Verify the fourth-order convergence of the Störmer–Verlet method with respect to the modified Hamiltonian $\tilde{H}_2$ numerically. Take, for example, initial conditions $\boldsymbol{q} = (1, 0)^T$ and $\boldsymbol{p} = (0, 1)^T$.

7. *BCH formula.* Verify the first two terms in the BCH formula (5.36) for the composition of two vector field exponentials

$$\boldsymbol{\Psi}_{\Delta t} = \boldsymbol{\Phi}_{\Delta t, \boldsymbol{f}} \circ \boldsymbol{\Phi}_{\Delta t, \boldsymbol{g}} = \mathrm{e}^{\Delta t L_g} \mathrm{e}^{\Delta t L_f} \mathbf{id}.$$

Also verify that the next term in the expansion is equal to

$$\frac{\Delta t^2}{12}\left([L_f, [L_f, L_g]] + [L_g, [L_g, L_f]]\right).$$

7. *Lie derivatives.* Verify formula (5.38), i.e. show that

$$L_{\{H_1,H_2\}}f = L_{H_1}[L_{H_2}f] - L_{H_2}[L_{H_1}f] = (L_{H_1}L_{H_2} - L_{H_2}L_{H_1})f,$$

by using $L_H f = \{f, H\}$, $\{f, g\} = -\{g, f\}$, and the Jacobi identity [7]

$$\{\{f, g\}, h\} + \{\{g, h\}, f\} + \{\{h, f\}, g\} = 0.$$

8. *Conservation of modified energy.* If one is only interested in monitoring the conservation of a more accurate energy expression, then one does not necessarily have to use a modified Hamiltonian $\tilde{H}(q, p; \Delta t)$ as previously discussed. This has been pointed out by SKEEL AND HARDY [177] who gave a practical construction for high-order modified energies. Below we describe a somewhat similar approach which has been used by MOORE AND REICH [137] to verify numerical energy conservation laws for Hamiltonian PDEs. The idea has also been used by HAIRER AND LUBICH to prove conservation of energy for the Störmer–Verlet and related methods [81].

We write the Störmer–Verlet method in its leapfrog form

$$\frac{\boldsymbol{q}^{n+1} - 2\boldsymbol{q}^n + \boldsymbol{q}^{n-1}}{\Delta t^2} = -\nabla_q V(\boldsymbol{q}).$$

a. Show that

$$\frac{q(t_n + \Delta t) - 2q(t_n) + q(t_n - \Delta t)}{\Delta t^2} = \ddot{q}(t_n) + \frac{\Delta t^2}{12} q^{(4)}(t_n) + \mathcal{O}(\Delta t^4),$$

where $q(t)$ is sufficiently smooth.

b. We use the result to formulate the following modified equation

$$\frac{d^2}{dt^2}\boldsymbol{q} + \frac{\Delta t^2}{12}\frac{d^4}{dt^4}\boldsymbol{q} = -\nabla_q V(\boldsymbol{q}) \tag{5.42}$$

for the leapfrog method. Show that

$$E = \frac{1}{2}\|\dot{\boldsymbol{q}}\|^2 + V(\boldsymbol{q}) + \frac{\Delta t^2}{12}\left[\dot{\boldsymbol{q}} \cdot \boldsymbol{q}^{(3)} - \frac{1}{2}\|\ddot{\boldsymbol{q}}\|^2\right] \tag{5.43}$$

is conserved along the solutions of the modified equations (5.42).

c. Show that the modified equation (5.42) is the Euler–Lagrange equation for the Lagrangian functional

$$\mathcal{L}[\boldsymbol{q}] = \int_{t_0}^{t_1} \left[\frac{1}{2}\|\dot{\boldsymbol{q}}\|^2 - V(\boldsymbol{q}) - \frac{\Delta t^2}{24}\|\ddot{\boldsymbol{q}}\|^2\right] dt.$$

d. How would you verify numerically that the modified energy (5.43) is preserved to fourth-order along numerical solutions $\{\boldsymbol{q}^n\}$? Hint: See [137].

6

Higher-order methods

In Chapter 4, we introduced several first- and second-order symplectic integration methods for Hamiltonian systems. In this chapter, we will discuss the construction of "higher-order" symplectic methods (with order greater than two), focusing in particular on those types of schemes that have been found to be most useful for practical computations. In traditional practice, higher-order integrators are employed for solving problems with relatively smooth solutions, such as gravitational simulations (solar system simulations, satellite trajectories). They are also traditionally used for many types of computations when *very high accuracy* (for example near the rounding error of the computer) is desired.

As we have seen in Chapter 2, the appropriateness of a given numerical method for a given computational task is a complicated issue. In some cases, the principles of geometric integration are in contradiction with the demand for high accuracy. If the purpose of simulation is to reconstruct, as exactly as possible, a particular trajectory segment, it may not matter what sort of qualitative features the integrator possesses: the efficiency of the integration method in terms of solution error per unit work is instead of paramount importance. Since the development of symplectic integrators adds a number of additional constraints on the design of the method, such schemes typically sacrifice something in efficiency compared with their nonsymplectic counterparts at similar accuracy, for example requiring an extra force evaluation or two at each timestep. Thus the problem of correctly determining the precise entry point and time instant that a space probe arrives at the Martian atmosphere is a task best handled by a standard integration method, for example, a high-order multistep integrator (for example, DIVA [99]) or explicit Runge–Kutta method (for example, RKSUITE [29]). These methods are well-suited to short time integration at extreme high accuracy.

On the other hand, it is often the case that integrations are carried out on terrifically long time intervals, or with respect to a large set of initial data. The goal is to accurately represent the qualitative long-term dynamics of the model, but some tolerance of individual trajectory error may be acceptable. Examples where this is the case include studies of the stability of the solar system, planet formation, and stellar cluster evolution. In some of the outer solar system simulations of SUSSMAN AND WISDOM for example, the computation was performed on a time interval of 100 million years [187], and more recent simulations have pushed the time intervals into the billions of years. On such time intervals, even the trajectories of the relatively slow-moving outer planets cannot be resolved to very high accuracy. Instead, the goal of these types of calculations is generally to obtain a useful, qualitatively correct result regarding a robust pattern in the long-term evolution of the system. The particular initial conditions are chosen arbitrarily within some set, and the questions asked concern the *generic behavior* for initial conditions within this set. In this setting, the most natural and efficient scheme is often some sort of geometric integrator, for example a symplectic or time-symmetric method.

To clarify this discussion, consider the challenge of predicting collisions of the earth with near-earth asteroids using simulation. One might ask either of the following questions, which are answerable in part using simulation: (i) when will a given object next collide with the earth and with what impact force, or (ii) on average how often will the earth be hit by such an object and with what average impact force. If the problem at hand is similar to the first one, then the best method for simulation is likely to be a traditional, high-order integrator with optimized choice of integration parameters. If the problem is more like the second one, the use of a high-order symplectic method may be indicated. Note that each of the two types of problems also requires a suitable choice of data: very accurate initial data and parameters for questions of type (i) and perhaps a sampling from an appropriate range of suitable initial data and model parameters in the case (ii).

In summary, accurate simulation of nonlinear systems generally calls for a high-order method. On the other hand, the efficient solution of nonlinear systems over very long time intervals or with many initial points typically benefits from the preservation of geometric properties such as the symplectic structure or a reversing symmetry. In this chapter, we will explain how to develop methods to provide us with some of the best aspects of both worlds: high order to improve the accuracy of the method when high accuracy is needed, as well as conservation of available geometric structure.

6.1 Construction of higher-order methods

The design of a symplectic integrator of order $p \geq 1$ involves, roughly speaking, two steps. First a stepsize dependent family of symplectic maps,

$$z^{n+1} = \boldsymbol{\Psi}_{\Delta t}(z^n; \gamma_1, \gamma_2, \ldots, \gamma_k),$$

is defined in terms of k arbitrary parameters. The method is then "fine tuned" by adjusting the parameters so that the local error of the method is $O(\Delta t^{p+1})$

$$\boldsymbol{\Psi}_{\Delta t}(\boldsymbol{z}) - \boldsymbol{\Phi}_{\Delta t}(\boldsymbol{z}) = \mathcal{O}(\Delta t^{p+1}).$$

In this chapter, we will discuss higher-order methods for general canonical Hamiltonian systems

$$\frac{d}{dt}\boldsymbol{q} = \nabla_{\boldsymbol{p}} H(\boldsymbol{q}, \boldsymbol{p}), \tag{6.1}$$

$$\frac{d}{dt}\boldsymbol{p} = -\nabla_{\boldsymbol{q}} H(\boldsymbol{q}, \boldsymbol{p})), \tag{6.2}$$

with an important special class of canonical systems obtained for Hamiltonian functions that are separable into kinetic and potential energy, i.e. of type

$$H(\boldsymbol{q}, \boldsymbol{p}) = T(\boldsymbol{p}) + V(\boldsymbol{q}).$$

We cover several approaches to the derivation of higher-order methods: composition methods, classical Runge–Kutta (RK) as well as partitioned Runge–Kutta (PRK) methods, and methods based on generating functions.

Composition methods offer an attractive and inexpensive way to obtain higher-order symplectic methods. They only require either (1) a splitting of the Hamiltonian into two or more explicitly solvable subproblems or, if this is not available, (2) a symmetric second-order symplectic approximation. The latter approach can, of course, be applied to any Hamiltonian problem, whether or not a splitting can be identified. Composition methods are not restricted to Hamiltonian problems and provide a general, systematic way to construct higher-order methods that preserve underlying geometric properties of the analytic problem.

On the other hand, there is a very well-developed order theory for RK and PRK methods and we "only" have to identify those schemes within these classes that are symplectic. This and the mathematical elegance of RK methods have contributed to their great popularity, especially among numerical analysts. However, it appears that, for many problems, symplectic RK methods are necessarily implicit and therefore more expensive in their implementation and are outperformed by composition methods unless very high precision computations are required.

6.2 Composition methods

In this section, we introduce the composition method approach to higher-order integration of Hamiltonian systems. We begin our treatment with systems in separable Hamiltonian form. The reader is referred to the survey article by McLachlan and Quispel [132] for a general discussion of splitting and composition methods.

We also wish to point to NERI [145], FOREST AND RUTH [62], YOSHIDA [205], and SUZUKI [188]) for early contributions to the theory of composition methods and their application to symplectic integration.

6.2.1 Composition methods for separable Hamiltonian systems

Consider a separable Hamiltonian system

$$\frac{d}{dt}\boldsymbol{q} = \nabla_{\boldsymbol{p}} T(\boldsymbol{p}),$$
$$\frac{d}{dt}\boldsymbol{p} = -\nabla_{\boldsymbol{q}} V(\boldsymbol{q}),$$

and observe that each of the two Hamiltonian subsystems

$$\frac{d}{dt}\boldsymbol{q} = \mathbf{0},$$
$$\frac{d}{dt}\boldsymbol{p} = -\nabla_{\boldsymbol{q}} V(\boldsymbol{q}),$$

and

$$\frac{d}{dt}\boldsymbol{q} = \nabla_{\boldsymbol{p}} T(\boldsymbol{p}),$$
$$\frac{d}{dt}\boldsymbol{p} = \mathbf{0}$$

is exactly solvable. Let us denote the corresponding flow maps by $\boldsymbol{\Psi}_{t,V}$ and $\boldsymbol{\Psi}_{t,T}$, respectively. First- and second-order methods can be obtained by simple compositions of these flow maps, as discussed in Chapters 4 and 5.

Here we will generalize this approach to methods of arbitrarily high order. For this purpose, we consider a multi-composition of the form

$$\boldsymbol{\Psi}_{\Delta t} := \boldsymbol{\Phi}_{c_s \Delta t, V} \circ \boldsymbol{\Phi}_{d_s \Delta t, T} \circ \cdots \circ \boldsymbol{\Phi}_{c_1 \Delta t, V} \circ \boldsymbol{\Phi}_{d_1 \Delta t, T} \circ \boldsymbol{\Phi}_{c_0 \Delta t, V}, \tag{6.3}$$

with $\{c_i\}_{i=0,\ldots,s}$ and $\{d_i\}_{i=1,\ldots,s}$ representing appropriate weight factors. The integer s is termed the *number of stages* of the concatenation method.

A method is said to be consistent if the order is at least one, i.e., if the local error is at least $O(\Delta t^2)$. Let us determine the requirement on the method parameters in a composition method for consistency. Observe that the terms in (6.3) commute to second order, so we have, for example

$$\boldsymbol{\Phi}_{d_1 \Delta t, T} \circ \boldsymbol{\Phi}_{c_1 \Delta t, V} = \boldsymbol{\Phi}_{c_1 \Delta t, V} \circ \boldsymbol{\Phi}_{d_1 \Delta t, T} + O(\Delta t^2),$$

We apply this fact recursively s times and use the group property associated to the flow maps on T and V, for example

$$\Phi_{\alpha T} \circ \Phi_{\beta T} = \Phi_{(\alpha+\beta)T},$$

for any scalar α, β. It follows that

$$\boldsymbol{\Psi}_{\Delta t} = \boldsymbol{\Phi}_{(d_1+d_2+\ldots+d_s)\Delta t,T} \circ \boldsymbol{\Phi}_{(c_0+c_1+c_2+\ldots+c_s)\Delta t,V} + O(\Delta t^2).$$

Since consistency is to be a property of the method, and hence should hold independent of T and V, the following *consistency conditions* result

$$\sum_{i=0}^{s} c_i = \sum_{i=1}^{s} d_i = 1\,.$$

Furthermore, if a consistent composition (6.3) is symmetric, i.e.

$$c_i = c_{s-i} \qquad \text{and} \quad d_i = d_{s+1-i},$$

$i = 1, \ldots, s$, then the method is automatically second order.

The Störmer–Verlet method (2.16)–(2.18) of Section 2.4 is obtained by choosing $s = 1$, $c_0 = 1/2$, $d_1 = 1$, $c_1 = 1/2$. The method is clearly consistent and symmetric; in fact, we have already pointed out in earlier chapters that this is a second-order method.

The conditions on the weight factors c_i and d_i to obtain order $p > 1$ can, for example, be obtained by a recursive application of the Baker–Campbell–Hausdorff formula (see Section 5.4). Another approach to derive order conditions uses an extension of Runge–Kutta rooted trees to composition methods [142].

Once the algebraic-order conditions are developed, it becomes an interesting challenge to find solutions to the resulting nonlinear system. Typically, one chooses the number of stages high enough so that, at a given order, there are too many parameters for the number of order conditions; one can then impose some additional conditions (such as minimization of some coefficient or coefficients appearing in the leading term of the error expansion) so that a (locally) unique and optimal solution is obtained.

In Table 6.1, we summarize two popular methods by providing the coefficients. Both methods have an even number of stages and are symmetric and, hence, we only have to give the first $s/2$ coefficients. References are also given in this table to the publication where the method appeared. These methods can be applied to any canonical Hamiltonian system with separable Hamiltonian $H = T(\boldsymbol{p}) + V(\boldsymbol{q})$. In fact, the listed methods can also be applied to any splitting $H = H_1 + H_2$ with explicitly solvable Hamiltonian functions H_1 and H_2.

Since the number of order conditions for general composition methods (6.3) rapidly grows with the desired order of the method, one may attempt to restrict the search for higher-order methods to methods based on the concatenation using a second-order symmetric method as discussed below.

Table 6.1 Symmetric composition methods for separable Hamiltonian systems

stages	order	coefficients		reference
6	4	$c_0 = 0.0792036964311957$ $c_1 = 0.353172906049774$ $c_2 = -0.0420650803577195$ $c_3 = 1 - 2(c_0 + c_1 + c_2)$	$d_1 = 0.209515106613362$ $d_2 = -0.143851773179818$ $d_3 = 1/2 - d_1 - d_2$	Blanes & Moan[23]
10	6	$c_0 = 0.0502627644003922$ $c_1 = 0.413514300428344$ $c_2 = 0.0450798897943977$ $c_3 = -0.188054853819569$ $c_4 = 0.541960678450780$ $c_5 = 1 - 2\sum_{i=0}^{4} c_i$	$d_1 = 0.148816447901042$ $d_2 = -0.132385865767784$ $d_3 = 0.067307604692185$ $d_4 = 0.432666402578175$ $d_5 = 1/2 - \sum_{i=1}^{4} d_i$	Blanes & Moan[23]

6.2.2 Composition methods based on second-order symmetric methods

It was observed by Yoshida [205] and Suzuki [188]) that an efficient way to obtain higher-order methods for separable Hamiltonian systems is to restrict the search to methods based on the concatenation of a second-order symmetric method with stepsize $w\Delta t$. For example, in the case of a separable Hamiltonian system, we might base this on

$$\boldsymbol{\Psi}_{w\Delta t} := \boldsymbol{\Phi}_{w\Delta t/2,T} \circ \boldsymbol{\Phi}_{w\Delta t,V} \circ \boldsymbol{\Phi}_{w\Delta t/2,T}. \tag{6.4}$$

Here $w \neq 0$ is a free parameter. Note that $\boldsymbol{\Psi}_{w\Delta t}$ is a second-order approximation to the exact flow map $\boldsymbol{\Phi}_{t,H}$ with $t = w\Delta t$.

Higher-order methods are constructed by s-fold concatenating $\boldsymbol{\Psi}_{w\Delta t}$ with different values for w, i.e.

$$\widetilde{\boldsymbol{\Psi}}_{\Delta t} = \boldsymbol{\Psi}_{w_s\Delta t} \circ \boldsymbol{\Psi}_{w_{s-1}\Delta t} \circ \cdots \circ \boldsymbol{\Psi}_{w_2\Delta t} \circ \boldsymbol{\Psi}_{w_1\Delta t}, \tag{6.5}$$

with $\{w_i\}_{i=1,\ldots,s}$ as free parameters. We also require that the concatenation method (6.5) is symmetric, i.e.

$$w_{s+1-i} = w_i, \qquad i = 1, \ldots, s,$$

and only consider odd values of s. The reason for considering symmetric methods based on the composition of symmetric methods is twofold: (i) symmetric methods are always of even order (see Theorem 1 in Section 4.3.3) and time reversible, (ii) the odd power terms in the Taylor expansion of the local error vanish and, therefore, the order conditions simplify. We present three methods of type (6.5) in Table 6.2. Because of symmetry, we again only provide the $(s+1)/2$ first coefficients w_i.

Table 6.2 Symmetric composition methods based on a symmetric second-order method

stages	order	coefficients	reference
5	4	$w_1 = 0.28$ $w_2 = 0.62546642846767004501$ $w_3 = 1 - 2\sum_{i=1}^{2} w_i$	McLachlan[130]
7	6	$w_1 = 0.78451361047755726382$ $w_2 = 0.23557321335935813368$ $w_3 = -1.17767998417887100695$ $w_4 = 1 - 2\sum_{i=1}^{3} w_i$	Yoshida[205]
9	6	$w_1 = 0.39216144400731413928$ $w_2 = 0.33259913678935943860$ $w_3 = -0.70624617255763935981$ $w_4 = 0.08221359629355080023$ $w_5 = 1 - 2\sum_{i=1}^{4} w_i$	Kahan and Li[97]
15	8	$w_1 = .74167036435061295345$ $w_2 = -.40910082580003159400$ $w_3 = .19075471029623837995$ $w_4 = -.57386247111608226666$ $w_5 = .29906418130365592384$ $w_6 = .33462491824529818378$ $w_7 = .31529309239676659663$ $w_8 = 1 - 2\sum_{i=1}^{7} w_i$	McLachlan[130]

For the concatenation methods (6.5) it is not essential that the method (6.4) is used. The necessary requirement is that the method is of one-step form, is second order and symmetric. For example, one could replace the method (6.4) by the standard Störmer–Verlet method (which is obtained from (6.4) by interchanging V and T) or by the implicit midpoint rule. Thus, in contrast to (6.3), this concatenation method can also be used to obtain higher-order methods for general Hamiltonian systems (6.1)–(6.2) as long as we are able to identify a second-order and symmetric method $\boldsymbol{\Psi}_{\Delta t}$ for the problem at hand.

6.2.3 Post-processing of composition methods

Composition methods of type (6.5) can be enhanced by the idea of postprocessing [117, 204, 131]. Introduce a coordinate transformation

$$\hat{z} = \boldsymbol{\psi}(z),$$

and design a one-step method

$$\hat{z}^{n+1} = \widehat{\boldsymbol{\Psi}}_{\Delta t}(\hat{z}^n), \tag{6.6}$$

such that the composed map

$$\widetilde{\Psi}_{\Delta t} = \psi^{-1} \circ \widehat{\Psi}_{\Delta t} \circ \psi \tag{6.7}$$

is a higher-order method for the given differential equation. Observe that

$$\begin{aligned} \widetilde{\Psi}^2_{\Delta t} &= \psi^{-1} \circ \widehat{\Psi}_{\Delta t} \circ \psi\psi^{-1} \circ \widehat{\Psi}_{\Delta t} \circ \psi \\ &= \psi^{-1} \circ \widehat{\Psi}^2_{\Delta t} \circ \psi, \end{aligned}$$

and, similarly

$$\widetilde{\Psi}^m_{\Delta t} = \psi^{-1} \circ \widehat{\Psi}^m_{\Delta t} \circ \psi.$$

Thus time integration is done in terms of the variable $\widehat{z}$ with ψ only applied infrequently whenever an output value is needed in terms of the original variable z. Because of this, we can afford for ψ to be relatively expensive to evaluate as long as $\widehat{\Psi}_{\Delta t}$ is "cheap."

Now assume that we are given a second-order symmetric method

$$z^{n+1} = \Psi_{\Delta t}(z^n).$$

As in Section 6.2.2, define a symmetric composition method of type (6.5) by (s odd)

$$\widehat{\Psi}_{\Delta t} = \Psi_{w_1\Delta t} \circ \Psi_{w_2\Delta t} \circ \cdots \circ \Psi_{w_2\Delta t} \circ \Psi_{w_1\Delta t}, \tag{6.8}$$

with $\{w_i\}_{i=1,\ldots,(s+1)/2}$ free parameters; i.e. the transformed variable $\widehat{z}$ is updated according to (6.6) using (6.8). We still need the transformation ψ. Here we use

$$\psi = \Psi_{-c_m\Delta t} \circ \Psi_{-c_{m-1}\Delta t} \circ \cdots \circ \Psi_{-c_1\Delta t} \circ \Psi_{c_m\Delta t} \circ \Psi_{c_{m-1}\Delta t} \circ \cdots \circ \Psi_{c_2\Delta t} \circ \Psi_{c_1\Delta t}, \tag{6.9}$$

with $\{c_i\}_{i=1,\ldots,m}$ as free parameters. Note that the inverse of ψ is easy to compute since the basic method $\Psi_{\Delta t}$ is symmetric, i.e.

$$\psi^{-1} = \Psi_{-c_1\Delta t} \circ \Psi_{-c_2\Delta t} \circ \cdots \circ \Psi_{-c_m\Delta t} \circ \Psi_{c_1\Delta t} \circ \Psi_{c_2\Delta t} \circ \cdots \circ \Psi_{c_{m-1}\Delta t} \circ \Psi_{c_m\Delta t}.$$

In Table 6.3 we give the coefficients for a sixth-order method due to BLANES [22]. Numerical comparisons of these methods can be found in Section 6.5.

6.3 Runge–Kutta methods

An alternative approach to higher-order integration is provided by the class of Runge–Kutta (RK) methods. It is well-known that certain implicit RK methods of Radau type (generalizing the implicit Euler method) are useful in the context

Table 6.3 Symmetric composition methods with post-processing

stages	order	coefficients		reference
7	6	$w_1 = 0.513910778424374$ $w_2 = 0.364193022833858$ $w_3 = -0.867423280969274$ $w_4 = 1 - 2(w_1 + w_2 + w_3)$	$c_1 = -(c_2 + c_3 + c_4 + c_5)$ $c_2 = -0.461165940466494$ $c_3 = -0.074332422810238$ $c_4 = 0.384998538774070$ $c_5 = 0.375012038697862$	BLANES[22]

of systems with strong dissipation, such as electronic circuits or chemical reaction dynamics. For these so-called *stiff systems*, implicit methods are often much more efficient than explicit methods. This is due to the inherent stepsize restriction of explicit methods in the context of strong dissipation (see, for example, [82, 84]). The situation is different for Hamiltonian systems for which, in general, implicit methods are found to be less competitive compared with the best explicit methods.

The situation is slightly different for partitioned Runge–Kutta methods where explicit methods exist for separable Hamiltonian systems. However, these methods can also be discussed in the context of splitting methods [148, 172].

On the other hand, implicit Runge–Kutta and partitioned Runge–Kutta methods often provide the only reasonable means for symplectic treatment of *non-separable* Hamiltonian systems. Below, we give a brief summary of the relevant issues for classical and partitioned Runge–Kutta methods.

6.3.1 Implicit Runge–Kutta methods

The general formula of an s-stage Runge–Kutta method applied to a Hamiltonian system (6.1)–(6.2) is

$$\boldsymbol{Q}_i = \boldsymbol{q}^n + \Delta t \sum_{j=1}^{s} a_{ij} \boldsymbol{F}_j, \qquad i = 1, \ldots, s, \tag{6.10}$$

$$\boldsymbol{P}_i = \boldsymbol{p}^n + \Delta t \sum_{j=1}^{s} a_{ij} \boldsymbol{G}_j, \qquad i = 1, \ldots, s, \tag{6.11}$$

$$\boldsymbol{q}^{n+1} = \boldsymbol{q}^n + \Delta t \sum_{i=1}^{s} b_i \boldsymbol{F}_i, \tag{6.12}$$

$$\boldsymbol{p}^{n+1} = \boldsymbol{p}^n + \Delta t \sum_{i=1}^{s} b_i \boldsymbol{G}_i. \tag{6.13}$$

with $s \geq 1$ the number of stages, $(\boldsymbol{Q}_i, \boldsymbol{P}_i)$, $i = 1, \ldots, s$, the internal stage variables, and the abbreviations

$$\boldsymbol{F}_i = +\nabla_{\boldsymbol{p}} H(\boldsymbol{Q}_i, \boldsymbol{P}_i), \quad i = 1, \ldots, s,$$
$$\boldsymbol{G}_i = -\nabla_{\boldsymbol{q}} H(\boldsymbol{Q}_i, \boldsymbol{P}_i), \quad i = 1, \ldots, s.$$

Recall that we called a Runge–Kutta method "explicit" in Section 2.4 if the coefficients $\{a_{ij}\}$ satisfy $a_{ij} = 0$ for $j \geq i$ and *implicit* otherwise. For implicit methods, the equations (6.10)–(6.11) are, in general, a system of nonlinear equations which must be solved by some iterative method, such as Newton's method or fixed-point iteration. Because implicit methods are based on iteration, and several force (or vector field) evaluations are needed at each timestep, they are typically more costly to implement than explicit methods. However, good starting approximations can greatly enhance the efficiency of the necessary iterative method. This has been discussed by HAIRER, LUBICH AND WANNER [80] and by CALVO AND PORTILLO [41].

An example of a symplectic Runge–Kutta method is provided by the implicit midpoint rule, which can be written as a one-stage ($s = 1$) RK method:

$$\boldsymbol{Q}_1 = \boldsymbol{q}^n + \frac{\Delta t}{2} \nabla_{\boldsymbol{p}} H(\boldsymbol{Q}_1, \boldsymbol{P}_1),$$
$$\boldsymbol{P}_1 = \boldsymbol{p}^n - \frac{\Delta t}{2} \nabla_{\boldsymbol{q}} H(\boldsymbol{Q}_1, \boldsymbol{P}_1),$$
$$\boldsymbol{q}^{n+1} = \boldsymbol{q}^n + \Delta t \nabla_{\boldsymbol{p}} H(\boldsymbol{Q}_1, \boldsymbol{P}_1),$$
$$\boldsymbol{p}^{n+1} = \boldsymbol{p}^n - \Delta t \nabla_{\boldsymbol{q}} H(\boldsymbol{Q}_1, \boldsymbol{P}_1).$$

The corresponding coefficients in (6.10)–(6.13) are $a_{11} = 1/2$ and $b_1 = 1$. For a general (implicit) s-stage RK method we have s^2 parameters $\{a_{ij}\}$ and s parameters $\{b_i\}$. Conditions on the parameters $\{a_{ij}\}$ and $\{b_i\}$ that insure a certain order of the corresponding method can be obtained by Taylor series expansion of the exact time-Δt-flow map and its approximation generated by the RK method followed by a matching of the terms in Δt up to the desired order of the method. Since the complexity of the resulting equations, or *order conditions*, increases rapidly with the desired order of the method, simplifying conditions are often introduced. This leads to special classes of RK methods like the *Gauss-Legendre methods*, the *Radau methods*, and the *Lobatto methods* which are all based on the idea of *collocation*. We refer the reader to the textbooks by HAIRER, NORSETT AND WANNER [82] and HAIRER AND WANNER [84] for a detailed exposition of this material.

Further restrictions apply if we demand that the RK method is symplectic. As first shown by SANZ-SERNA [170], LASAGNI [106], and SURIS [185] the conditions on the parameters $\{a_{ij}\}$ and $\{b_i\}$ are:

CONDITIONS OF SYMPLECTICNESS FOR RUNGE–KUTTA METHODS

$$b_i a_{ij} + b_j a_{ji} - b_i b_j = 0\,, \qquad i,j = 1,\dots,s. \tag{6.14}$$

Proof: It is convenient to first express the Runge–Kutta method in more compact form. We associate with any s-stage Runge–Kutta method the $s \times s$ matrix $\boldsymbol{A} = \{a_{ij}\}$ and the column vector $\boldsymbol{b} = \{b_i\}$. We also introduce tensor product notation. The tensor product $\boldsymbol{T} \otimes \boldsymbol{S}$, of an arbitrary $m \times m$ matrix $\boldsymbol{T} = \{t_{ij}\}$ and an arbitrary $n \times n$ matrix $\boldsymbol{S} = \{s_{ij}\}$, is a $k \times k$ matrix, $k = mn$, defined by

$$\boldsymbol{T} \otimes \boldsymbol{S} = \begin{bmatrix} t_{11}\boldsymbol{S} & t_{12}\boldsymbol{S} & \cdots & t_{1m}\boldsymbol{S} \\ t_{21}\boldsymbol{S} & t_{22}\boldsymbol{S} & \cdots & t_{2m}\boldsymbol{S} \\ \vdots & \vdots & & \vdots \\ t_{m1}\boldsymbol{S} & t_{m2}\boldsymbol{S} & \dots & t_{mm}\boldsymbol{S} \end{bmatrix}.$$

An account of the properties of tensor products is given by LANCHESTER [103]. For our purposes it is sufficient to know that

$$(\boldsymbol{R} \otimes \boldsymbol{S})(\boldsymbol{T} \otimes \boldsymbol{S}) = (\boldsymbol{R}\boldsymbol{T}) \otimes \boldsymbol{S},$$

whenever the products are defined.

Using tensor product notation, a RK method (6.10)–(6.13) can be compactly rewritten as

$$\begin{aligned} \boldsymbol{q}^{n+1} &= \boldsymbol{q}^n + \Delta t(\boldsymbol{b}^T \otimes \boldsymbol{I})\boldsymbol{F}, \\ \boldsymbol{p}^{n+1} &= \boldsymbol{p}^n + \Delta t(\boldsymbol{b}^T \otimes \boldsymbol{I})\boldsymbol{G}, \\ \boldsymbol{Q} &= \boldsymbol{e} \otimes \boldsymbol{q}^n + \Delta t(\boldsymbol{A} \otimes \boldsymbol{I})\boldsymbol{F}, \\ \boldsymbol{P} &= \boldsymbol{e} \otimes \boldsymbol{p}^n + \Delta t(\boldsymbol{A} \otimes \boldsymbol{I})\boldsymbol{G}, \end{aligned}$$

where $\boldsymbol{e} = [1, 1, \dots, 1]^T \in \mathbb{R}^s$, $\boldsymbol{I} \in \mathbb{R}^{d\times d}$ is the identity matrix, and

$$\boldsymbol{Q} = \begin{bmatrix} \boldsymbol{Q}_1 \\ \boldsymbol{Q}_2 \\ \vdots \\ \boldsymbol{Q}_s \end{bmatrix} \in \mathbb{R}^{sd}, \qquad \boldsymbol{F} = \begin{bmatrix} \boldsymbol{F}_1 \\ \boldsymbol{F}_2 \\ \vdots \\ \boldsymbol{F}_s \end{bmatrix} \in \mathbb{R}^{sd},$$

etc. The linearization of a RK method (6.10)–(6.13) is now implicitly defined by the system

$$d\boldsymbol{q}^{n+1} = d\boldsymbol{q}^n + \Delta t(\boldsymbol{b}^T \otimes \boldsymbol{I})d\boldsymbol{F}, \tag{6.15}$$

$$d\boldsymbol{p}^{n+1} = d\boldsymbol{p}^n + \Delta t(\boldsymbol{b}^T \otimes \boldsymbol{I})d\boldsymbol{G}, \tag{6.16}$$

$$d\boldsymbol{Q} = \boldsymbol{e} \otimes d\boldsymbol{q}^n + \Delta t(\boldsymbol{A} \otimes \boldsymbol{I})d\boldsymbol{F}, \tag{6.17}$$

$$d\boldsymbol{P} = \boldsymbol{e} \otimes d\boldsymbol{p}^n + \Delta t(\boldsymbol{A} \otimes \boldsymbol{I})d\boldsymbol{G}. \tag{6.18}$$

where $dF = F_Q dQ + F_P dP$ etc. It is important to keep in mind that, because we started from a Hamiltonian system

$$F_Q = -G_P^T, \quad F_P = F_P^T, \quad G_Q = G_Q^T. \tag{6.19}$$

The first two equations (6.15)–(6.16) are combined to give

$$\begin{aligned}\frac{dq^{n+1}\wedge dp^{n+1} - dq^n \wedge dp^n}{\Delta t} &= dq^n \wedge (b^T \otimes I)dG - dp^n \wedge (b^T \otimes I)dF + \\ &\quad \Delta t(b^T \otimes I)dF \wedge (b^T \otimes I)dG \\ &= dq^n \wedge (b^T \otimes I)dG - dp^n \wedge (b^T \otimes I)dF + \\ &\quad \Delta t dF \wedge (bb^T \otimes I)dG.\end{aligned}$$

Next we introduce the diagonal matrix $B \in \mathbb{R}^{s\times s}$ via $Be = b$. Hence we obtain from (6.17)–(6.18) the identities

$$\begin{aligned}dQ \wedge (B\otimes I)dG &= e \otimes dq^n \wedge (B\otimes I)dG + \Delta t(A\otimes I)dF \wedge (B\otimes I)dG \\ &= dq^n \wedge (b^T \otimes I)dG + \Delta t dF \wedge (A^T B \otimes I)dG,\end{aligned}$$

and

$$\begin{aligned}dP \wedge (B\otimes I)dF &= e \otimes dp^n \wedge (B\otimes I)dF + \Delta t(A\otimes I)dG \wedge (B\otimes I)dF \\ &= dp^n \wedge (b^T \otimes I)dF - \Delta t dF \wedge (B^T A \otimes I)dG.\end{aligned}$$

These three equations, obtained so far, can be combined into a single equation

$$\begin{aligned}\frac{dq^{n+1}\wedge dp^{n+1} - dq^n \wedge dp^n}{\Delta t} &= \Delta t dF \wedge \left(\left\{bb^T - BA - A^T B\right\} \otimes I\right) dG + \\ &\quad + dQ \wedge (B\otimes I)dG - dP \wedge (B\otimes I)dF.\end{aligned}$$

The symplecticity condition (6.14) is equivalent to

$$BA + A^T B - bb^T = 0.$$

Furthermore, since (6.19)

$$\begin{aligned}dQ \wedge (B\otimes I)dG - dP \wedge (B\otimes I)dF &= dQ \wedge (B\otimes I)G_P dP - \\ &\quad - dP \wedge (B\otimes I)F_Q dQ \\ &= -dP \wedge (B\otimes I)\left\{F_Q + G_P^T\right\} dQ \\ &= 0.\end{aligned}$$

Table 6.4 Coefficients for the two-stage Gauss–Legendre method of orders 4 and 6

4th order/2 stage			
$c_1 = \frac{1}{2} - \frac{\sqrt{3}}{6}$	$a_{11} = \frac{1}{4}$	$a_{12} = \frac{1}{4} - \frac{\sqrt{3}}{6}$	
$c_2 = \frac{1}{2} + \frac{\sqrt{3}}{6}$	$a_{21} = \frac{1}{4} + \frac{\sqrt{3}}{6}$	$a_{22} = \frac{1}{4}$	
	$b_1 = \frac{1}{2}$	$b_2 = \frac{1}{2}$	
6th order/3 stage			
$c_1 = \frac{1}{2} - \frac{\sqrt{15}}{10}$	$a_{11} = \frac{5}{36}$	$a_{12} = \frac{2}{9} - \frac{\sqrt{15}}{15}$	$a_{13} = \frac{5}{36} - \frac{\sqrt{15}}{30}$
$c_2 = \frac{1}{2}$	$a_{21} = \frac{5}{36} + \frac{\sqrt{15}}{24}$	$a_{22} = \frac{2}{9}$	$a_{23} = \frac{5}{36} - \frac{\sqrt{15}}{24}$
$c_3 = \frac{1}{2} + \frac{\sqrt{15}}{10}$	$a_{31} = \frac{5}{36} + \frac{\sqrt{15}}{30}$	$a_{32} = \frac{2}{9} + \frac{\sqrt{15}}{15}$	$a_{33} = \frac{5}{36}$
	$b_1 = \frac{5}{18}$	$b_2 = \frac{4}{9}$	$b_3 = \frac{5}{18}$

Thus

$$d\boldsymbol{q}^{n+1} \wedge d\boldsymbol{p}^{n+1} = d\boldsymbol{q}^n \wedge d\boldsymbol{p}^n,$$

as desired. □

Upon setting $i = j$ in (6.14), we obtain $2a_{ii} - b_i = 0$ for $b_i \neq 0$, which implies that $a_{ii} \neq 0$. Thus we obtain the following important restriction on symplectic RK methods:

Lemma 1 *Symplectic Runge–Kutta methods are necessarily implicit, i.e.,* $a_{ij} \neq 0$ *for some* $i, j \in \{1, \ldots, s\}$, $j \geq i$. □

Among implicit RK methods, the Gauss–Legendre methods can be shown to satisfy (6.14) [39]. For example, the two-stage Gauss–Legendre method given in Table 6.4 is fourth order and symplectic. The additional coefficients c_i are needed when solving time-dependent Hamiltonian problems. Specifically, each stage variable $(\boldsymbol{Q}_i, \boldsymbol{P}_i)$ is an approximation to the exact solution at time $t_i = t_n + c_i\Delta t$.

More generally, Gauss–Legrendre methods with $s \geq 1$ stages are of order $p = 2s$. This is the optimal order obtainable for a given number of stages among all possible symplectic collocation Runge–Kutta methods.

Implicit RK methods require the solution of the nonlinear equations (6.10)–(6.11). In general, these nonlinear equations can be solved by Newton's method, but this task requires the computation of the Jacobian and its inverse. This can be avoided provided the stepsize Δt is small enough. In this case, the following fixed-point iteration can be used instead of Newton's method:

FIXED-POINT ITERATION FOR IMPLICIT RK METHODS

For $l = 1, \ldots,$ till convergence

$$\boldsymbol{Q}_i^l = \boldsymbol{q}^n + \Delta t \sum_{j=1}^{s} a_{ij} \boldsymbol{F}_j^{l-1}, \quad i = 1, \ldots, s, \tag{6.20}$$

$$\boldsymbol{P}_i^l = \boldsymbol{p}^n + \Delta t \sum_{j=1}^{s} a_{ij} \boldsymbol{G}_j^{l-1}, \quad i = 1, \ldots, s, \tag{6.21}$$

with

$$\boldsymbol{F}_j^{l-1} = +\nabla_{\boldsymbol{p}} H(\boldsymbol{Q}_j^{l-1}, \boldsymbol{P}_j^{l-1}),$$
$$\boldsymbol{G}_j^{l-1} = -\nabla_{\boldsymbol{q}} H(\boldsymbol{Q}_j^{l-1}, \boldsymbol{P}_j^{l-1})$$

and initial values $\boldsymbol{Q}_j^0 = \boldsymbol{q}^n$, $\boldsymbol{P}_j^0 = \boldsymbol{p}^n$.

It is worth pointing out that the condition (6.14) also implies that the corresponding Runge–Kutta method exactly preserves any quadratic first integral of the form

$$I = \boldsymbol{z}^T \boldsymbol{C} \boldsymbol{z} + \boldsymbol{d}^T \boldsymbol{z}, \quad \boldsymbol{z} = (\boldsymbol{q}, \boldsymbol{p})^T,$$

$\boldsymbol{C} \in \mathbb{R}^{2d \times 2d}$ a symmetric matrix and $\boldsymbol{d} \in \mathbb{R}^{2d}$ a vector. For example, the Gauss–Legendre methods exactly preserve total linear and angular momentum for Hamiltonian systems with pairwise distance-dependent interactions. They also conserve the Hamiltonian of linear systems since the Hamiltonian H is quadratic in $\boldsymbol{q}$ and $\boldsymbol{p}$. The proof of this result has been given by COOPER [46] and is analogous to a result given below for integrals of partitioned Runge–Kutta methods.

6.3.2 Partitioned Runge–Kutta methods

Runge–Kutta methods were designed for general differential equations without assuming that the phase space variable $\boldsymbol{z}$ can be partitioned into two sets of variables $\boldsymbol{q}$ and $\boldsymbol{p}$ as is the case for canonical Hamiltonian systems (6.1)–(6.2). Examining (6.10)–(6.13), it seems natural to use a different set of parameters $\{a_{ij}\}$ and $\{b_i\}$ for each variable $\boldsymbol{q}$ and $\boldsymbol{p}$. Let us denote these parameter sets by $\{\widehat{a}_{ij}\}$, $\{\widehat{b}_i\}$ and $\{\widetilde{a}_{ij}\}$, $\{\widetilde{b}_i\}$, respectively. This leads us to the class of partitioned Runge–Kutta (PRK) methods (already introduced in Section 2.5)

$$\boldsymbol{Q}_i = \boldsymbol{q}^n + \Delta t \sum_{j=1}^{s} \widehat{a}_{ij} \boldsymbol{F}_j, \qquad i = 1, \ldots, s, \tag{6.22}$$

$$\boldsymbol{P}_i = \boldsymbol{p}^n + \Delta t \sum_{j=1}^{s} \widetilde{a}_{ij} \boldsymbol{G}_j, \qquad i = 1, \ldots, s, \tag{6.23}$$

$$\boldsymbol{q}^{n+1} = \boldsymbol{q}^n + \Delta t \sum_{i=1}^{s} \widehat{b}_i \boldsymbol{F}_i, \tag{6.24}$$

$$\boldsymbol{p}^{n+1} = \boldsymbol{p}^n + \Delta t \sum_{i=1}^{s} \widetilde{b}_i \boldsymbol{G}_i, \tag{6.25}$$

with s again the number of stages and $\boldsymbol{F}_i$ and $\boldsymbol{G}_i$ defined as before for RK methods. Thus we now have $2s(s+1)$ free parameters to obtain a certain order and to guarantee symplecticness. Order conditions for PRK methods are derived in the same way as for implicit RK methods and we again refer the interested reader to the textbooks by HAIRER, NORSETT, AND WANNER [82] and HAIRER AND WANNER [84].

The conditions for a PRK method to be symplectic are:

CONDITION OF SYMPLECTICNESS FOR PARTITIONED RK METHODS

$$\widehat{b}_i \widetilde{a}_{ij} + \widetilde{b}_j \widehat{a}_{ji} - \widehat{b}_i \widetilde{b}_j = 0, \qquad i, j = 1, \ldots, s, \tag{6.26}$$

$$\widehat{b}_i - \widetilde{b}_i = 0, \qquad i = 1, \ldots, s. \tag{6.27}$$

The proof of this condition is very similar to the proof given in the previous section for Runge–Kutta methods, and is omitted.

As first shown by SUN [184], proper combination of Lobatto IIIA and Lobatto IIIB Runge–Kutta methods gives rise to symplectic PRK methods. It should be noted that neither of the two classes of RK methods are symplectic, in general, when considered individually! An example of symplectic PRK method, that is based on a Lobatto IIIA–IIIB pair, is the second-order generalized leapfrog scheme.

GENERALIZED LEAPFROG METHOD

$$\boldsymbol{p}^{n+1/2} = \boldsymbol{p}^n - \frac{\Delta t}{2} \nabla_{\boldsymbol{q}} H(\boldsymbol{p}^{n+1/2}, \boldsymbol{q}^n), \tag{6.28}$$

$$\boldsymbol{q}^{n+1} = \boldsymbol{q}^n + \frac{\Delta t}{2} \left[\nabla_{\boldsymbol{p}} H(\boldsymbol{p}^{n+1/2}, \boldsymbol{q}^n) + \nabla_{\boldsymbol{p}} H(\boldsymbol{p}^{n+1/2}, \boldsymbol{q}^{n+1}) \right], \tag{6.29}$$

$$\boldsymbol{p}^{n+1} = \boldsymbol{p}^{n+1/2} - \frac{\Delta t}{2} \nabla_{\boldsymbol{q}} H(\boldsymbol{p}^{n+1/2}, \boldsymbol{q}^{n+1}). \tag{6.30}$$

Table 6.5 Coefficients for the three-stage Lobatto IIIA–IIIB pair of order 4

$\widehat{c}_1 = 0$	$\widehat{a}_{11} = 0$	$\widehat{a}_{12} = 0$	$\widehat{a}_{13} = 0$
$\widehat{c}_2 = 1/2$	$\widehat{a}_{21} = 5/24$	$\widehat{a}_{22} = 1/3$	$\widehat{a}_{23} = -1/24$
$\widehat{c}_3 = 1$	$\widehat{a}_{31} = 1/6$	$\widehat{a}_{32} = 2/3$	$\widehat{a}_{33} = 1/6$
	$\widehat{b}_1 = 1/6$	$\widehat{b}_2 = 2/3$	$\widehat{b}_3 = 1/6$
$\widetilde{c}_1 = 0$	$\widetilde{a}_{11} = 1/6$	$\widetilde{a}_{12} = -1/6$	$\widetilde{a}_{13} = 0$
$\widetilde{c}_2 = 1/2$	$\widetilde{a}_{21} = 1/6$	$\widetilde{a}_{22} = 1/3$	$\widetilde{a}_{23} = 0$
$\widetilde{c}_3 = 1$	$\widetilde{a}_{31} = 1/6$	$\widetilde{a}_{32} = 5/6$	$\widetilde{a}_{13} = 0$
	$\widetilde{b}_1 = 1/6$	$\widetilde{b}_2 = 2/3$	$\widetilde{b}_3 = 1/6$

In fact, the generalized leapfrog method is the combination of the trapezoidal rule (Lobatto IIIA) for the $\boldsymbol{q}$ variable with a variant of the midpoint rule (Lobatto IIIB) for the $\boldsymbol{p}$ variable. The generalized leapfrog method is ideal for low-order integration of non-separable Hamiltonian problems and can be used as a basic method for higher-order symmetric compositions methods as described in Sections 6.2.2 and 6.2.3.

In Table 6.5, we give the coefficients for a three-stage fourth-order Lobatto IIIA–IIIB PRK method.

Higher-order symplectic PRK methods are implicit when applied to general Hamiltonian systems of type (6.1)–(6.2) and can be solved by a fixed-point iteration similar to the one discussed for implicit RK methods, or by Newton iteration.

The situation changes for systems with a *separable Hamiltonian*,

$$H(\boldsymbol{q}, \boldsymbol{p}) = T(\boldsymbol{p}) + V(\boldsymbol{q}).$$

While symplectic RK methods are still necessarily implicit, explicit PRK methods can be found. However, this class of explicit PRK methods is equivalent to the class of composition methods and so we do not discuss them here in any further detail [148, 172].

We close this section with a proof of the fact that the symplecticness conditions (6.26)–(6.27) also imply exact preservation of any first integral of the form

$$I = \boldsymbol{q}^T W \boldsymbol{p} + \boldsymbol{d}_1^T \boldsymbol{q} + \boldsymbol{d}_2^T \boldsymbol{p}, \tag{6.31}$$

$W \in \mathbb{R}^{d \times d}$ a constant matrix, $\boldsymbol{d}_1 \in \mathbb{R}^d$ and $\boldsymbol{d}_2 \in \mathbb{R}^d$ two vectors. Recall from Section 5.3 that first integrals of this type arise from the symmetry of H under a linear canonical point transformation. For example, symplectic PRK methods exactly preserve total linear and angular momentum for systems with pairwise distance-dependent interactions.

Using the same tensor product notation and the abbreviations introduced in the previous section, a partitioned RK method (6.22)–(6.25) can be rewritten as

$$q^{n+1} = q^n + \Delta t(\widehat{b}^T \otimes I)F, \tag{6.32}$$
$$p^{n+1} = p^n + \Delta t(\widetilde{b}^T \otimes I)G, \tag{6.33}$$
$$Q = e \otimes q^n + \Delta t(\widehat{A} \otimes I)F, \tag{6.34}$$
$$P = e \otimes p^n + \Delta t(\widetilde{A} \otimes I)G, \tag{6.35}$$

where $\widehat{A} = \{\widehat{a}_{ij}\}$, $\widetilde{A} = \{\widetilde{a}_{ij}\}$, $\widehat{b} = \{\widehat{b}_i\}$, and $\widetilde{b} = \{\widetilde{b}_i\}$ are the coefficient matrices, vectors respectively, of the PRK method. The conditions of symplecticness (6.26)–(6.27) are equivalent to

$$\widehat{B}\widetilde{A} + \widehat{A}^T\widetilde{B} - \widehat{b}\widetilde{b}^T = 0, \qquad \widehat{b} = \widetilde{b}.$$

In the sequel we will assume $b = \widehat{b} = \widetilde{b}$ and denote the associated diagonal $s \times s$ matrix by B.

Let us assume that the given Hamiltonian ODE has a first integral of the from (6.31) with, for simplicity, $d_1 = d_2 = 0$. We now check conservation of such first integrals under timestepping by symplectic PRK methods.

The first two equations (6.32)–(6.33) can be combined to yield

$$\frac{(q^{n+1})^T W p^{n+1} - (q^n)^T W p^n}{\Delta t} = (q^n)^T (b^T \otimes W)G + F^T(b \otimes W)p^n$$
$$+ \Delta t F^T (bb^T \otimes W)G.$$

Similarly, equations (6.34)–(6.35) give rise to

$$Q^T(B \otimes W)G = (q^n)^T(b \otimes W)G + \Delta t F^T(\widehat{A}^T B \otimes W)G,$$

and

$$F^T(B \otimes W)P = F^T(b \otimes W)p^n + \Delta t F^T(B\widetilde{A} \otimes W)G.$$

These formulas may be combined to yield

$$\frac{(q^{n+1})^T W p^{n+1} - (q^n)^T W p^n}{\Delta t} = \Delta t F^T(\left\{bb^T - B\widetilde{A} - \widehat{A}^T B\right\} \otimes W)G$$
$$+ Q^T(B \otimes W)G + F^T(B \otimes W)P.$$

Note that $dI/dt = 0$ implies that

$$Q^T(B \otimes W)G + F^T(B \otimes W)P = 0.$$

Hence symplectic PRK methods preserve first integrals arising from an invariance of the Hamiltonian H under linear point transformations. The same statement is of course true for symplectic classical RK methods where $\widehat{A} = \widetilde{A}$.

6.4 Generating functions

Composition methods offer a systematic way to the construction of symplectic maps; they rely on a splitting of the Hamiltonian into explicitly solvable Hamiltonian problems and the group property of symplectic maps under composition. In contrast, the symplectic RK methods appear more as a pure strike of good luck. However, symplectic RK methods and other available symplectic methods can be derived from a *generating function* [7, 73] that automatically guarantees the symplecticness of the associated method. We will give a brief summary of generating function methods in this section.

Generating functions are scalar valued functions S that can be used to derive symplectic maps from $(\boldsymbol{q}, \boldsymbol{p})$ to $(\bar{\boldsymbol{q}}, \bar{\boldsymbol{p}})$ in a systematic manner. There exists several types of generating funtions. We focus here on the following two: (i) $S_1(\boldsymbol{q}, \bar{\boldsymbol{p}})$ and (ii) $S_2(\boldsymbol{q}, \bar{\boldsymbol{q}})$. The associated symplectic map $\boldsymbol{\psi} : (\boldsymbol{q}, \boldsymbol{p}) \to (\bar{\boldsymbol{q}}, \bar{\boldsymbol{p}})$ is defined implicitly by either

$$\boldsymbol{p} = \nabla_{\boldsymbol{q}} S_1(\boldsymbol{q}, \bar{\boldsymbol{p}}), \quad \bar{\boldsymbol{q}} = \nabla_{\bar{\boldsymbol{p}}} S_1(\boldsymbol{q}, \bar{\boldsymbol{p}}), \tag{6.36}$$

or

$$\boldsymbol{p} = -\nabla_{\boldsymbol{q}} S_2(\boldsymbol{q}, \bar{\boldsymbol{q}}), \quad \bar{\boldsymbol{p}} = \nabla_{\bar{\boldsymbol{q}}} S_2(\boldsymbol{q}, \bar{\boldsymbol{q}}), \tag{6.37}$$

respectively. The symplecticness of $\boldsymbol{\psi}$ is easily verified (see the Exercises).

Let us focus first on generating functions of type S_1. The identity map corresponds to $S_1 = \boldsymbol{q}^T \bar{\boldsymbol{p}}$. One can now unfold the identity by going to a parameter dependent family of generating functions:

$$S_1(\boldsymbol{q}, \bar{\boldsymbol{p}}(\tau), \tau) = \boldsymbol{q}^T \bar{\boldsymbol{p}}(\tau) + \tau \Delta S_1(\boldsymbol{q}, \bar{\boldsymbol{p}}(\tau), \tau). \tag{6.38}$$

According to (6.36), this yields an associated family of symplectic maps,

$$\begin{bmatrix} \bar{\boldsymbol{q}}(\tau) \\ \bar{\boldsymbol{p}}(\tau) \end{bmatrix} = \boldsymbol{\psi}_\tau \left(\begin{bmatrix} \boldsymbol{q} \\ \boldsymbol{p} \end{bmatrix} \right).$$

A simple example of (6.38) is provided by

$$S_1(\boldsymbol{q}, \bar{\boldsymbol{p}}(\tau), \tau) = \boldsymbol{q}^T \bar{\boldsymbol{p}}(\tau) + \tau H(\boldsymbol{q}, \bar{\boldsymbol{p}}(\tau)),$$

and (6.36) results in

$$\boldsymbol{p} = \bar{\boldsymbol{p}}(\tau) + \tau \nabla_{\boldsymbol{q}} H(\boldsymbol{q}, \bar{\boldsymbol{p}}(\tau)), \quad \bar{\boldsymbol{q}}(\tau) = \boldsymbol{q} + \tau \nabla_{\bar{\boldsymbol{p}}} H(\boldsymbol{q}, \bar{\boldsymbol{p}}(\tau)).$$

If we set $\tau = \Delta t$ and identify $(\boldsymbol{q}, \boldsymbol{p})$ with $(\boldsymbol{q}^n, \boldsymbol{p}^n)$ and $(\bar{\boldsymbol{q}}, \bar{\boldsymbol{p}})$ with $(\boldsymbol{q}^{n+1}, \boldsymbol{p}^{n+1})$, respectively, the symplectic Euler-B method emerges.

Perhaps more importantly, a fundamental result in classical mechanics states that there is a generating function S_1 of type (6.38) such that the associated transformation $\boldsymbol{\psi}_\tau$ is equivalent to the exact flow map $\boldsymbol{\Phi}_{\tau,H}$ of a canonical Hamiltonian system with Hamiltonian H. The generating function S_1 satisfies the Hamilton–Jacobi equation [7, 8]

$$\partial_\tau S_1(\boldsymbol{q}, \bar{\boldsymbol{p}}, \tau) - H(\nabla_{\bar{\boldsymbol{p}}} S_1(\boldsymbol{q}, \bar{\boldsymbol{p}}, \tau), \bar{\boldsymbol{p}}) = 0. \tag{6.39}$$

Hence one can try to obtain symplectic methods by approximately solving the Hamilton–Jacobi equation (6.39) near the identity transformation and to use the approximative S_1 to define a symplectic integration method via (6.36). Indeed, this approach was followed by Feng [58] and Channel and Scovel [43, 44] and inspired early investigations of symplectic methods. Later it was shown by Lasagni [106, 107] and Suris [186] that symplectic (partitioned) Runge–Kutta methods can be derived from generating functions of both types S_1 and S_2, respectively. These results are important since they guarantee the existence of globally defined modified Hamiltonian functions in backward error analysis [16, 80] (see the Exercises).

Generating functions of type S_2 are also very useful because of their close connection to Largange's equation (recall Section 3.2) and the Lagrangian variational principle [7, 73]. We illustrate the basic idea by a simple example. Take the Lagrangian function

$$L(q, \dot{q}) = \frac{1}{2}\dot{q}^2 - V(q)$$

of a single degree-of-freedom particle system with mass equal to one and a potential energy function $V(q)$. One can replace the time derivative in L by a discrete approximation and use this approximation as a generating function of type S_2, e.g.

$$S_2(q, \bar{q}; \tau) = \tau L\left(q, \frac{\bar{q} - q}{\tau}\right) = \frac{\tau}{2}\left(\frac{\bar{q} - q}{\tau}\right)^2 - \tau V(q).$$

The associated symplectic transformation is given by

$$p = -\frac{\partial S_2}{\partial q} = -\frac{q - \bar{q}}{\tau} + \tau V'(q), \qquad \bar{p} = \frac{\partial S_2}{\partial \bar{q}} = \frac{\bar{q} - q}{\tau}.$$

After sorting variables, we obtain the explicit expression

$$\bar{p} = p - \tau V'(q), \qquad \bar{q} = q + \tau \bar{p},$$

which turns again into the symplectic Euler-B method with $(q, p) = (q^n, p^n)$, $(\bar{q}, \bar{p}) = (q^{n+1}, p^{n+1})$, and $\tau = \Delta t$. See the survey article by Marsden and West [125] for a detailed discussion of this particular generating function approach to symplectic integration and the closely related discrete variational principle. See also the Exercises.

6.5 Numerical experiments

6.5.1 Arenstorf orbits

Let us consider the gravitational system consisting of a couple of massive bodies and one light one moving in the (x, y)-plane. We assume that the orbit of the two heavy bodies is essentially circular, and that the mass of the light body is so small that we can neglect the influence of the light body (for example, asteroid) on the motion of the two heavy ones. It is convenient to pass to a moving reference frame which rotates with the angular velocity of the planetary system around the center of mass of the two heavy bodies. We also choose units of length and time and mass so that the angular velocity of rotation, the sum of the masses of the heavy bodies, and the gravitational constant are all equal to one. The equations of motion for the light body are then given by

$$\ddot{x} = +2\dot{y} - V_x(x, y), \tag{6.40}$$

$$\ddot{y} = -2\dot{x} - V_y(x, y), \tag{6.41}$$

where the potential V is given by

$$V(x, y) = -\frac{x^2 + y^2}{2} - \frac{\mu_1}{r_1} - \frac{\mu_2}{r_2}.$$

Here μ_1 is the mass of the first (lighter) heavy body, $\mu_2 = 1 - \mu_1$ is the mass of the second heavy body, and

$$r_1 = \sqrt{(x - \mu_2)^2 + y^2}, \quad r_2 = \sqrt{(x + \mu_1)^2 + y^2}$$

are the distances from the light body to the two heavy bodies which (in the moving coordinate frame) are fixed at points $(\mu_2, 0)$ and $(-\mu_1, 0)$. This problem is also called the *restricted three-body problem*. See [8] for further details.

The restricted three-body problem possesses interesting periodic solutions often called *Arenstorf orbits*. For example, $\mu_1 = 0.012277471$ and initial data

$$x = 0.994, \quad \dot{x} = 0,$$
$$y = 0, \quad \dot{y} = -2.001585106379082$$

lead to a periodic orbit with period

$$T = 17.0652165601579625\text{5}.$$

The numerical computation of this orbit faces difficulties because of the two singularities in V at $(x, y) = (\mu_2, 0)$ and $(x, y) = (-\mu_1, 0)$. Note that the periodic orbit gets very close to the first singularity at $t = k \cdot T$, $k = 0, \pm 1, \pm 2, \ldots$, which makes this a challenging problem for numerical computations. See Fig. 6.1 for a

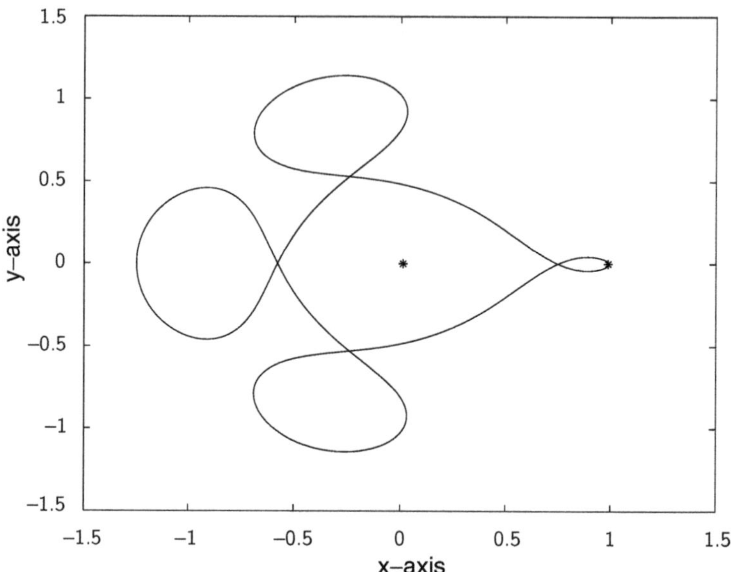

Figure 6.1 Arenstorf orbit of the restricted three-body problem. The two (fixed) heavy bodies are marked by $*$.

graphical presentation and [82] for the numerical computation of the Arenstorf orbit using "standard" numerical methods.

Let us discuss the Hamiltonian structure of the restricted three-body problem. In fact, the equations (6.40)–(6.41) can be brought into the form of the equations of motion for a particle in a magnetic field. We define $\boldsymbol{q} = (x, y, 0)^T$ and $\boldsymbol{p} = (\dot{x}, \dot{y}, 0)^T$ and obtain the Hamiltonian

$$H = \frac{1}{2}||\boldsymbol{p}||^2 + V(\boldsymbol{q}), \quad V(\boldsymbol{q}) = V(x, y).$$

The restricted three-body equations of motion are then equivalent to

$$\frac{d}{dt}\boldsymbol{q} = \boldsymbol{p},$$
$$\frac{d}{dt}\boldsymbol{p} = \boldsymbol{b} \times \boldsymbol{p} - \nabla_q V(\boldsymbol{q}),$$

with $\boldsymbol{b} = (0, 0, -2)^T$. Both Scovel's method from Section 4.5.2 and the implicit midpoint method can now be used to obtain symplectic integration methods.

We apply the higher-order composition methods from the previous subsections to the Arenstorf orbit of the restricted three-body problem. Scovel's method is used as the basic second-order integrator, i.e.

$$\boldsymbol{\Psi}_{w\Delta t} = \boldsymbol{\Phi}_{w\Delta t/2,V} \circ \boldsymbol{\Phi}_{w\Delta t,T} \circ \boldsymbol{\Phi}_{w\Delta t/2,V}.$$

The equations of motion are integrated over one period T and the error in the coordinates $(x(t), y(t))$ of the light body at $t = T$, i.e.

$$e(T) = \max(x(T) - 0.994, y(T)),$$

is recorded as a function of the necessary number of force evaluations per unit time interval. See Fig. 6.2 for a comparison of the various methods. We find that higher accuracy of a method pays off in terms of efficiency over a fixed time interval (here one period). In particular, both the sixth-order method of Table 6.1 and the eighth-order method of Table 6.2 work very well. We mention that this problem offers significant scope for adaptive integration. We will come back to this issue in Chapter 9.

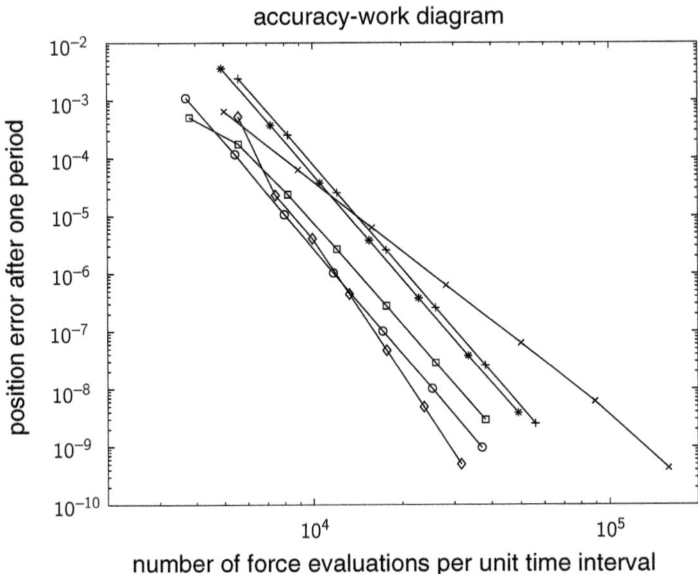

Figure 6.2 Accuracy versus efficiency comparison of fourth-order (x) and sixth-order (o) composition methods of Table 6.1, the sixth-order 7-stages (+) and 9-stages (∗) composition methods of Table 6.2, the sixth-order (□) post-processed method of Table 6.3, and the eighth-order (◇) composition method of Table 6.2 using the Arenstorf orbit of the restricted three-body problem as a numerical test case.

6.5.2 Solar system

To compare the relative efficiencies of the different methods a model of the solar system was constructed as follows. The model used was based on ten bodies: Mercury, Venus, Earth, Mars, Jupiter, Saturn, Uranus, Neptune, Pluto, and the

Sun with the entire system rotating around its center of gravity, and uses an all-body potential. The integrator ignores all but gravitational forces.

The Hamiltonian equation, where m_i is the mass of the ith planet and G is the gravitational constant is

$$H(q,p) = \frac{1}{2}\sum_{i=1}^{10}\frac{||p_i||^2}{m_i} - \sum_{i=1}^{9}\sum_{j=i+1}^{10}\frac{Gm_im_j}{||q_i - q_j||}.$$

Then the differential equations are

$$\dot{q}_i = \frac{p_i}{m_i}, \quad \dot{p}_i = \sum_{j=1,j\neq i}^{10}\frac{Gm_im_j(q_i - q_j)}{||q_i - q_j||^3}, \quad i = 1, 2, \ldots, N.$$

This model treats the Sun as one of the bodies and the entire system orbits around the centre of mass of the system. Initial data were taken from the DE118 18-digit state vectors provided by the NASA Jet Propulsion Laboratory, and represents the state of the Solar System at Julian date 2440400.50 (00:00 on June 28, 1969).

All of the composition and post-processing schemes were based on the second-order Störmer–Verlet method. The methods compared were as follows:

label	order	method type	table	entry
2nd	2	Verlet		
4thM	4	Composition	6.2	1
4thRK	4	Runge–Kutta	6.4	1
6thM	6	Composition	6.2	3
6thPP	6	Post-processing	6.3	1
6thRK	6	Runge–Kutta	6.4	2
6thY	6	Composition	6.2	2
8thM	8	Composition	6.2	4

The methods have been compared by plotting the maximum energy error over a ten year simulation against the number of force evaluations done, as shown in Fig. 6.3. From the graph it is clear that the high-order explicit methods dominate for small maximum errors, with the post-processing and eighth-order methods providing the greatest efficiency. The implicit Runge–Kutta methods fare less well in the comparison.

In this model each step of the Runge–Kutta method was resolved to a fixed arbitrary value which was small enough so that there was no influence on the maximum energy error. This approach gives rise to the odd curves in the work–energy diagrams at large stepsize, as, at the large stepsizes corresponding to the larger maximum energy error, the simplified nonlinear solver becomes inefficient and increasing numbers of iterations are needed to resolve the solution to the specified accuracy.

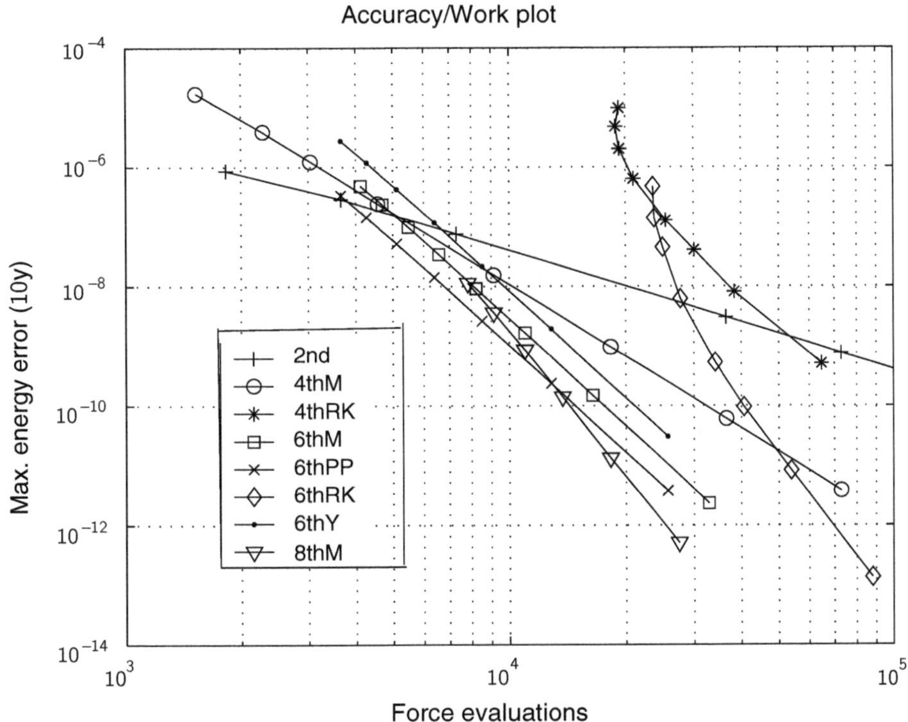

Figure 6.3 Comparison of method efficiency for the solar system model.

Longer simulations demonstrate the procession in Pluto's orbit; this can be seen in Fig. 6.4 which represents the trajectories in the Solar System over one million years.

6.6 Exercises

1. *Numerical comparison for Arenstorf orbit.* Implement Scovel's second-order splitting method (compare Section 4.5.2) for the restricted three-body problem of Section 6.5.1. Using Scovel's method, also implement the higher-order composition methods as described in Sections 6.2.2–6.2.3. Reproduce the data shown in Fig. 6.2. Note that the work per unit time interval is defined as the number of timesteps taken to approximate the solution over one period

Figure 6.4 One million year simulation of the solar system. The cloud of points illustrates the precession in the orbit of Pluto.

T multiplied by the number of stages of the method and divided by the period T. The accuracy is defined in terms of the absolute error in the computed positions after one period, i.e.

$$e(T) = \max(x(T) - 0.994, y(T)).$$

2. *Higher-order composition methods.* Let $\boldsymbol{\Psi}_{\Delta t}$ be a symmetric and symplectic method of order two (for example the Störmer–Verlet method). Show that the composed method

$$\hat{\boldsymbol{\Psi}}_{\Delta t} = \boldsymbol{\Psi}_{w_1 \Delta t} \circ \boldsymbol{\Psi}_{w_2 \Delta t} \circ \boldsymbol{\Psi}_{w_1 \Delta t}$$

is fourth order if

$$2w_1 + w_2 = 1, \qquad 2w_1^3 + w_2^3 = 0.$$

Hint: Since $\boldsymbol{\Psi}_{\Delta t}$ is symplectic, there exists a modified Hamiltonian $\tilde{H}(\Delta t)$ whose corresponding time-Δt-flow map is "identical" to $\boldsymbol{\Psi}_{\Delta t}$ (see Chapter 5). Because $\boldsymbol{\Psi}_{\Delta t}$ is also symmetric, the Taylor expansion of $\tilde{H}(\Delta t)$ contains only terms of even order in Δt (see Exercises in Chapter 5). Use the Taylor series expansion

$$\tilde{H}(\Delta t) = H + \Delta t^2 \delta H_2 + \Delta t^4 \delta H_4 + \ldots$$

to derive a modified Hamiltonian for the (symplectic and symmetric) maps $\boldsymbol{\Psi}_{w_i\Delta t}$. Finally, use the BCH formula of Section 5.4 to derive the modified Hamiltonian for the method $\hat{\boldsymbol{\Psi}}_{\Delta t}$. Show that $2w_1 + w_2 = 1$ implies second order. If also $2w_1^3 + w_2^3 = 0$, then the method is fourth order.

3. *Symplectic methods from generating functions.* Show that a map $\boldsymbol{\psi} : (\boldsymbol{q}, \boldsymbol{p}) \to (\bar{\boldsymbol{q}}, \bar{\boldsymbol{p}})$ defined by either (6.36) or (6.37) is symplectic. Find a generating function that yields the symplectic Euler-A method.

4. *Hamilton–Jacobi equation.* Given a generating function (6.38) that satisfies the Hamilton–Jacobi equation (6.39), show that the associated symplectic map (6.36) is equivalent to the flow map $\boldsymbol{\Phi}_{\tau,H}$. Hint: Differentiate

$$\boldsymbol{p} = \nabla_{\boldsymbol{q}} S_1(\boldsymbol{q}, \bar{\boldsymbol{p}}(\tau), \tau),$$

with respect to τ and compare with the gradient of (6.39) with respect to $\boldsymbol{q}$. Repeat the same calculation for

$$\bar{\boldsymbol{q}}(\tau) = \nabla_{\bar{\boldsymbol{p}}} S_1(\boldsymbol{q}, \bar{\boldsymbol{p}}(\tau), \tau).$$

Also note that

$$\begin{bmatrix} \bar{\boldsymbol{q}}(\tau) \\ \bar{\boldsymbol{p}}(\tau) \end{bmatrix} = \boldsymbol{\Phi}_{\tau,H}\left(\begin{bmatrix} \boldsymbol{q} \\ \boldsymbol{p} \end{bmatrix} \right)$$

is equivalent to

$$\frac{d}{d\tau}\bar{\boldsymbol{p}} = -\nabla_{\bar{\boldsymbol{q}}} H(\bar{\boldsymbol{q}}, \bar{\boldsymbol{p}}), \qquad \frac{d}{d\tau}\bar{\boldsymbol{q}} = \nabla_{\bar{\boldsymbol{p}}} H(\bar{\boldsymbol{q}}, \bar{\boldsymbol{p}}).$$

5. *Symplectic methods from generating functions.* Given a Lagrangian functions $L(\boldsymbol{q}, \dot{\boldsymbol{q}}) = \|\dot{\boldsymbol{q}}\|^2/2 - V(\boldsymbol{q})$. What symplectic method do you obtain from the generating function

$$S(\boldsymbol{q}, \bar{\boldsymbol{q}}, \tau) = \tau L\left(\frac{\boldsymbol{q} + \bar{\boldsymbol{q}}}{2}, \frac{\bar{\boldsymbol{q}} - \boldsymbol{q}}{\tau} \right) = \frac{\tau}{2}\|\frac{\bar{\boldsymbol{q}} - \boldsymbol{q}}{\tau}\|^2 - \tau V\left(\frac{\bar{\boldsymbol{q}} + \boldsymbol{q}}{2} \right)?$$

One can clearly apply this approach to more general Lagrangian functions $L(\boldsymbol{q}, \dot{\boldsymbol{q}})$. Take, for example, the Lagrangian function (3.17) given in Section 3.2 to describe the motion of a charged particle in a magnetic field. Discuss several options to obtain a generating function and compare the associated numerical methods with the ones used in Section 4.5.2.

6. *Modified Hamiltonian functions via generating functions.* Let S_H denote the generating function of type (6.38) for the exact flow map $\boldsymbol{\Phi}_{\tau,H}$. Show that

$$S_H(\boldsymbol{q}, \bar{\boldsymbol{p}}, \tau) = \boldsymbol{q}^T \bar{\boldsymbol{p}} + \tau H(\boldsymbol{q}, \bar{\boldsymbol{p}}) + \frac{\tau^2}{2} \nabla_{\boldsymbol{q}} H(\boldsymbol{q}, \bar{\boldsymbol{p}})^T \nabla_{\bar{\boldsymbol{p}}} H(\boldsymbol{q}, \bar{\boldsymbol{p}}) + \mathcal{O}(\tau^3),$$

via explicit Taylor expansion of (6.39) for $S_1 = S_H$.

The symplectic Euler-B method $\boldsymbol{\Psi}_{\Delta t}$ can be obtained from the generating function

$$S(\boldsymbol{q}, \bar{\boldsymbol{p}}, \Delta t) = \boldsymbol{q}^T \bar{\boldsymbol{p}} + \Delta t H(\boldsymbol{q}, \bar{\boldsymbol{p}}).$$

Show that the first-order modified Hamiltonian is given by

$$\tilde{H}_1 = H + \Delta t \delta H_1,$$

with

$$\delta \tilde{H}_1(\boldsymbol{q}, \boldsymbol{p}) = \lim_{\tau \to 0} \frac{S(\boldsymbol{q}, \boldsymbol{p}, \tau) - S_H(\boldsymbol{q}, \boldsymbol{p}, \tau)}{\tau^2}.$$

Can this approach be generalized to higher-order corrections? See also [80].

7. *Discrete variational mechanics.* Consider the Lagrangian function $L(q, \dot{q})$ and the action integral

$$\mathcal{L}[q] = \int_{t_0}^{t_1} L(q(t), \dot{q}(t))\, dt.$$

We replace the integral by a finite sum

$$\mathcal{L}[\{q^n\}] = \sum_n L\left(q^n, \frac{q^{n+1} - q^n}{\Delta t}\right) \Delta t$$

and find the local minimizer from the condition

$$\frac{\partial}{\partial q^n} \mathcal{L}[\{q^n\}] = 0.$$

What numerical scheme do you obtain by explicitly evaluating the formula for $L(q, \dot{q}) = \dot{q}^2/2 - V(q)$? The derivation is a simple example of the discrete variational principle [125].

7

Constrained mechanical systems

In this chapter, we discuss the problem of simulating a mechanical system subject to one or several constraints. This subject is rather broad, and there are many theoretical issues which arise in the most general settings concerning the formulation of the equations of motion and the properties of solutions. To simplify the discussion, we primarily restrict ourselves to the treatment of constraints which can be described by algebraic relations among the position variables of the system, i.e., defined by equations of the form

$$g_i(\boldsymbol{q}) = 0, \qquad i = 1, \ldots, m,$$

for smooth functions g_i. A mechanical system subject to such constraints is typically termed *holonomic*. Derivation of equations of motion for a holonomically constrained mechanical system is not very much more complicated than for an unconstrained system.

When it comes to numerical discretization, however, the constraints introduce a few challenges. For one thing, the propagation of errors in numerical algorithms for a constrained differential equation is more complicated than for ordinary differential equations. For another, the constraints are often an intrinsic component of the modeling of the system, and the configuration manifold (the set of points for which the constraints are satisfied) is an essential part of the extension of the concept of symplecticness. For these reasons, it seems important that the constraints are accurately resolved at each step. This is in contrast to the case for unconstrained systems, where a certain error growth must generally be tolerated.

In this chapter, we will introduce several approaches for numerical discretization of holonomically constrained systems, focusing initially on the SHAKE method due to RYCKAERT, CICCOTTI, AND BERENDSEN [167] and RATTLE [5] variant, as these have proven very popular in molecular dynamics. We then develop more general type of methods. Recently, schemes like those discussed in this chapter have seen widespread use in computer graphics and gaming applications.

7.1 *N*-body systems with holonomic constraints

In the introduction of this book, we have already encountered simple examples of constrained Hamiltonian systems (the bead on a wire, the spherical pendulum). It was shown that D'Alembert's principle enabled us to write equations of motion with respect to a single constraint $g(\boldsymbol{q}) = 0$ by introducing a constraint force that always acts in the direction of the normal to the constraint surface.

The extension of the Newtonian formulation to systems with multiple particles and multiple holonomic constraints is straightforward. Given m algebraic constraints $g_i(\boldsymbol{q}) = 0$, $i = 1, \ldots, m$, on a multiparticle system, it can be shown that the constraint forces due to each constraint act in the normal direction to the corresponding surface. Newton's Second Law then suggests the following form for the equations of motion:

N-BODY SYSTEMS WITH HOLONOMIC CONSTRAINTS

$$\frac{d}{dt}\boldsymbol{q} = \boldsymbol{v}, \tag{7.1}$$

$$\boldsymbol{M}\frac{d}{dt}\boldsymbol{v} = -\nabla_{\boldsymbol{q}} V(\boldsymbol{q}) - \sum_{i=1}^{m} \nabla_{\boldsymbol{q}} g_i(\boldsymbol{q})\lambda_i, \tag{7.2}$$

$$0 = g_i(\boldsymbol{q}), \qquad i = 1, 2, \ldots, m. \tag{7.3}$$

Here, as in previous chapters, $\boldsymbol{M}$ represents a positive definite and symmetric (typically diagonal) mass matrix. We typically impose the assumption that the gradients of the constraint functions $\nabla_{\boldsymbol{q}} g_i(\boldsymbol{q})$ form a linearly independent set.

We define the *configuration manifold* $\mathcal{M}$ as the space of all positions subject to the position constraints:

$$\mathcal{M} = \{\boldsymbol{q} \in \mathbb{R}^d \mid g_i(\boldsymbol{q}) = 0, \quad i = 1, \ldots, m\}.$$

Let $\bar{\boldsymbol{q}}$ be a point of the configuration manifold and consider the set of all smooth parameterized curves containing $\bar{\boldsymbol{q}}$ and lying in $\mathcal{M}$. Each such parameterized curve $\boldsymbol{q}(t)$ (with, say, $\boldsymbol{q}(t_0) = \bar{\boldsymbol{q}}$) has a certain *velocity vector* $\bar{\boldsymbol{v}}$ at $t = t_0$, $\bar{\boldsymbol{v}} = \dot{\boldsymbol{q}}(t_0)$. Obviously, because of $g_i(\boldsymbol{q}(t)) = 0$ for all t, we must have that

$$\frac{d}{dt} g_i(\boldsymbol{q}(t)) = \nabla_{\boldsymbol{q}} g_i(\boldsymbol{q}(t)) \cdot \dot{\boldsymbol{q}}(t) = \nabla_{\boldsymbol{q}} g_i(\boldsymbol{q}(t)) \cdot \boldsymbol{v}(t) = 0, \tag{7.4}$$

and, in particular

$$\nabla_{\boldsymbol{q}} g_i(\bar{\boldsymbol{q}}) \cdot \bar{\boldsymbol{v}} = 0. \tag{7.5}$$

The set of all possible velocity vectors at the point $\bar{\boldsymbol{q}}$ is a linear vector space

$$T_{\bar{\boldsymbol{q}}}\mathcal{M} = \{\bar{\boldsymbol{v}} \in \mathbb{R}^d \mid (\nabla_{\boldsymbol{q}} g_i(\bar{\boldsymbol{q}})) \cdot \bar{\boldsymbol{v}} = 0, \quad i = 1, \ldots, m\},$$

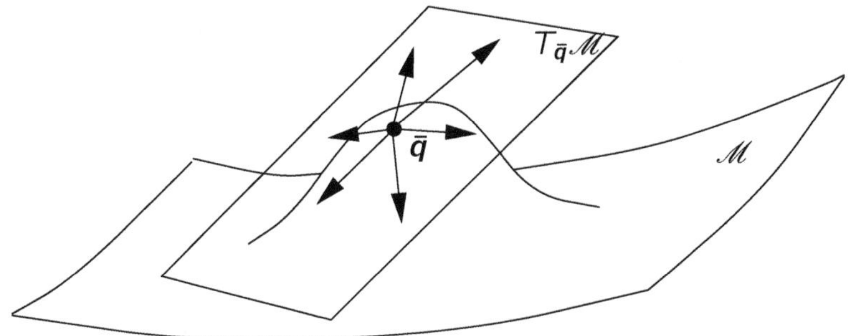

Figure 7.1 The tangent space.

called the *tangent space* at $\bar{\boldsymbol{q}}$ (Fig. 7.1). The *tangent bundle* of $\mathcal{M}$ is the space of all pairs $(\boldsymbol{q}, \boldsymbol{v})$ with $\boldsymbol{q}$ lying in $\mathcal{M}$ and $\boldsymbol{v}$ lying in $T_{\boldsymbol{q}}\mathcal{M}$. The tangent bundle is denoted $T\mathcal{M}$.

The system (7.1)–(7.3) can be written more compactly by introducing the vector function $\boldsymbol{g}(\boldsymbol{q}) = (g_1(\boldsymbol{q}), \ldots, g_m(\boldsymbol{q}))^T$, denoting its Jacobian matrix by

$$\boldsymbol{G}(\boldsymbol{q}) = \boldsymbol{g}_{\boldsymbol{q}}(\boldsymbol{q}), \quad \text{or, equivalently,} \quad \boldsymbol{G}(\boldsymbol{q})^T = \nabla_{\boldsymbol{q}}\boldsymbol{g}(\boldsymbol{q}),$$

and letting $\boldsymbol{\lambda}$ represent the m-vector of multipliers $\boldsymbol{\lambda} = (\lambda_1, \ldots, \lambda_m)^T$. Then the equations of motion become

$$\frac{d}{dt}\boldsymbol{q} = \boldsymbol{v}, \tag{7.6}$$

$$\boldsymbol{M}\frac{d}{dt}\boldsymbol{v} = -\nabla_{\boldsymbol{q}}V(\boldsymbol{q}) - \boldsymbol{G}(\boldsymbol{q})^T\boldsymbol{\lambda}, \tag{7.7}$$

$$\boldsymbol{0} = \boldsymbol{g}(\boldsymbol{q}). \tag{7.8}$$

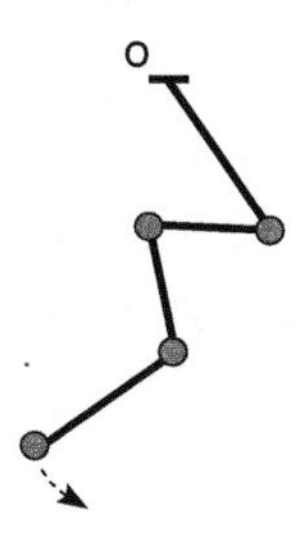

Example 1 Consider a simple *constraint chain* in the plane made up of N-point particles with masses $m_1, m_2, \ldots, m_N$, with each successive pair of particles joined by a length constraint, say with lengths $l_1, l_2, \ldots, l_{N-1}$. If, moreover, the first point is linked to the origin by a similar length constraint (say with length l_0), then we have a *multiple pendulum* (see diagram). These problems are important model problems used in biochemistry and biophysics (polymers exhibit such a chain structure) as well as in engineering.

We might assume that the system moves in some applied uniform potential energy field $V = \sum_{i=1}^{N} \phi(\boldsymbol{q}_i)$, so that the external force acting on the ith particle is

$$\boldsymbol{F}_i = -\nabla_{\boldsymbol{q}_i}\phi(\boldsymbol{q}_i).$$

Let us number the multipliers in accordance with the indexing of the constraints, then following the discussion of this section, the equations of motion are, for any of the internal nodes of the chain

$$\frac{d}{dt}\boldsymbol{q}_i = \boldsymbol{v}_i,$$

$$m_i \frac{d}{dt}\boldsymbol{v}_i = \boldsymbol{F}_i - \lambda_{i-1}(\boldsymbol{q}_i - \boldsymbol{q}_{i-1}) - \lambda_i(\boldsymbol{q}_i - \boldsymbol{q}_{i+1}),$$

whereas the first node obeys

$$\frac{d}{dt}\boldsymbol{q}_1 = \boldsymbol{v}_1,$$

$$m_1 \frac{d}{dt}\boldsymbol{v}_1 = \boldsymbol{F}_1 - \lambda_0 \boldsymbol{q}_1 - \lambda_1(\boldsymbol{q}_1 - \boldsymbol{q}_2),$$

and the last node moves according to

$$\frac{d}{dt}\boldsymbol{q}_N = \boldsymbol{v}_N,$$

$$m_N \frac{d}{dt}\boldsymbol{v}_N = \boldsymbol{F}_N - \lambda_{N-1}(\boldsymbol{q}_N - \boldsymbol{q}_{N-1}).$$

Here the constraints take the form

$$g_i(\boldsymbol{q}) = \frac{1}{2}(\|\boldsymbol{q}_i - \boldsymbol{q}_{i+1}\|^2 - l_i^2) = 0,$$

for $i = 1, \ldots, N-1$, while $g_0(\boldsymbol{q}) = \frac{1}{2}(\|\boldsymbol{q}_1\|^2 - l_0^2) = 0$. The transpose of the constraint Jacobian matrix, $\boldsymbol{G}(\boldsymbol{q})^T$, here is

$$\begin{bmatrix} \boldsymbol{q}_1 & \boldsymbol{q}_1 - \boldsymbol{q}_2 & \boldsymbol{0} & \cdots & \boldsymbol{0} & \boldsymbol{0} & \boldsymbol{0} \\ \boldsymbol{0} & \boldsymbol{q}_2 - \boldsymbol{q}_1 & \boldsymbol{q}_2 - \boldsymbol{q}_3 & \cdots & \boldsymbol{0} & \boldsymbol{0} & \boldsymbol{0} \\ \boldsymbol{0} & \boldsymbol{0} & \boldsymbol{q}_3 - \boldsymbol{q}_2 & \cdots & \boldsymbol{0} & \boldsymbol{0} & \boldsymbol{0} \\ \vdots & \vdots & \vdots & \vdots & \vdots & \vdots & \vdots \\ \boldsymbol{0} & \boldsymbol{0} & \boldsymbol{0} & \cdots & \boldsymbol{0} & \boldsymbol{q}_{N-1} - \boldsymbol{q}_{N-2} & \boldsymbol{q}_{N-1} - \boldsymbol{q}_N \\ \boldsymbol{0} & \boldsymbol{0} & \boldsymbol{0} & \cdots & \boldsymbol{0} & \boldsymbol{0} & \boldsymbol{q}_N - \boldsymbol{q}_{N-1} \end{bmatrix}.$$

□

Just as for the one-particle, one-constraint case, the multipliers λ_i in the constrained Newton equations (7.6)–(7.8) could be eliminated by differentiating the velocity constraints (7.4) one more time with respect to t, resulting, after an exercise in calculus, in a system of unconstrained differential equations

$$\frac{d}{dt}\boldsymbol{q} = \boldsymbol{v}, \tag{7.9}$$

$$\boldsymbol{M}\frac{d}{dt}\boldsymbol{v} = -\nabla_q V(\boldsymbol{q}) - \boldsymbol{G}(\boldsymbol{q})^T \boldsymbol{\Lambda}(\boldsymbol{q}, \boldsymbol{v}), \tag{7.10}$$

with $\boldsymbol{\Lambda}(\boldsymbol{q}, \boldsymbol{v})$ given by

$$\boldsymbol{\Lambda}(\boldsymbol{q}, \boldsymbol{v}) = -(\boldsymbol{G}\boldsymbol{M}^{-1}\boldsymbol{G}^T)^{-1}[\boldsymbol{G}\boldsymbol{M}^{-1}\nabla_q V(\boldsymbol{q}) + \boldsymbol{g}_{qq} < \boldsymbol{v}, \boldsymbol{v} >]. \qquad (7.11)$$

Here we have used the shorthand $\boldsymbol{G} = \boldsymbol{G}(\boldsymbol{q})$ and the symbol $\boldsymbol{g}_{qq} < \boldsymbol{v}, \boldsymbol{v} >$ is to be understood to represent a vector of m components with the ith component given by

$$(g_i)_{qq} < \boldsymbol{v}, \boldsymbol{v} >= \sum_{k,l=1}^{d} \frac{\partial^2 g_i}{\partial q_k \partial q_l} v_k v_l.$$

These equations are well defined provided the square $m \times m$ matrix $\boldsymbol{G}\boldsymbol{M}^{-1}\boldsymbol{G}^T$ is nonsingular along the configuration manifold $\mathcal{M}$, i.e., when $\boldsymbol{g}(\boldsymbol{q}) = \boldsymbol{0}$. Note that this is always the case if the matrix $\boldsymbol{G}(\boldsymbol{q})$ has full rank for $\boldsymbol{q} \in \mathcal{M}$ (has linearly independent rows) and $\boldsymbol{M}$ is positive definite (see Exercises). For prescribed initial conditions $(\boldsymbol{q}_0, \boldsymbol{v}_0) \in T\mathcal{M}$, the associated solution $(\boldsymbol{q}(t), \boldsymbol{v}(t))$ of (7.9)–(7.11) stays in $T\mathcal{M}$ and, hence, such a solution $(\boldsymbol{q}(t), \boldsymbol{v}(t))$ also satisfies (7.6)–(7.8). For that reason, one often refers to the system of differential and algebraic equations (7.6)–(7.8) as a description of an ordinary differential equation (ODE) on a manifold (here $T\mathcal{M}$) and (7.9)–(7.11) is called the underlying ODE.

7.2 Numerical methods for constraints

We now consider the development of numerical integration methods suitable for integrating a constrained mechanical system. The obvious geometric property that we would like to preserve is the constraint $\boldsymbol{g}(\boldsymbol{q}) = \boldsymbol{0}$. However, any such trajectory will also need to satisfy the tangency condition (7.5), thus we seek methods for solving (7.1)–(7.3) which preserve the tangent bundle $T\mathcal{M}$.

There are many approaches to this problem. We have already seen in previous chapters that the use of different formulations of the equations can have a profound impact on the numerical integration process. Roughly speaking, methods for constrained integration can be divided into two classes: (i) methods based on integration of some related (unconstrained) ordinary differential equation, e.g., the underlying ODE (7.9)–(7.11), and (ii) methods based on direct discretization of the constrained equations of motion (7.1)–(7.3). In this section, we consider methods of the second class. Schemes of the other type are mentioned at the end of this chapter.

7.2.1 Direct discretization: SHAKE and RATTLE

Experience indicates that the best results can generally be obtained using a direct discretization of the equations of motion. The idea in direct discretization is to apply the ideas used to in constructing approximation methods for unconstrained differential equations to the constrained equations of motion. This requires some generalization of the standard types of methods. In this section, we will introduce popular methods which generalize the Störmer–Verlet method introduced in Chapter 2, and we comment on issues such as accuracy and stability. The results of this subsection will be generalized to more general constrained Hamiltonian systems in the next section.

SHAKE discretization was proposed in 1976 by RYCKAERT, CICCOTTI, AND BERENDSEN [167]. A paper of ANDERSEN [5] later introduced a related formulation, which he called RATTLE. LEIMKUHLER AND SKEEL [113] analyzed and compared the two methods, discovering, apparently for the first time, that they were (i) equivalent to each other and (ii) symplectic. Our treatment is based on the exposition of [113].

Upon rewriting (7.6)–(7.7) as a single second-order equation in $\boldsymbol{q}$ and applying the standard leapfrog discretization (2.19) of Section 2.5, one quite naturally arrives at the SHAKE discretization

$$\boldsymbol{M}\frac{\boldsymbol{q}^{n+1} - 2\boldsymbol{q}^n + \boldsymbol{q}^{n-1}}{\Delta t^2} = -\nabla_{\boldsymbol{q}} V(\boldsymbol{q}^n) - \boldsymbol{G}(\boldsymbol{q}^n)^T \boldsymbol{\lambda}^n, \tag{7.12}$$

$$\boldsymbol{0} = \boldsymbol{g}(\boldsymbol{q}^{n+1}). \tag{7.13}$$

The method can be implemented as follows. We solve (7.12) for $\boldsymbol{q}^{n+1}$ and insert in (7.13), resulting in a system of m equations in the m unknown Lagrange multipliers $\boldsymbol{\lambda}^n = \{\lambda_i^n\}$

$$\boldsymbol{0} = \tilde{\boldsymbol{g}}(\boldsymbol{\lambda}^n) := \boldsymbol{g}(\bar{\boldsymbol{q}}^{n+1} - \Delta t^2 \boldsymbol{M}^{-1}\boldsymbol{G}(\boldsymbol{q}^n)^T \boldsymbol{\lambda}^n),$$

where

$$\bar{\boldsymbol{q}}^{n+1} := 2\boldsymbol{q}^n - \boldsymbol{q}^{n-1} - \Delta t^2 \boldsymbol{M}^{-1}\nabla_{\boldsymbol{q}} V(\boldsymbol{q}^n)$$

represents an unconstrained step with the leapfrog method. Some possibilities for the treatment of these nonlinear equations will be discussed later in this section.

The equations (7.12)–(7.13) can be recast in a position-velocity formulation by setting $\boldsymbol{v}^{n+1/2} = (\boldsymbol{q}^{n+1} - \boldsymbol{q}^n)/\Delta t$ and defining $\boldsymbol{v}^n = \frac{1}{2}(\boldsymbol{v}^{n-1/2} + \boldsymbol{v}^{n+1/2})$, resulting in:

SHAKE IN POSITION-VELOCITY FORM

$$q^{n+1} = q^n + \Delta t v^{n+1/2}, \tag{7.14}$$
$$Mv^{n+1/2} = Mv^{n-1/2} - \Delta t \nabla_q V(q^n) - \Delta t G(q^n)^T \lambda^n, \tag{7.15}$$
$$0 = g(q^{n+1}), \tag{7.16}$$
$$v^n = \frac{1}{2}\left(v^{n+1/2} + v^{n-1/2}\right). \tag{7.17}$$

SHAKE can be viewed as a mapping of $\mathcal{M}$, but it does not define a mapping of the tangent bundle $T\mathcal{M}$. Yet, SHAKE is algebraically equivalent to another method, RATTLE, which is a mapping of the tangent bundle $T\mathcal{M}$. The idea behind RATTLE is to correct the SHAKE solution so that it lies on $T\mathcal{M}$ through appropriate projection of the velocity v^{n+1} on to the tangency constraint (7.5), the result being the scheme:

RATTLE DISCRETIZATION

$$q^{n+1} = q^n + \Delta t v^{n+1/2}, \tag{7.18}$$
$$Mv^{n+1/2} = Mv^n - \frac{\Delta t}{2}\nabla_q V(q^n) - \frac{\Delta t}{2} G(q^n)^T \lambda^n_{(r)}, \tag{7.19}$$
$$0 = g(q^{n+1}), \tag{7.20}$$
$$Mv^{n+1} = Mv^{n+1/2} - \frac{\Delta t}{2}\nabla_q V(q^{n+1}) - \frac{\Delta t}{2} G(q^{n+1})^T \lambda^{n+1}_{(v)}, \tag{7.21}$$
$$0 = G(q^{n+1}) v^{n+1}. \tag{7.22}$$

The multipliers $\lambda^n_{(r)}$ are chosen in order to enforce the position constraints (7.20), while $\lambda^{n+1}_{(v)}$ relates to the velocity constraints (7.22) and is determined at time level t_n by the linear system

$$\left[G(q^n) M^{-1} G(q^n)^T\right] \lambda^n_{(v)} = G(q^n)\left(\frac{2}{\Delta t} v^{n-1/2} - M^{-1}\nabla_q V(q^n)\right). \tag{7.23}$$

Combining equations (7.19) and

$$Mv^n = Mv^{n-1/2} - \frac{\Delta t}{2}\nabla_q V(q^n) - \frac{\Delta t}{2} G(q^n)^T \lambda^n_{(v)}$$

results in

$$M\boldsymbol{v}^{n+1/2} = M\boldsymbol{v}^{n-1/2} - \Delta t \nabla_{\boldsymbol{q}} V(\boldsymbol{q}^n) - \frac{\Delta t}{2} \boldsymbol{G}(\boldsymbol{q}^n)^T (\boldsymbol{\lambda}^n_{(\mathrm{r})} + \boldsymbol{\lambda}^n_{(\mathrm{v})}), \tag{7.24}$$

where $\boldsymbol{\lambda}^n_{(\mathrm{v})}$ is determined by (7.23) and is assumed to be computed at this stage. On the other hand, $\boldsymbol{\lambda}^n_{(\mathrm{r})}$ is chosen so that

$$\boldsymbol{g}(\boldsymbol{q}^{n+1}) = \boldsymbol{g}(\boldsymbol{q}^n + \Delta t \boldsymbol{v}^{n+1/2}) = \boldsymbol{0}$$

at the next time level.

We recognize that (7.24) is equivalent to the SHAKE update (7.15) with $\boldsymbol{\lambda}^n$ replaced by $\frac{1}{2}(\boldsymbol{\lambda}^n_{(\mathrm{r})} + \boldsymbol{\lambda}^n_{(\mathrm{v})})$. Thus RATTLE and SHAKE are formally equivalent when viewed as iterations from $(\boldsymbol{q}^n, \boldsymbol{v}^{n-\frac{1}{2}})$ to $(\boldsymbol{q}^{n+1}, \boldsymbol{v}^{n+\frac{1}{2}})$. In other words, a proper initialization of RATTLE would produce a sequence of approximations identical to that produced by SHAKE, except that the two solutions would differ in the velocity approximation $\boldsymbol{v}^n$ at the end of each timestep.

Example 2 Let us illustrate the use of SHAKE with the familiar example of the planar pendulum. For the pendulum, applying SHAKE results in the discrete equations

$$\begin{aligned}
x_{n+1} &= x_n + \Delta t u_{n+1/2}, \\
y_{n+1} &= y_n + \Delta t v_{n+1/2}, \\
m u_{n+1/2} &= m u_{n-1/2} - \Delta t x_n \lambda_n, \\
m v_{n+1/2} &= m v_{n-1/2} - \Delta t m g - \Delta t y_n \lambda_n, \\
L^2 &= (x_{n+1})^2 + (y_{n+1})^2,
\end{aligned}$$

where we have violated our usual convention of writing time step index as a superscript rather than a subscript in order to avoid ambiguity with the multiplicative power.

The SHAKE method can be interpreted as follows: we first ignore the constraints and take a step of the Störmer–Verlet method, resulting for the pendulum in

$$\begin{aligned}
\bar{x}_{n+1} &= x_n + \Delta t u_{n-1/2}, \\
\bar{y}_{n+1} &= y_n + \Delta t v_{n-1/2} - \Delta t^2 g.
\end{aligned}$$

(This is always a point of a quadratic arc through the previous step.)

Next, we compute the oblique projection of the unconstrained step on to the configuration manifold along the direction of the normal to the constraint at the *previous timestep* t_n. For the pendulum, this means finding the intersection of a certain straight line with the circle by solving the following quadratic equation for λ_n

$$(\bar{x}_{n+1} - \frac{1}{m}\Delta t^2 x_n \lambda_n)^2 + (\bar{y}_{n+1} - \frac{1}{m}\Delta t^2 y_n \lambda_n)^2 = L^2.$$

The situation is diagrammed in Fig. 2. Note that there are, in general, two real solutions to this equation, corresponding to the two points of intersection with the circle. For small timesteps, the correct choice of multiplier will be obvious (the appropriate solution will be small in magnitude). As a simple rule, we might always choose the solution corresponding to the smaller value of the multiplier, but if the stepsize is sufficiently large, this approach could lead to an incorrect choice. In general, we should consider the smooth continuation of the solution in the parameter Δt from $\Delta t = 0$ and $\lambda^n(0) = 0$ to the final value $\lambda(\Delta t)$.

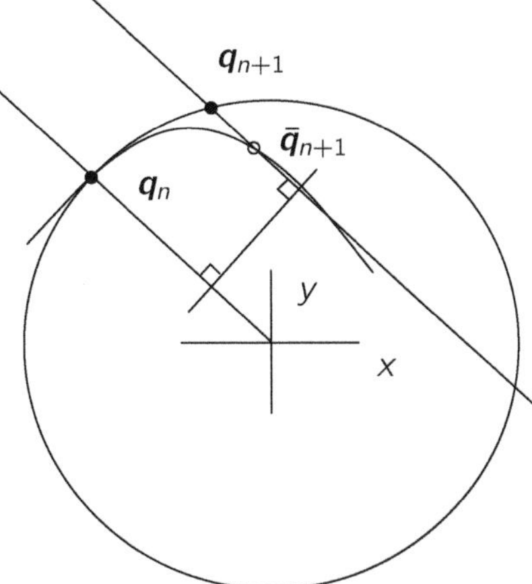

Figure 7.2 SHAKE discretization applied to the planar pendulum. Starting from a point on the circle ($q_n = (x_n, y_n)$), an unconstrained step is taken (following a parabolic arc in the parameter Δt) to $\bar{q}_{n+1} = (\bar{x}_{n+1}, \bar{y}_{n+1})$. Next, we find the projection on to the circle of radius L along the direction q_n. □

These same considerations enter into the discussion of the general case (7.14)–(7.17). The correct choice of the multiplier can be identified by smoothly continuing the solution as a function of the parameter Δt starting from $\Delta t = 0$ and $\boldsymbol{\lambda}^n(0) = \mathbf{0}$. For small timesteps, the correct choice of multiplier $\boldsymbol{\lambda}^n$ will thus be small in magnitude, so that a Newton iteration started from the zero initial guess will tend to find the correct solution. For larger timesteps, though, caution must be exercised to make certain that the solution obtained is the correct one.

We next briefly comment on the error growth in these methods. Since SHAKE and RATTLE are formally equivalent, the propagation of errors is identical for the two methods. The modification of the end of timestep velocity in the RATTLE step does not effect the stability or global convergence of the method in any way.

It is relatively straightforward to derive a formula for the local error introduced in a single timestep using SHAKE or RATTLE based on comparison of the Taylor expansions of the solution and numerical solution in a step of size Δt, starting from some given point $(\boldsymbol{q}^n, \boldsymbol{v}^n) \in T\mathcal{M}$. For both methods, this calculation shows that the local error is of order Δt^3. Thus these methods are second-order accurate. (Higher-order schemes are discussed below, in the context of Hamiltonian systems.)

7.2.2 Implementation

At every step of the SHAKE (or RATTLE) discretization, we must solve a system of nonlinear equations of dimension equal to the number of constraints. In some cases, these constraints can be dealt with very easily. For example, if the constraints are linear, the work involved is usually a step of Gaussian elimination or the use of some other linear solver, and unless the dimension of the system is exceptionally large, or the equations poorly conditioned, this computation will be easy to implement. In other cases, the constraints may admit a decoupling that enables their simplified solution.

Example 3 Consider a system of N particles, each of mass m, attached to the surface of the unit sphere and interacting in some homogeneous two-body potential ϕ. The energy takes the form

$$E = \frac{1}{2}\sum_{i=1}^{N} m_i \|\boldsymbol{v}_i\|^2 + V(\boldsymbol{q}_1, \ldots, \boldsymbol{q}_N),$$

where $V(\boldsymbol{q}) = V(\boldsymbol{q}_1, \ldots, \boldsymbol{q}_N) = \sum_{i=1}^{N-1}\sum_{j=i+1}^{N} \phi(\|\boldsymbol{q}_i - \boldsymbol{q}_j\|)$. The equations of motion for the ith particle, $i = 1, \ldots, N$, are

$$\begin{aligned}
\dot{\boldsymbol{q}}_i &= \boldsymbol{v}_i, \\
m\dot{\boldsymbol{v}}_i &= -\nabla_{\boldsymbol{q}_i} V(\boldsymbol{q}) - \boldsymbol{q}_i \lambda_i, \\
\|\boldsymbol{q}_i\|^2 &= 1,
\end{aligned}$$

and the equations of the ith and jth particles are coupled only through the potential energy. Applying SHAKE discretization results in

$$\boldsymbol{q}_i^{n+1} = \boldsymbol{q}_i^n + \Delta t \boldsymbol{v}_i^{n+1/2}, \tag{7.25}$$

$$m\boldsymbol{v}_i^{n+1/2} = m\boldsymbol{v}_i^{n-\frac{1}{2}} - \Delta t \nabla_{\boldsymbol{q}_i} V(\boldsymbol{q}^n) - \Delta t \boldsymbol{q}_i^n \lambda_i^n, \tag{7.26}$$

$$\|\boldsymbol{q}_i^{n+1}\|^2 = 1, \tag{7.27}$$

Introducing equation (7.26) into (7.25), then combining the resulting formula with the constraint (7.27) results in N independent quadratic equations to be solved for the N multipliers λ_i^n, $i = 1, \ldots, N$. □

On the other hand, in most cases, such as for example the constraint chain, one has to solve a system of nonlinear equations for the multipliers $\boldsymbol{\lambda}^n = \{\lambda_i^n\}_{i=1,\ldots,m}$ of the form

$$\boldsymbol{g}\left(\bar{\boldsymbol{q}}^{n+1} - \Delta t^2 \boldsymbol{M}^{-1} \sum_{i=1}^{m} \boldsymbol{G}(\boldsymbol{q}^n)^T \boldsymbol{\lambda}^n\right) = \boldsymbol{0}, \tag{7.28}$$

where $\bar{\boldsymbol{q}}^{n+1}$ represents an *unconstrained* step using Störmer–Verlet/leapfrog. The paper [167] describing the SHAKE discretization also provided an iterative solver for the nonlinear equations (7.28); the term SHAKE is used typically to refer to the combined procedure of time discretization together with the iterative (*coordinate resetting*) algorithm for constraints. We divide these here into two procedures, namely *SHAKE iteration* for coordinate resetting and *SHAKE discretization* (7.14)–(7.16) .

In SHAKE iteration, we cycle through the constraints, adjusting one multiplier at each iteration. If g_i is the ith component of g, we denote by $\boldsymbol{G}_i = \nabla_{\boldsymbol{q}} g_i(\boldsymbol{q})$ the ith row of the constraint Jacobian matrix (i.e. the gradient of the ith constraint function), then the iteration is as follows: First, we initialize

$$\boldsymbol{Q} := \bar{\boldsymbol{q}}^{n+1} = \boldsymbol{q}^n + \Delta t \boldsymbol{v}^{n-1/2} - \Delta t^2 \boldsymbol{M}^{-1} \nabla_{\boldsymbol{q}} V(\boldsymbol{q}^n),$$

which is equivalent to taking $\boldsymbol{\lambda}^n$ to be zero in (7.28).

Next, we cycle through the list of constraints and correct each constraint one after another by using the following procedure. For $i = 1, \ldots, m$, compute an offset $\Delta\Lambda_i$ in order to satisfy the ith linearized constraint equation

$$\Delta\Lambda_i := \frac{g_i(\boldsymbol{Q})}{\boldsymbol{G}_i(\boldsymbol{Q})\boldsymbol{M}^{-1}\boldsymbol{G}_i(\boldsymbol{q}^n)},$$

and update $\boldsymbol{Q}$ by

$$\boldsymbol{Q} := \boldsymbol{Q} - \boldsymbol{M}^{-1}\boldsymbol{G}_i(\boldsymbol{q}^n)^T \Delta\Lambda_i.$$

This cycle is repeated until all constraint residuals $g_i(\boldsymbol{Q})$ are smaller than some prescribed tolerance (usually a multiple of the unit rounding error). At this point we set $\boldsymbol{q}^{n+1} = \boldsymbol{Q}$ and continue with the next timestep.

As was shown in [13], SHAKE iteration is really a variant of nonlinear Gauss–Seidel–Newton iteration and its convergence can be justified and analyzed in the framework of Ortega and Rheinboldt [150]. In particular, it can be shown that, given a good enough initial guess for the multiplier, or a small enough step size Δt, this iterative method eventually converges.

One can consider a variety of improvements to SHAKE iteration. Noting that SHAKE is essentially a nonlinear Gauss–Seidel iteration, it seems natural to consider the use of an SOR-type technique. For this purpose, we could introduce a *parameter* ω, changing the offset $\Delta\Lambda_i$ at each step of iteration to $\omega\Delta\Lambda_i$.

The parameter ω can be a fixed value obtained through some preliminary experiment, or it can be obtained automatically during integration by a simple adaptive algorithm. We stress that this is a cost-free enhancement of SHAKE and it can lead to substantial speedups in the constraint-solving portion of a timestepping algorithm. Because the method is just an alternative nonlinear equation solver to SHAKE iteration, the converged numerical solution will not differ from that obtained by SHAKE iteration.

A second alternative to SHAKE iteration was also considered in [13]: we could use a Newton iteration, or variant thereof, to compute successive updates to the vector $\boldsymbol{\lambda}^n$. In particular, we could apply a true Newton iteration to the original nonlinear equations and use *sparse matrix techniques* to solve the resulting linear equations. This turns out to be somewhat expensive for most problems because the factorization of a matrix – even a sparse matrix – is relatively costly. A better alternative is to use a quasi-Newton iteration described below.

All these methods iteratively improve the whole vector of offsets $\boldsymbol{\Delta\Lambda} = \{\Delta\Lambda_i\}$

$$\boldsymbol{\Delta\Lambda} := \boldsymbol{R}^{-1}\boldsymbol{g}(\boldsymbol{Q}),$$

where for the true Newton iteration (NIP) we have $\boldsymbol{R} = \boldsymbol{G}(\boldsymbol{Q})\boldsymbol{M}^{-1}\boldsymbol{G}(\boldsymbol{q}^n)^T$ with $\boldsymbol{Q}$ the latest approximation obtained from

$$\boldsymbol{Q} := \boldsymbol{Q} - \boldsymbol{M}^{-1}\boldsymbol{G}(\boldsymbol{q}^n)^T\boldsymbol{\Delta\Lambda}.$$

For the quasi-Newton iteration we have $\boldsymbol{R} = \boldsymbol{G}\boldsymbol{M}^{-1}\boldsymbol{G}^T$, with $\boldsymbol{G} = \boldsymbol{G}(\boldsymbol{q}^k)$ for some $\boldsymbol{q}^k$ computed at a previous timestep t_k ($\boldsymbol{G}$ is updated as needed for convergence). Note that the quasi-Newton iteration matrix $\boldsymbol{R}$ is symmetric and positive-definite and, hence, can be factorized in a very efficient manner [13].

SHAKE and RATTLE are particularly useful in the context of molecular dynamics. Further details will be provided in Chapter 11. We only mention here that the iterative techniques described in this section have been integrated into the CHARMM molecular dynamics software package [34] and successfully applied to several examples: C_{60}, a box of water, Myoglobin, BPTI; detailed results are reported in [13]. Roughly speaking, we can summarize the results as follows: (1) the use of an SOR parameter can improve SHAKE iteration convergence by a factor of two to three in protein dynamics simulations for no additional cost or loss of robustness, (2) the adaptive scheme for determining optimal ω in [13], although not robust, does demonstrate the feasibility of computing the SOR parameter adaptively, and (3) the symmetric adaptive Newton method may show improvement over SHAKE iteration at large step size Δt for problems with high bond connectivity, e.g. the *buckminsterfullerene* C_{60}.

7.2.3 Numerical experiment

To conclude this section, we apply the RATTLE method to simulate a small model system consisting of six unit-mass particles linked by six rigid unit length constraints. The hexagonal starting configuration is shown in Fig. 7.3.

Note there are six multipliers and constraint equations, and, in the planar case that we consider here, a total of $2 \times 6 - 6 = 6$ degrees of freedom. The next nearest neighbors are linked by springs with rest length $\sqrt{3}$. Pairs numbered (1,4), (2,5), and (3,6) are linked by springs with rest length 2 (see Fig. 7.3). In this way the system tends to retain the shape of a hexagon during simulation.

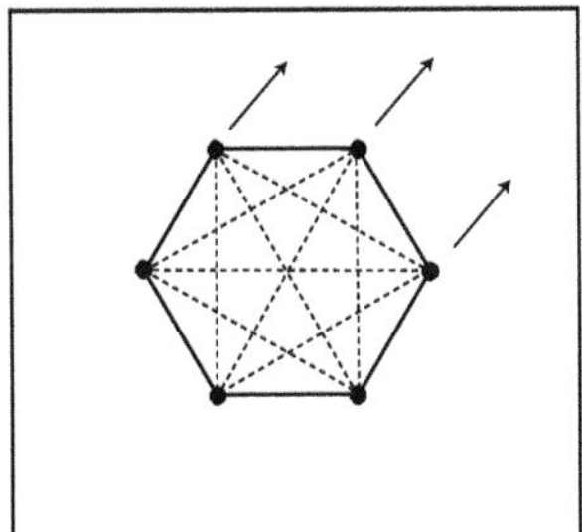

Figure 7.3 Diagram of a planar constraint loop consisting of six particles of unit mass.

We placed the system in a square box with sides at $x = \pm 2$, $y = \pm 2$. The interaction with the boundary of the box was introduced with the following soft wall potential

$$V_{\mathrm{wall}} = \frac{\epsilon}{6} \sum_{i=1}^{6} \left[(x_i + 2)^{-6} + (x_i - 2)^{-6} + (y_i + 2)^{-6} + (y_i - 2)^{-6} \right].$$

We chose $\epsilon = 0.01$, strongly localizing collisions at nearly the point of contact with the boundary of the region.

It is important to make sure that the initial velocities are consistent with the constraint (7.5). This can always be achieved by a simple projection step. In our case, we simply initialized the positions at the global minimum of potential energy (regular hexagon configuration) and gave each atom of the system the same initial velocity, propelling the entire object rigidly toward the boundary. Following the first impact, the system begins chaotic tumbling and oscillatory vibration.

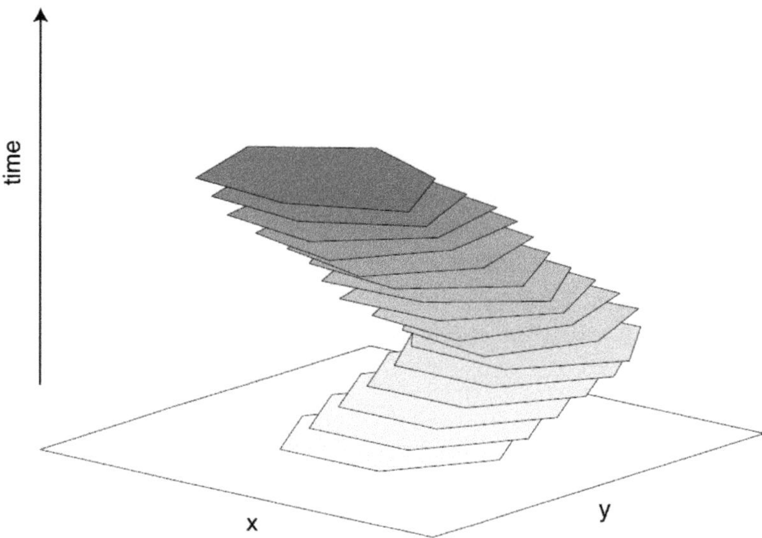

Figure 7.4 Snapshots of the motion of the six particle chain.

We applied RATTLE to simulate the constraint chain over a time interval $[0, 100]$. Some frames of the motion are shown in Fig. 7.4.

We compare energy errors for various simulations. In the first run, the stepsize was set to $\Delta t = 0.08$. The result, shown in the upper panel of Fig. 7.5, is not very encouraging. The energy seems to hop about in an almost random pattern. The jumps in energy coincide with close approaches of a particle of the chain to the walls of the box, i.e. "collisions." Here the r^{-6} potential gives rise to a strong restraining force, propelling the particle away at relatively high velocity. At these points, the stability of the method breaks down for this stepsize: in essence, the "perturbed Hamiltonian" expansion discussed in Chapter 5 does not really exist at these stepsizes; the exponentially small error term mentioned in Chapter 5 is not really small at all. When the stepsize is reduced, say to $\Delta t = 0.04$, the energy is stabilized at nearly the correct value (see lower panel of Fig. 7.5).[1]

The second-order convergence of the RATTLE method can be verified numerically by computing trajectories with several different stepsizes, then graphing the error v. stepsize in logarithmic scale (see Fig. 7.6).

[1]That this stabilization can be achieved for some well-chosen stepsize is of course one of the remarkable features of symplectic methods. In each simulation with a symplectic method, typically the first step is to identify the maximum stepsize for which energy stabilizes; it is hoped that in the case of a symplectic method the energy stability acts as a sort of gauge for the physicality of a simulation.

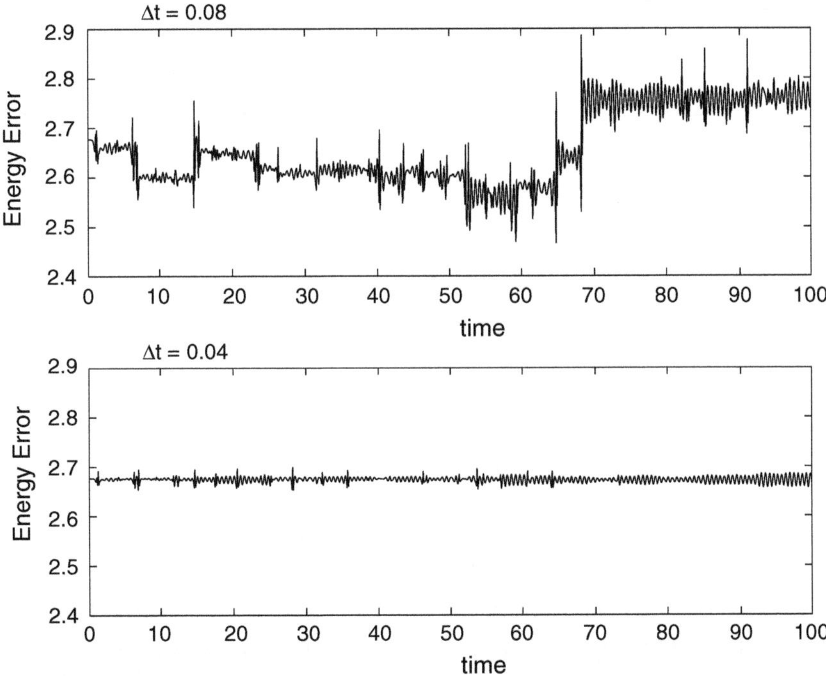

Figure 7.5 Energies against time for two runs of RATTLE on the six-particle chain, upper panel $\Delta t = 0.08$, lower panel $\Delta t = 0.04$.

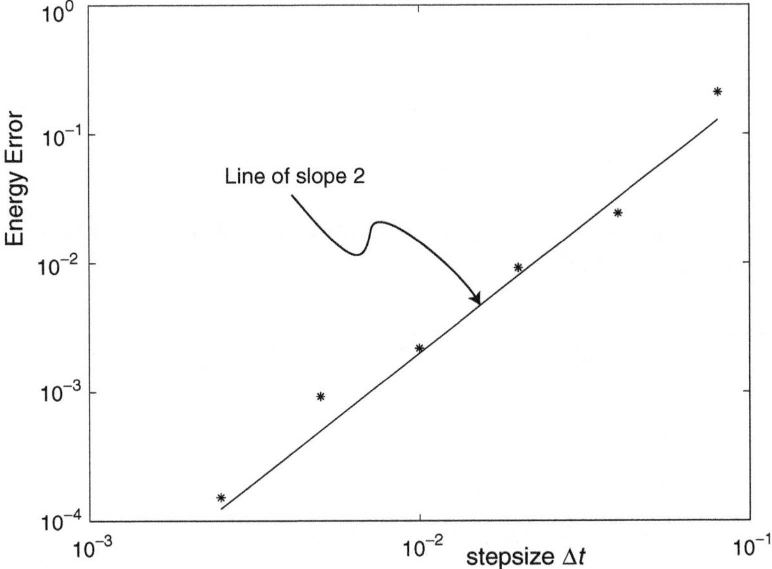

Figure 7.6 Verification of second-order convergence of RATTLE.

7.3 Transition to Hamiltonian mechanics

We now turn to the Hamiltonian formalism with constraints. As done before in Chapter 3, we formally introduce conjugate momenta $\boldsymbol{p} = \boldsymbol{M}\dot{\boldsymbol{q}}$ and rewrite (7.1)–(7.3) as

$$\frac{d}{dt}\boldsymbol{q} = \boldsymbol{M}^{-1}\boldsymbol{p},$$
$$\frac{d}{dt}\boldsymbol{p} = -\nabla_{\boldsymbol{q}} V(\boldsymbol{q}) - \boldsymbol{G}(\boldsymbol{q})^T\boldsymbol{\lambda},$$
$$\boldsymbol{0} = \boldsymbol{g}(\boldsymbol{q}).$$

The first equation is the gradient of the augmented Hamiltonian

$$\tilde{H} = \frac{1}{2}\boldsymbol{p}^T\boldsymbol{M}^{-1}\boldsymbol{p} + V(\boldsymbol{q}) + \boldsymbol{g}(\boldsymbol{q})^T\boldsymbol{\lambda},$$

with respect to $\boldsymbol{p}$, while the second equation is equal to the negative gradient of $\tilde{H}$ with respect to $\boldsymbol{q}$ with $\boldsymbol{\lambda}$ treated as a constant.

The form of the Hamiltonian $\tilde{H}$ suggests a correspondence between Hamiltonian formulations with and without constraints. Let H be the energy of a certain unconstrained mechanical system, so the equations of motion are just

$$\frac{d}{dt}\boldsymbol{q} = +\nabla_{\boldsymbol{p}} H(\boldsymbol{q}, \boldsymbol{p}),$$
$$\frac{d}{dt}\boldsymbol{p} = -\nabla_{\boldsymbol{q}} H(\boldsymbol{q}, \boldsymbol{p}).$$

Similar to the situation above, introduce the augmented Hamiltonian

$$\tilde{H}(\boldsymbol{q}, \boldsymbol{p}) = H(\boldsymbol{q}, \boldsymbol{p}) + \boldsymbol{g}(\boldsymbol{q})^T\boldsymbol{\lambda},$$

and write the equations of motion (the formulation of a constrained Hamiltonian system in natural coordinates):

HAMILTONIAN FORMULATION OF A HOLONOMIC SYSTEM

$$\frac{d}{dt}\boldsymbol{q} = \nabla_{\boldsymbol{p}} H(\boldsymbol{q}, \boldsymbol{p}), \tag{7.29}$$
$$\frac{d}{dt}\boldsymbol{p} = -\nabla_{\boldsymbol{q}} H(\boldsymbol{q}, \boldsymbol{p}) - \boldsymbol{G}(\boldsymbol{q})^T\boldsymbol{\lambda}, \tag{7.30}$$
$$\boldsymbol{0} = \boldsymbol{g}(\boldsymbol{q}). \tag{7.31}$$

Example 4 The constrained Hamiltonian formulation (7.29)–(7.31) can be derived from a constrained Lagrangian formulation. Recall from Chapter 3 that Lagrange's equations for a mechanical system with Lagrangian function $L(\boldsymbol{q}, \dot{\boldsymbol{q}})$ are given by

$$\frac{d}{dt}\nabla_{\dot{q}} L(\boldsymbol{q}, \dot{\boldsymbol{q}}) - \nabla_q L(\boldsymbol{q}, \dot{\boldsymbol{q}}) = \boldsymbol{0}.$$

These are the Euler–Lagrange equations minimizing the action integral

$$\mathcal{L}[\boldsymbol{q}] = \int_{t_0}^{t_1} L(\boldsymbol{q}(t), \dot{\boldsymbol{q}}(t))\, dt.$$

To obtain the constrained Lagrange's equation we consider the augmented action integral

$$\mathcal{L}[\boldsymbol{q}, \boldsymbol{\lambda}] = \int_{t_0}^{t_1} \left[L(\boldsymbol{q}(t), \dot{\boldsymbol{q}}(t)) - \boldsymbol{\lambda}(t)^T \boldsymbol{g}(\boldsymbol{q}(t))\right]\, dt,$$

and, upon taking variations in $\boldsymbol{q}$ and $\boldsymbol{\lambda}$, derive the associated Euler–Lagrange equations

$$\frac{d}{dt}\nabla_{\dot{q}} L(\boldsymbol{q}, \dot{\boldsymbol{q}}) = \nabla_q L(\boldsymbol{q}, \dot{\boldsymbol{q}}) + \boldsymbol{G}(\boldsymbol{q})^T \boldsymbol{\lambda}, \tag{7.32}$$

$$\boldsymbol{0} = \boldsymbol{g}(\boldsymbol{q}). \tag{7.33}$$

The transition to the Hamiltonian formulation can be achieved by introducing the canonical momenta

$$\boldsymbol{p} = \nabla_{\dot{q}} L(\boldsymbol{q}, \dot{\boldsymbol{q}}),$$

and by employing the standard Legendre transformation

$$\tilde{H}(\boldsymbol{q}, \boldsymbol{p}, \boldsymbol{\lambda}) = \boldsymbol{p} \cdot \dot{\boldsymbol{q}} - \left[L(\boldsymbol{q}, \dot{\boldsymbol{q}}) - \boldsymbol{\lambda}^T \boldsymbol{g}(\boldsymbol{q})\right] = H(\boldsymbol{q}, \boldsymbol{p}) + \boldsymbol{\lambda}^T \boldsymbol{g}(\boldsymbol{q}).$$

For example, the N-body Lagrangian

$$L(\boldsymbol{q}, \dot{\boldsymbol{q}}) = \frac{1}{2}\dot{\boldsymbol{q}}^T \boldsymbol{M} \dot{\boldsymbol{q}} - V(\boldsymbol{q})$$

gives rise to the Hamiltonian

$$\tilde{H} = \frac{1}{2}\boldsymbol{p}^T \boldsymbol{M}^{-1} \boldsymbol{p} + V(\boldsymbol{q}) + \boldsymbol{g}(\boldsymbol{q})^T \boldsymbol{\lambda}.$$

□

It is easily verified that the constrained equations (7.29)–(7.31) can be reduced to the Lagrangian equations (7.32)–(7.33) and that both formulations lead to identical solutions curves $\boldsymbol{q}(t)$. However, while $(\boldsymbol{q}(t), \dot{\boldsymbol{q}}(t)) \in T\mathcal{M}$, the natural coordinates $(\boldsymbol{q}, \boldsymbol{p})$ must satisfy the equation (7.31) and the corresponding momentum-level constraint

$$\frac{d}{dt}\boldsymbol{g}(\boldsymbol{q}) = \boldsymbol{G}(\boldsymbol{q})\nabla_{\boldsymbol{p}} H(\boldsymbol{q}, \boldsymbol{p}) = \boldsymbol{0}. \tag{7.34}$$

The subset of $\mathbb{R}^{2d}$ defined by (7.31) and (7.34) we term the *phase space* and denote by $\mathscr{P}$.

Observe that the energy H of the unconstrained system remains a first integral of (7.29)–(7.31). To see this, first observe that, along solutions

$$\begin{aligned}\frac{d}{dt}H(\boldsymbol{q}, \boldsymbol{p}) &= H_{\boldsymbol{q}}(\boldsymbol{q}, \boldsymbol{p})\, \dot{\boldsymbol{q}} + H_{\boldsymbol{p}}(\boldsymbol{q}, \boldsymbol{p})\dot{\boldsymbol{p}} \\ &= H_{\boldsymbol{q}}(\boldsymbol{q}, \boldsymbol{p})\, \nabla_{\boldsymbol{p}} H(\boldsymbol{q}, \boldsymbol{p}) - H_{\boldsymbol{p}}(\boldsymbol{q}, \boldsymbol{p}) \left[\nabla_{\boldsymbol{q}} H(\boldsymbol{q}, \boldsymbol{p}) + \boldsymbol{G}(\boldsymbol{q})^T \boldsymbol{\lambda}\right] \\ &= -\boldsymbol{\lambda}^T \left[\boldsymbol{G}(\boldsymbol{q})\, \nabla_{\boldsymbol{p}} H(\boldsymbol{q}, \boldsymbol{p})\right],\end{aligned}$$

then note that the latter expression vanishes along the hidden constraint (7.34).

7.4 The symplectic structure with constraints

In this section, we show that (7.29)–(7.31) is indeed a generalization of unconstrained canonical Hamiltonian systems as discussed in Chapter 3. According to a perspective due to DIRAC [50], the flow of a Hamiltonian system on a manifold can be embedded in the flow of an *unconstrained* Hamiltonian system. Following this idea, the solutions of (7.29)–(7.31) can simply be viewed as evolving in the phase space of a standard canonical Hamiltonian system with extended Hamiltonian

$$\hat{H} = H(\boldsymbol{q}, \boldsymbol{p}) + \boldsymbol{g}(\boldsymbol{q})^T \boldsymbol{\Lambda}(\boldsymbol{q}^{-1}, \boldsymbol{p}),$$

and $\boldsymbol{\Lambda}(\boldsymbol{q}, \boldsymbol{v})$ given by (7.11) if $H = \boldsymbol{p}^T \boldsymbol{M}^{-1}\boldsymbol{p}/2 + V(\boldsymbol{q})$ or by a proper generalization for more general unconstrained Hamiltonian functions H. The manifold $\mathscr{P}$ is an invariant manifold for the solution operator $\boldsymbol{\Phi}_{t,\hat{H}}$. For a discussion of Dirac's method, see MARSDEN AND RATIU [124]; for a discussion of numerical methods based on this, see LEIMKUHLER AND REICH [111].

Dirac's approach allows us to consider the symplectic structure on the phase space $\mathscr{P}$ as the restriction of the canonical symplectic structure $d\boldsymbol{q} \wedge d\boldsymbol{p}$ on $\mathbb{R}^{2d}$ to the phase space (submanifold) $\mathscr{P}$. We outline the derivation of this restricted symplectic structure and its invariance under the flow operator of (7.29)–(7.31) in more detail in the remainder of this section. Contrary to this global approach, one can also introduce a symplectic structure on phase space by considering local parametrizations of $\mathscr{P}$. This is the more traditional approach and a short outline will be provided later in this chapter.

In keeping with the notation introduced earlier in Chapters 3 and 4, we start with differential one-forms $d\boldsymbol{q}$ and $d\boldsymbol{p}$ defined over the unconstrained space $\mathbb{R}^{2d}$. We now restrict these one-forms to the solution space by requiring that

$$\boldsymbol{G}(\boldsymbol{q})d\boldsymbol{q} = \boldsymbol{0} \quad \text{and} \quad \boldsymbol{f}_q(\boldsymbol{q},\boldsymbol{p})d\boldsymbol{q} + \boldsymbol{f}_p(\boldsymbol{q},\boldsymbol{p})d\boldsymbol{p} = \boldsymbol{0}, \tag{7.35}$$

for all $(\boldsymbol{q},\boldsymbol{p}) \in \mathscr{P}$, where

$$\boldsymbol{f}(\boldsymbol{q},\boldsymbol{p}) = \boldsymbol{G}(\boldsymbol{q})\nabla_{\boldsymbol{p}} H(\boldsymbol{q},\boldsymbol{p})$$

is the hidden constraint function. In more abstract terms, the differential one-forms $(d\boldsymbol{q}, d\boldsymbol{p})$ are now elements of the cotangent space $T^*\mathscr{P}$ of $\mathscr{P}$ (imbedded into $\mathbb{R}^{2d}$) [124].

The variational equations corresponding to (7.29)–(7.31) can be obtained by straightforward linearization and the resulting equations for the differential one-forms are

$$\frac{d}{dt}d\boldsymbol{q} = \boldsymbol{H}_{pq}(\boldsymbol{q},\boldsymbol{p})d\boldsymbol{q} + \boldsymbol{H}_{pp}(\boldsymbol{q},\boldsymbol{p})d\boldsymbol{p}, \tag{7.36}$$

$$\frac{d}{dt}d\boldsymbol{p} = -\boldsymbol{H}_{qq}(\boldsymbol{q},\boldsymbol{p})d\boldsymbol{q} - \boldsymbol{H}_{qp}(\boldsymbol{q},\boldsymbol{p})d\boldsymbol{p} - d\left(\boldsymbol{G}(\boldsymbol{q})\boldsymbol{\lambda}\right), \tag{7.37}$$

$$\boldsymbol{0} = \boldsymbol{G}(\boldsymbol{q})d\boldsymbol{q}. \tag{7.38}$$

It is easily verified by differentiation of (7.38) with respect to time that the solutions of (7.36)–(7.38) indeed satisfy (7.35). Hence we are left with the task to show that the canonical wedge product $d\boldsymbol{q} \wedge d\boldsymbol{p}$, restricted to (7.35), is an invariant of (7.36)–(7.38). We need the following:

Lemma 1 *Assume that the differential one-form $d\boldsymbol{q}$ satisfies (7.38), then*

$$d\boldsymbol{q} \wedge d(\boldsymbol{G}(\boldsymbol{q})^T\boldsymbol{\lambda}) = \boldsymbol{0}.$$

Proof:

$$d\boldsymbol{q} \wedge d(\boldsymbol{G}(\boldsymbol{q})^T\boldsymbol{\lambda}) = d\boldsymbol{q} \wedge \boldsymbol{G}(\boldsymbol{q})^T d\boldsymbol{\lambda} + \sum_{i=0}^{m} \lambda_i d\boldsymbol{q} \wedge \boldsymbol{\Gamma}_i d\boldsymbol{q},$$

where the components of $\boldsymbol{\lambda}$ have been indexed by a subscript and

$$\boldsymbol{\Gamma}_i = \left\{ \frac{\partial^2 g_i}{\partial q_k \partial q_l}(\boldsymbol{q}) \right\}$$

is the (symmetric!) Hessian matrix of the ith constraint function. Now

$$d\boldsymbol{q} \wedge \boldsymbol{G}(\boldsymbol{q})^T d\boldsymbol{\lambda} = \boldsymbol{G}(\boldsymbol{q})d\boldsymbol{q} \wedge d\boldsymbol{\lambda} = \boldsymbol{0},$$

for any values of $d\boldsymbol{\lambda}$ because of (7.38). Finally, each of the terms of the summation can be eliminated since $d\boldsymbol{q} \wedge \boldsymbol{\Gamma}_i d\boldsymbol{q} = \mathbf{0}$. □

We now easily verify (dropping arguments to simplify notations) that

$$\begin{aligned}\frac{d}{dt} d\boldsymbol{q} \wedge d\boldsymbol{p} &= d\dot{\boldsymbol{q}} \wedge d\boldsymbol{p} + d\boldsymbol{q} \wedge d\dot{\boldsymbol{p}} \\ &= (\boldsymbol{H}_{pq} d\boldsymbol{q} + \boldsymbol{H}_{pp} d\boldsymbol{p}) \wedge d\boldsymbol{p} - d\boldsymbol{q} \wedge \left[\boldsymbol{H}_{qq} d\boldsymbol{q} + \boldsymbol{H}_{qp} d\boldsymbol{p} + d\left(\boldsymbol{G}^T \boldsymbol{\lambda}\right)\right] \\ &= \boldsymbol{H}_{pq} d\boldsymbol{q} \wedge d\boldsymbol{p} - d\boldsymbol{q} \wedge \boldsymbol{H}_{qp} d\boldsymbol{p} + \boldsymbol{H}_{pp} d\boldsymbol{p} \wedge d\boldsymbol{p} - d\boldsymbol{q} \wedge \boldsymbol{H}_{qq} d\boldsymbol{q}.\end{aligned}$$

These terms all vanish using the properties of the wedge product given in Chapter 3. Thus the canonical wedge product is preserved along solutions of (7.36)–(7.38).

In most textbooks on classical mechanics, mechanical systems subject to a holonomic constraint manifold $\mathcal{M}$ are formulated as Hamiltonian systems on the cotangent space $T^*\mathcal{M}$ (see ARNOLD [7] and MARSDEN AND RATIU [124]). Here we have defined constrained Hamiltonian systems over a different phase space $\mathcal{P}$. However, it turns out, not entirely unexpectedly, that the cotangent space $T^*\mathcal{M}$ and the phase space $\mathcal{P}$ are diffeomorphic under the symplectic transformation

$$(\boldsymbol{q}, \boldsymbol{p}) \in \mathcal{P} \to (\boldsymbol{q}, \bar{\boldsymbol{p}}) \in T^*\mathcal{M}$$

defined by

$$\bar{\boldsymbol{p}} = \boldsymbol{p} + \boldsymbol{G}(\boldsymbol{q})^T \boldsymbol{\mu}, \qquad \boldsymbol{G}(\boldsymbol{q})\bar{\boldsymbol{p}} = \mathbf{0}.$$

Here we have identified $T^*\mathcal{M}$ with the tangent space $T\mathcal{M}$ as defined earlier in Section 7.1. Indeed, upon applying Lemma 1, it is easy to show that

$$d\boldsymbol{q} \wedge d\boldsymbol{p} = d\boldsymbol{q} \wedge d\bar{\boldsymbol{p}}.$$

Hence the symplectic structure on $T^*\mathcal{M}$ is again given by the restriction of the standard canonical structure to $T^*\mathcal{M}$. See also MCLACHLAN AND SCOVEL [133].

7.5 Direct symplectic discretization

We come to the direct symplectic integration of general Hamiltonian systems (7.29)–(7.30) subject to holonomic constraints (7.31). We first discuss general second-order methods and then briefly outline the derivation of higher-order methods by composition.

7.5.1 Second-order methods

In LEIMKUHLER AND SKEEL [113], the symplecticness of the SHAKE and RATTLE methods was first explained. Here we will introduce a generalization of this framework, following the ideas of REICH [155]. Let us consider a Hamiltonian system

$$\tilde{H} = H(\boldsymbol{q}, \boldsymbol{p}) + \boldsymbol{g}(\boldsymbol{q})^T \boldsymbol{\lambda},$$

subject to the constraint $\boldsymbol{g}(\boldsymbol{q}) = \mathbf{0}$. Let us also assume that a second-order, symplectic, and symmetric integration method is known for the unconstrained system

$$\frac{d}{dt}\boldsymbol{q} = \nabla_p H(\boldsymbol{q}, \boldsymbol{p}), \qquad \frac{d}{dt}\boldsymbol{p} = -\nabla_q H(\boldsymbol{q}, \boldsymbol{p}).$$

For $H = T(\boldsymbol{p}) + V(\boldsymbol{q})$, this could, for example, be the Störmer–Verlet method. For a general, non-separable, Hamiltonian, one could apply the implicit midpoint rule or the generalized Störmer–Verlet/leapfrog method. In any case, let us denote the chosen method by $\boldsymbol{\Psi}_{\Delta t}$. Then the following method is a constraint-preserving, symplectic, symmetric, and second-order method for the constrained Hamiltonian system with Hamiltonian $\tilde{H}$:

SECOND-ORDER CONSTRAINT-PRESERVING SYMPLECTIC INTEGRATOR

$$\bar{\boldsymbol{p}}^n = \boldsymbol{p}^n - \frac{\Delta t}{2} G(\boldsymbol{q}^n)^T \boldsymbol{\lambda}^n_{(r)}, \tag{7.39}$$

$$(\boldsymbol{q}^{n+1}, \bar{\boldsymbol{p}}^{n+1}) = \boldsymbol{\Psi}_{\Delta t}(\boldsymbol{q}^n, \bar{\boldsymbol{p}}^n), \tag{7.40}$$

$$\mathbf{0} = \boldsymbol{g}(\boldsymbol{q}^{n+1}), \tag{7.41}$$

$$\boldsymbol{p}^{n+1} = \bar{\boldsymbol{p}}^{n+1} - \frac{\Delta t}{2} G(\boldsymbol{q}^{n+1})^T \boldsymbol{\lambda}^{n+1}_{(v)}, \tag{7.42}$$

$$\mathbf{0} = G(\boldsymbol{q}^{n+1}) \nabla_p H(\boldsymbol{q}^{n+1}, \boldsymbol{p}^{n+1}). \tag{7.43}$$

Note that this method reduces to RATTLE for $H = \boldsymbol{p}^T \boldsymbol{M}^{-1} \boldsymbol{p}/2 + V(\boldsymbol{q})$ and taking the Störmer–Verlet method as $\boldsymbol{\Psi}_{\Delta t}$. This correspondence should guide us to the implementation of the method for a more general $\boldsymbol{\Psi}_{\Delta t}$.

Let us now briefly verify the various geometric properties of (7.39)–(7.43). (i) The method obviously conserves phase space $\mathcal{P}$. However, without changing the propagation in the $\boldsymbol{q}$-variables, one can also enforce the cotangent space

$T^*\mathscr{M}$ by replacing the momentum-constraint equation (7.43) by

$$\mathbf{0} = \mathbf{G}(\mathbf{q}^{n+1})\mathbf{p}^{n+1}.$$

See also McLachlan and Scovel [133]. (ii) The method is symplectic. Indeed, by the symplecticness of $\boldsymbol{\Psi}_{\Delta t}$ we know that

$$d\mathbf{q}^n \wedge d\bar{\mathbf{p}}^n = d\mathbf{q}^{n+1} \wedge d\bar{\mathbf{p}}^{n+1}.$$

On the other hand, both momentum maps (7.39) and (7.42) are also symplectic. This is a consequence of Lemma 1 of Section 7.4. (iii) The method is also symmetric as easily shown by replacing Δt by $-\Delta t$ and $(\mathbf{q}^n, \mathbf{p}^n, \boldsymbol{\lambda}^n_{(\mathrm{r})})$ by $(\mathbf{q}^{n+1}, \mathbf{p}^{n+1}, \boldsymbol{\lambda}^{n+1}_{(\mathrm{v})})$. (iv) The method is certainly consistent. Since it is also symmetric, second-order convergence follows.

Let us give an explicit example. We take the generalized Störmer–Verlet/leapfrog method as $\boldsymbol{\Psi}_{\Delta t}$. This yields the method

$$\begin{aligned}
\mathbf{p}^{n+1/2} &= \mathbf{p}^n - \frac{\Delta t}{2}\nabla_q H(\mathbf{p}^{n+1/2}, \mathbf{q}^n) - \frac{\Delta t}{2}\mathbf{G}(\mathbf{q}^n)\boldsymbol{\lambda}^n_{(\mathrm{r})}, \\
\mathbf{q}^{n+1} &= \mathbf{q}^n + \frac{\Delta t}{2}\left[\nabla_p H(\mathbf{p}^{n+1/2}, \mathbf{q}^n) + \nabla_p H(\mathbf{p}^{n+1/2}, \mathbf{q}^{n+1})\right], \\
\mathbf{0} &= \mathbf{g}(\mathbf{q}^{n+1}), \\
\mathbf{p}^{n+1} &= \mathbf{p}^{n+1/2} - \frac{\Delta t}{2}\nabla_q H(\mathbf{p}^{n+1/2}, \mathbf{q}^{n+1}) - \frac{\Delta t}{2}\mathbf{G}(\mathbf{q}^{n+1})\boldsymbol{\lambda}^{n+1}_{(\mathrm{v})}, \\
\mathbf{0} &= \mathbf{G}(\mathbf{q}^{n+1})\nabla_p H(\mathbf{q}^{n+1}, \mathbf{p}^{n+1}).
\end{aligned}$$

This method has first been proposed by Jay [95].

Let us briefly comment on the conservation of first integrals. Assume we are given a function F such that $\{F, \tilde{H}\} = 0$ and, furthermore, both the unconstrained Hamiltonian H as well as the constraints $\mathbf{g}(\mathbf{q})$ are invariant under the Hamiltonian flow map of F. Then (7.39)–(7.43) will preserve F if and only if the unconstrained method $\boldsymbol{\Psi}_{\Delta t}$ conserves F.

7.5.2 Higher-order methods

Higher-order methods can now be obtained in two ways. One option is to generalize partitioned Runge–Kutta methods based on Lobbatto IIIA–IIIB quadrature rules to constrained Hamiltonian systems. For details see Jay [95]. However, the methods suffer from the same drawbacks as mentioned in Chapter 6 for general symplectic (partitioned) Runge–Kutta methods. The second option is to apply the idea of composition. Denote the second-order symmetric method

(7.39)–(7.43) by $\hat{\boldsymbol{\Psi}}_{\Delta t}$. Then we consider the concatenation method

$$\tilde{\boldsymbol{\Psi}}_{\Delta t} = \hat{\boldsymbol{\Psi}}_{w_s \Delta t} \circ \hat{\boldsymbol{\Psi}}_{w_{s-1}\Delta t} \circ \cdots \circ \hat{\boldsymbol{\Psi}}_{w_2 \Delta t} \circ \hat{\boldsymbol{\Psi}}_{w_1 \Delta t}, \tag{7.44}$$

with $\{w_i\}_{i=1,\ldots,s}$ free parameters that can be chosen exactly as described in Chapter 6. For example, the coefficients for a sixth-order method, as given in Chapter 6, can be applied to (7.44) and, again, yield a sixth-order (constrained) method. The crucial point is that the error analysis for such composition methods is the same independent of the fact that $(\boldsymbol{q}^n, \boldsymbol{p}^n) \in \mathcal{P}$ instead of $(\boldsymbol{q}^n, \boldsymbol{p}^n) \in \mathbb{R}^{2d}$. See REICH [155, 157] for further details.

7.6 Alternative approaches to constrained integration

In this section, we summarize some alternative approaches to integration of the constrained problem. Generally speaking, approaches can be divided into two classes of methods according to how they maintain the constraints: one class of schemes relies on a local parametrization of the constraint manifold, while the other uses projections.

The traditional treatment of constrained systems is based on a reduction to an unconstrained system in some minimal set of variables. The phase space of the problem then becomes a (flat) Euclidean space, but the parameters are defined as complicated nonlinear functions of the natural coordinates. The appeal of this approach is that the ordinary differential equations can then be treated directly by a standard numerical method. If the original constrained system is a holonomic Hamiltonian system, we show here that the reduced problem can also be taken to be Hamiltonian, hence a symplectic method can be used for discretization. However, we also point out some pitfalls with this approach and we do not recommend it for general integrations.

7.6.1 Parametrization of manifolds – local charts

Consider a constrained Hamiltonian system (7.1)–(7.3). The full rank assumption on $\boldsymbol{G}(\boldsymbol{q})$ assumed in the formulation of the Euler–Lagrange equations implies, via the implicit function theorem, that the components of the phase space variable $\boldsymbol{z} = (\boldsymbol{q}, \boldsymbol{p}) \in \mathbb{R}^{2d}$ can at any point be separated into two subvectors $\boldsymbol{z}_1$ and $\boldsymbol{z}_2$, of dimensions $2m$ and $2(d-m)$, respectively, and the constraints locally solved for $\boldsymbol{z}_1$ smoothly in terms of $\boldsymbol{z}_2$. We say that the *dimension* of the manifold $T\mathcal{M}$ is $2(d-m)$ because of this fact. The local partitioning of the variables into $\boldsymbol{z}_1$ and $\boldsymbol{z}_2$ is one example of a *parameterization* of the constraint. More generally, we can introduce any set of independent variables (*parameters*) together with appropriate functions which allow us to describe, at least locally, the constraint surface in terms of those parameters.

Example 5 Consider the plane pendulum which has the position constraint $x^2+y^2 = L^2$ and the tangency constraint $x\dot{x} + y\dot{y} = 0$. As long as $y > 0$, we can solve the position constraint for y in terms of x

$$y = \sqrt{L^2 - x^2},$$

and solve the tangency condition $x\dot{x} + y\dot{y} = 0$ for $v = \dot{y}$ in terms of x, and $u = \dot{x}$

$$v = -xu/\sqrt{L^2 - x^2}.$$

After differentiating with respect to time, it is formally possible to rewrite the pendulum equations of motion in terms of x and u only. A similar formulation could be used to described the case $y < 0$, and we could as well rewrite the equations in terms of y, v (with appropriate domain restrictions). □

Parameterizations are, at first glance, appealing mechanisms for describing constrained mechanical motion, however they may introduce a number of complications in practice. For instance, the set of parameters is usually only valid locally, i.e. within a finite region of space; for realistic mechanical problems we often need a set of parameterizations, defined in overlapping regions of space, which together include the entire phase space of the problem (the tangent bundle of $\mathcal{M}$). The parameterizations are then typically taken to map parameters from an open subspace of Euclidean space (of dimension $2(d-m)$) on to the tangent bundle. Such a family of parameterizations is called an *atlas* of the manifold, and the individual parameterizations are referred to as *local charts*.

7.6.2 The Hamiltonian case

The concept of local chart carries over to the Hamiltonian setting. If we introduce a parameterization of the position constraint in terms of a parameter $\boldsymbol{\xi}$, then the corresponding canonical momenta $\boldsymbol{\nu}$ are defined by a proper generalization of the canonical lift transformation introduced in Chapter 3. We summarize the important result in the following theorem.

Theorem 2 *Consider the general mechanical system (7.29)–(7.31) subject to $m < d$ smooth constraint $\mathbf{g}(\mathbf{q}) = \mathbf{0}$. Assume that $\mathbf{G}(\mathbf{q})$ is of full rank (has linearly independent rows) along the configuration manifold $\mathcal{M}$. Suppose that $\boldsymbol{\phi} : \mathbb{R}^{d-m} \to \mathbb{R}^d$ defines a local coordinatization of the configuration space in terms of a new variable $\boldsymbol{\xi} \in \mathbb{R}^{d-m}$ by*

$$\mathbf{q} = \boldsymbol{\phi}(\boldsymbol{\xi}).$$

Then the canonical momenta $\boldsymbol{\nu}$ are defined by the equations

$$\boldsymbol{\phi}_{\xi}(\boldsymbol{\xi})^T \mathbf{p} = \boldsymbol{\nu},$$

and the Hamiltonian corresponding to the new variables is

$$\tilde{H}(\boldsymbol{\xi}, \boldsymbol{\nu}) = H(\boldsymbol{\phi}(\boldsymbol{\xi}), \left(\boldsymbol{\phi}_{\xi}(\boldsymbol{\xi})^T\right)^+ \boldsymbol{\nu}),$$

where $\boldsymbol{A}^+ = \boldsymbol{A}^T(\boldsymbol{A}\boldsymbol{A}^T)^{-1}$ *represents a right pseudo-inverse of the matrix* $\boldsymbol{A}$.

Because the differential equations (7.29)–(7.31) can be expressed as an unconstrained Hamiltonian system in the parameters, we know that the two-form $d\boldsymbol{\xi} \wedge d\boldsymbol{\nu}$ will be preserved by the phase flow. This could be taken as a definition of the canonical two-form for the constrained system.

We have already seen that, along the constraint, the canonical two-form $d\boldsymbol{q} \wedge d\boldsymbol{p}$ is conserved. These two expressions are in fact equivalent, as the following theorem shows.

Theorem 3 *If* $d\boldsymbol{q}$ *and* $d\boldsymbol{p}$ *represent a solution of the constrained variational equations (7.36)–(7.37), and* $\boldsymbol{\xi}$ *and* $\boldsymbol{\nu}$ *are defined as in Theorem 2, then*

$$d\boldsymbol{q} \wedge d\boldsymbol{p} = d\boldsymbol{\xi} \wedge d\boldsymbol{\nu}.$$

Proof. The proof of this theorem is straightforward: first, note that

$$\begin{aligned} d\boldsymbol{q} \wedge d\boldsymbol{p} &= (\boldsymbol{\phi}_{\xi}(\boldsymbol{\xi})d\boldsymbol{\xi}) \wedge d\boldsymbol{p} \\ &= d\boldsymbol{\xi} \wedge (\boldsymbol{\phi}_{\xi}(\boldsymbol{\xi}))^T d\boldsymbol{p}. \end{aligned}$$

Now

$$d\boldsymbol{\nu} = (\boldsymbol{\phi}_{\xi}(\boldsymbol{\xi}))^T d\boldsymbol{p} + \sum_{i=1}^{d} p_i \phi_{i,\xi\xi} d\boldsymbol{\xi},$$

where ϕ_i is the ith component of the vector function $\boldsymbol{\phi}$ and $\phi_{i,\xi\xi}$ its Hessian. Therefore

$$\begin{aligned} d\boldsymbol{\xi} \wedge d\boldsymbol{\nu} &= d\boldsymbol{\xi} \wedge (\boldsymbol{\phi}_{\xi}(\boldsymbol{\xi}))^T d\boldsymbol{p} + \sum_{i=1}^{d} p_i d\boldsymbol{\xi} \wedge \phi_{i,\xi\xi}(\boldsymbol{\xi}) d\boldsymbol{\xi} \\ &= d\boldsymbol{\xi} \wedge (\boldsymbol{\phi}_{\xi}(\boldsymbol{\xi}))^T d\boldsymbol{p}, \end{aligned}$$

which concludes the proof. □

Although the above discussion is local in the sense that it assumes that the parameters are defined by a single coordinate chart, there is nothing to prevent extending the idea to a family of charts defining a "symplectic atlas."

7.6.3 Numerical methods based on local charts

A straightforward approach to discretization can be based on a symplectic atlas: simply parameterize the phase space $\mathscr{P}$ in local charts, and then solve the resulting (unconstrained) Hamiltonian systems in the parameters using a symplectic integrator.

Several problems with this approach surface rapidly when one attempts to use it to perform a simulation. These include the difficulty of identifying the boundaries of the local charts for the manifold in a general purpose numerical integration framework and the potential additional computational complexity that may be introduced in the parametric vector fields.

In the Hamiltonian setting, there is an additional difficulty introduced by parameterization. According to the results of the last section, the Hamiltonian that arises via a canonical parameterization is typically nonseparable. This generally occurs in the presence of a nonlinear constraint regardless of whether the original Hamiltonian was separable. Nonseparable Hamiltonian functions are, typically much more difficult to integrate than separable Hamiltonian functions because they require the use of an implicit scheme (whereas separable Hamiltonian functions can be treated with explicit integrators).

However, as explained by LEIMKUHLER AND REICH [111], there is still one more serious drawback to using parameterizations to integrate a Hamiltonian system, having to do with the "nearby Hamiltonian" introduced in Chapter 5. Recall that a key feature of a symplectic integrator is that it generates the exact flow, up to an exponentially small error term, of a nearby Hamiltonian system. Thus taking many steps with the same symplectic integrator is the same as taking many steps along the flow of the perturbed Hamiltonian system. This implies the existence of a conserved quantity, not much different from the exact energy, and confers a certain long-term stability on the numerical simulation and, consequently, a certain structural stability of the system under the process of discretization.

When the numerical solution is obtained by integrating the differential equations in local minimal coordinates (local charts), we find that the symplectic map associated to the numerical simulation changes whenever we switch to a new chart: the realization as the flow of a single Hamiltonian system therefore does not apply. Thus the perturbed conserved quantity is lost, and the result observed in practice is typically a loss of long-term stability. The only exception to this rule is provided if one can find local coordinates in which one can integrate the equations of motion exactly. Then the modified Hamiltonian is the same as the original energy function and a change of charts does not affect conservation of energy. This obvious statement has some importance for the implementation of splitting methods in local coordinates (see BENETTIN, CHERUBINI, AND FASSÒ [17]).

In summary, as a general proposition, use of local coordinate charts as a basis for numerical integration is unwise for the following reasons:

- Identifying the boundaries of local charts and switching between separate coordinate charts in the course of a simulation can introduce a costly computational overhead. Missed chart changes may lead to singularities in the solutions of the differential equations.
- The use of a parametrization often adds significant computational complexity, e.g. by introducing transcendental functions into a vector field where only simple polynomials may otherwise be needed.
- Finally, the use of parameterized equations may greatly limit the possibilities for obtaining efficient geometric numerical integrators respecting other invariants of the flow.

Despite these pessimistic comments, parameterizations may sometimes be used effectively in special applications for which the constraint geometry is well understood, or for which the choice of appropriate parameters greatly simplifies the description of the vector field.

7.6.4 Methods based on projection

Because the flow of the *underlying ordinary differential equations* (7.9)–(7.10) has the tangent bundle $T\mathcal{M}$ as an invariant submanifold, this formulation is sometimes proposed as the basis for numerical simulation. However, it is important to note that $T\mathcal{M}$ will typically *not* be an invariant submanifold for the discretization method applied to those equations. This means that the numerical solution, started from initial conditions in $T\mathcal{M}$ may drift gradually off the manifold and into the larger Euclidean space in which the flow is embedded. One way around this problem is to *project* the numerical solution back on to $T\mathcal{M}$ at the end of each timestep [84] or to stabilize the manifold [10].

7.7 Exercises

1. *Local charts.* It is also possible to describe the pendulum in terms of just two Euclidean charts as follows (refer to Fig. 7.7). In the first chart, we identify a point $Q(x, y)$ of the unit circle with a point on the real line ξ by passing a line through Q and the point at the top of the unit circle N and letting R denote the point of intersection with the x-axis. This works for all points Q on the unit circle except N itself. In the other chart, we identify points of the

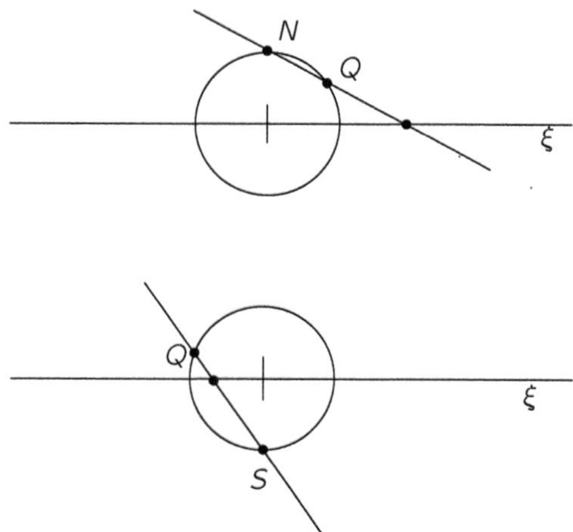

Figure 7.7 Parametrization of the circle in two charts on the real line.

unit circle to points of the x-axis in the same way, except that we use the "south pole" S instead of N as a point of the straight line.

a. Write out formulas for these two parameterizations and determine the canonical equations for the pendulum problem in each chart.

b. Discretize the equations using the implicit midpoint rule and device a proper criterion to switch from one chart to the other. Perform a numerical experiment with $g = 0$ and $(q_1(0), q_2(0), p_1(0), p_2(0)) = (1, 0, 0, -2)$. Monitor the total energy.

c. Repeat the experiment under (b) with a SHAKE/RATTLE discretization of the constrained formulation for the pendulum.

2. *Regularity of constraints.* Let $\boldsymbol{M}$ be a symmetric positive definite matrix. Show that the matrix $\boldsymbol{G}\boldsymbol{M}^{-1}\boldsymbol{G}^T$ is nonsingular provided the matrix $\boldsymbol{G}$ has full rank (i.e. has linearly independent rows).

3. *Time-reversibility.* Show that SHAKE and RATTLE are time-reversible.

4. *Linear stability analysis.* Consider the *linearly constrained* system with equations of motion

$$\begin{aligned}
\dot{x} &= u, \\
\dot{y} &= v, \\
\dot{u} &= -\omega_1^2 x - \alpha\lambda, \\
\dot{v} &= -\omega_2^2 y - \beta\lambda, \\
\alpha x + \beta y &= 0,
\end{aligned}$$

where α, β, ω_1 and ω_2 are real numbers, and we will assume the normalization $\alpha^2 + \beta^2 = 1$.

a. Eliminate the multiplier (by twice differentiating the constraint relationship with respect to time) to obtain

$$\lambda = \alpha\omega_1^2 x + \beta\omega_2^2 y.$$

b. Next, assume $\beta \neq 0$, and solve the constraint for y in terms of x (this is a global parameterization). Show that the equations of motion reduce to

$$\dot{x} = u,$$
$$\dot{u} = -\left((1-\alpha^2)\omega_1^2 + \alpha^2\omega_2^2\right))x,$$

which describes a new harmonic oscillator with frequency

$$\bar{\omega} = \sqrt{(1-\alpha^2)\omega_1^2 + \alpha^2\omega_2^2}.$$

c. Apply the RATTLE discretization to the constrained system, resulting in

$$\begin{aligned} x^{n+1} &= x^n + \Delta t u^{n+1/2}, \\ y^{n+1} &= y^n + \Delta t v^{n+1/2}, \\ u^{n+1/2} &= u^n - \frac{1}{2}\Delta t\omega_1^2 x^n - \frac{1}{2}\Delta t\alpha\lambda_{(r)}^n, \\ v^{n+1/2} &= v^n - \frac{1}{2}\Delta t\omega_2^2 y^n - \frac{1}{2}\Delta t\beta\lambda_{(r)}^n, \\ \alpha x^{n+1} + \beta y^{n+1} &= 0, \end{aligned}$$

and

$$\begin{aligned} u^{n+1} &= u^{n+1/2} - \frac{1}{2}\Delta t\omega_1^2 x^{n+1} - \frac{1}{2}\Delta t\alpha\lambda_{(v)}^{n+1}, \\ v^{n+1} &= v^{n+1/2} - \frac{1}{2}\Delta t\omega_2^2 y^{n+1} - \frac{1}{2}\Delta t\beta\lambda_{(v)}^{n+1}, \\ \alpha u^{n+1} + \beta v^{n+1} &= 0. \end{aligned}$$

Assume $\alpha x^n + \beta y^n = 0$ and also $\alpha u^n + \beta v^n = 0$, so that

$$\lambda_{(r)}^n = \alpha\omega_1^2 x^n + \beta\omega_2^2 y^n.$$

Find a similar expression for $\lambda_{(v)}^{n+1}$.

Show that RATTLE reduces in the coordinates x^n, u^n, to a Störmer–Verlet discretization of the harmonic oscillator with frequency $\bar{\omega}$. What can you conclude about possible stepsize restrictions?

e. Repeat the discussion for RATTLE replaced by the implicit midpoint method as the basic integration method in (7.39)–(7.43).

5. *Constraint chains.* The iteration matrix

$$R = G(q^n)M^{-1}G(q^n)^T$$

appearing in a quasi-Newton method for SHAKE takes a particularly simple form for problems only involving length constraints. We can write these constraints in the following compact form

$$g_i = \frac{1}{2}(q^T S_i q - L_i^2),$$

where L_i is the length of the constraint, and S_i is a $d \times d$ *stamp matrix* with 3×3 blocks. This matrix is zero in all but the following components: if the ith constraint links particles k and l, then the kk and ll blocks of S_i are I_3 (3×3 identity matrix) and the kl and lk blocks are $-I_3$. With this notation it is easy to show that the ij element R_{ij} of R has the form

$$R_{ij} = q^T S_i M^{-1} S_j q.$$

This implies in turn that

$$R_{ii} = (\frac{1}{m_{\text{left}(i)}} + \frac{1}{m_{\text{right}(i)}})L_i^2,$$

where left(i) and right(i) represent the two particle indices associated to constraint i. Now R_{ij} is zero unless constraints i and j share a common particle. If right(i) = left(j), then we can write

$$\begin{aligned} R_{ij} &= \frac{1}{m_{\text{left}(j)}}(q_{\text{left}(i)} - q_{\text{right}(i)})^T(q_{\text{left}(j)} - q_{\text{right}(j)}) \\ &= \frac{1}{m_{\text{left}(j)}} r_i r_j \cos\theta_{ij} \end{aligned}$$

with $r_i = \|q_{\text{left}(i)} - q_{\text{right}(i)}\|$, $r_j = \|q_{\text{left}(j)} - q_{\text{right}(j)}\|$ and θ_{ij} is the angle between the two constraints i and j. Now along solutions of the constrained equations, we know that $r_i \equiv L_i$, etc.

a. Consider a system of three particles of equal mass connected by two rigid length constraints of equal length. Find an explicit expression for the associated 2×2 matrix R.

b. Implement the SHAKE time discretization and apply (i) a quasi-Newton method and (ii) SHAKE iteration to solve the nonlinear constraint equations. Compare the two approaches in terms of efficiency.

c. Consider a system of three particles of equal mass connected by three rigid length constraints of equal length. Repeat exercises (a) and (b).

8

Rigid body dynamics

In this chapter, we discuss formulation issues and symplectic integration methods for simulating the motion of a rigid body. Rigid bodies arise frequently in engineering, chemistry, and physics. For example, they occur in molecular simulation when the flexibility of small polyatomic units such as the water molecule, or CH_4 is ignored. Cogwheels, space vehicles, and the planets are some other objects that are commonly modeled by rigid bodies.

Even in the absence of external applied forces, any rigid body more complicated than a uniform sphere will exhibit complicated motion, as defined by the moments of inertia of the body. A hint of the potential complexity of the motion is provided by the classic illustration using a hardbound book, which typically has three unequal moments of inertia $I_1 < I_2 < I_3$ with I_1 corresponding to an axis drawn along the binding, I_2 to an axis across the cover, and I_3 to an axis through the pages of the book (see Fig. 8.1). As the book is tossed up and spinning around each of the axes, the following dynamics are observed: around the first and third axes, the motion combines a stable periodic rotation with the rising and falling motion due to gravity, whereas the rotation with respect to the middle axis is much more complicated. (See, for example, [124] for more explanation.) (It helps to place a rubber band around the book's cover to keep it closed while conducting experiments.)

Developing a method to simulate general rigid body motions, especially for long-term integration, proves an interesting and challenging task. The first issue we must confront is the selection of a set of coordinates that describe body orientation and spatial position. Since a rigid body has six degrees of freedom, the positional description must include at least this many configuration variables. For example, a rigid body can be formulated in terms of three Euler angles describing the orientation in space relative to a reference configuration defined in terms of a certain sequence of axial rotations together with the three coordinates of the center of mass. However, if these variables, together with the corresponding momenta, are selected as the basis for a canonical description of the motion of

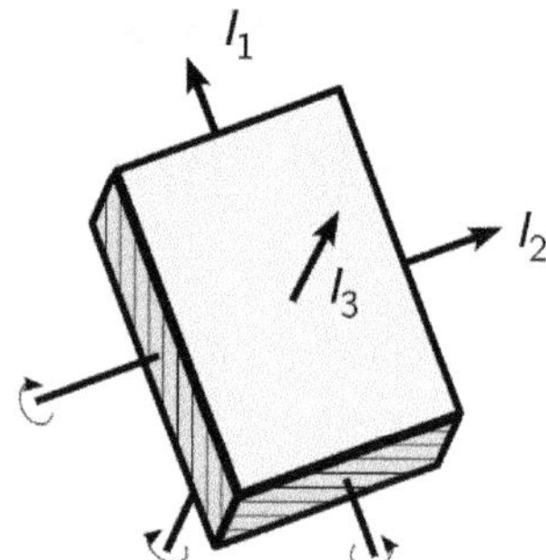

Figure 8.1 A hardbound book tossed up and spinning about one of its axes of inertia provides a simple illustration of rigid body motion.

the body, the resulting equations of motion are found to have singularities which impede their numerical solution.

To overcome this defect of the Euler angles, it is possible – and often desirable – to introduce additional redundant variables together with constraints. We will develop various approaches by relating them to a natural canonical formulation of the rigid body dynamics in terms of *rotation matrices:* the Lie group $SO(3)$ consisting of 3×3 orthogonal matrices with unit determinant.[1] This is a *global coordinatization* in the language of the last chapter. The rotational kinetic energy term of the Lagrangian can be written in these variables simply as

$$\mathcal{L}_{\mathrm{rot}} = \frac{1}{2}\mathrm{trace}(\dot{\boldsymbol{Q}}\boldsymbol{R}\dot{\boldsymbol{Q}}^T), \tag{8.1}$$

where $\boldsymbol{Q}$ is subject to an orthogonality constraint $\boldsymbol{Q}^T\boldsymbol{Q} = I_3$, I_3 the 3×3 identity matrix, and $\boldsymbol{R}$ is a constant 3×3 matrix closely related to the "inertia tensor" of the body.

In addition to the rotation matrix formulation, other popular choices of parameterization of rigid body motion include Hamilton's quaternions (Cayley–Klein parameters) and cartesian (particle) models. Another formulation is based on the Euler equations, which can be obtained by a formal reduction procedure. In principle, any of these formulations can be used with a variety of timestepping methods. However, *the choice of parameterization and the design of a discretization method are not independent.* We will see that some sets of parameters facilitate efficient symplectic/reversible discretization while others may make this task difficult or render the resulting schemes useless because of the computational expense involved.

[1]A Lie group is a set which possesses simultaneously the structure of an algebraic group and that of a smooth manifold. We do not assume any knowledge of Lie group theory here, although such knowledge will undoubtedly enable the reader to gain more from the discussion.

8.1 Rigid bodies as constrained systems

We now consider the formulation of the equations of motion for a single rigid body acted on by a (possibly nonlinear) potential field. The basic ideas naturally extend to systems of interacting rigid bodies.

The definition of a rigid body (see [104, 7, 2]) is simply a set of particles of prescribed mass whose positions are relatively fixed in space. Assume for the moment that the number of particles is finite, and that their instantaneous positions are $\boldsymbol{q}_i$, $i = 1, \ldots, k$. The kinetic energy of this collection is given by $T = \sum_{i=1}^{k} m_i \|\dot{\boldsymbol{q}}_i\|^2$, and the potential by $V = V(\boldsymbol{q}_1, \boldsymbol{q}_2, \ldots, \boldsymbol{q}_k)$.

Define by $\boldsymbol{q}_{\mathrm{cm}}$ the center of mass of the set of particles:

$$\boldsymbol{q}_{\mathrm{cm}} = \frac{\sum_{i=1}^{k} m_i \boldsymbol{q}_i}{\sum_{i=1}^{k} m_i}.$$

In a fixed *reference configuration* $(\boldsymbol{r}_1^0, \boldsymbol{r}_2^0, \ldots, \boldsymbol{r}_k^0)$ the center of mass is located at the origin. Any possible configuration of the particles of the rigid body can be expressed in terms of two types of elementary motions applied to some reference configuration: a common rotation of each of the particles of the body about the center of mass followed by a common translation in space of each particle. Let $\boldsymbol{Q}$ denote an orthogonal 3×3 matrix with unit determinant, i.e. a rotation matrix with respect to vectors in $\mathbb{R}^3$, and let $\boldsymbol{d} \in \mathbb{R}^3$ be a displacement, then we may write

$$\boldsymbol{q}_i = \boldsymbol{Q}\boldsymbol{r}_i^0 + \boldsymbol{d}.$$

If we view the points of the body as moving in space along some smooth trajectory, we can imagine a corresponding trajectory in the orientation $\boldsymbol{Q}$ and spatial position $\boldsymbol{d}$, i.e.

$$\boldsymbol{q}_i(t) = \boldsymbol{Q}(t)\boldsymbol{r}_i^0 + \boldsymbol{d}(t).$$

Since $\boldsymbol{q}_i(t) = \boldsymbol{q}_{\mathrm{cm}}(t)$ for $\boldsymbol{r}_i^0 = \boldsymbol{0}$, we have $\boldsymbol{d}(t) = \boldsymbol{q}_{\mathrm{cm}}(t)$ and the velocities therefore satisfy

$$\frac{d}{dt}\boldsymbol{q}_i(t) = \dot{\boldsymbol{q}}_{\mathrm{cm}}(t) + \dot{\boldsymbol{Q}}(t)\boldsymbol{r}_i^0,$$

so the kinetic energy can be written as

$$T = \frac{1}{2}\sum_{i=1}^{k} m_i \|\dot{\boldsymbol{q}}_{\mathrm{cm}}(t) + \dot{\boldsymbol{Q}}(t)\boldsymbol{r}_i^0\|^2.$$

Expanding the squared two-norm using the properties of the inner product, we find

$$T = \frac{1}{2}\sum_{i=1}^{k} m_i (\|\dot{\boldsymbol{q}}_{\mathrm{cm}}(t)\|^2 + 2\dot{\boldsymbol{q}}_{\mathrm{cm}}(t) \cdot [\dot{\boldsymbol{Q}}(t)\boldsymbol{r}_i^0] + \|\dot{\boldsymbol{Q}}(t)\boldsymbol{r}_i^0\|^2).$$

Now the middle term in this expression vanishes, since

$$\sum_{i=1}^{k} m_i \dot{\boldsymbol{q}}_{\mathrm{cm}}(t) \cdot [\dot{\boldsymbol{Q}}(t)\boldsymbol{r}_i^0] = \dot{\boldsymbol{q}}_{\mathrm{cm}}(t) \cdot [\dot{\boldsymbol{Q}}(t) \sum_{i=1}^{k} m_i \boldsymbol{r}_i^0] = 0,$$

as $(\boldsymbol{r}_1^0, \ldots, \boldsymbol{r}_k^0)$ is a reference configuration, i.e. $\sum_i m_i \boldsymbol{r}_i^0 = \boldsymbol{0}$.

We can now define translation and rotational parts of the kinetic energy by

$$T_{\mathrm{trans}} = \frac{1}{2} \sum_{i=1}^{k} m_i \|\dot{\boldsymbol{q}}_{\mathrm{cm}}\|^2, \tag{8.2}$$

and

$$T_{\mathrm{rot}} = \frac{1}{2} \sum_{i=1}^{k} m_i \|\dot{\boldsymbol{Q}}\boldsymbol{r}_i^0\|^2.$$

To express the rotational kinetic energy more compactly, observe the following identity of linear algebra: for any pair of vectors $\boldsymbol{u}, \boldsymbol{v} \in \mathbb{R}^k$

$$\boldsymbol{u}^T \boldsymbol{v} = \mathrm{tr}\left(\boldsymbol{u}\boldsymbol{v}^T\right),$$

where $\mathrm{tr}(\boldsymbol{A})$ represents the *trace* (sum of diagonal elements) of a matrix $\boldsymbol{A}$. Using this fact, we have

$$\begin{aligned}
T_{\mathrm{rot}} &= \frac{1}{2} \sum_{i=1}^{k} m_i \|\dot{\boldsymbol{Q}}\boldsymbol{r}_i^0\|^2 \\
&= \frac{1}{2} \sum_{i=1}^{k} m_i \left[\dot{\boldsymbol{Q}}\boldsymbol{r}_i^0\right]^T \left[\dot{\boldsymbol{Q}}\boldsymbol{r}_i^0\right] \\
&= \frac{1}{2} \sum_{i=1}^{k} m_i \,\mathrm{tr}\left(\left[\dot{\boldsymbol{Q}}\boldsymbol{r}_i^0\right] \left[\dot{\boldsymbol{Q}}\boldsymbol{r}_i^0\right]^T\right) \\
&= \frac{1}{2} \sum_{i=1}^{k} m_i \,\mathrm{tr}\left(\dot{\boldsymbol{Q}}\boldsymbol{r}_i^0 [\boldsymbol{r}_i^0]^T \dot{\boldsymbol{Q}}^T\right) \\
&= \frac{1}{2} \mathrm{tr}\left(\dot{\boldsymbol{Q}}\boldsymbol{R}\dot{\boldsymbol{Q}}^T\right),
\end{aligned}$$

where we have defined the symmetric matrix $\boldsymbol{R}$ by

$$\boldsymbol{R} = \sum_{i=1}^{k} m_i \boldsymbol{r}_i^0 [\boldsymbol{r}_i^0]^T.$$

The matrix $\boldsymbol{R}$ can be viewed as a mass-weighted sum of projections along the displacements of points in the reference configuration. We will refer to $\boldsymbol{R}$ as the *mass tensor* of the body.[2]

Since $\boldsymbol{R}$ is symmetric one can always find a reference coordinate system $\bar{\boldsymbol{e}}_i$, $i = 1, 2, 3$, so that $\boldsymbol{R}$ is diagonal with diagonal entries r_{ii}, $i = 1, 2, 3$. Note that the coordinate system is held fixed on the rigid body contrary to a laboratory coordinate system which is fixed in space. The mass tensor $\boldsymbol{R}$ can now be written as

$$\boldsymbol{R} = \sum_{i=1}^{3} r_{ii} \bar{\boldsymbol{e}}_i \bar{\boldsymbol{e}}_i^T = \begin{bmatrix} r_{11} & 0 & 0 \\ 0 & r_{22} & 0 \\ 0 & 0 & r_{33} \end{bmatrix}. \tag{8.3}$$

We will see later in this chapter that this particular choice of the reference system is equivalent to the statement that the inertia tensor of the rigid body is also diagonal with the principal moments of inertia equal to $I_1 = r_{22} + r_{33}$, $I_2 = r_{11} + r_{33}$, and $I_3 = r_{11} + r_{22}$. From now on we will assume that all matrices are expressed with respect to the reference coordinate system $\bar{\boldsymbol{e}}_i$, $i = 1, 2, 3$.

In the given variables, the Lagrangian becomes

$$L = T_{\mathrm{rot}}(\dot{\boldsymbol{Q}}) + T_{\mathrm{trans}}(\dot{\boldsymbol{q}}_{\mathrm{cm}}) - V_{\mathrm{ext}}(\boldsymbol{Q}, \boldsymbol{q}_{\mathrm{cm}}), \tag{8.4}$$

where V_{ext} is the potential energy function expressed in terms of the center of mass of the collection of particles and the rotation matrix. We must bear in mind, however, that this Lagrangian is subject to a holonomic constraint: the condition that $\boldsymbol{Q}$ must be an orthogonal matrix!

In order to formulate the constraint and augmented Lagrangian, we note that the orthogonality condition on $\boldsymbol{Q}$

$$\boldsymbol{Q}^T \boldsymbol{Q} = I_3,$$

gives rise to six independent constraints on the matrix $\boldsymbol{Q}$, so six multipliers are needed. These can be introduced in the form of the six independent elements of a symmetric matrix $\boldsymbol{\Lambda}$, with the augmented Lagrangian having the simple

[2]We may also think of replacing larger groups of rigidly constrained point masses by small groups of "pseudo" particles so that the resulting rigid body has the same mass tensor as the original one, and therefore identical dynamical properties. Such "pseudo" particle formulations for arbitrary rigid bodies are described in the classic book of ROUTH [164], and has been used as a basis for simulation, for example, in [15]. The essential idea is that any constrained particle system can be treated numerically by the SHAKE or RATTLE method described in Chapter 7.

expression

$$\tilde{L} = T_{\rm rot}(\dot{\boldsymbol{Q}}) + T_{\rm trans}(\dot{\boldsymbol{q}}_{\rm cm}) - V_{\rm ext}(\boldsymbol{Q}, \boldsymbol{q}_{\rm cm}) - \operatorname{tr}((\boldsymbol{Q}^T\boldsymbol{Q} - \boldsymbol{I}_3)\boldsymbol{\Lambda}). \quad (8.5)$$

8.1.1 Hamiltonian formulation

Recall that the gradient $\nabla_{\boldsymbol{q}} V(\boldsymbol{q})$ of a scalar-valued function $V(\boldsymbol{q})$ is defined by

$$\langle \nabla_{\boldsymbol{q}} V(\boldsymbol{q}), \boldsymbol{\delta q} \rangle = \lim_{\varepsilon \to 0} \frac{V(\boldsymbol{q} + \varepsilon \boldsymbol{\delta q}) - V(\boldsymbol{q})}{\varepsilon}.$$

Since we will now work with matrices, we have to introduce an appropriate inner product. Throughout this chapter we will use

$$\langle \boldsymbol{A}, \boldsymbol{B} \rangle = \operatorname{tr}(\boldsymbol{A}\boldsymbol{B}^T) \quad (8.6)$$

for any two 3×3 matrices $\boldsymbol{A}$ and $\boldsymbol{B}$.

Once the (unconstrained) Lagrangian (8.4) has been found, we may introduce the canonical momentum in the usual way by a matrix

$$\boldsymbol{P} = \nabla_{\dot{\boldsymbol{Q}}} L(\boldsymbol{Q}, \dot{\boldsymbol{Q}}).$$

One finds that this matrix can be expressed compactly as $\boldsymbol{P} = \dot{\boldsymbol{Q}}\boldsymbol{R}$. It is important to keep in mind that the two matrices $\boldsymbol{Q}$ and $\boldsymbol{R}$ do *not* commute. Hence the ordering of the two variables in the definition of $\boldsymbol{P}$ is crucial.

It must be pointed out that there are certain rigid bodies for which the matrix $\boldsymbol{R}$ is singular; specifically linear and planar bodies have this feature. For the moment we will simply assume that $\boldsymbol{R}$ is nonsingular; we will correct the equations for the case of planar bodies in Section 8.1.2.

The momentum corresponding to the center of mass has the expression

$$\boldsymbol{p}_{\rm cm} = M\dot{\boldsymbol{q}}_{\rm cm},$$

with $M = \sum_{i=1}^k m_i$ the total mass of the rigid body.

The constrained Hamiltonian formulation is now found exactly in the same manner as outlined in the previous chapter and we obtain:

CONSTRAINED HAMILTONIAN FORMULATION FOR A NONPLANAR BODY

$$\tilde{H} = \frac{1}{2}\mathrm{tr}(\boldsymbol{P}\boldsymbol{R}^{-1}\boldsymbol{P}^T) + \frac{1}{2M}\|\boldsymbol{p}_{\mathrm{cm}}\|^2 + V_{\mathrm{ext}}(\boldsymbol{Q}, \boldsymbol{q}_{\mathrm{cm}}) + \mathrm{tr}((\boldsymbol{Q}^T\boldsymbol{Q} - \boldsymbol{I}_3)\boldsymbol{\Lambda}), \quad (8.7)$$

$$\frac{d}{dt}\boldsymbol{q}_{\mathrm{cm}} = M^{-1}\boldsymbol{p}_{\mathrm{cm}}, \quad (8.8)$$

$$\frac{d}{dt}\boldsymbol{p}_{\mathrm{cm}} = -\nabla_{\boldsymbol{q}_{\mathrm{cm}}} V_{\mathrm{ext}}(\boldsymbol{Q}, \boldsymbol{q}_{\mathrm{cm}}), \quad (8.9)$$

$$\frac{d}{dt}\boldsymbol{Q} = \boldsymbol{P}\boldsymbol{R}^{-1}, \quad (8.10)$$

$$\frac{d}{dt}\boldsymbol{P} = -\nabla_{\boldsymbol{Q}} V_{\mathrm{ext}}(\boldsymbol{Q}, \boldsymbol{q}_{\mathrm{cm}}) - 2\boldsymbol{Q}\boldsymbol{\Lambda}, \quad (8.11)$$

$$\boldsymbol{I}_3 = \boldsymbol{Q}^T\boldsymbol{Q}. \quad (8.12)$$

The holonomic constraint (8.12) implies that $\boldsymbol{Q}$ is a rotation matrix provided the initial $\boldsymbol{Q}(0) = \boldsymbol{Q}_0$ is a rotation matrix; i.e. $\boldsymbol{Q}_0^T\boldsymbol{Q}_0 = \boldsymbol{I}_3$ and $\det \boldsymbol{Q}_0 = 1$. The set of all rotation matrices $\boldsymbol{Q} \in \mathbb{R}^3$ forms the group $SO(3)$.

Differentiate the constraint $\boldsymbol{Q}^T\boldsymbol{Q} = \boldsymbol{I}_3$

$$\boldsymbol{Q}^T\dot{\boldsymbol{Q}} + \dot{\boldsymbol{Q}}^T\boldsymbol{Q} = \boldsymbol{0},$$

hence, from (8.10) and using the symmetry of $\boldsymbol{R}$

$$\boldsymbol{Q}^T\boldsymbol{P}\boldsymbol{R}^{-1} + \boldsymbol{R}^{-1}\boldsymbol{P}^T\boldsymbol{Q} = \boldsymbol{0}. \quad (8.13)$$

Thus, when viewed as a constrained system, the equations (8.10)–(8.12) constitute a Hamiltonian system on the manifold

$$\mathcal{P} = \{(\boldsymbol{Q}, \boldsymbol{P}) \in \mathbb{R}^{3\times3} \times \mathbb{R}^{3\times3} : \boldsymbol{Q}^T\boldsymbol{Q} = \boldsymbol{I}_3, \boldsymbol{Q}^T\boldsymbol{P}\boldsymbol{R}^{-1} + \boldsymbol{R}^{-1}\boldsymbol{P}^T\boldsymbol{Q} = \boldsymbol{0}\}.$$

The symplectic structure on $\mathcal{P}$ is given by the restriction of the canonical symplectic structure on $\mathbb{R}^{3\times3} \times \mathbb{R}^{3\times3}$ to $\mathcal{P}$. The canonical structure on $\mathbb{R}^{3\times3} \times \mathbb{R}^{3\times3}$, in turn is obtained by viewing 3×3 matrices as vectors in $\mathbb{R}^9$.

Note that the manifold $\mathcal{P}$ is not the cotangent bundle $T^*SO(3)$, which can be identified with the manifold

$$T^*SO(3) = \{(\boldsymbol{Q}, \bar{\boldsymbol{P}}) \in \mathbb{R}^{3\times3} \times \mathbb{R}^{3\times3} : \boldsymbol{Q}^T\boldsymbol{Q} = \boldsymbol{I}_3, \boldsymbol{Q}^T\bar{\boldsymbol{P}} + \bar{\boldsymbol{P}}^T\boldsymbol{Q} = \boldsymbol{0}\}.$$

However, we can relate any element $(\boldsymbol{Q}, \boldsymbol{P}) \in \mathcal{P}$ to an element $(\boldsymbol{Q}, \bar{\boldsymbol{P}}) \in T^*SO(3)$ via the transformation

$$\bar{\boldsymbol{P}} = \boldsymbol{P} - \boldsymbol{Q}\boldsymbol{\Gamma},$$

where the symmetric matrix $\boldsymbol{\Gamma} \in \mathbb{R}^{3\times 3}$ is defined by

$$\boldsymbol{\Gamma} = \frac{1}{2}\left(\boldsymbol{Q}^T\boldsymbol{P} + \boldsymbol{P}^T\boldsymbol{Q}\right).$$

This is equivalent to

$$\bar{\boldsymbol{P}} = \frac{1}{2}\left(\boldsymbol{P} - \boldsymbol{Q}\boldsymbol{P}^T\boldsymbol{Q}\right). \tag{8.14}$$

It can be shown that this map is symplectic (this is equivalent to showing that a map from $(\boldsymbol{q}, \boldsymbol{p})$ to $(\boldsymbol{q}, \bar{\boldsymbol{p}})$ defined by

$$\bar{\boldsymbol{p}} = \boldsymbol{p} - \boldsymbol{G}(\boldsymbol{q})^T\boldsymbol{\lambda}$$

is symplectic provided $\boldsymbol{g}(\boldsymbol{q}) = \boldsymbol{0}$, as pointed out in Lemma 1 of Chapter 7). We conclude that the equations (8.10)–(8.12) combined with (8.14) define a Hamiltonian system on $T^*SO(3)$.

Let us also briefly discuss the evaluation of $\nabla_{\boldsymbol{Q}} V_{\text{ext}}(\boldsymbol{Q}, \boldsymbol{q}_{\text{cm}})$. In many applications, for example a force action on a fixed reference location $\boldsymbol{r}^0$ on the rigid body, we are led to a potential energy term of the form

$$V(\boldsymbol{q}) = V(\boldsymbol{q}_{\text{cm}} + \boldsymbol{Q}\boldsymbol{r}^0) = V_{\text{ext}}(\boldsymbol{Q}, \boldsymbol{q}_{\text{cm}}).$$

Then the definition

$$\text{tr}(\nabla_{\boldsymbol{Q}} V_{\text{ext}}(\boldsymbol{Q}, \boldsymbol{q}_{\text{cm}})\,\delta\boldsymbol{Q}^T) = \lim_{\varepsilon\to 0} \frac{V_{\text{ext}}(\boldsymbol{Q} + \varepsilon\delta\boldsymbol{Q}, \boldsymbol{q}_{\text{cm}}) - V_{\text{ext}}(\boldsymbol{Q}, \boldsymbol{q}_{\text{cm}})}{\varepsilon} \tag{8.15}$$

leads to the expression

$$\nabla_{\boldsymbol{Q}} V_{\text{ext}}(\boldsymbol{Q}, \boldsymbol{q}_{\text{cm}}) = \nabla_{\boldsymbol{q}} V(\boldsymbol{q})\,[\boldsymbol{r}^0]^T. \tag{8.16}$$

See the Exercises.

8.1.2 Linear and planar bodies

In the case of linear and planar bodies, the above discussion must be amended. For a linear body (a pendulum), in which all the points of the body lie along a straight line, the orientation in space is defined by a single unit vector $\boldsymbol{u}$, so there is just one multiplier. The mass tensor $\boldsymbol{R}$ will have only its $(1,1)$ element nonzero.

For a planar body, the orientation is defined by two vectors, say $\boldsymbol{u}_1$ and $\boldsymbol{u}_2$. There are three constraints (unit length for each of the two vectors and orthogonality of $\boldsymbol{u}_1$ and $\boldsymbol{u}_2$) and therefore three multipliers which can be cast in the form of a 2×2 symmetric matrix.

A unified treatment of the equations for linear, planar, and three-dimensional bodies is possible in the formulation (8.7) and (8.8)–(8.12) if we view $\boldsymbol{Q}$ and $\boldsymbol{P}$ as lying in $\mathbb{R}^{3\times k}$, with k the dimension of the body, and suppose $\boldsymbol{R}$ and $\boldsymbol{\Lambda}$ to be $k \times k$ (rather than 3×3) symmetric matrices. (See the Exercises.)

8.1.3 Symplectic discretization using SHAKE

Let us now consider the symplectic integration of the rigid body. For simplicity, we will initially restrict our treatment to the case where the center of mass of the body is fixed in space, so that only rotational motions, governed by equations (8.10)–(8.12), need be considered.

As first pointed out by McLachlan and Scovel [133] and Reich [154, 156], the obvious approach is to apply the SHAKE/RATTLE method to the constrained equations. This yields a symplectic method that propagates the variable $(\boldsymbol{Q}, \boldsymbol{P})$ on $\mathcal{P}$. The discrete equations for the RATTLE discretization can be written as follows:

RATTLE DISCRETIZATION OF RIGID BODY MOTION

$$\boldsymbol{Q}^{n+1} = \boldsymbol{Q}^n + \Delta t \boldsymbol{P}^{n+1/2} \boldsymbol{R}^{-1}, \tag{8.17}$$

$$\boldsymbol{P}^{n+1/2} = \boldsymbol{P}^n - \frac{\Delta t}{2} \nabla_{\boldsymbol{Q}} V_{\text{ext}}(\boldsymbol{Q}^n) - \Delta t \boldsymbol{Q}^n \boldsymbol{\Lambda}^n_{(r)}, \tag{8.18}$$

$$\boldsymbol{I}_3 = [\boldsymbol{Q}^{n+1}]^T \boldsymbol{Q}^{n+1}, \tag{8.19}$$

$$\boldsymbol{P}^{n+1} = \boldsymbol{P}^{n+1/2} - \frac{\Delta t}{2} \nabla_{\boldsymbol{Q}} V_{\text{ext}}(\boldsymbol{Q}^{n+1}) - \Delta t \boldsymbol{Q}^{n+1} \boldsymbol{\Lambda}^{n+1}_{(v)}, \tag{8.20}$$

$$\boldsymbol{0} = [\boldsymbol{Q}^{n+1}]^T \boldsymbol{P}^{n+1} \boldsymbol{R}^{-1} + \boldsymbol{R}^{-1} [\boldsymbol{P}^{n+1}]^T \boldsymbol{Q}^{n+1}. \tag{8.21}$$

The implementation of the RATTLE method requires that we solve a nonlinear system at each timestep. Setting

$$\bar{\boldsymbol{Q}}^{n+1} := \boldsymbol{Q}^n + \Delta t \boldsymbol{P}^n \boldsymbol{R}^{-1} - \frac{\Delta t^2}{2} \nabla_{\boldsymbol{Q}} V_{\text{ext}}(\boldsymbol{Q}^n) \, \boldsymbol{R}^{-1},$$

and writing $\boldsymbol{Q}$ for $\boldsymbol{Q}^{n+1}$ as well as $\boldsymbol{\Lambda}$ for $\boldsymbol{\Lambda}^n_{(r)}$, the equations become

$$\boldsymbol{Q} = \bar{\boldsymbol{Q}}^{n+1} - \Delta t^2 \boldsymbol{Q}^n \boldsymbol{\Lambda} \boldsymbol{R}^{-1},$$
$$\boldsymbol{I}_3 = \boldsymbol{Q}^T \boldsymbol{Q}.$$

Substituting the first equation into the second, we obtain

$$\boldsymbol{I}_3 = [\bar{\boldsymbol{Q}}^{n+1}]^T \bar{\boldsymbol{Q}}^{n+1} - \Delta t^2 \left\{ \boldsymbol{R}^{-1} \boldsymbol{\Lambda} [\boldsymbol{Q}^n]^T \bar{\boldsymbol{Q}}^{n+1} + [\bar{\boldsymbol{Q}}^{n+1}]^T \boldsymbol{Q}^n \boldsymbol{\Lambda} \boldsymbol{R}^{-1} \right\} + \\ + \Delta t^4 \boldsymbol{R}^{-1} \boldsymbol{\Lambda} \boldsymbol{\Lambda} \boldsymbol{R}^{-1},$$

where we have made use of the symmetry of $\boldsymbol{R}$ and $\boldsymbol{\Lambda}$. To work out a simple iterative solution scheme, we first neglect terms of order Δt^3 and higher. Taking note of $[\bar{\boldsymbol{Q}}^{n+1}]^T \boldsymbol{Q}^n = \boldsymbol{I}_3 + \mathcal{O}(\Delta t)$, we obtain the linear relation

$$\boldsymbol{0} \approx ([\bar{\boldsymbol{Q}}^{n+1}]^T \bar{\boldsymbol{Q}}^{n+1} - \boldsymbol{I}_3) - \Delta t^2 (\boldsymbol{\Lambda} \boldsymbol{R}^{-1} + \boldsymbol{R}^{-1} \boldsymbol{\Lambda}),$$

which we have to solve for $\bar{\boldsymbol{\Lambda}} = \Delta t^2 \boldsymbol{\Lambda}$.

We now recall the special form (8.3) of the mass tensor $\boldsymbol{R}$ which allows us to easily solve any system of type

$$\mathbf{0} = \boldsymbol{M} - (\bar{\boldsymbol{\Lambda}}\boldsymbol{R}^{-1} + \boldsymbol{R}^{-1}\bar{\boldsymbol{\Lambda}}) \tag{8.22}$$

for $\bar{\boldsymbol{\Lambda}}$. Specifically, multiplication of (8.22) by $\bar{\boldsymbol{e}}_i^T$ from the right and by $\bar{\boldsymbol{e}}_j$ from the left yields $0 = m_{ij} - \bar{\lambda}_{ij}/r_{jj} - \bar{\lambda}_{ij}/r_{ii}$. This relation is easily inverted to express the solution $\bar{\boldsymbol{\Lambda}} = \{\bar{\lambda}_{ij}\}$ explicitly as

$$\bar{\lambda}_{ij} = \frac{r_{ii}r_{jj}}{r_{ii} + r_{jj}} m_{ij} \qquad (i, j = 1, 2, 3). \tag{8.23}$$

In our case, we set

$$\boldsymbol{M} = [\bar{\boldsymbol{Q}}^{n+1}]^T \bar{\boldsymbol{Q}}^{n+1} - \boldsymbol{I}_3$$

and obtain, in first approximation

$$\boldsymbol{Q}_{(1)} = \bar{\boldsymbol{Q}}^{n+1} - \boldsymbol{Q}^n \bar{\boldsymbol{\Lambda}}_{(1)} \boldsymbol{R}^{-1},$$

with $\bar{\boldsymbol{\Lambda}}_{(1)} = \{\bar{\lambda}_{ij}\}$ determined by (8.23). This guides us to the simple quasi-Newton iteration

$$\boldsymbol{Q}_{(k)} = \boldsymbol{Q}_{(k-1)} - \boldsymbol{Q}^n \bar{\boldsymbol{\Lambda}}_{(k)} \boldsymbol{R}^{-1},$$
$$\mathbf{0} = (\boldsymbol{Q}_{(k-1)}^T \boldsymbol{Q}_{(k-1)} - \boldsymbol{I}_3) - (\bar{\boldsymbol{\Lambda}}_{(k)} \boldsymbol{R}^{-1} + \boldsymbol{R}^{-1} \bar{\boldsymbol{\Lambda}}_{(k)}),$$

for $k > 1$ till convergence at which point we set $\boldsymbol{Q}^{n+1} = \boldsymbol{Q}_{(k+1)}$. During the iteration we also use the computed $\bar{\boldsymbol{\Lambda}}_{(k)}$ to update the momentum matrix $\boldsymbol{P}^{n+1/2}$. Compare the discussion in Section 7.2.2. A description of other Newton-type iteration schemes can be found in [98].

The momentum constraint (8.21) leads to a linear equation in $\boldsymbol{\Lambda}_{(v)}^{n+1}$ of type (8.22) and, hence, is easy to enforce.

8.1.4 Numerical experiment: a symmetric top

We perform a numerical experiment involving an axially symmetric rigid body with its center of mass held fixed ($\boldsymbol{q}_{\rm cm} = \mathbf{0}$). We further suppose that a conservative force is applied to a fixed reference point $\boldsymbol{r}^0$ on the axis of symmetry.

The rigid body is given moments of inertia $I_1 = 4$, $I_2 = 4$ and $I_3 = 1$, making it a long, thin object, symmetric around the third axis of inertia with mass tensor

$$\boldsymbol{R} = \begin{bmatrix} 0.5 & 0 & 0 \\ 0 & 0.5 & 0 \\ 0 & 0 & 3.5 \end{bmatrix}.$$

The body is attached by a spring (coefficient $K = 5$) from the fixed reference point $\boldsymbol{r}^0 = (0, 0, 1)^T$ on the rigid body to a point $\boldsymbol{q}^0 = (0, 0, -1)^T$ fixed in the laboratory frame. The associated potential energy expression is given by

$$V_{\text{ext}}(\boldsymbol{Q}) = \frac{K}{2}\|\boldsymbol{Q}\boldsymbol{r}^0 - \boldsymbol{q}^0\|^2.$$

The object is initially oriented so that the third axis of inertia coincides with the z-axis in the laboratory frame implying $\boldsymbol{Q}(0) = \boldsymbol{I}_3$.[3] Initial body angular momenta applied are $\pi_1 = 0$, $\pi_2 = 0.1$ and $\pi_3 = 2$. This means that the body is initially set spinning around its axis of symmetry and is provided a small downward impetus. The associated initial momentum matrix $\boldsymbol{P}$ is given as the solution of

$$\widehat{\Pi} = \begin{bmatrix} 0 & -2 & 0.1 \\ 2 & 0 & 0 \\ -0.1 & 0 & 0 \end{bmatrix} = \boldsymbol{P} - \boldsymbol{P}^T,$$
$$\boldsymbol{0} = \boldsymbol{P}\boldsymbol{R}^{-1} + \boldsymbol{R}^{-1}\boldsymbol{P}^T,$$

which is equivalent to

$$\boldsymbol{P}\boldsymbol{R}^{-1} + \boldsymbol{R}^{-1}\boldsymbol{P} = \boldsymbol{R}^{-1}\widehat{\boldsymbol{\pi}}.$$

The origin of these equations is explained in the following section. The explicit solution is

$$\boldsymbol{P}(0) = \begin{bmatrix} 0 & -1 & 0.0875 \\ 1 & 0 & 0 \\ -0.0125 & 0 & 0 \end{bmatrix}.$$

Under these circumstances, the body spirals gradually away from the upright position, eventually swinging down (this phase is quite rapid) before returning again to the upright position. Although it is a slight misuse of terminology, let us refer to such a motion as a "quasi-period."

The variation in speed of oscillation over one quasi-period makes this a somewhat challenging integration problem. The results of a numerical experiment using a RATTLE integrator and $\Delta t = 0.1$ are shown in Fig. 8.2. The motion $\boldsymbol{q}(t) = \boldsymbol{Q}(t)\boldsymbol{r}^0$ of the symmetry axis of the body, as seen in the laboratory frame, is plotted on the left. On the right are shown the fluctuations in the z-component of $\boldsymbol{q}(t)$ v. time, above the plot of energy error v. time. The rapid changes in the energy error are associated with the times of rapid motion as the top briefly swings down.

[3] From a computational point of view it is often useful to set $\boldsymbol{Q}(0)$ equal to the identity matrix, i.e. to identify the laboratory frame with the body reference coordinate system $\bar{e}_i$, $i = 1, 2, 3$, at time $t = 0$.

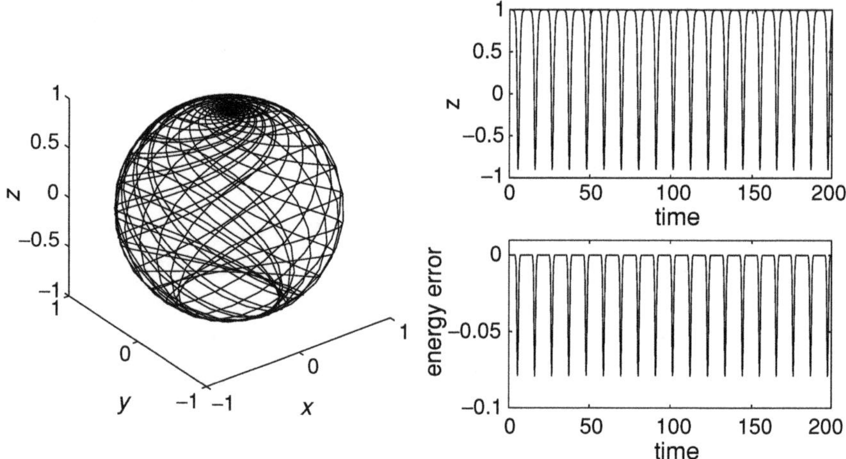

Figure 8.2 Motion of the symmetric top, as computed using RATTLE.

8.2 Angular momentum and the inertia tensor

A common alternative description for the rotational kinetic energy is based on the *body angular momentum* vector $\boldsymbol{\pi} \in \mathbb{R}^3$. Introduction of this concept will pave the way for another very useful approach to symplectic integration based on the rigid body Euler equations. We start the derivation by introducing the *spatial angular velocity* vector $\boldsymbol{\omega}$ via the classical definition

$$\dot{\boldsymbol{r}}_i = \boldsymbol{\omega} \times \boldsymbol{r}_i,$$

where $\boldsymbol{r}_i = \boldsymbol{q}_i - \boldsymbol{q}_{\mathrm{cm}}$ is a displacement vector. Using the setting from Section 8.1, the standard (spatial) *angular momentum* vector $\boldsymbol{m}$ can then be expressed as

$$\begin{aligned} \boldsymbol{m} &= \sum_{i=1}^{k} m_i \boldsymbol{r}_i \times \dot{\boldsymbol{r}}_i \\ &= \sum_{i=1}^{k} m_i \boldsymbol{r}_i \times (\boldsymbol{\omega} \times \boldsymbol{r}_i) \\ &= \sum_{i=1}^{k} m_i \left[\|\boldsymbol{r}_i\|^2 - \boldsymbol{r}_i \boldsymbol{r}_i^T \right] \boldsymbol{\omega}. \end{aligned}$$

Since

$$\boldsymbol{r}_i = \boldsymbol{Q} \boldsymbol{r}_i^0,$$

we obtain

$$\boldsymbol{m} = \boldsymbol{Q} \left(\sum_{i=1}^{k} m_i \left[\|\boldsymbol{r}_i^0\|^2 - \boldsymbol{r}_i^0 \left[\boldsymbol{r}_i^0 \right]^T \right] \right) \boldsymbol{Q}^T \boldsymbol{\omega},$$

in terms of a fixed reference configuration. The *body angular momentum* vector $\boldsymbol{\pi}$ is now defined by $\boldsymbol{\pi} = \boldsymbol{Q}^T \boldsymbol{m}$. Similarly the *body angular velocity* vector is $\boldsymbol{\omega}^{\text{body}} = \boldsymbol{Q}^T \boldsymbol{\omega}$ and we obtain the relation

$$\boldsymbol{\pi} = \boldsymbol{T}\boldsymbol{\omega}^{\text{body}},$$

with the *inertial tensor* given by

$$\boldsymbol{T} = \sum_{i=1}^{k} m_i \left[\|\boldsymbol{r}_i^0\|^2 - \boldsymbol{r}_i^0 \left[\boldsymbol{r}_i^0\right]^T \right] = \sum_{i=1}^{k} m_i \|\boldsymbol{r}_i^0\|^2 \boldsymbol{I}_3 - \boldsymbol{R}.$$

If the mass tensor $\boldsymbol{R}$ is diagonalized and represented in the form (8.3), then

$$\sum_{i=1}^{k} m_i \|\boldsymbol{r}_i^0\|^2 = \sum_{i=1}^{3} r_{ii} \|\bar{\boldsymbol{e}}_i\|^2 = r_{11} + r_{22} + r_{33},$$

and

$$\boldsymbol{T} = \begin{bmatrix} r_{22} + r_{33} & 0 & 0 \\ 0 & r_{11} + r_{33} & 0 \\ 0 & 0 & r_{11} + r_{22} \end{bmatrix},$$

with the principal moments of inertia $I_i = \sum_{j \neq i} r_{jj}$.

Let us link the angular momentum and angular velocity vectors to the canonical variables in the constrained formulation (8.8)–(8.12). Since $\boldsymbol{r}_i = \boldsymbol{Q}\boldsymbol{r}_i^0$, we obtain

$$\dot{\boldsymbol{Q}}\boldsymbol{r}_i^0 = \boldsymbol{\omega} \times \boldsymbol{Q}\boldsymbol{r}_i^0. \tag{8.24}$$

Next we make use of the fact that for any vector $\boldsymbol{u} \in \mathbb{R}^3$, we can write

$$\widehat{\boldsymbol{\omega}}\boldsymbol{u} = \boldsymbol{\omega} \times \boldsymbol{u},$$

where $\widehat{\boldsymbol{\omega}}$ is the 3×3 skew-symmetric matrix

$$\widehat{\boldsymbol{\omega}} = \begin{bmatrix} 0 & -\omega_3 & \omega_2 \\ \omega_3 & 0 & -\omega_1 \\ -\omega_2 & \omega_1 & 0 \end{bmatrix},$$

$\boldsymbol{\omega} = (\omega_1, \omega_2, \omega_3)^T$.[4] Hence (8.24) implies the relation

$$\widehat{\boldsymbol{\omega}} = \dot{\boldsymbol{Q}}\boldsymbol{Q}^T = \boldsymbol{P}\boldsymbol{R}^{-1}\boldsymbol{Q}^T.$$

[4]Similar expression will be frequently used in the rest of this chapter. We remind the reader of the following convention. If $\boldsymbol{b}$ is a 3-vector, then the associated skew-symmetric 3×3 matrix is denoted by $\widehat{\boldsymbol{b}}$. We use capital letters for the skew matrix forms of body angular momentum and angular velocity since these are standard notation often found in the literature.

It is also easy to verify from (8.24) and

$$Q^T \dot{Q} r_i^0 = Q^T \left[\omega \times Q r_i^0 \right] = (Q^T \omega) \times r_i^0$$

that

$$\widehat{\omega}^{\text{body}} = Q^T \dot{Q} = Q^T P R^{-1}.$$

Furthermore, we have the important identity:

$$\widehat{\pi} = \widehat{\omega}^{\text{body}} R - R \left[\widehat{\omega}^{\text{body}} \right]^T = Q^T P - P^T Q,$$

for the skew matrix form of the body angular momentum vector π, which can be verified by explicit computation using (8.3). Specifically, multiplication from the left by $\bar{e}_i^T$ and by $\bar{e}_j$ from the right yields

$$\widehat{\pi}_{ij} = (r_{ii} + r_{jj}) \widehat{\omega}_{ij}^{\text{body}},$$

which is equivalent to $\pi_k = I_k \omega_k^{\text{body}}$, $k = 1, 2, 3$. (See also the exercises.)

From $\widehat{\omega} = \dot{Q} Q^T$ and $\widehat{\omega}^{\text{body}} = Q^T \dot{Q}$, we see that $\widehat{\omega} = Q \widehat{\omega}^{\text{body}} Q^T$. The same transformation applies to the angular momentum vectors and we obtain the expression

$$\widehat{m} = Q \widehat{\pi} Q^T = P Q^T - Q P^T. \qquad (8.25)$$

This expression is useful when setting up the initial conditions for a constrained rigid body formulation. In particular, given an initial angular momentum vector m (or an initial π) and an initial orientation matrix Q the unique initial momentum matrix P is determined by (8.25) subject to $(Q, P) \in \mathcal{P}$.

With this we have completed our brief discussion on how the classical definitions of angular velocity and angular momentum relate to the canonical variables used in the constrained rigid body formulation (8.8)–(8.12).

8.3 The Euler equations of rigid body motion

We next show that the equations of a rigid body with its center of mass held fixed at $q_{\text{cm}} = \mathbf{0}$ can be reduced to a system of three differential equations for the angular momenta and associated equations for the orientation of the rigid body. These differential equations have a special generalized (noncanonical) Hamiltonian structure which can be treated using an explicit Hamiltonian splitting method.

We start our discussion from the angular momentum identity $m = Q\pi = \sum_i m_i r_i \times \dot{r}_i$. Differentiation with respect to time yields

$$\dot{m} = \dot{Q}\pi + Q\dot{\pi} = \sum_i m_i r_i \times \ddot{r}_i,$$

which, multiplied by $\boldsymbol{Q}^T$, is equivalent to

$$\dot{\boldsymbol{\pi}} = -\boldsymbol{Q}^T\dot{\boldsymbol{Q}}\boldsymbol{\pi} + \sum_i m_i\boldsymbol{q}_i^0 \times \boldsymbol{Q}^T\ddot{\boldsymbol{r}}_i.$$

We now take note of

$$\boldsymbol{Q}^T\dot{\boldsymbol{Q}}\boldsymbol{\pi} = \widehat{\boldsymbol{\omega}}^{\text{body}}\boldsymbol{\pi} = (\widehat{\boldsymbol{T}^{-1}\boldsymbol{\pi}})\,\boldsymbol{\pi} = \boldsymbol{T}^{-1}\boldsymbol{\pi}\times\boldsymbol{\pi} = -\boldsymbol{\pi}\times\boldsymbol{T}^{-1}\boldsymbol{\pi} = -\widehat{\boldsymbol{\pi}}\boldsymbol{T}^{-1}\boldsymbol{\pi},$$

and assume an external force $m_i\ddot{\boldsymbol{r}}_i = \boldsymbol{F}_i$ acting on the rigid body at a point $\boldsymbol{r}_i = \boldsymbol{Q}\boldsymbol{r}_i^0$, then we obtain the *Euler equation*

$$\dot{\boldsymbol{\pi}} = \widehat{\boldsymbol{\pi}}\boldsymbol{T}^{-1}\boldsymbol{\pi} + \sum_i \boldsymbol{r}_i^0 \times \boldsymbol{Q}^T\boldsymbol{F}_i \tag{8.26}$$

for forced rigid body motion. We call

$$\boldsymbol{N} = \sum_i \boldsymbol{r}_i^0 \times \boldsymbol{Q}^T\boldsymbol{F}_i = \boldsymbol{Q}^T\left(\sum_i \boldsymbol{r}_i \times \boldsymbol{F}_i\right)$$

the *applied torque in body coordinates.*

The differential equation for the rotation matrix $\boldsymbol{Q}$ had already been derived previously as

$$\dot{\boldsymbol{Q}} = \boldsymbol{Q}\widehat{\boldsymbol{\omega}}^{\text{body}} = \boldsymbol{Q}(\widehat{\boldsymbol{T}^{-1}\boldsymbol{\pi}}). \tag{8.27}$$

Let us concentrate for a moment on the free rigid body, i.e. $\boldsymbol{F}_i = \boldsymbol{0}$. Then the equation (8.26) can be solved independently of (8.27). The *free rigid body Euler equation*

$$\dot{\boldsymbol{\pi}} = \widehat{\boldsymbol{\pi}}\boldsymbol{T}^{-1}\boldsymbol{\pi}$$

is an example of a noncanonical Hamiltonian system (actually a Lie–Poisson system) with Hamiltonian function (kinetic energy)

$$\mathcal{T}(\boldsymbol{\pi}) = \frac{1}{2}\boldsymbol{\pi}^T\boldsymbol{T}^{-1}\boldsymbol{\pi} = \frac{1}{2}\left(\frac{\pi_1^2}{I_1} + \frac{\pi_2^2}{I_2} + \frac{\pi_3^2}{I_3}\right) \tag{8.28}$$

and *Lie–Poisson structure matrix*

$$\boldsymbol{J}(\boldsymbol{\pi}) = \widehat{\boldsymbol{\pi}} = \begin{pmatrix} 0 & -\pi_3 & \pi_2 \\ \pi_3 & 0 & -\pi_1 \\ -\pi_2 & \pi_1 & 0 \end{pmatrix}.$$

In other words, the Euler equation (8.26) is equivalent to the more abstract formulation

$$\frac{d}{dt}\boldsymbol{\pi} = \boldsymbol{J}(\boldsymbol{\pi})\nabla_{\boldsymbol{\pi}}\mathcal{T}(\boldsymbol{\pi}). \tag{8.29}$$

The Lie–Poisson bracket $\{F, G\}$ of two functions $F(\boldsymbol{\pi})$ and $G(\boldsymbol{\pi})$ is defined by

$$\{F, G\}(\boldsymbol{\pi}) = [\nabla_{\boldsymbol{\pi}} F(\boldsymbol{\pi})]^T \, J(\boldsymbol{\pi}) \, \nabla_{\boldsymbol{\pi}} G(\boldsymbol{\pi}).$$

The Lie–Poisson bracket is anti-symmetric and satisfies the Jacobi identity [149, 124, 7]. In this sense the Euler equation (8.29) is a generalization of the Hamiltonian formulations we have encountered so far. Note, however, that the matrix $\boldsymbol{J}$ is *not* invertible hence we *cannot* define a symplectic form $\boldsymbol{J}^{-1} d\boldsymbol{\pi} \wedge d\boldsymbol{\pi}$. On the other hand, it is easily verified that the Euclidean norm of the momentum vector $\boldsymbol{\pi}$ is a first integral of (8.29) for any choice of the Hamiltonian $\mathcal{T}$. This is due to the fact that the structure matrix $\boldsymbol{J}$ is singular and of rank two. Functions that are first integrals of a system for any choice of the Hamiltonian are also called *Casimir functions.* They are always linked to the fact that the structure matrix $\boldsymbol{J}$ is singular. We also point out that the rigid body structure matrix $\boldsymbol{J} = \widehat{\boldsymbol{\pi}}$ is invertible on the level sets of $||\boldsymbol{\pi}|| =$ constant. This allows us to define a (canonical) symplectic structure on each of these two-dimensional level sets. See OLVER [149], MARSDEN AND RATIU [124], or ARNOLD [7] for more details.

Let us next have a closer look at the equation (8.27). We can view the rotation matrix $\boldsymbol{Q} \in SO(3)$ as the collection of three column vectors $\boldsymbol{s}_j \in \mathbb{R}^3$ that are orthogonal to each other and that have unit length; i.e.

$$\boldsymbol{Q}^T = [\boldsymbol{s}_1, \boldsymbol{s}_2, \boldsymbol{s}_3] .$$

Then equation (8.27) is rewritten as

$$\frac{d}{dt}\boldsymbol{Q}^T = -(\widehat{\boldsymbol{T}^{-1}\boldsymbol{\pi}})\boldsymbol{Q}^T,$$

which is equivalent to

$$\frac{d}{dt}\boldsymbol{s}_j = -(\boldsymbol{T}^{-1}\boldsymbol{\pi}) \times \boldsymbol{s}_j = \boldsymbol{s}_j \times (\boldsymbol{T}^{-1}\boldsymbol{\pi}) = \widehat{\boldsymbol{s}}_j \nabla_{\boldsymbol{\pi}} \mathcal{T}(\boldsymbol{\pi}), \qquad (j = 1, 2, 3). \quad (8.30)$$

The system (8.29)–(8.30) constitutes a Hamiltonian system in the variable

$$\boldsymbol{z} = (\boldsymbol{s}_1, \boldsymbol{s}_2, \boldsymbol{s}_3, \boldsymbol{\pi})^T \in \mathbb{R}^{12}$$

and with the skew-symmetric Lie–Poisson structure matrix[5]

$$J(\boldsymbol{z}) = \begin{pmatrix} \boldsymbol{0} & \boldsymbol{0} & \boldsymbol{0} & \widehat{\boldsymbol{s}}_1 \\ \boldsymbol{0} & \boldsymbol{0} & \boldsymbol{0} & \widehat{\boldsymbol{s}}_2 \\ \boldsymbol{0} & \boldsymbol{0} & \boldsymbol{0} & \widehat{\boldsymbol{s}}_3 \\ \widehat{\boldsymbol{s}}_1 & \widehat{\boldsymbol{s}}_2 & \widehat{\boldsymbol{s}}_3 & \widehat{\boldsymbol{\pi}} \end{pmatrix} \in \mathbb{R}^{12\times 12}. \qquad (8.31)$$

[5]The matrix $\boldsymbol{J}$ is skew-symmetric but depends on the state variable $\boldsymbol{z}$. Thus we would also have to verify that $\boldsymbol{J}$ satifies the Jacobi identity to show that $\boldsymbol{J}$ indeed defines a Poisson bracket $\{F, G\} = (\nabla_{\boldsymbol{z}} F)^T \boldsymbol{J} \nabla_{\boldsymbol{z}} G$.

The equations of motion of a free rigid body take now the abstract form

$$\frac{d}{dt}z = J(z)\nabla_z \mathcal{H}(z),$$

where

$$\mathcal{H} = \mathcal{T} = \frac{1}{2}\boldsymbol{\pi}^T \boldsymbol{T}^{-1}\boldsymbol{\pi}.$$

Using this general Hamiltonian framework, one can consider systems where the Hamiltonian $\mathcal{H}$ depends on the state variable $\boldsymbol{z}$ in an arbitrary way. Typically, the Hamiltonian $\mathcal{H}$ is however of the form

$$\mathcal{H} = \frac{1}{2}\boldsymbol{\pi}^T \boldsymbol{T}^{-1}\boldsymbol{\pi} + \mathcal{V}(\boldsymbol{s}_1, \boldsymbol{s}_2, \boldsymbol{s}_3).$$

Example 1 Consider a rigid body with fixed center of mass $\boldsymbol{q}_{\rm cm} = \boldsymbol{0}$ and an external force $\boldsymbol{F}$ acting on a (fixed) point $\boldsymbol{r}^0$ in the body reference configuration. In other words, the force is assumed to be given by

$$\boldsymbol{F}(\boldsymbol{r}) = -\nabla_r V(\boldsymbol{r}), \quad \boldsymbol{r} = \boldsymbol{Q}\boldsymbol{r}^0.$$

We call this force the *spatial force*. The rigid body potential energy is

$$\mathcal{V}(\boldsymbol{s}_1, \boldsymbol{s}_2, \boldsymbol{s}_3) = V(\boldsymbol{Q}\boldsymbol{r}^0),$$

and the corresponding equations of motion are given by

$$\frac{d}{dt}\boldsymbol{\pi} = \boldsymbol{\pi} \times (\boldsymbol{T}^{-1}\boldsymbol{\pi}) + \sum_{j=1}^{3} \boldsymbol{s}_j \times \nabla_{\boldsymbol{s}_j} \mathcal{V}(\boldsymbol{s}_1, \boldsymbol{s}_2, \boldsymbol{s}_3)$$

and

$$\frac{d}{dt}\boldsymbol{s}_j = \boldsymbol{s}_j \times (\boldsymbol{T}^{-1}\boldsymbol{\pi}),$$

$j = 1, 2, 3$. Let us write $\boldsymbol{r}$ in terms of the unit coordinate vectors $\boldsymbol{e}_j$ (fixed laboratory frame with $\boldsymbol{e}_j^T\boldsymbol{Q} = \boldsymbol{s}_j^T$), i.e.

$$\boldsymbol{r} = \sum_{j=1}^{3} (\boldsymbol{s}_j^T \boldsymbol{r}^0)\, \boldsymbol{e}_j,$$

since $\boldsymbol{e}_j^T\boldsymbol{r} = \boldsymbol{e}_j^T\boldsymbol{Q}\boldsymbol{r}^0 = \boldsymbol{s}_j^T\boldsymbol{r}^0$. Then

$$\begin{aligned}
\sum_{j=1}^{3} \boldsymbol{s}_j \times \nabla_{\boldsymbol{s}_j} \mathcal{V}(\boldsymbol{s}_1, \boldsymbol{s}_2, \boldsymbol{s}_3) &= -\sum_{j=1}^{3} \boldsymbol{s}_j \times \left((\boldsymbol{F}(\boldsymbol{r})^T \boldsymbol{e}_j)\, \boldsymbol{r}^0\right) \\
&= -\sum_{j=1}^{3} \left((\boldsymbol{F}(\boldsymbol{r})^T \boldsymbol{e}_j)\, \boldsymbol{s}_j\right) \times \boldsymbol{r}^0 \\
&= \boldsymbol{r}^0 \times (\boldsymbol{Q}^T \boldsymbol{F}(\boldsymbol{r}))
\end{aligned}$$

is the standard formula for the applied torque in body coordinates (compare (8.26)). This procedure generalizes to any number of particle locations

$$r_i = \sum_{j=1}^{3} (s_j^T r_i^0)\, e_j, \qquad i = 1, \ldots, k,$$

and potential energies $V = V(r_1, \ldots, r_k)$. □

8.3.1 Symplectic discretization of the Euler equations

We have already seen that SHAKE/RATTLE discretization can be used to treat the constrained formulation (8.17)–(8.21). This leads to a system of nonlinear equations that needs to be solved at each timestep.

An attractive alternative for simulating free rigid body motion is based on Euler equations (8.29)–(8.30) and the idea of splitting the Hamiltonian into explicitly solvable components. This was proposed by McLachlan [128], Reich [153, 154], and Touma and Wisdom [193]. The Hamiltonian of a free rigid body is given by (8.28), which we write as

$$\mathcal{T}(\boldsymbol{\pi}) = \mathcal{T}_1(\pi_1) + \mathcal{T}_2(\pi_2) + \mathcal{T}_3(\pi_3), \tag{8.32}$$

where $\mathcal{T}_j(\pi_j) = \pi_j^2/(2I_j), j = 1, 2, 3$. The crucial observation is that each entry $\mathcal{T}_j$ gives rise to equations of motion which can be solved exactly. Let us demonstrate this for $\mathcal{T}_1$. The corresponding equations of motion can be written as a system of linear differential equations

$$\frac{d}{dt}\boldsymbol{\pi} = \boldsymbol{A}_1 \boldsymbol{\pi}, \qquad \frac{d}{dt} s_j = \boldsymbol{A}_1 s_j, \quad (j = 1, 2, 3),$$

where

$$\boldsymbol{A}_1 = \frac{1}{I_1} \begin{pmatrix} 0 & 0 & 0 \\ 0 & 0 & \pi_1 \\ 0 & -\pi_1 & 0 \end{pmatrix}.$$

Note that

$$\frac{d}{dt}\pi_1 = 0,$$

which implies that the solutions are simply given by

$$\boldsymbol{\pi}(t) = e^{t\boldsymbol{A}_1}\boldsymbol{\pi}(0), \qquad s_j(t) = e^{t\boldsymbol{A}_1} s_j(t), \quad (j = 1, 2, 3), \tag{8.33}$$

with

$$e^{t\boldsymbol{A}_1} = \begin{pmatrix} 1 & 0 & 0 \\ 0 & \cos\omega_1 t & \sin\omega_1 t \\ 0 & -\sin\omega_1 t & \cos\omega_1 t \end{pmatrix}, \quad \omega_1 = \frac{\pi_1}{I_1}.$$

The equations (8.33) provide an explicit expression for the exact flow map $\boldsymbol{\Phi}_{t,\mathcal{T}_1}$ corresponding to the system

$$\frac{d}{dt}\mathbf{z} = J(\mathbf{z})\nabla_{\mathbf{z}}\mathcal{T}_1(\pi_1),$$

$\mathbf{z} = (\mathbf{s}_1, \mathbf{s}_2, \mathbf{s}_3, \boldsymbol{\pi})^T$. Similar expressions can be found for the flow maps corresponding to the entries $\mathcal{T}_2$ and $\mathcal{T}_3$ in (8.32).

A second-order symplectic method is now obtained by subsequent concatenation of the exact solutions corresponding to the five terms in

$$\mathcal{T} = \frac{1}{2}\mathcal{T}_1 + \frac{1}{2}\mathcal{T}_2 + \mathcal{T}_3 + \frac{1}{2}\mathcal{T}_2 + \frac{1}{2}\mathcal{T}_1,$$

over a time interval $t = \Delta t$. To be more precise. We define a numerical method

$$\mathbf{z}^{n+1} = \boldsymbol{\Psi}_{\Delta t,\mathcal{T}}(\mathbf{z}^n)$$

by the composition of flow maps

$$\boldsymbol{\Psi}_{\Delta t,\mathcal{T}} = \boldsymbol{\Phi}_{\Delta t/2,\mathcal{T}_1} \circ \boldsymbol{\Phi}_{\Delta t/2,\mathcal{T}_2} \circ \boldsymbol{\Phi}_{\Delta t,\mathcal{T}_3} \circ \boldsymbol{\Phi}_{\Delta t/2,\mathcal{T}_2} \circ \boldsymbol{\Phi}_{\Delta t/2,\mathcal{T}_1}. \tag{8.34}$$

This method is obviously symplectic since each flow map preserves $\|\boldsymbol{\pi}\|$ and is symplectic on the level sets of constant $\|\boldsymbol{\pi}\|$. The method also conserves the spatial angular momentum vector $\mathbf{I}$ exactly. This follows again from the exact conservation under each flow map. This splitting and its implementation can be improved. For example, one could replace exact matrix exponentials by the Cayley transform, i.e., for any of three skew-symmetric matrices $\mathbf{A}_i$ we can use

$$e^{t\mathbf{A}_i} \approx (\mathbf{I} - t\mathbf{A}_i)^{-1}(\mathbf{I} + t\mathbf{A}_i) \in SO(3).$$

There are also other splitting of $\mathcal{T}$ possible. See the discussion in the following subsection and TOUMA AND WISDOM [193], DULLWEBER, LEIMKUHLER, AND MCLACHLAN. [54] as well as BUSS [38] and FASSO [56].

This method can now also be used to integrate more general rigid body systems with Hamiltonian

$$\mathcal{H} = \frac{1}{2}\boldsymbol{\pi}^T\mathbf{T}^{-1}\boldsymbol{\pi} + \mathcal{V}(\mathbf{s}_1, \mathbf{s}_2, \mathbf{s}_3). \tag{8.35}$$

Similar to the Störmer–Verlet method, we split the total Hamiltonian into the kinetic energy $\mathcal{T} = 1/2\boldsymbol{\pi}^T\mathbf{T}^{-1}\boldsymbol{\pi}$ and the potential energy $\mathcal{V}$. A second-order method can then be derived based on

$$\mathcal{H} = \frac{1}{2}\mathcal{V} + \mathcal{T} + \frac{1}{2}\mathcal{V}, \tag{8.36}$$

The equations of motion corresponding to the potential energy $\mathcal{V}$ are given by

$$\frac{d}{dt}\boldsymbol{\pi} = \sum_{j=1}^{3} \mathbf{s}_j \times \nabla_{\mathbf{s}_j}\mathcal{V}(\mathbf{s}_1, \mathbf{s}_2, \mathbf{s}_3), \qquad \frac{d}{dt}\mathbf{s}_j = 0, \quad (j = 1, 2, 3),$$

and can be solved exactly. In particular, since

$$\boldsymbol{N} = \sum_{j=1}^{3} \mathbf{s}_j \times \nabla_{\mathbf{s}_j} \mathcal{V}(\mathbf{s}_1, \mathbf{s}_2, \mathbf{s}_3) = \text{const.},$$

we obtain $\boldsymbol{\pi}(t') = \boldsymbol{\pi}(t) + (t'-t)\boldsymbol{N}$ and $\mathbf{s}_j(t') = \mathbf{s}_j(t)$. Denote the corresponding flow map by $\boldsymbol{\Phi}_{t,\mathcal{V}}$. Then, based on the splitting (8.36) and upon using (8.34) to propagate the kinetic energy part, we obtain the second-order symplectic method

$$\boldsymbol{\Psi}_{\Delta t} = \boldsymbol{\Phi}_{\Delta t/2,\mathcal{V}} \circ \boldsymbol{\Psi}_{\Delta t,\mathcal{T}} \circ \boldsymbol{\Phi}_{\Delta t/2,\mathcal{V}}, \tag{8.37}$$

for the generalized Euler system

$$\frac{d}{dt} z = J(z) \nabla_z \mathcal{H}(z),$$

with Hamiltonian (8.35).

8.3.2 Numerical experiment: the Lagrangian top

Below we show that the Hamiltonian for a Lagrangian top is of the form (8.35) and conduct a numerical experiment to test the symplectic splitting method. See [7, 73] for a detailed analysis of the Lagrangian top.

The matrix $\boldsymbol{T}$ is defined by the three principal moments of inertia of the (axially symmetric) top. Symmetry of the top implies that $I_1 = I_2$. The potential energy function $\mathcal{V}$ is obtained by introducing the fixed reference vector $\boldsymbol{r}^0$ to point from the stationary point of contact to the center of mass, i.e. $\boldsymbol{r}^0 = (0, 0, L)^T$. We set the distance from the fixed point to the center of mass equal to one, i.e., $L = 1$. Then the potential energy becomes

$$\mathcal{V}(\boldsymbol{Q}) = -c\boldsymbol{k}^T \boldsymbol{Q} \boldsymbol{r}^0,$$

where $\boldsymbol{k} = (0, 0, -1)^T$ gives the direction of gravity and $c = mg$ is the product of the mass m of the top and the gravitational constant g. The orthogonal matrix $\boldsymbol{Q}(t)$ describes the rotation of the rigid body about its stationary point of contact and $\boldsymbol{Q}^T = [\mathbf{s}_1, \mathbf{s}_2, \mathbf{s}_3]$ as before. Note that the potential energy can be written as a function of $\mathbf{s}_3$ alon

$$\mathcal{V}(\mathbf{s}_3) = -c\left[\boldsymbol{r}^0\right]^T \boldsymbol{Q}^T \boldsymbol{k} = c\left[\boldsymbol{r}^0\right]^T \mathbf{s}_3.$$

Employing the above outlined abstract Hamiltonian framework, we obtain the equations of motion

$$\frac{d}{dt}\boldsymbol{\pi} = \boldsymbol{\pi} \times \boldsymbol{T}^{-1}\boldsymbol{\pi} + c\mathbf{s}_3 \times \boldsymbol{r}^0, \tag{8.38}$$

$$\frac{d}{dt}\mathbf{s}_i = \mathbf{s}_i \times \boldsymbol{T}^{-1}\boldsymbol{\pi}, \qquad (i = 1, 2, 3). \tag{8.39}$$

These equations can be integrated using the composition method (8.37) with the kinetic energy part integrated by the standard splitting (8.34). However, one can solve the free Euler equation for a symmetric rigid body exactly. To do so, we split the kinetic energy into two contributions [193, 54]

$$\mathcal{T} = \frac{1}{2I}\boldsymbol{\pi}^T\boldsymbol{\pi} + \frac{1}{2}\left(\frac{1}{I_3} - \frac{1}{I}\right)\pi_3^2,$$

$I = I_1 = I_2$, instead of three which yields the modified composition method

$$\boldsymbol{\Psi}_{\Delta t,\mathcal{T}} = \boldsymbol{\Phi}_{\Delta t,\tilde{\mathcal{T}}_2} \circ \boldsymbol{\Phi}_{\Delta t,\tilde{\mathcal{T}}_1},$$

where

$$\tilde{\mathcal{T}}_1 = \frac{1}{2I}\boldsymbol{\pi}^T\boldsymbol{\pi},$$

and

$$\tilde{\mathcal{T}}_2 = \frac{1}{2}\left(\frac{1}{I_3} - \frac{1}{I}\right)\pi_3^2.$$

Note that $\boldsymbol{\pi}(t) = const.$ for the Hamiltonian $\tilde{\mathcal{T}}_1$ which allows for the exact integration of the equation of motion for $\mathbf{s}_i$, $i = 1, 2, 3$; i.e.,

$$\frac{d}{dt}\boldsymbol{\pi} = \mathbf{0}, \qquad \frac{d}{dt}\mathbf{s}_i = -\frac{1}{I}\widehat{\boldsymbol{\pi}}\mathbf{s}_i.$$

Even more importantly, the two Hamiltonian functions $\mathcal{T}_1$, $\mathcal{T}_2$ commute and, hence, the composition method $\boldsymbol{\Psi}_{\Delta t,\mathcal{T}}$ is equal to the exact flow map:

$$\boldsymbol{\Phi}_{\Delta t,\mathcal{T}} = \boldsymbol{\Phi}_{\Delta t,\tilde{\mathcal{T}}_1} \circ \boldsymbol{\Phi}_{\Delta t,\tilde{\mathcal{T}}_2}. \tag{8.40}$$

The exact propagator $\boldsymbol{\Phi}_{\Delta t,\mathcal{T}}$ for the kinetic energy part of the Lagrangian top's Hamiltonian is now used in (8.37) to obtain a second-order symplectic method for the spinning top.

We conduct a numerical experiments with $c = 1$ and

$$\boldsymbol{Q}(0) = \begin{bmatrix} \cos 0.6 & 0 & -\sin 0.6 \\ 0 & 1 & 0 \\ \sin 0.6 & 0 & \cos 0.6 \end{bmatrix}.$$

This corresponds to an initial $\mathbf{s}_3 = (\sin 0.6, 0, \cos 0.6)^T$. The equations of motion are integrated with the above splitting methods using a stepsize of $\Delta t = 0.1$. The motion of the center of mass $\boldsymbol{q}_{\rm cm}(t) = \boldsymbol{Q}\boldsymbol{r}^0$ is given in Fig. 8.3 for $I_1 = I_2 = 5$, $I_3 = 1$, and the initial body angular momentum $\boldsymbol{\pi} = (0, 0, 5)^T$. The error in energy and the motion of the z-component of $\boldsymbol{q}_{\rm cm}(t)$ are shown on the right

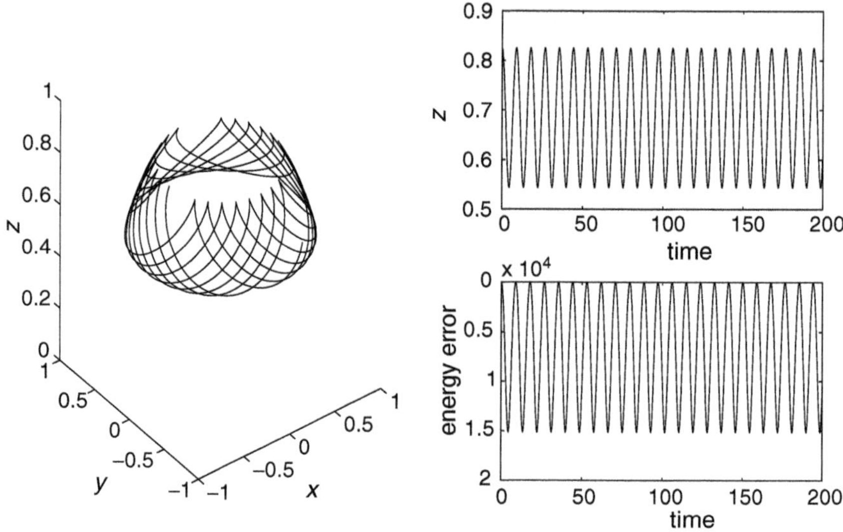

Figure 8.3 Motion of the Lagrangian top, as computed with an explicit splitting method.

of Fig. 8.3. Note that this experiment corresponds to a Lagrangian top initially spinning only about its axis of symmetry and which is released from that position without applying any additional torque. See also [7, 73] and problem 4 in the Exercises.

As a second test we implemented the splitting algorithm (8.37) for the symmetric top example of Section 8.1.4 with the purpose of comparing with our previously computed RATTLE solution. For a stepsize of $\Delta t = 0.1$ and with the exact propagator (8.40) used in (8.37), the numerical trajectories are essentially indistinguishable from those obtained using RATTLE. However, the splitting scheme preserves the energy better by a factor of about 40. Furthermore, the splitting method is much easier to implement and more efficient than the RATTLE approach and emerges here as the clear winner.

The situation changes when the exact propagator (8.40) is replaced by the standard kinetic energy splitting (8.34). The results, using an identical timestep of $\Delta t = 0.1$ are shown in Fig. 8.4. There are some significant differences with the results using RATTLE (compare with Fig. 8.2). First, the new trajectory appears to be somewhat less regular, related to the appearance for the first time of sharp peaks in the evolution of the energy error. In Fig. 8.5 a projection of a very

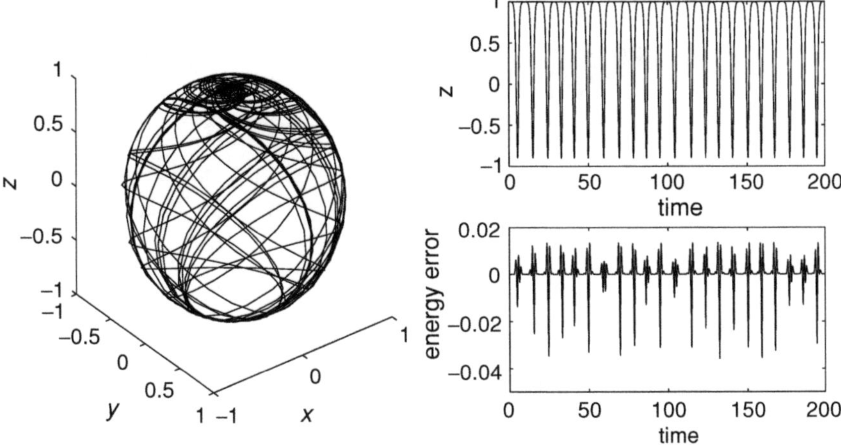

Figure 8.4 Motion of the symmetric sprung top, as computed using the standard reduction/splitting scheme.

long time numerical solution in the neighborhood of the upright body position is shown for each method, dramatizing the improved regularity of the RATTLE orbit. Even more striking, perhaps, is the fact that the new computation finds too many quasi-periods (22 instead of 19) in the integration on $[0, 200]$. In fact 19 is very nearly the correct figure for this time interval, and this is only achieved by an eight-fold reduction in the integration timestep. On the other hand, the standard reduction/splitting method exhibits better energy conservation, even at the large timestep, than the RATTLE approach. (A careful comparison of methods must include attention to the efficient implementation of the RATTLE nonlinear solver, which can add to the work.)

The experiment highlights the necessity for a careful choice of the splitting. It is often advisable to use a splitting of the rigid body kinetic energy into an entirely symmetric contribution $\mathcal{T}_s = \frac{1}{2I}\boldsymbol{\pi}^T\boldsymbol{\pi}$ and a "perturbation," i.e.

$$\mathcal{T} = \frac{1}{2I}\boldsymbol{\pi}^T\boldsymbol{\pi} + \frac{1}{2}\left(\frac{1}{I_1} - \frac{1}{I}\right)\pi_1^2 + \frac{1}{2}\left(\frac{1}{I_2} - \frac{1}{I}\right)\pi_2^2 + \frac{1}{2}\left(\frac{1}{I_3} - \frac{1}{I}\right)\pi_3^2.$$

A natural choice is to set $I = I_i$ for some appropriate index $i \in \{1, 2, 3\}$ (see [56] for a detailed comparison).

8.3.3 Integrable discretization: RATTLE and the scheme of Moser and Veselov

The free rigid body is completely integrable. It was first pointed out that a completely integrable discretization for the free rigid body is possible in an article of

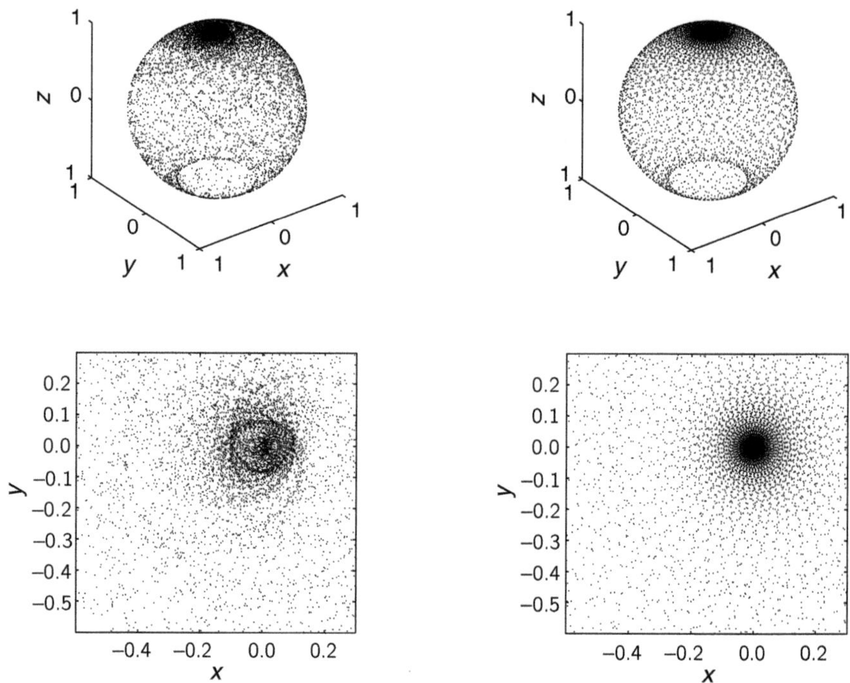

Figure 8.5 Motion of the symmetric sprung top, as computed using the standard reduction/splitting scheme (left) and using the RATTLE approach (right).

MOSER AND VESELOV [141] (see also LEWIS AND SIMO [114]). Perhaps as remarkable in our context is the fact, noted by MCLACHLAN AND ZANNA [127], that the RATTLE method when applied to the free rigid body is actually equivalent to the Moser–Veselov algorithm.

To see this, introduce into the RATTLE method (8.17)–(8.21) the auxiliary variables $\boldsymbol{Y}^n = \boldsymbol{P}^n[\boldsymbol{Q}^n]^T, \boldsymbol{Y}^{n+1/2} = \boldsymbol{P}^{n+1/2}[\boldsymbol{Q}^n]^T$ and $\boldsymbol{Y}^{n+1} = \boldsymbol{P}^{n+1}[\boldsymbol{Q}^{n+1}]^T$. By construction, one has $\boldsymbol{Y}^n = \boldsymbol{R}\widehat{\boldsymbol{\omega}}^n$ and hence $\widehat{\boldsymbol{\pi}}^n = \boldsymbol{Y}^n - [\boldsymbol{Y}^n]^T$. Moreover, $\boldsymbol{Y}^{n+1/2} = \boldsymbol{Y}^n - \frac{1}{2}\Delta t \boldsymbol{Q}^n[\boldsymbol{P}^{n+1/2}]^T \boldsymbol{R}^{-1}\boldsymbol{P}^{n+1/2}[\boldsymbol{Q}^n]^T = \boldsymbol{Y}^n + \Delta t \boldsymbol{S}$, where $\boldsymbol{S}$ is a symmetric matrix. Introducing $\boldsymbol{Q} = \boldsymbol{Q}^{n+1}[\boldsymbol{Q}^n]^T = \boldsymbol{I}_3 + \Delta t \boldsymbol{R}^{-1}\boldsymbol{P}^{n+1/2}[\boldsymbol{Q}^n]^T = \boldsymbol{I}_3 + \Delta t \boldsymbol{R}^{-1}\boldsymbol{Y}^{n+1/2}$, it is immediately verified that $\boldsymbol{Q}^{n+1}$ is orthogonal if and only if $\boldsymbol{Q}$ is orthogonal.

Now, we rewrite $\Delta t \widehat{\boldsymbol{\pi}}^n = \Delta t(\boldsymbol{Y}^n - [\boldsymbol{Y}^n]^T)$ in terms of $\boldsymbol{Q}$. We have $\Delta t \widehat{\boldsymbol{\pi}}^n = \boldsymbol{R}(\boldsymbol{Q} - \boldsymbol{I}_3) - \Delta t^2 \boldsymbol{S} - (\boldsymbol{Q}^T - \boldsymbol{I}_3)\boldsymbol{R} + \Delta t^2 \boldsymbol{S}^T = \boldsymbol{R}\boldsymbol{Q} - \boldsymbol{Q}^T\boldsymbol{R}$, since $\boldsymbol{S}$ is symmetric. The equation $\Delta t \widehat{\boldsymbol{\pi}}^n = \boldsymbol{R}\boldsymbol{Q} - \boldsymbol{Q}^T\boldsymbol{R}$ is precisely the Moser–Veselov equation that one has to solve for an orthogonal matrix $\boldsymbol{Q}$. Similarly, $\Delta t \widehat{\boldsymbol{\pi}}^{n+1} = \boldsymbol{Q}\boldsymbol{R} - \boldsymbol{R}\boldsymbol{Q}^T =$

$\Delta t(\boldsymbol{Y}^{n+1} - [\boldsymbol{Y}^{n+1}]^T)$ coincides with the second step of the Moser–Veselov algorithm.

8.4 Order 4 and order 6 variants of RATTLE for the free rigid body

The observation mentioned above due to McLachlan and Zanna [127] of the exact integrability of the RATTLE method has ramifications for the design of higher-order schemes. When applied with constant step size, it follows from the integrability result that the RATTLE scheme solves exactly the modified equation

$$\frac{d}{dt}\boldsymbol{\pi} = (1 + \Delta t^2 \tau_3 + \Delta t^4 \tau_5 + \cdots)\boldsymbol{\pi} \times \boldsymbol{T}^{-1}\boldsymbol{\pi},$$

where the τ_{2i+1} are constants depending on the Hamiltonian, the inertial tensor $\boldsymbol{T}$, the mass tensor $\boldsymbol{R}$ and the Casimir $\|\boldsymbol{\pi}\|$. This modified equation is a time reparametrization of the original free rigid body equation. Hence, by rescaling the initial condition $\boldsymbol{\pi}_0$ it is possible to improve the order of RATTLE.

Introduce the family of constants

$$C_{R,i,j} = r_{11}^i r_{22}^j + r_{11}^i r_{33}^j + r_{22}^i r_{33}^j, \qquad i, j = 1, 2, \ldots,$$

associated to the mass tensor $\boldsymbol{R}$ and set

$$C_{R,i} = C_{R,i,i}, \qquad C_R = C_{R,1}.$$

Denote as

$$\mathcal{H}_2 = r_{11}^2 \pi_1^2 + r_{22}^2 \pi_2^2 + r_{33}^2 \pi_3^2,$$

the constant which is a linear combination of the Hamiltonian $\mathcal{H}$ and the Casimir $\|\boldsymbol{\pi}\|$ of the free rigid body. Consider the constants

$$\tau_3 = \frac{1}{6\det(\boldsymbol{T})^2}\left((3\det(\boldsymbol{R})\mathrm{tr}(\boldsymbol{R}) + C_{R,2})\|\boldsymbol{\pi}\|^2 + (3C_R + \mathrm{tr}(\boldsymbol{R}^2))\mathcal{H}_2\right),$$

and

$$\begin{aligned}\tau_5 = \frac{1}{40\det(\boldsymbol{T})^4}\Big(&(3\mathrm{tr}(\boldsymbol{R}^4) + 27C_{R,2} + 15\mathrm{tr}(\boldsymbol{R}^2)C_R + 45\det(\boldsymbol{R})\mathrm{tr}(\boldsymbol{R}))\mathcal{H}_2^2\\ &+ (10C_{R,3} + 50\det(\boldsymbol{R})\mathrm{tr}(\boldsymbol{R})C_R\\ &+ 10\det(\boldsymbol{R})\mathrm{tr}(\boldsymbol{R})\mathrm{tr}(\boldsymbol{R}^2) + 2C_{R,2}\mathrm{tr}(\boldsymbol{R}^2)\\ &-28\det(\boldsymbol{R}^2))\|\boldsymbol{\pi}\|^2\mathcal{H}_2\\ &+ (60\det(\boldsymbol{R}^2)C_R + 3C_{R,4} + 27\det(\boldsymbol{R}^2)\mathrm{tr}(\boldsymbol{R}^2)\\ &+ 15\det(\boldsymbol{R})(C_{R,2,3} + C_{R,3,2}))\|\boldsymbol{\pi}\|_2^4\Big).\end{aligned}$$

Scaling the initial condition $\boldsymbol{\pi}_0$ to $\boldsymbol{\pi}_0/(1+\Delta t^2\tau_3)$, applying the RATTLE scheme, and then multiplying back by $(1+\Delta t^2\tau_3)$, raises the order of the RATTLE scheme from 2 to 4, and preserves integrability and symplecticity.

Similarly, scaling the initial condition $\boldsymbol{\pi}_0$ to $\boldsymbol{\pi}_0/(1+\Delta t^2\tau_3+\Delta t^4(\tau_5-2\tau_3^2))$, applying RATTLE and multiplying back by the same scaling factor, improves the order of RATTLE to 6. For more details on these methods and accompanying numerical experiments, see [127].

8.5 Freely moving rigid bodies

So far, we have considered a single rigid body with its center of mass held fixed. We now relax this condition and consider a freely moving rigid body. Let us assume that a spatial coordinate system is given and that the represention of a vector in spatial coordinates is denoted by $\boldsymbol{q} \in \mathbb{R}^3$. The particular spatial vector that points from the origin of the laboratory frame to the center of mass of the rigid body is denoted by $\boldsymbol{q}_{cm}$. We consider also a fixed reference vector $\boldsymbol{r}^0$ that points from the center of mass to a given (fixed) point on the rigid body. Then the position of this point is given by the spatial vector

$$\boldsymbol{q}(t) = \boldsymbol{q}_{cm}(t) + \boldsymbol{Q}(t)\boldsymbol{r}^0 = \boldsymbol{q}_{cm}(t) + \sum_{j=1}^{3} \left(\boldsymbol{s}_j(t)^T \boldsymbol{r}^0\right) \boldsymbol{e}_j, \tag{8.41}$$

where $\boldsymbol{e}_j$, $j=1,2,3$, are (fixed) coordinate vectors in the laboratory frame.[6] Thus the motion of any fixed point on the rigid body is completely characterized by the evolution of the center of mass $\boldsymbol{q}_{cm}$ and the three vectors $\boldsymbol{s}_j$. Let us assume for a moment that no external forces act on the rigid body. Then the equations of motion for the center of mass and the rotation matrix decouple and are simply given by

$$\frac{d}{dt}\boldsymbol{\pi} = \boldsymbol{\pi} \times \boldsymbol{T}^{-1}\boldsymbol{\pi}, \qquad \frac{d}{dt}\boldsymbol{s}_j = \boldsymbol{s}_j \times \boldsymbol{T}^{-1}\boldsymbol{\pi}, \quad (j=1,2,3),$$

and

$$\frac{d}{dt}\boldsymbol{q}_{cm} = \frac{1}{M}\boldsymbol{p}_{cm}, \qquad \frac{d}{dt}\boldsymbol{p}_{cm} = \boldsymbol{0}.$$

M is the total mass of the rigid body and $\boldsymbol{p}_{cm} \in \mathbb{R}^3$ the momentum of the center of mass. The combined system can be propagated in time by solving the center of mass equations of motion exactly and by applying the second-order symplectic

[6] These coordinate vectors should be carefully distinguished from the fixed coordinate vectors $\bar{\boldsymbol{e}}_j$ in the body reference frame as defined in (8.3).

integration method for the Euler equations as suggested in the previous section. Let us denote the resulting time-Δt-propagator by $\boldsymbol{\Psi}_{\Delta t,\mathcal{T}}$, where $\mathcal{T}$ stands for the kinetic energy

$$\mathcal{T} = \frac{1}{2M}\|\boldsymbol{p}_{cm}\|^2 + \frac{1}{2}\boldsymbol{\pi}^T \boldsymbol{T}^{-1}\boldsymbol{\pi},$$

of the unforced rigid body.

Consider now a rigid body with an external force $\boldsymbol{F}$ acting at a (fixed) reference point $\boldsymbol{r}^0$ on the rigid body. In other words, the force is assumed to be given in spatial coordinates by

$$\boldsymbol{F}(\boldsymbol{q}) = -\nabla_q V(\boldsymbol{q}), \quad \boldsymbol{q} = \boldsymbol{Q}\boldsymbol{r}^0 + \boldsymbol{q}_{cm}.$$

Then, upon applying (8.41), we formally obtain

$$\mathcal{V}(\boldsymbol{q}_{cm}, \mathbf{s}_1, \mathbf{s}_2, \mathbf{s}_3) = V(\boldsymbol{q}) = V\left(\boldsymbol{q}_{cm} + \sum_{j=1}^{3}\left(\mathbf{s}_j^T \boldsymbol{r}^0\right)\boldsymbol{e}_j\right).$$

It is easily verified (see Example 1) that the equations of motion corresponding to the potential energy $\mathcal{V}$ are given by

$$\frac{d}{dt}\boldsymbol{\pi} = \boldsymbol{r}^0 \times \sum_{j=1}^{3}\left(\boldsymbol{e}_j^T \boldsymbol{F}(\boldsymbol{q})\right)\mathbf{s}_j, \qquad \frac{d}{dt}\mathbf{s}_j = \mathbf{0}, \quad (j = 1, 2, 3),$$

and

$$\frac{d}{dt}\boldsymbol{q}_{cm} = \mathbf{0}, \qquad \frac{d}{dt}\boldsymbol{p}_{cm} = \boldsymbol{F}(\boldsymbol{q}).$$

A symplectic splitting method can be based on

$$\mathcal{H} = \frac{1}{2}\mathcal{V} + \mathcal{T} + \frac{1}{2}\mathcal{V}. \tag{8.42}$$

We have already derived several second-order symplectic methods $\boldsymbol{\Psi}_{\Delta t,\mathcal{T}}$ for the integration of the Hamiltonian $\mathcal{H} = \mathcal{T}$ (free rigid body). The equations of motion corresponding to $\mathcal{H} = \mathcal{V}$ can be integrated exactly since $d\boldsymbol{q}_{cm}/dt = d\mathbf{s}_i/dt = d\boldsymbol{q}/dt = \mathbf{0}$. Let us denote the corresponding flow map by $\boldsymbol{\Phi}_{t,\mathcal{V}}$. Then a second-order symplectic method for the integration of (8.42) is provided by the composition method

$$\boldsymbol{\Psi}_{\Delta t} = \boldsymbol{\Phi}_{\Delta t/2,\mathcal{V}} \circ \boldsymbol{\Psi}_{\Delta t,\mathcal{T}} \circ \boldsymbol{\Phi}_{\Delta t/2,\mathcal{V}}. \tag{8.43}$$

This splitting approach to the integration of rigid bodies was proposed by TOUMA AND WISDOM in [193] and by REICH in [156].

Example 2 Consider a completely symmetric rigid body; i.e. $I = I_1 = I_2 = I_3$, with a spring attached to it at a point $\boldsymbol{r}^0$ on the rigid body. The spring is assumed to have a rest length L, a spring constant K, and its second end being attached to the origin of the coordinate system. The potential energy of the spring is

$$V(\boldsymbol{q}) = \frac{K}{2}(r - L)^2,$$

where $r = \|\boldsymbol{q}\|$, $\boldsymbol{q} = \boldsymbol{q}_{cm} + \boldsymbol{Q}\boldsymbol{r}^0$, and the spatial force is

$$\boldsymbol{F}(\boldsymbol{q}) = -K\frac{r - L}{r}\boldsymbol{q}.$$

The force $\boldsymbol{F}_{cm}$ acting on the center of mass is

$$\boldsymbol{F}_{cm} = \boldsymbol{F}(\boldsymbol{q}) = -K\frac{r - L}{r}\left(\boldsymbol{q}_{cm} + \sum_{j=1}^{3}(\boldsymbol{s}_j^T\boldsymbol{r}^0)\boldsymbol{e}_j\right),$$

and the applied torque $\boldsymbol{N}$ in body coordinates is

$$\boldsymbol{N} = K\frac{r - l}{r}\sum_{j=1}^{3}(\boldsymbol{q}^T\boldsymbol{e}_j)\,\boldsymbol{s}_j \times \boldsymbol{r}^0.$$

The kinetic energy of the rigid body is

$$\mathcal{T} = \frac{1}{2I}\boldsymbol{\pi}^T\boldsymbol{\pi} + \frac{1}{2M}\boldsymbol{p}_{cm}^T\boldsymbol{p}_{cm},$$

with associated free rigid body equations of motion

$$\frac{d}{dt}\boldsymbol{\pi} = \boldsymbol{0}, \qquad \frac{d}{dt}\boldsymbol{s}_j = -I^{-1}\widehat{\boldsymbol{\pi}}\boldsymbol{s}_j, \quad (j = 1, 2, 3),$$

and

$$\frac{d}{dt}\boldsymbol{p}_{cm} = \boldsymbol{0}, \qquad \frac{d}{dt}\boldsymbol{q}_{cm} = M^{-1}\boldsymbol{p}_{cm},$$

which can be solved exactly. Thus a symplectic integrator can be based on the splitting

$$\mathcal{H} = \frac{1}{2}\mathcal{V} + \mathcal{T} + \frac{1}{2}\mathcal{V},$$

with each of the three entries being explicitly integrable.

We conduct a numerical experiment for the set of parameters $I = K = M = L = 1$ and $\boldsymbol{r}^0 = (0, 0, 1)$. The initial conditions are $\boldsymbol{q}_c = \boldsymbol{0}$, $\boldsymbol{Q} = \boldsymbol{I}$, $\boldsymbol{\pi} = (1, 1, 2)^T$, and $\boldsymbol{p}_c = (1, 0, 0)^T$. See Fig. 8.6 for the numerical results obtained for $\Delta t = 0.01$. □

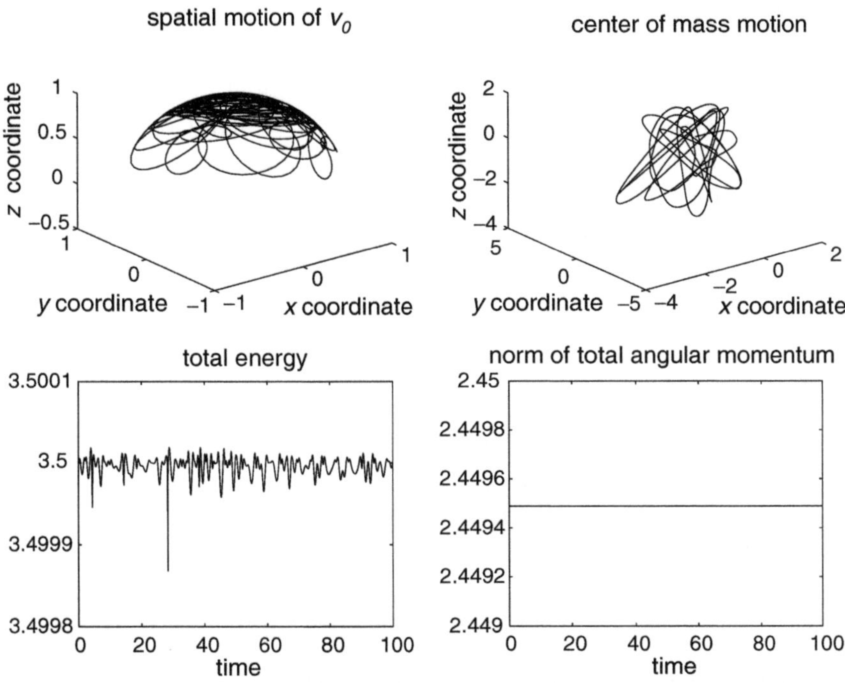

Figure 8.6 Rigid body motion under an external force field.

The approach described so far easily generalizes to systems of interacting rigid bodies. The associated algorithm has been successfully applied to rigid body molecular dynamics by DULLWEBER, LEIMKUHLER, AND MCLACHLAN [54].

We mention that one can also treat rigid bodies that are linked by constraints. Take, for example, two rigid bodies that are connected by a rigid rod linking a material point $\boldsymbol{r}_1^0$ on the first rigid body and a material point $\boldsymbol{r}_2^0$ on the second body. The associated (holonomic) constraint can be formulated as

$$0 = r_{12} - L,$$

where L is the length of the rod and $r_{12} = \|\boldsymbol{r}_{12}\|$

$$\boldsymbol{r}_{12} = \boldsymbol{q}_{cm,1} + \boldsymbol{Q}_1 \boldsymbol{r}_1^0 - \boldsymbol{q}_{cm,2} - \boldsymbol{Q}_2 \boldsymbol{r}_2^0.$$

The Hamiltonian of the systems is

$$\mathcal{H} = \mathcal{T}_1 + \mathcal{T}_2 + \lambda(r_{12} - L).$$

Although such systems can be treated numerically by a natural extension of the splitting methods considered in this chapter, it is probably conceptually easier to apply an approach based on a RATTLE discretization in terms of rotation matrices.

8.6 Other formulations for rigid body motion

We next briefly outline other popular choices for formulating rigid body dynamics. The following two approaches can be viewed as defining a set of generalized coordinates $\boldsymbol{\xi}$ in some appropriate configuration space, together with a map from the parameter space to the 3×3 orthogonal matrices.

While each of these choices of parameters can be used as the basis for numerical simulation, they have certain disadvantages compared with the methods outlined in the previous sections.

8.6.1 Euler angles

It was observed by EULER (ca. 1776) that the orientation of a body in space can be specified in terms of three successive rotations with respect to some set of coordinate axes *fixed in the body*. The specific sequence of rotations used by Euler is as follows:

1. A counterclockwise rotation about the z-axis through an angle α.
2. A counterclockwise rotation about the x-axis through an angle β.
3. A counterclockwise rotation about the z-axis through an angle γ.

The rotation matrix $\boldsymbol{Q}$ can be expressed as a product of these three planar rotations according to

$$\boldsymbol{Q} = \begin{bmatrix} \cos\gamma & \sin\gamma & 0 \\ -\sin\gamma & \cos\gamma & 0 \\ 0 & 0 & 1 \end{bmatrix} \begin{bmatrix} 1 & 0 & 0 \\ 0 & \cos\beta & \sin\beta \\ 0 & -\sin\beta & \cos\beta \end{bmatrix} \begin{bmatrix} \cos\alpha & \sin\alpha & 0 \\ -\sin\alpha & \cos\alpha & 0 \\ 0 & 0 & 1 \end{bmatrix}.$$

There is nothing particularly special about the choice of axes for rotations (z, then x, then z) used by Euler: the only restriction is that no two successive axes of the sequence may be the same. The Euler angles provide a parameterized description of the space of rotation matrices, and associated equations of motion can be derived. However these equations do not cover all possible orientations of a rigid body, and singularities in the equations defining the parameters complicate numerical integration of the resulting equations of motion, not just at the point of singularity, but also in the vicinity of it. One way of resolving the problem is to switch to a new set of angle variables (defined by a different sequence of axes) whenever integration proceeds into the vicinity of a singular point, but this approach is algorithmically cumbersome and suffers from the several other problems mentioned for local charts in Chapter 7. In particular, it is very difficult to obtain a satisfactory geometric method based on this approach.

8.6.2 Quaternions

Because of problems with the Euler angles, an alternative set of parameters is more frequently used in simulations, based on Hamilton's quaternions. This approach is essentially equivalent to the use of "Euler parameters" or "Cayley–Klein variables," although implementations may vary slightly.

Hamilton's quaternions are a quadruple of parameters $\boldsymbol{\sigma} = (\sigma_0, \sigma_1, \sigma_2, \sigma_3)$ subject to the constraint $\sigma_0^2 + \sigma_1^2 + \sigma_2^2 + \sigma_3^2 = 1$. The rotation matrix is defined in terms of the parameters by

$$\boldsymbol{Q} = \boldsymbol{I}_3 + 2\sigma_0\widehat{\boldsymbol{e}} + 2\widehat{\boldsymbol{e}}^2, \qquad \widehat{\boldsymbol{e}} = \begin{bmatrix} 0 & -\sigma_3 & \sigma_2 \\ \sigma_3 & 0 & -\sigma_1 \\ -\sigma_2 & \sigma_1 & 0 \end{bmatrix}. \tag{8.44}$$

The quaternions obey coupled differential equations involving the body angular momenta $\boldsymbol{\pi}$. These differential equations take the form

$$\frac{d}{dt}\boldsymbol{\sigma} = \boldsymbol{B}(\boldsymbol{\sigma})\boldsymbol{T}^{-1}\boldsymbol{\pi}; \qquad \frac{d}{dt}\boldsymbol{\pi} = \widehat{\boldsymbol{\pi}}\boldsymbol{T}^{-1}\boldsymbol{\pi} + \boldsymbol{N},$$

where $\boldsymbol{T} = \mathrm{diag}(I_1, I_2, I_3)$ is the (diagonalized) inertial tensor,

$$\boldsymbol{B}(\boldsymbol{\sigma}) = \begin{bmatrix} -\sigma_1 & -\sigma_2 & -\sigma_3 \\ \sigma_0 & -\sigma_3 & \sigma_2 \\ \sigma_3 & \sigma_0 & -\sigma_1 \\ -\sigma_2 & \sigma_1 & \sigma_0 \end{bmatrix},$$

and $\boldsymbol{N}$ is the applied torque in body coordinates.

The equations of motion can be shown to be Hamiltonian with a noncanonical Lie–Poisson structure similar to that of the Euler equation. In fact, the same splitting ideas as developed in Section 8.3.1 can be applied and lead essentially to identical propagation schemes. The only difference is that one either works directly with the rotation matrix $\boldsymbol{Q} \in \mathbb{R}^{3\times 3}$ or with $\boldsymbol{\sigma} \in \mathbb{R}^4$ and employs the relation (8.44).

A constrained Hamiltonian version of the quaternionic description is also possible by viewing the quaternions as a set of generalized coordinates subject to the holonomic constraint $\|\boldsymbol{\sigma}\|^2 = 1$. Using (8.44), we introduce $\boldsymbol{\sigma}$ and $\dot{\boldsymbol{\sigma}}$ into the rigid body Lagrangian (8.5) and determine the canonical momenta through the formula

$$\boldsymbol{p}_\sigma = \frac{\partial L}{\partial \dot{\boldsymbol{\sigma}}}.$$

Although symplectic integration methods based on such an approach can be formulated, the RATTLE approach based on the rotation matrix $\boldsymbol{Q}$ is found to be more efficient and conceptional easier to implement.

8.7 Exercises

1. *Lagrange and Hamiltonian equations for rigid bodies.* Verify the gradient expression (8.16) using the definition (8.15). Also show that

$$\boldsymbol{P}\boldsymbol{R}^{-1} = \frac{1}{2}\nabla_{\boldsymbol{P}}\mathrm{tr}(\boldsymbol{P}\boldsymbol{R}^{-1}\boldsymbol{P}^T).$$

Essentially the same result can be used to derive the constrained Lagrange equation from the Lagrangian function (8.5). (Hint: You may find the identities

$$\mathrm{tr}(\boldsymbol{A}) = \mathrm{tr}(\boldsymbol{A}^T), \quad \mathrm{tr}(\boldsymbol{A}\boldsymbol{B}) = \mathrm{tr}(\boldsymbol{B}\boldsymbol{A}), \quad \boldsymbol{u}^T\boldsymbol{v} = \mathrm{tr}(\boldsymbol{u}\boldsymbol{v}^T)$$

useful.)

2. *Planar rigid bodies.* Let us introduce a set of orthogonal unit vectors $\boldsymbol{t}_i \in \mathbb{R}^3$, $i = 1, 2, 3$, such that

$$\boldsymbol{Q} = [\boldsymbol{t}_1, \boldsymbol{t}_2, \boldsymbol{t}_3] \in \mathbb{R}^{3\times 3}.$$

We also write

$$\dot{\boldsymbol{Q}} = [\dot{\boldsymbol{t}}_1, \dot{\boldsymbol{t}}_2, \dot{\boldsymbol{t}}_3]$$

for the time derivative. The rotational kinetic energy of a rigid body can then be expressed as

$$T_{\mathrm{rot}} = \frac{1}{2}\mathrm{tr}(\dot{\boldsymbol{Q}}\boldsymbol{R}\dot{\boldsymbol{Q}}^T) = \frac{1}{2}\sum_{i=1}^{3} r_{ii}\|\dot{\boldsymbol{t}}_i\|^2.$$

We have $r_{33} = 0$ for a planar rigid body and the rotational kinetic energy reduces to

$$T_{\mathrm{rot}} = \frac{1}{2}\sum_{i=1}^{2} r_{ii}\|\dot{\boldsymbol{t}}_i\|^2.$$

Since the third unit vector $\boldsymbol{t}_3$ is no longer part of the equations, we also redefine the matrix $\boldsymbol{Q}$ by

$$\boldsymbol{Q} = [\boldsymbol{t}_1, \boldsymbol{t}_2] \in \mathbb{R}^{3\times 2}.$$

This matrix still satisfies $\boldsymbol{Q}^T\boldsymbol{Q} = \boldsymbol{I}_2$ but $\boldsymbol{I}_2$ is now the 2×2 identity matrix.

a. Verify that

$$T_{\mathrm{rot}} = \frac{1}{2}\sum_{i=1}^{2} r_{ii}\|\dot{\boldsymbol{t}}_i\|^2 = \frac{1}{2}\mathrm{tr}(\dot{\boldsymbol{Q}}\boldsymbol{R}\dot{\boldsymbol{Q}}^T),$$

with $\dot{\boldsymbol{Q}} = [\dot{\boldsymbol{t}}_1, \dot{\boldsymbol{t}}_2] \in \mathbb{R}^{3\times 2}$ and

$$\boldsymbol{R} = \begin{bmatrix} r_{11} & 0 \\ 0 & r_{22} \end{bmatrix}.$$

b. The augmented Lagrangian takes the expression

$$\tilde{L} = T_{\mathrm{rot}}(\dot{\boldsymbol{Q}}) + T_{\mathrm{trans}}(\dot{\boldsymbol{q}}_{\mathrm{cm}}) - V_{\mathrm{ext}}(\boldsymbol{Q}, \boldsymbol{q}_{\mathrm{cm}}) - \mathrm{tr}((\boldsymbol{Q}^T\boldsymbol{Q} - \boldsymbol{I}_2)\boldsymbol{\Lambda}).$$

Derive the associated Hamiltonian equations of motion and suggest a numerical implementation.

3. *Mass tensor and inertia tensor.* Let $\boldsymbol{R}$ be the diagonal mass tensor of a free rigid body and $\boldsymbol{T}$ the corresponding diagonal inertia tensor. We have the following relation between the diagonal elements of these two matrices

$$t_{ii} = \sum_{j\neq i} r_{jj}.$$

Show that this relation implies that

$$\widehat{\boldsymbol{b}}\boldsymbol{R} + \boldsymbol{R}\widehat{\boldsymbol{b}} = \widehat{\boldsymbol{T}\boldsymbol{b}}$$

holds for any vector $\boldsymbol{b}$ and associated skew-symmetric matrix $\widehat{\boldsymbol{b}}$.

4. *Lagrangian top.* Show that the third component $m_3 = \boldsymbol{s}_3^T\boldsymbol{\pi}$ of the spatial angular momentum vector $\boldsymbol{m} = \boldsymbol{Q}\boldsymbol{\pi}$ is preserved under the equations (8.38)–(8.39) for the symmetric top. Show that the third component π_3 of the body angular momentum is also preserved. Denote the energy of the Lagrangian top by E and introduce the four constants (compare [7, 73])

$$\alpha = \frac{2E - \pi_3^2/m_3}{I_1},$$

$$\beta = \frac{c}{I_1},$$

$$a = \frac{\pi_3}{I_1},$$

$$b = \frac{m_3}{I_1}.$$

Provided that the distance between the stationary point of contact and the center of mass is equal to one, it can be shown that the third component $u(t) = q_3(t)$ of the center of mass motion $\boldsymbol{q}_{\rm cm}(t) = \boldsymbol{Q}(t)\boldsymbol{r}^0$ satisfies the differential equation

$$\dot{u}^2 = (1-u^2)(\alpha - \beta u) - (b - au)^2.$$

Of particular importance are the two roots of

$$0 = (1-u_i^2)(\alpha - \beta u_i) - (b - au_i)^2,$$

$i = 1, 2$, in the interval $u \in [-1, 1]$. With $u_1 < u_2$ the motion in $u(t)$ satisfies $u_1 \le u(t) \le u_2$. Compute the two roots for the data given in Section 8.3.2 and compare with the computed motion in $u(t) = q_3(t)$ as displayed in Fig. 8.3. Note also that the initial body angular momentum vector implies that $u_2 = b/a$.

5. *Structure matrix and Casimir functions.* Find the structure matrix $\boldsymbol{J}$ corresponding to the Hamiltonian equations of motion for a single rigid body with kinetic energy

$$\mathcal{T}(\boldsymbol{p}_{cm}, \boldsymbol{\pi}) = \frac{1}{2m}\boldsymbol{p}_{cm}^T\boldsymbol{p}_{cm} + \frac{1}{2}\boldsymbol{\pi}^T\boldsymbol{T}^{-1}\boldsymbol{\pi},$$

moving in $\mathbb{R}^3$ under a potential energy $\mathcal{V}(\boldsymbol{q}_{cm}, \boldsymbol{s}_1, \boldsymbol{s}_2, \boldsymbol{s}_3)$. What is the rank of J? Find the associated Casimir functions.

6. *Angular momentum conservation.* Show that total angular momentum

$$\boldsymbol{m}_{\rm tot} = \boldsymbol{q}_{cm} \times \boldsymbol{p}_{cm} + \boldsymbol{Q}\boldsymbol{\pi},$$

is conserved for any single rigid body with a potential energy of the form

$$\mathcal{V} = \psi(\|\boldsymbol{q}\|), \qquad \boldsymbol{q} = \boldsymbol{q}_{cm} + \boldsymbol{Q}\boldsymbol{r}^0.$$

7. *Numerical conservation of angular momentum.* Consider the RATTLE algorithm (8.17)–(8.21) for a free rigid body; i.e. $V_{\rm ext} = 0$. Show that the spatial angular momentum matrix $\widehat{\boldsymbol{m}}^n = \boldsymbol{P}^n[\boldsymbol{Q}^n]^T - \boldsymbol{Q}^n[\boldsymbol{P}^n]^T$ is exactly conserved along numerical solutions. Hint: use the fact that $\widehat{\boldsymbol{m}}^n$ is skew symmetric.

8. *Coupled rigid bodies.* Write the equations of motion for two identical rigid bodies with total mass m and tensor of inertia $\boldsymbol{T}$ interacting through a harmonic spring with rest length L and force constant K. The spring is attached to the rigid bodies at the material points $\boldsymbol{r}_i^0$, $i = 1, 2$. Give a second-order composition method for the case $I = I_1 = I_2 \neq I_3$ (symmetric rigid body).

9. *Quaternions.* Given a unit vector $\boldsymbol{n} \in \mathbb{R}^3$ and a real number ω, verify RODRIGUES' formula

$$\boldsymbol{Q} = \mathrm{e}^{\omega \hat{\boldsymbol{n}}},$$

where the rotation matrix $\boldsymbol{Q}$ is given by (8.44) with $\sigma_0 = \cos(\omega/2)$ and

$$[\sigma_1, \sigma_2, \sigma_3] = \sin(\omega/2)\,[n_1, n_2, n_2].$$

RODRIGUES' formula is useful for implementing a splitting method for symmetric or nearly symmetric rigid bodies. Note that σ_0 can be replaced by $\sigma_0 = (1 - \omega^2/4)/(1 + \omega^2/4)$ and

$$[\sigma_1, \sigma_2, \sigma_3] = \frac{\omega}{1 + \omega^2/4}\,[n_1, n_2, n_2],$$

which provides an inexpensive approximation to the exact matrix exponential for $\omega \ll 1$.

9

Adaptive geometric integrators

It is well known that during the integration of nonlinear systems of ordinary differential equations changes in the character of the solutions may demand corresponding changes in the integration timestep. In the context of Hamiltonian systems this need develops from fluctuations in the forces along a solution curve. For motivation we need look no further than the historic Kepler problem which was introduced in Chapter 3. Kepler's first two laws tell us (i) that a body in bound gravitational motion moves along an ellipse with the fixed point at one focus and that (ii) the orbit sweeps out equal sections of the ellipse in equal times. This is diagrammed in the illustration below (Fig. 9.1), which shows a Keplerian orbit and several points equally spaced in time along the orbit.

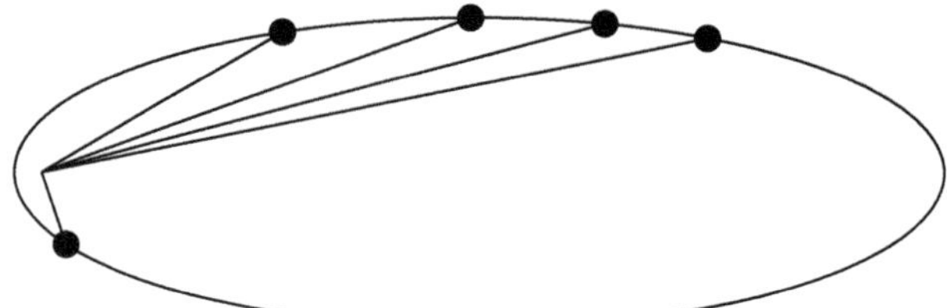

Figure 9.1 An orbit of the Kepler problem. The moving body sweeps out equal areas in equal times, moving most rapidly in the vicinity of the fixed body. The points shown along the Kepler ellipse are encountered at uniform intervals in time.

Recall that the force acting on the moving body in the Kepler system is inversely proportional to the square of the separation from the fixed body. When the orbits in the Kepler problem (or, more generally, in a gravitational N-body problem) are highly eccentric (as they are for example for Comets in our Solar System) the magnitude of the force along orbits can vary considerably. It is clear that the dynamics in the vicinity of close approach will be challenging to resolve accurately. For this reason, some sort of adaptivity in time is essential.[1]

[1] Of course it must not be overlooked that the Kepler problem is exactly integrable, and methods that are intended to be effective for N-body gravitation should inevitably take advantage of the

The traditional approach to varying stepsize in ordinary differential equation solvers is based on selecting a stepsize so that an error estimate is kept below a prescribed tolerance. Because computing the global error is essentially impossible in general purpose integration, the standard approaches use a local error estimate. For example, it is common when integrating with a Runge–Kutta method to add an additional stage to the method so that a higher-order approximation is provided; the difference between the two approximations is then used as an error estimate. Schemes like this can be designed to work with symplectic methods, but the results obtained in numerical experiments are wholly unsatisfying. The observation has been made that, for long term stability, *fixed stepsize symplectic methods actually outperform their variable stepsize counterparts when compared on an efficiency basis.* This raised a quandary when it first was discovered by GLADMAN, DUNCAN, AND CANDY [72], CALVO AND SANZ-SERNA [42], SKEEL AND GEAR [176] and others in the early 1990s. After all, the symplectic maps form a group, hence the composition of symplectic integrators with various stepsizes

$$\boldsymbol{\Psi} = \boldsymbol{\Psi}_{\Delta t_1}\boldsymbol{\Psi}_{\Delta t_2}\ldots\boldsymbol{\Psi}_{\Delta t_N},$$

remains a symplectic map. Why should the performance be so poor? The same poor behaviour was observed whether or not the stepsize sequence was obtained from a series of computations based on the solution or was provided a priori.

The best explanation emerged with the development of the backward error analysis. In order to construct the so-called perturbed Hamiltonian expansion for a symplectic method, it is necessary to assume that the symplectic maps used in each timestep are identical. If the stepsize varies, we have the problem that we generate approximations to one, then another perturbed Hamiltonian as the stepsize changes; there is no reason to expect that this sort of process would produce a stable long-term propagation. It seems from this that the prospects for getting good results from symplectic methods with traditional variable stepsize techniques are dim.

9.1 Sundman and Poincaré transformations

The key to effective geometric integration with variable stepsize lies in the use of a time transformation

$$\frac{dt}{d\tau} = g(\boldsymbol{q}, \boldsymbol{p}), \tag{9.1}$$

nearly Keplerian character of many of the trajectories. However, in this instance we are merely using the unmodified Kepler problem as a device to illustrate a general approach to variable stepsize.

where g is a smooth, positive scalar-valued function of positions and momenta. We refer to this as a *Sundman transformation*. If this transformation is applied to an autonomous system of differential equations, there results a new system of differential equations whose solutions evolve in the rescaled, or "fictive," time. In fact, the actual orbits of the system do not change; they are just traversed at a different speed. If the original Hamiltonian is $H(\boldsymbol{q}, \boldsymbol{p})$, the system that results from application of (9.1) is

$$\frac{d}{d\tau}\boldsymbol{q} = \frac{d\boldsymbol{q}}{dt}\frac{dt}{d\tau} = g(\boldsymbol{q}, \boldsymbol{p})\nabla_{\boldsymbol{p}} H(\boldsymbol{q}, \boldsymbol{p}), \tag{9.2}$$

$$\frac{d}{d\tau}\boldsymbol{p} = \frac{d\boldsymbol{p}}{dt}\frac{dt}{d\tau} = -g(\boldsymbol{q}, \boldsymbol{p})\nabla_{\boldsymbol{q}} H(\boldsymbol{q}, \boldsymbol{p}). \tag{9.3}$$

This system is, in general, no longer Hamiltonian. (It is Hamiltonian in very special cases, such as when the function g is a function of a first integral of the problem – for example the energy – but this choice appears to be of little practical interest.) On the other hand, if $g(\boldsymbol{q}, -\boldsymbol{p}) = g(\boldsymbol{q}, \boldsymbol{p})$ and $g(\boldsymbol{q}, \boldsymbol{p}) > 0$, then *time-reversibility* can be maintained by the Sundman transformation. Sundman transformations had actually been in frequent use in computational astronomy from the mid 1960s for varying the timestep (albeit without maintaining the preservation of time-reversal symmetry under subsequent discretization).

Now consider the fixed-stepsize integration of (9.2)–(9.3) with a stepsize $\Delta\tau$ in fictive time, generating a discrete trajectory $\{(\boldsymbol{q}^n, \boldsymbol{p}^n)\}$. Because of the time-transformation, each fixed step in τ corresponds to a time-step in t of length $\Delta t_n \approx g(\boldsymbol{q}^n, \boldsymbol{p}^n)\Delta\tau$. This relation is not exact; in fact, to resolve the time variable we need to numerically solve the differential equation (9.1) in tandem with the system in order to determine the evolution of time. STOFFER [178, 179] pointed out that the system (9.2)–(9.3),(9.1) could be solved by a reversible integrator, yielding a reversible variable stepsize method. A related scheme, also based on reversible Sundman transformation, was suggested in 1995 by the astronomers HUT, MAKINO AND MCMILLAN as a modification of the Störmer–Verlet method [90]. The methods mentioned so far are all implicit, a consequence of the coupling the various components of the vector field via the Sundman transformation. This implicitness is undesirable in many of the applications where variable stepsize methods are most likely to be needed, such as gravitational dynamics. A semi-explicit approach ("Adaptive Verlet") was eventually developed by HUANG AND LEIMKUHLER [89].

Before turning our attention to this scheme, let us consider the symplectic alternative. Along with Sundman transformations, there is another class of time transformations which can be used as the foundation for geometric integration. From a given Hamiltonian $H(\boldsymbol{q}, \boldsymbol{p})$ construct a new Hamiltonian by first subtracting the energy E of a trajectory, then multiplying this difference by a positive, smooth scaling function $g(\boldsymbol{q}, \boldsymbol{p})$

$$\tilde{H}(\boldsymbol{q}, \boldsymbol{p}) = g(\boldsymbol{q}, \boldsymbol{p})(H(\boldsymbol{q}, \boldsymbol{p}) - E). \tag{9.4}$$

Noting that the energy is constant along trajectories, we can infer that if at some initial point on the trajectory $H(\boldsymbol{q}^0, \boldsymbol{p}^0) = E$, then this same relation is satisfied at *all* points of the trajectory, thus, as well, $\tilde{H} \equiv 0$ along this trajectory. The equations of motion for the modified Hamiltonian (9.4) take the form

$$\frac{d}{d\tau}\boldsymbol{q} = g(\boldsymbol{q}, \boldsymbol{p})\nabla_{\boldsymbol{p}} H(\boldsymbol{q}, \boldsymbol{p}) + (H(\boldsymbol{q}, \boldsymbol{p}) - E)\nabla_{\boldsymbol{p}} g(\boldsymbol{q}, \boldsymbol{p}), \tag{9.5}$$

$$\frac{d}{d\tau}\boldsymbol{p} = -g(\boldsymbol{q}, \boldsymbol{p})\nabla_{\boldsymbol{q}} H(\boldsymbol{q}, \boldsymbol{p}) - (H(\boldsymbol{q}, \boldsymbol{p}) - E)\nabla_{\boldsymbol{q}} g(\boldsymbol{q}, \boldsymbol{p}). \tag{9.6}$$

These equations can be seen to reduce to (9.2)–(9.3) along the energy surface, thus the Poincaré and Sundman transformations are formally equivalent, in the absence of perturbation. The transformation (9.4) has the same effect as (9.1) with respect to the exact solutions. This suggests that we could implement a variable stepsize integration strategy by first applying the Poincaré transformation and then discretizing the resulting system using a fixed stepsize symplectic integrator, an idea independently suggested by HAIRER [78] and REICH [158]. In this case, our numerical method would consist of the iteration of a single symplectic map, so standard backward error analysis (see Chapter 5) could be used to justify the resulting method.

In the context of numerical integrations, however, it should be emphasized that this approach raises some new issues. First, the energy is unlikely to be exactly conserved, and this destroys the direct correspondence between time and fictive time when the Poincaré transformation is used. From (9.4) we can see that if an error $\Delta\tilde{H}$ is introduced in the Hamiltonian $\tilde{H}$ at some stage of numerical integration, then the true energy will satisfy

$$\Delta H = \frac{\Delta\tilde{H}}{g},$$

thus in situations where g becomes very small (where the timesteps needed for integration are small), the true energy error can be large, and the equations (9.5)–(9.6) will differ significantly from the Sundman-transformed equations (9.2)–(9.3); hence the correspondence between time and fictive time will be destroyed at points where the computation is most difficult.

A second problem raised by the Poincaré transformation has to do with the numerical implementation of the method. We know from earlier chapters that the available methods for integrating Hamiltonian systems with a nonseparable Hamiltonian are all implicit. Since (9.4) couples the positions and momenta, even when the original Hamiltonian H is separable, the methods may not be practical for many applications. On the other hand, there are settings where such a symplectic approach may be reasonable: for example, if the original Hamiltonian is of low dimension or is given in a nonseparable form, or if certain combinations of numerical method and time transformation are employed. We will briefly return to discuss the symplectic approach later in this chapter.

9.2 Reversible variable stepsize integration

We now consider the application of reversible adaptive methods. Eventually, we will see that this approach allows for construction of efficient higher-order variable stepsize methods, and we will see how this reversible adaptive framework can be combined with more sophisticated methods such as those used to regularize few-body close approaches in N-body gravitation.

Recall that a differential equation system

$$\frac{d}{dt}z = \boldsymbol{f}(z), \qquad z \in \mathbb{R}^d$$

is reversible with respect to a linear involution $\boldsymbol{S}$ (a mapping of $\mathbb{R}^d$ with $\boldsymbol{S}^2 = \boldsymbol{I}$) if

$$-\boldsymbol{S}\boldsymbol{f}(\boldsymbol{S}z) = \boldsymbol{f}(z).$$

In particular, a canonical Hamiltonian system is reversible with respect to $\boldsymbol{p} \to -\boldsymbol{p}$ if $H(\boldsymbol{q}, -\boldsymbol{p}) = H(\boldsymbol{q}, \boldsymbol{p})$. A system of differential equations

$$\frac{d}{dt}\boldsymbol{q} = \boldsymbol{F}(\boldsymbol{q}, \boldsymbol{p}),$$
$$\frac{d}{dt}\boldsymbol{p} = \boldsymbol{G}(\boldsymbol{q}, \boldsymbol{p})$$

is reversible under the same involution if $\boldsymbol{F}(\boldsymbol{q}, -\boldsymbol{p}) = -\boldsymbol{F}(\boldsymbol{q}, \boldsymbol{p})$, and $\boldsymbol{G}(\boldsymbol{q}, -\boldsymbol{p}) = \boldsymbol{G}(\boldsymbol{q}, \boldsymbol{p})$.

From the definition, it is easy to see that applying a Sundman transformation (9.1) with positive scaling g and $g(\boldsymbol{S}z) = g(z)$ to a reversible system $\frac{d}{dt}z = \boldsymbol{f}(z)$ will yield another reversible system. If we discretize the rescaled system

$$\frac{dz}{d\tau} = g(z)\boldsymbol{f}(z)$$

by some numerical method we obtain a discrete trajectory $\{z^n\}$, and each point on this trajectory will correspond to a point τ_n in the rescaled time. Evidently, we must solve (9.1) along with the differential equations for the phase variables in order to resolve the temporal variable.

It remains to find a suitable discretization of the system (9.2)–(9.3). One approach, which is usually acceptable for low-dimensional problems, is just to apply the implicit midpoint method (or some other reversible integrator) to the rescaled equations. This has the advantage of simplicity and generality. We will illustrate this general approach to adaptive integration with a simple example: the soft impact oscillator with one degree-of-freedom Hamiltonian

$$H(q,p) = \frac{p^2}{2} + \frac{1}{q^2} + \frac{k}{2}q^2.$$

The orbits of this problem are periodic. A phase plane orbit (for $k = 0.1$) is shown below (Fig. 9.2), along with graphs of each coordinate against time. Note that the momenta must change very rapidly on a short time interval as the material point approaches the origin, and then gradually on the remainder of the period; this problem can benefit from some sort of adaptivity in the timestep.

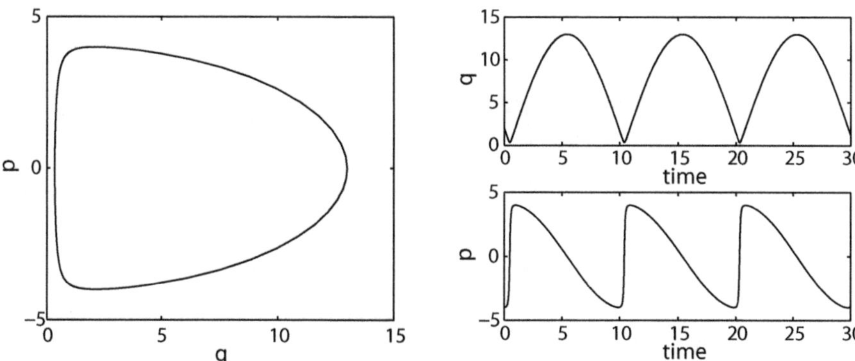

Figure 9.2 An orbit of the "impact-oscillator" for $k = 0.1$. The figure to the left shows the motion in the phase plane; in the right, the positions and momenta are graphed against time. Note the sharp change in the momenta as q approaches the origin.

We solve this system numerically for the following initial condition $q(0) = 2$, $p(0) = -4$ for $t \in [0, 100]$ using the following three related second-order methods: (1) the trapezoidal rule (abbreviated "TR"), (2) an implementation of the trapezoidal rule with a third-order error estimate (TR23),[2] and (3) Trapezoidal Rule applied to the rescaled equations

$$\frac{d}{dt}q = g(q)p,$$
$$\frac{d}{dt}p = g(q)(2q^{-3} - kq),$$

[2]This solver is implemented in MATLAB 6 as **ode23t** and is specifically recommended for problems with little or no damping.

with the rescaling function

$$g(q) = \frac{1}{1 + q^{-2}}$$

(abbreviated TRS).

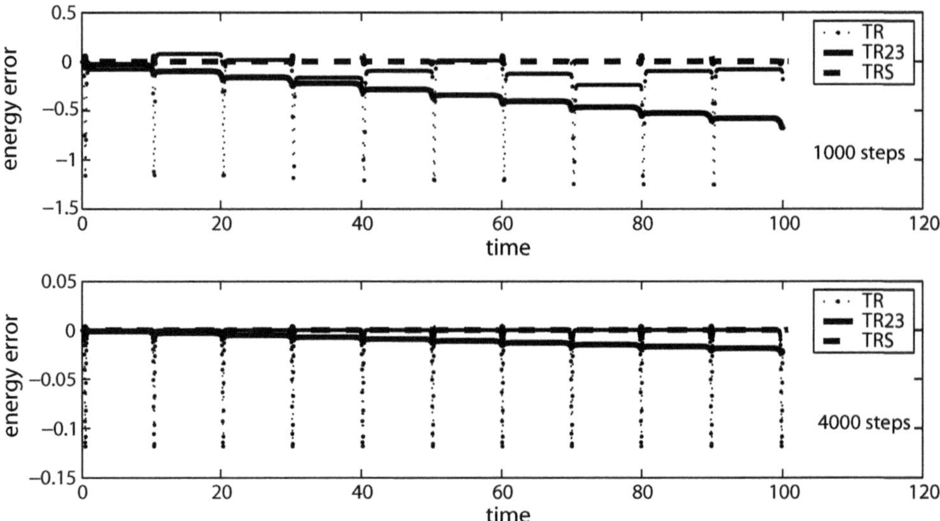

Figure 9.3 Energy errors in simulations of the impact oscillator using the three variants of trapezoidal rule. In the upper figure the simulation timestep, tolerance, and fictive timestep were all tuned so that a total of 1000 timesteps were taken to cover the time interval $[0, 100]$. In the lower figure, 4000 timesteps were used.

The energy errors for representative simulations are shown in Fig. 9.3. In order to perform comparable experiments, we tuned the integration parameters (timestep, tolerance, or fictive timestep) so that all methods required the same number of trapezoidal rule steps to integrate to time 100. Observe that the energy of both the trapezoidal rule and the traditional variable stepsize methods fluctuate substantially, in the vicinity of the "collision." On the other hand, the reversible variable stepsize method shows a stable, relatively mild energy variation with no secular drift. Comparing the upper and lower panels of Fig. 9.3, we see that the results are qualitatively similar regardless of the number of timesteps used in the simulation.

The mechanism which leads to the energy drift in TR23 becomes clearer when we look at the close-up of the energy in the vicinity of collision (Fig. 9.4).

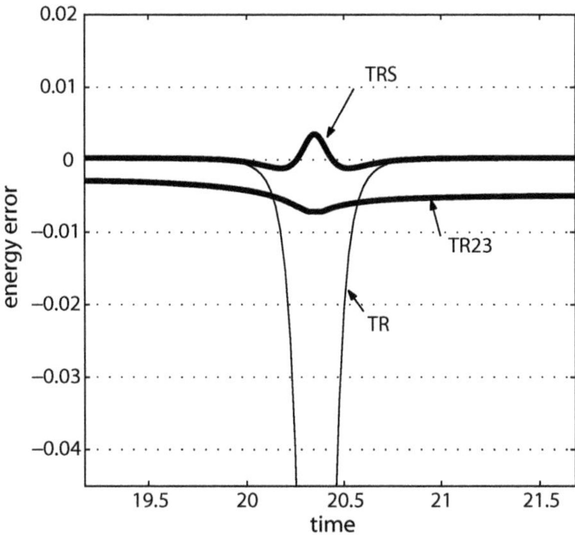

Figure 9.4 Close-up of the energy evolution in the vicinity of a collision. The TR and TRS schemes return to the pre-collisional value. By contrast, the TR23 method loses a substantial amount of energy during the collision.

For the TR and TRS methods, which are time reversible, the energy error graph can be seen to be geometrically symmetric around a vertical line through the time of collision. Because the energy is nearly exactly maintained far from the collision, the symmetry in energy implies that it must go back to the correct value as we leave the point of collision. By contrast, the TR23 method, which actually controls the local error in energy at the point of collision somewhat better than TRS, does not possess the symmetric energy behavior of the other methods. As a consequence, even on the relatively modest interval considered, the reversible variable stepsize method easily wins this competition, when compared on an efficiency basis (with efficiency measured by energy error per timestep); on longer time intervals or in larger problems, one can expect the results to be even more dramatic. The important symmetry property for the evolution of the energy error is not guaranteed to be present in all simulations. It can be shown to hold for two-body problems with central forces [109]. The benefits of the reversible adaptive framework are less apparent in the presence of interactions between three or more bodies. We will return to this issue in the section on Coulombic problems, below.

The choice of the time-rescaling function is critical to the success of the symmetric adaptive methods. We discuss this problem in the following and subsequent sections of this chapter.

9.2.1 Local error control as a time transformation

Given a symmetric one-step method $\boldsymbol{\Psi}_{\Delta t}$, an adaptive method can be generated by solving the equations

$$z^{n+1} = \boldsymbol{\Psi}_{\Delta t_n}(z^n),$$
$$0 = P(\Delta t_n, z^n, z^{n+1}).$$

If the scalar function P is symmetric with respect to transposition of its last two arguments, then the resulting method will be symmetric.

As suggested by STOFFER [179] (see also [83]) we can base P on an estimate for the local error using various schemes, such as finite differences. In the case where the underlying method $\boldsymbol{\Psi}_{\Delta t}$ is a symmetric Runge–Kutta or Partitioned Runge–Kutta method, a traditional embedded error estimate can be developed using the stages of the method; the only restriction is that the formula for the error estimate we use should be symmetric.

To illustrate, suppose we start with the trapezoidal rule,

$$z^{n+1} = z^n + \frac{\Delta t}{2}\left[\boldsymbol{f}(z^n) + \boldsymbol{f}(z^{n+1})\right]. \tag{9.7}$$

The local error in one step of Trapezoidal Rule is

$$\mathbf{le} = \frac{\Delta t^3}{12}\frac{d^3}{dt^3}z(\xi).$$

This error can be estimated in a variety of ways. We can also use the differential equation itself to define $\frac{d^3}{dt^3}z$ along the solution. In a one-dimensional case we would have

$$\frac{d}{dt}z = f \Rightarrow \frac{d^2}{dt^2}z = f'f \Rightarrow \frac{d^3}{dt^3}z = f''f^2 + f'^2 f$$

(or in higher dimensions, a more complicated version of this formula). The local error in the nth step of trapezoidal rule can then be approximated by

$$le \approx \frac{\Delta t^3}{12}(f''f^2 + f'^2 f).$$

Setting the magnitude of this estimate to a prescribed tolerance `tol` yields

$$\Delta t = \left[\frac{12\,\mathtt{tol}}{|f''f^2 + f'^2 f|}\right]^{1/3}.$$

Note that the stepsize has been written as a function of `tol` and the solution itself. We could, for example, evaluate the estimate at $z^{n+1/2} = (z^n + z^{n+1})/2$, then introduce the expression for Δt directly into (9.7), the result being a second-order implicit adaptive integrator. The parameter `tol` acts in the usual way to allow a refinement of the approximate solution. In practice we would need to modify the stepsize formula to avoid singularities and to ensure the smoothness of the timestep map.

Alternatively, if we define $\Delta\tau = \mathtt{tol}^{1/3}$, we see that we are simply solving a certain Sundman rescaling of the original problem, so that any of the schemes mentioned in this chapter could be used. This example is a special case of a more general observation that symmetric methods based on error estimates can be viewed as reversible discretizations of a Sundman-transformed problem (see [179]).

Controlling local error makes sense if the goal of computation is a numerical solution with a small trajectory error, but this is not necessarily the situation we confront when using a geometric integrator. On the other hand, there are likely to be situations where some combination of various types of stepsize controls, including one based on local error, should be used in simulation.

In the following subsections, we consider the construction of efficient methods for implementing reversible variable stepsize based on time reparameterization. The schemes described above are implicit. In the sequel we show how semi-explicit and even fully explicit methods can be developed.

9.2.2 Semi-explicit methods based on generalized leapfrog

We consider a reparameterization of a separable Hamiltonian system

$$\frac{d\boldsymbol{q}}{d\tau} = g(\boldsymbol{q}, \boldsymbol{p})\boldsymbol{M}^{-1}\boldsymbol{p}, \tag{9.8}$$

$$\frac{d\boldsymbol{p}}{d\tau} = -g(\boldsymbol{q}, \boldsymbol{p})\nabla_{\boldsymbol{q}} V(\boldsymbol{q}). \tag{9.9}$$

The system is not solvable by the standard Verlet method.

On the other hand, if we apply the generalized leapfrog method, we obtain the following system

$$\boldsymbol{q}^{n+1} = \boldsymbol{q}^n + \frac{\Delta\tau}{2}(g(\boldsymbol{q}^n, \boldsymbol{p}^{n+\frac{1}{2}}) + g(\boldsymbol{q}^{n+1}, \boldsymbol{p}^{n+\frac{1}{2}}))\boldsymbol{M}^{-1}\boldsymbol{p}^{n+\frac{1}{2}}, \tag{9.10}$$

$$\boldsymbol{p}^{n+\frac{1}{2}} = \boldsymbol{p}^n - \frac{\Delta\tau}{2} g(\boldsymbol{q}^n, \boldsymbol{p}^{n+\frac{1}{2}})\nabla_{\boldsymbol{q}} V(\boldsymbol{q}^n), \tag{9.11}$$

$$\boldsymbol{p}^{n+1} = \boldsymbol{p}^{n+\frac{1}{2}} - \frac{\Delta\tau}{2} g(\boldsymbol{q}^{n+1}, \boldsymbol{p}^{n+\frac{1}{2}})\nabla_{\boldsymbol{q}} V(\boldsymbol{q}^{n+1}). \tag{9.12}$$

The equations (9.10)–(9.12) are implicit, requiring the solution of a nonlinear system at each step, but the fact that g is scalar means that a relatively efficient Newton solver based on rank-one updates is possible, as pointed out in [89]. Moreover, in case g is a function only of $\boldsymbol{q}$ (or $\boldsymbol{p}$) the scheme actually becomes semi-explicit, in the sense that the timestepping can be performed with only one force evaluation per timestep. Specifically, if $g = g(\boldsymbol{q})$, then (9.11) gives $\boldsymbol{p}^{n+\frac{1}{2}}$ explicitly. If we evaluate g on both sides of (9.10) and set $\gamma^{n+1} = g(\boldsymbol{q}^{n+1})$, then we arrive at

$$\gamma^{n+1} = g(\boldsymbol{q}^n + \frac{\Delta\tau}{2}(\gamma^n + \gamma^{n+1})M^{-1}\boldsymbol{p}_{n+\frac{1}{2}}),$$

but since $\boldsymbol{p}^{n+1/2}$ is now known, this is just a scalar nonlinear equation for γ^{n+1}, readily solvable by Newton iteration. This scheme will be efficient provided the evaluation of g and its gradient are inexpensive. For example if the system involves long-ranged forces, but the time reparameterization is short-ranged – a typical case – then the per timestep cost of incorporating adaptivity in this way will be neglible.

9.2.3 Differentiating the control

We next describe an approach based on introducing the stepsize control as a new variable along with its own differential equation. After discretization, this results in an algorithm requiring evaluation of the Hessian of the potential energy function, or rather its products with certain vectors. Differentiating $\gamma = g(\boldsymbol{q}, \boldsymbol{p})$ with respect to the reparameterized time τ yields

$$\begin{aligned}\frac{d}{d\tau}\gamma &= (\nabla_{\boldsymbol{q}} g(\boldsymbol{q}, \boldsymbol{p}))^T \frac{d}{d\tau}\boldsymbol{q} + (\nabla_{\boldsymbol{p}} g(\boldsymbol{q}, \boldsymbol{p}))^T \frac{d}{d\tau}\boldsymbol{p} \\ &= g(\boldsymbol{q}, \boldsymbol{p}) \left[(\nabla_{\boldsymbol{q}} g(\boldsymbol{q}, \boldsymbol{p}))^T \boldsymbol{M}^{-1}\boldsymbol{p} - (\nabla_{\boldsymbol{p}} g(\boldsymbol{q}, \boldsymbol{p}))^T \nabla_{\boldsymbol{q}} V(\boldsymbol{q})\right].\end{aligned}$$

We then look for a discretization for the coupled system in $(\boldsymbol{q}, \boldsymbol{p}, \gamma)$. We will illustrate for the choice

$$g(\boldsymbol{q}, \boldsymbol{p}) = (\boldsymbol{p}^T \boldsymbol{M}^{-2}\boldsymbol{p} + \|\nabla_{\boldsymbol{q}} V(\boldsymbol{q})\|^2)^{-\frac{1}{2}}.$$

Since $\nabla_{\boldsymbol{p}} g(\boldsymbol{q}, \boldsymbol{p}) = -g(\boldsymbol{q}, \boldsymbol{p})^{-3}\boldsymbol{M}^{-2}\boldsymbol{p}$ and $\nabla_{\boldsymbol{q}} g(\boldsymbol{q}, \boldsymbol{p}) = -g(\boldsymbol{q}, \boldsymbol{p})^{-3} V_{\boldsymbol{qq}}(\boldsymbol{q}) \nabla_{\boldsymbol{q}} V(\boldsymbol{q})$,

$$\frac{d}{d\tau}\gamma = \frac{\boldsymbol{p}^T \boldsymbol{M}^{-1} V_{\boldsymbol{qq}}(\boldsymbol{q}) \nabla_{\boldsymbol{q}} V(\boldsymbol{q}) - \boldsymbol{p}^T \boldsymbol{M}^{-2} \nabla_{\boldsymbol{q}} V(\boldsymbol{q})}{g(\boldsymbol{q}, \boldsymbol{p})^2} =: G(\boldsymbol{q}, \boldsymbol{p}).$$

Next, we partition the variables as $\boldsymbol{q}$ and $(\boldsymbol{p}, \gamma)$ and discretize using the generalized leapfrog

$$\boldsymbol{q}^{n+1} = \boldsymbol{q}^n + \Delta\tau g^{n+\frac{1}{2}} \boldsymbol{M}^{-1} \boldsymbol{p}^{n+\frac{1}{2}}, \tag{9.13}$$

$$\gamma^{n+\frac{1}{2}} = \gamma^n + \frac{\Delta\tau}{2} G(\boldsymbol{q}^n, \boldsymbol{p}^{n+\frac{1}{2}}), \tag{9.14}$$

$$\boldsymbol{p}^{n+\frac{1}{2}} = \boldsymbol{p}^n - \frac{\Delta\tau}{2} g^{n+\frac{1}{2}} \nabla_{\boldsymbol{q}} V(\boldsymbol{q}^n), \tag{9.15}$$

$$\gamma^{n+1} = \gamma^{n+1/2} + \frac{\Delta\tau}{2} G(\boldsymbol{q}^{n+1}, \boldsymbol{p}^{n+\frac{1}{2}}), \tag{9.16}$$

$$\boldsymbol{p}^{n+1} = \boldsymbol{p}^{n+1/2} - \frac{\Delta\tau}{2} g^{n+\frac{1}{2}} \nabla_{\boldsymbol{q}} V(\boldsymbol{q}^{n+1}), \tag{9.17}$$

$$t_{n+1} = t_n + \Delta\tau\gamma^{n+\frac{1}{2}}. \tag{9.18}$$

It can be shown that these equations reduce to solving a cubic polynomial for $\gamma^{n+1/2}$, after which all steps are explicit. For details see [89], which also includes numerical experiments with the method. This scheme works well in practice. For the application of this idea in the context of gravitational dynamics, see [134].

Note that, for N-body systems, efficient schemes are typically available which allow the rapid computation of matrix–vector products involving the Hessian matrix in tandem with the force evaluation [118].

9.2.4 The Adaptive Verlet method

We now describe a reversible adaptive timestepping scheme which admits both semi-explicit and fully explicit variants. The idea is to write the Sundman-transformed equations of motion for a Newtonian mechanical system as a constrained differential equation system, in the form

$$\frac{d}{d\tau}\boldsymbol{q} = \gamma \boldsymbol{M}^{-1}\boldsymbol{p},$$

$$\frac{d}{d\tau}\boldsymbol{p} = \gamma \nabla_{\boldsymbol{q}} V(\boldsymbol{q}),$$

$$\gamma = g(\boldsymbol{q}, \boldsymbol{p}).$$

This approach broadens the possibilities for methods, since we may now evaluate g at different points than the vector field. A broad class of second-order symmetric methods can be written in the form

$$\boldsymbol{q}^{n+1/2} = \boldsymbol{q}^n + \frac{1}{2}\Delta t \gamma^n \boldsymbol{M}^{-1}\boldsymbol{p}^{n+1/2},$$

$$\boldsymbol{p}^{n+1/2} = \boldsymbol{p}^n + \frac{1}{2}\Delta t \gamma^n \nabla_{\boldsymbol{q}} V(\boldsymbol{q}^n),$$

$$0 = R(\boldsymbol{q}^n, \boldsymbol{q}^{n+1}, \boldsymbol{q}^{n+1/2}, \boldsymbol{p}^n, \boldsymbol{p}^{n+1}, \boldsymbol{p}^{n+1/2}, \gamma^n, \gamma^{n+1}),$$

$$\boldsymbol{q}^{n+1} = \boldsymbol{q}^{n+1/2} + \frac{1}{2}\Delta t \gamma^{n+1} \boldsymbol{M}^{-1}\boldsymbol{p}^{n+1/2},$$

$$\boldsymbol{p}^{n+1} = \boldsymbol{p}^{n+1/2} + \frac{1}{2}\Delta t \gamma^{n+1} \nabla_{\boldsymbol{q}} V(\boldsymbol{q}^{n+1}).$$

Here R is a smooth function depending on $g(\boldsymbol{q}, \boldsymbol{p})$, invariant under $\boldsymbol{q}^n \to \boldsymbol{q}^{n+1}$, $\boldsymbol{p}^n \to \boldsymbol{p}^{n+1}$, $\gamma^n \to \gamma^{n+1}$, and approximating to second-order accuracy the equation $\gamma = g(\boldsymbol{q}, \boldsymbol{p})$. Note that the method has the form of two separate Verlet half steps with a stepsize adjustment in between, but in some cases the stepsize adjustment could make the whole calculation fully implicit. These schemes are referred to collectively as the *Adaptive Verlet* method. The simplest explicit scheme of this type is defined by

$$\gamma^n + \gamma^{n+1} = 2g(\boldsymbol{q}^{n+1/2}, \boldsymbol{p}^{n+1/2}). \tag{9.19}$$

An alternative method, originally suggested by HUANG & LEIMKUHLER in [89], is defined by

$$\frac{1}{\gamma^{n+1}} + \frac{1}{\gamma^n} = \frac{1}{g(\boldsymbol{q}^{n+1/2}, \boldsymbol{p}^n)} + \frac{1}{g(\boldsymbol{q}^{n+1/2}, \boldsymbol{p}^{n+1})}. \tag{9.20}$$

When cost is measured in terms of function evaluations, the Adaptive Verlet method is usually much more efficient than the implicit schemes mentioned earlier. For example, when compared against the symmetric adaptive trapezoidal rule discretization (TRS in the previous section), applied to the same impact oscillator problem and with the same rescaling function, the Adaptive Verlet method requires about a quarter of the work required for the implicit scheme to compute solutions with the same accuracy. These considerations are likely to be ampified for larger systems.

9.3 Sundman transformations

Reversible variable stepsize methods require the specification of an auxiliary time-reparameterization function $g(\boldsymbol{q}, \boldsymbol{p})$. It is possible to choose this rescaling function based on a functional of the vector field; in this case the method becomes more or less problem independent, since the computations performed for varying stepsizes can be fully automated.

9.3.1 Arclength parameterization

In some sense the most natural rescaling is one which is provided by the vector field itself. Recall that the *arclength* s of a parameterized solution curve in the interval $[t_0, t_0 + \Delta t]$ is

$$s(\Delta t) = \int_{t_0}^{t_0+\Delta t} \|\frac{d}{dt} z(t)\| dt.$$

If we introduce a Sundman transformation of the form

$$\frac{dt}{d\tau} = \frac{1}{\|\frac{d}{dt} z\|},$$

the arclength along solution curves is normalized to the fictive timestep. We term this method "arclength reparameterization." Note that the Sundman transformation can be expressed as a function of the phase variables only by using the differential equations

$$\frac{dt}{d\tau} = \frac{1}{\|\boldsymbol{f}(z)\|}. \tag{9.21}$$

It is entirely possible that the vector field will vanish at a point on a solution curve, in which case the arclength reparameterization becomes singular. This can be corrected by introducing a regularizing parameter as discussed below in Section 9.3.3. Arclength reparameterization has the virtue of simplicity, but it is not usually the optimal choice on an efficiency basis.

9.3.2 Rescaling for the *N*-body problem

A particularly important application for variable stepsize methods arises in the simulation of Coulombic systems, including both gravitational dynamics and classical atomic models [19, 108]. In N-body systems, it is natural to choose the rescaling function as a homogeneous function of the distances. Here we show how a time transformation for two-body gravitational interactions can be naturally constructed as a consequence of a scaling symmetry, based on the treatment in [36]. This transformation is useful in Coulombic N-body applications.

Recall that the Kepler Hamiltonian is

$$H(\boldsymbol{q}, \boldsymbol{p}) = \frac{1}{2}\|\boldsymbol{p}\|^2 - \frac{1}{\|\boldsymbol{q}\|}, \tag{9.22}$$

with equations of motion

$$\frac{d}{dt}\boldsymbol{q} = \boldsymbol{p}, \tag{9.23}$$

$$\frac{d}{dt}\boldsymbol{p} = -\frac{\boldsymbol{q}}{\|\boldsymbol{q}\|^3}. \tag{9.24}$$

The system possesses angular momentum and energy as first integrals. (The existence of an additional first integral, the so-called Runge–Lenz vector, places the Kepler problem in the class of "super-integrable systems," a consequence of which is the existence of closed periodic orbits for a wide range of initial data.) Indeed, the Kepler problem admits three types of orbits expressed in terms of the projection of the phase variables on to the position plane, based on the value of the energy: (i) a closed ellipse for $H < 0$, (ii) a parabola for $H = 0$, (iii) a hyperbola for $H > 0$.

The Kepler problem possesses an important scaling symmetry. If we rescale the variables t, $\boldsymbol{q}$, and $\boldsymbol{p}$ as follows

$$t = \alpha\tilde{t}, \qquad \boldsymbol{q} = \alpha^{2/3}\tilde{\boldsymbol{q}}, \qquad \boldsymbol{p} = \alpha^{-1/3}\tilde{\boldsymbol{p}}, \tag{9.25}$$

then we find that the form of the system of equations (9.23)–(9.24) expressed in these new variables is the same as the original. This scaling symmetry reflects a natural invariance of physical systems with respect to changes of units.

Now consider a Sundman transformation of the form

$$\frac{dt}{d\tau} = \|\boldsymbol{q}\|^{3/2}, \tag{9.26}$$

which changes the Kepler equations to

$$\frac{d}{d\tau}\boldsymbol{q} = \|\boldsymbol{q}\|^{3/2}\boldsymbol{p}, \tag{9.27}$$

$$\frac{d}{d\tau}\boldsymbol{p} = -\|\boldsymbol{q}\|^{-3/2}\boldsymbol{q}. \tag{9.28}$$

Observe that the scaling symmetry for this new system becomes

$$\tau = \tilde{\tau}, \qquad E = \tilde{E}, \qquad \boldsymbol{q} = \alpha^{2/3}\tilde{\boldsymbol{q}}, \qquad \boldsymbol{p} = \alpha^{-1/3}\tilde{\boldsymbol{p}}. \tag{9.29}$$

The time variable has been made invariant under the action of the scaling symmetry. This choice of Sundman transformation has additional consequences. For example, it can be shown that with this choice, the collision event occurs in approximately the same fictive time regardless of the energy of the colliding particle [24, 109]. This choice is also found to be approximately optimal in numerical experiments.

9.3.3 Stepsize bounds

Regardless of which stepsize controls are used, it is usually necessary to modify the control to limit the stepsize to some range of values. Let us define $\Delta t_{\rm min}$ and $\Delta t_{\rm max}$ as minimum and maximum stepsizes, respectively. Let g be a given a stepsize control, such as one of those discussed above. Since the effective stepsize is $g\Delta\tau$, and $\Delta\tau$ is known and fixed at the start of simulation, the problem is just to define a monotonic function $g \to \hat{g} = \phi(g)$, so that the interval $[0, \infty)$ is remapped to $[a, b]$, where $a = \Delta t_{\rm min}/\Delta\tau$, and $b = \Delta t_{\rm max}/\Delta\tau$. This is easily accomplished. For example, let

$$\hat{g} = b\frac{g + a}{g + b}.$$

Alternative remappings are possible, but this simple heuristic works in most cases. Note that in Adaptive Verlet, stepsizes are allowed to oscillate slightly around the target value, so these are only approximate bounds on the actual stepsize that would be observed in simulation.

9.4 Backward error analysis

A criticism that may be made of time-reversible adaptive methods such as Adaptive Verlet is this: since the method does not define a symplectic map, we lose the powerful Hamiltonian backward error analysis (as considered in Chapter 5). However, it turns out that a meaningful backward error analysis is possible for time-reversible systems. Indeed, a theory of backward error analysis and preservation of approximate first integrals for reversible integrable systems has been developed without any recourse to the Hamiltonian structure (see the text [80] for a summary of the main results).

The idea first followed by HAIRER AND STOFFER [83] is to start with a perturbation series for the numerical flow, defining a sequence of interpolating vector fields $\tilde{f}_k(z;\Delta t)$, $k = 0, 1, \ldots$, which constitute approximate modified equations of a given order, and then to show that if the numerical method is reversible under involution $\boldsymbol{S}$, then the perturbed vector fields $\tilde{f}_k(z;\Delta t)$ satisfy the reversibility condition

$$-\boldsymbol{S}\tilde{f}_k(\boldsymbol{S}z) = \tilde{f}_k(z).$$

In this way we can view the numerical solution obtained from time-reversible discretization as approximately determined from the evolution of a reversible continuous dynamical system as already mentioned in Chapter 5.

As for symplectic methods, there are some limitations to this type of backward error analysis. First, one finds that the theory only guarantees in general that the best truncation of the modified equations accurately determines the numerical solution on an interval of length $\mathcal{O}(\Delta t^{-1})$. Work is in progress to try to determine estimates that would hold on much longer intervals (for example, $\mathcal{O}(e^{c/\Delta t})$), but the current efforts require that the system be close to integrable. At the time of this writing, it is not clear how this theory will develop. In many applications, particularly in large scale systems with chaotic solutions, the idea is often to use very large timesteps (just small enough to guarantee stability of the method) so that the accessible timescale is as large as possible, but in these cases there are serious questions concerning the practical relevance of backward error analysis. Nonetheless, backward error analysis is certainly valid for some types of simulations and its usefulness is likely only to increase as theoretical work continues on this topic.

We will restrict this discussion to the case of the Adaptive Verlet method, which is the most commonly used scheme. A backward error analysis of the Adaptive Verlet method is given in [45]. The method can be formally written as

$$\boldsymbol{q}^n = \tilde{\boldsymbol{q}}(\tau_n) + (-1)^n \hat{\boldsymbol{q}}(\tau_n), \tag{9.30}$$

$$\boldsymbol{p}^n = \tilde{\boldsymbol{p}}(\tau_n) + (-1)^n \hat{\boldsymbol{p}}(\tau_n), \tag{9.31}$$

$$\gamma^n = \tilde{\gamma}(\tau_n) + (-1)^n \hat{\gamma}(\tau_n), \tag{9.32}$$

where $\tau_n = n\Delta\tau$ and the modified equations are

$$\begin{aligned} \tilde{\boldsymbol{q}}' &= \tilde{\gamma} \boldsymbol{M}^{-1} \tilde{\boldsymbol{p}} + \Delta\tau^2 \tilde{\boldsymbol{Q}}_2(.) + \cdots, & \hat{\boldsymbol{q}} &= \Delta\tau^2 \hat{\boldsymbol{Q}}_2(.) + \cdots, \\ \tilde{\boldsymbol{p}}' &= \tilde{\gamma} \boldsymbol{F}(\tilde{\boldsymbol{q}}) + \Delta\tau^2 \tilde{\boldsymbol{P}}_2(.) + \cdots, & \hat{\boldsymbol{p}} &= \Delta\tau^2 \hat{\boldsymbol{P}}_2(.) + \cdots, \\ \tilde{\gamma} &= G(\tilde{\boldsymbol{q}}, \tilde{\boldsymbol{p}}) + \Delta\tau^2 \tilde{G}_2(.) + \cdots, & \hat{\gamma}' &= \hat{G}(.) + \Delta\tau^2 \hat{G}_2(.) + \cdots, \end{aligned}$$

with uniquely determined initial values satisfying

$$\boldsymbol{q}_0 = \tilde{\boldsymbol{q}}(0) + \hat{\boldsymbol{q}}(0), \qquad \boldsymbol{p}_0 = \tilde{\boldsymbol{p}}(0) + \hat{\boldsymbol{p}}(0), \qquad \gamma_0 = \tilde{\gamma}(0) + \hat{\gamma}(0).$$

All of the expansions are formal and in even powers of $\Delta\tau$. The functions in the above equations depend only on $\tilde{\boldsymbol{q}}, \tilde{\boldsymbol{p}}$ and $\hat{\gamma}$; importantly the formulas for $\hat{G}$, $\hat{G}_2$, Q_2, P_2 all contain $\hat{\gamma}$ as a factor. In addition it can be shown that $\hat{\gamma} = O(\Delta\tau^2)$ and this, with the above equations, gives $\hat{\boldsymbol{q}} = O(\Delta\tau^4)$, $\hat{\boldsymbol{p}} = O(\Delta\tau^4)$.

An instability can arise when the oscillating terms in the expansion for γ_n grow with time. The leading oscillatory term is of the form

$$\hat{\gamma}_2 = \hat{g}_2(\tau_n)\Delta\tau^2.$$

The presence of this artificial dynamic is the price we pay for a fully explicit scheme. In many applications, these wobbles are essentially invisible, but in certain applications, for example Coulombic problems, the oscillation may grow quite large at the most difficult points along the trajectory (points of close approach of two bodies). (In extreme cases, the oscillations can make the stepsize become negative, leading to a complete breakdown of the integrator.)

In general, the behaviour of $\hat{g}_2$ depends on the stepsize update formula and the differential equations, in a complicated way. For the important special case where $\hat{g}_2$ is a function of $\boldsymbol{q}$ only, it can be shown that for the stepsize update (9.19), we have, as we approach the collision,

$$\hat{g}_2(\tau) \sim K_1/\gamma,$$

while for the choice (9.20), the operative formula is

$$\hat{g}_2(\tau) \sim K_2\gamma.$$

Neither of these will present a significant problem in the case where the stepsize variation is moderate, but in applications such as gravitation, for which $\gamma \to 0$ in the vicinity of close approach, the stepsize formula (9.20) gives far better results than the other choice.

In practice, (9.20) is generally the safer choice, for the magnitude of γ would typically be inhibited in the design of the scaling function (see below), whereas its reciprocal would generally be allowed to grow substantially at difficult points on the trajectory. A scheme for eliminating wobble in Adaptive Verlet to leading order was also given in [45].

9.5 Generalized reversible adaptive methods

As discussed in [86] it is possible to generalize the variable stepsize apparatus to allow reversible adaptive integration of systems admitting a general linear reversing symmetry. Such methods can be constructed starting from the assumption of a certain base discretization method (not necessarily symmetric) $\boldsymbol{\Psi}_{\Delta t}$, together with its adjoint method $\boldsymbol{\Psi}^*_{\Delta t}$. Recall from Chapter 4 that we can construct a symmetric method by concatenation

$$\tilde{\boldsymbol{\Psi}}_{\Delta t} = \boldsymbol{\Psi}^*_{\frac{1}{2}\Delta t}\boldsymbol{\Psi}_{\frac{1}{2}\Delta t}.$$

A reversible variable stepsize method can also be constructed based on $\boldsymbol{\Psi}_{\Delta t}$

$$\begin{aligned} \boldsymbol{z}^{n+1/2} &= \boldsymbol{\Psi}_{\frac{1}{2}\gamma^n\Delta\tau}(\boldsymbol{z}^n), \\ \gamma^n + \gamma^{n+1} &= 2g(\boldsymbol{z}^{n+1/2}), \\ \boldsymbol{z}^{n+1} &= \boldsymbol{\Psi}^*_{\frac{1}{2}\gamma^{n+1}\Delta\tau}(\boldsymbol{z}^{n+1/2}). \end{aligned}$$

Clearly this method maps $(\boldsymbol{z}^n, \gamma^n))$ to $(\boldsymbol{z}^{n+1}, \gamma^{n+1})$. It is easy to compute the inverse of the associated map and to check the reversibility condition (with respect to the involution $\tau \to -\tau, \boldsymbol{z} \to \boldsymbol{S}\boldsymbol{z}, \gamma \to \gamma$. Because of symmetry, the method is of even order. The scheme is evidently convergent (since it is based on two applications of a convergent method), so it must be second order.

If we identify Δt^n with $\gamma^n\Delta\tau$, we could eliminate the variable γ^n and write the method in a more compact form. As we will shortly see, it turns out that it is also necessary in many cases to consider a different choice of the symmetric update for the time-rescaling parameter. In general we term this class of methods *symmetric adaptive composition* schemes:

<u>SYMMETRIC ADAPTIVE COMPOSITION</u>

Given $\boldsymbol{\Psi}_{\Delta t}$, a convergent integrator for $\frac{d}{dt}\boldsymbol{z} = \boldsymbol{f}(\boldsymbol{z})$, and $R(t_0, t_1, \boldsymbol{z}_0, \bar{\boldsymbol{z}}, \boldsymbol{z}_1, \tau)$ a function satisfying $R(t_0, t_1, \boldsymbol{z}_0, \bar{\boldsymbol{z}}, \boldsymbol{z}_1, \tau) \equiv R(t_1, t_0, \boldsymbol{z}_1, \bar{\boldsymbol{z}}, \boldsymbol{z}_0, \tau)$, $\frac{\partial R}{\partial t_1} \neq 0$, The following equations define a reversible method

$$\begin{aligned} \boldsymbol{z}^{n+1/2} &= \boldsymbol{\Psi}_{\frac{1}{2}\Delta t_n}(\boldsymbol{z}^n), \\ 0 &= R(\Delta t_n, \Delta t_{n+1}, \boldsymbol{z}^n, \boldsymbol{z}^{n+1/2}, \boldsymbol{z}^{n+1}, \Delta\tau), \\ \boldsymbol{z}^{n+1} &= \boldsymbol{\Psi}^*_{\frac{1}{2}\Delta t_{n+1}}(\boldsymbol{z}^{n+1/2}). \end{aligned}$$

This method reduces to Adaptive Verlet in case the base method is a symplectic Euler.

9.5.1 Switching

The idea of the variable stepsize method can be generalized to provide a tool for developing reversible methods with more complicated algorithmic features, such as variable order of accuracy or even adaptive replacement of one integrator by another. The idea of the Adaptive Verlet scheme is roughly this: we apply a given standard integrator with a particular stepsize, modify the stepsize in a symmetric way, then apply the adjoint of the integrator using the new stepsize. A similar idea can be used to modify the features of the method that is used or to allow additional forms of adaptivity in the numerical solution.

Consider the following problem. We have a vector field $\boldsymbol{f}$, and two integrators $\hat{\boldsymbol{\Psi}}^{(1)}_{\Delta t}$ and $\hat{\boldsymbol{\Psi}}^{(2)}_{\Delta t}$ preserving some given geometric structure associated to $\boldsymbol{f}$. Suppose further that we have an indicator (monitor) function $M(\boldsymbol{z})$ which quantifies some aspect of the solution, say the difficulty of computing it, or simply the usefulness of one of the two techniques compared with the other. We wish to develop a new integrator with the following properties:

1. It should automatically switch between $\hat{\boldsymbol{\Psi}}^{(1)}_{\Delta t}$ and $\hat{\boldsymbol{\Psi}}^{(2)}_{\Delta t}$ so that when $M(\boldsymbol{z}) < \beta$, say, we integrate with the first method, while when $M(\boldsymbol{z}) > \beta$ we use the second method, and

2. It should be smooth and preserve the same geometric structures as each of the two maps.

The idea is to introduce a smooth switching function, $\sigma(m)$ (sometimes called a *sigmoidal function*) on an interval containing β, smoothly passing from 1 to 0. Shifted scaled arctangent functions can be used for this purpose, but it is generally better for efficiency purposes to use a piecewise polynomial switch defined on a compact interval $[m_-, m_+]$ containing β. It is easy to construct a polynomial σ, with zero derivatives at m_- and m_+ of any desired degree, with the properties: $\sigma(m_-) = 1$, $\sigma(m_+) = 0$, σ monotone decreasing on $[m_-, m_+]$.

Now we construct a splitting of $\boldsymbol{f}$

$$\boldsymbol{f} = \boldsymbol{f}_1 + \boldsymbol{f}_2, \qquad \boldsymbol{f}_1(\boldsymbol{z}) = \sigma(M(\boldsymbol{z}))\boldsymbol{f}(\boldsymbol{z}), \qquad \boldsymbol{f}_2(\boldsymbol{z}) = (1 - \sigma(M(\boldsymbol{z})))\boldsymbol{f}(\boldsymbol{z}),$$

and recover the solution by iterating for example the following concatenation of maps: $\hat{\boldsymbol{\Psi}}_{\Delta t} = \hat{\boldsymbol{\Psi}}^{(1)}_{\sigma(M(\boldsymbol{z}))\Delta t} \circ \hat{\boldsymbol{\Psi}}^{(2)}_{(1-\sigma(M(\boldsymbol{z})))\Delta t}$. When $M(\boldsymbol{z}) < m_-$, the switch σ will be fully "on" (value 1), so that the effect will be to apply method $\hat{\boldsymbol{\Psi}}^{(1)}_{\Delta t}$ to $\boldsymbol{f}$. When $M(\boldsymbol{z}) > m_+$, the switch σ will be fully "off" (value 0), so that the the numerical solution will be propagated by applying method $\hat{\boldsymbol{\Psi}}^{(2)}_{\Delta t}$ to $\boldsymbol{f}$. Between m_- and m_+, both methods contribute to the numerical solution, by solving in each case some polynomially rescaled version of the original vector field. In other words, what we have done is to construct a *homotopy* of the two numerical integration maps $\hat{\Psi}^{(1)}_{\Delta t}$ and $\hat{\Psi}^{(2)}_{\Delta t}$.

The challenge is to implement, in an efficient manner, schemes based on this idea. In general, we cannot expect the structure $\sigma(M(\mathbf{z}))\mathbf{f}$ to match the structure of $\mathbf{f}$, for example.

One symmetric method that could be used in the general case is as follows

$$\begin{aligned} z^{n+1/2} &= \hat{\Psi}^{(2)}_{(1-\sigma(M(z^{n+1/2})))\Delta t/2} \circ \hat{\Psi}^{(1)}_{\sigma(M(z^n))\Delta t/2}(z^n), \\ z^{n+1} &= \hat{\Psi}^{(1)*}_{\sigma(M(z^{n+1}))\Delta t/2} \circ \hat{\Psi}^{(2)*}_{(1-\sigma(M(z^{n+1/2})))\Delta t/2}(z^{n+1/2}). \end{aligned}$$

The notation $\hat{\Psi}^{*}_{\Delta t}$ indicates the adjoint map of $\hat{\Psi}_{\Delta t}$, defined by $\hat{\Psi}^{*}_{\Delta t} = \hat{\Psi}^{-1}_{-\Delta t}$. The reversible switching integrator is symmetric. It is also second order. It switches the integrators completely. Its major drawback is that it is implicit, even if the two methods on which it is based are explicit.

Based on our experience with adaptive timestepping, the most natural approach to explicit schemes is to introduce a discrete variable σ_n which can be udpated according to a symmetric formula. However, such methods tend to introduce a small oscillatory component in the error in σ; since σ is zero in a substantial part of the domain, the numerical errors will tend to introduce an instability in the method. An explicit alternative is likely to be more successful if we change our perspective so that we switch based on the value of $M(\mathbf{z})$ rather than based on σ. Specifically, we could introduce a discrete monitor variable μ_n and employ an explicit scheme of the following sort

$$\begin{aligned} z^{n+1/2} &= \hat{\Psi}^{(2)}_{(1-\sigma(\mu_n))\Delta t/2} \circ \hat{\Psi}^{(1)}_{\sigma(\mu_n))\Delta t/2}(z^n), \\ \mu_{n+1} + \mu_n &= M(z^{n+1/2}), \\ z^{n+1} &= \hat{\Psi}^{(1)*}_{\sigma(\mu_{n+1}))\Delta t/2} \circ \hat{\Psi}^{(2)*}_{(1-\sigma(\mu_{n+1}))\Delta t/2}(z^{n+1/2}). \end{aligned}$$

If each of the two schemes involved is explicit with explicit adjoint, then the overall method is also explicit. The indicator function M should be designed so that $M(\mathbf{z})$ is bounded well away from zero, to avoid the possibility of μ becoming negative during the integration.

In practice, we expect that adaptivity must be introduced in a problem-specific way. For a detailed discussion of switching in the context of few body Coulombic systems, see [101].

9.6 Poincaré transformations

In this section, we briefly discuss the implementation of Poincaré transformations for symplectic variable stepsize integration. Recall that there are many explicit symplectic integrators for Hamiltonian systems in separated form, $H(\boldsymbol{q},\boldsymbol{p}) = T(\boldsymbol{p}) + V(\boldsymbol{q})$, but all known symplectic integrators which work for general H are implicit. In general, the Poincaré-transformed equations will require treatment with an implicit method, since the transformation couples the position and momenta variables.

Experience with reversible and symplectic methods in a variety of applications suggest that symplectic schemes often exhibit improved stability compared with their reversible counterparts. Comparisons of symplectic and reversible methods for small model problems also suggest that a symplectic scheme may be more reliable. On the other hand, efficiency considerations typically demand, even for moderate-sized applications, an explicit integrator.

It turns out that it is often the case that the time-reparameterization function g can be chosen to be a function of positions only, $g = g(\boldsymbol{q})$. In this case, a semi-explicit first-order method is possible.

The equations of motion after Poincaré transformation become

$$\frac{d}{d\tau}\boldsymbol{q} = g(\boldsymbol{q})\nabla_{\boldsymbol{p}}T(\boldsymbol{p}),$$
$$\frac{d}{d\tau}\boldsymbol{p} = -g(\boldsymbol{q})\nabla_{\boldsymbol{q}}V(\boldsymbol{q}) - (H(\boldsymbol{q},\boldsymbol{p}) - E)\nabla_{\boldsymbol{q}}g(\boldsymbol{q}).$$

If we discretize this system with the symplectic Euler method, there results

$$\boldsymbol{q}^{n+1} = \boldsymbol{q}^n + \Delta\tau g(\boldsymbol{q}^n)\nabla_{\boldsymbol{p}}T(\boldsymbol{p}^{n+1}), \tag{9.33}$$
$$\boldsymbol{p}^{n+1} = \boldsymbol{p}^n - \Delta\tau g(q^n)\nabla_{\boldsymbol{q}}V(\boldsymbol{q}^n) - \Delta\tau(T(\boldsymbol{p}^{n+1}) + V(\boldsymbol{q}^n) - E)\nabla_{\boldsymbol{q}}g(\boldsymbol{q}^n). \tag{9.34}$$

Define vectors

$$\boldsymbol{a} = \boldsymbol{p}^n - \Delta\tau g(q^n)\nabla_{\boldsymbol{q}}V(\boldsymbol{q}^n) - \Delta\tau(V(\boldsymbol{q}^n) - E)\nabla_{\boldsymbol{q}}g(\boldsymbol{q}^n),$$

and

$$\boldsymbol{b} = \nabla_{\boldsymbol{q}}g(\boldsymbol{q}^n).$$

Then we can write an implicit formula for the kinetic energy

$$\hat{T} := T(\boldsymbol{p}^{n+1}) = T(\boldsymbol{a} - \hat{T}\boldsymbol{b}).$$

Under normal circumstances, this is an inexpensive problem to solve for $\hat{T}$. For example, if $T(\boldsymbol{p}) = \frac{1}{2}\boldsymbol{p}^T\boldsymbol{M}^{-1}\boldsymbol{p}$, it becomes a scalar quadratic equation in $\hat{T}$. Once $\hat{T}$ is known, this can be inserted in (9.34) to yield $\boldsymbol{p}^{n+1}$, and then $\boldsymbol{q}^{n+1}$ can be computed from (9.33).

It would be desirable to find higher-order semi-explicit symplectic methods under these assumptions. A natural candidate for a second-order method would be based on the concatenation of (9.33)–(9.34) with its adjoint. The problem is therefore to find an efficient way to solve

$$\boldsymbol{q}^{n+1} = \boldsymbol{q}^n + \Delta\tau g(\boldsymbol{q}^{n+1})\nabla_{\boldsymbol{p}} T(\boldsymbol{p}^n), \tag{9.35}$$

$$\begin{aligned}\boldsymbol{p}^{n+1} = \boldsymbol{p}^n &- \Delta\tau g(q^{n+1})\nabla_{\boldsymbol{q}} V(\boldsymbol{q}^{n+1}) \\ &- \Delta\tau(T(\boldsymbol{p}^n) + V(\boldsymbol{q}^{n+1}) - E)\nabla_{\boldsymbol{q}} g(\boldsymbol{q}^{n+1}).\end{aligned} \tag{9.36}$$

The calculation is all explicit after $\boldsymbol{q}^{n+1}$ is obtained from (9.35), hence the challenge is to solve equations of the form

$$\hat{g} := g(\boldsymbol{q}^{n+1}) = g(\boldsymbol{a}' + \hat{g}\boldsymbol{b}'), \tag{9.37}$$

where $\boldsymbol{a}'$ and $\boldsymbol{b}'$ are fixed vectors. If this calculation can be performed efficiently, the entire method may be viable. In some cases, as when arclength reparameterization is used, the work required is similar to that of an implicit scheme for the full problem. However, in other cases, the work may be much less to solve (9.37) than it would be to apply an implicit method to the whole system. An example is where we know that the timestep is strongly dependent on only one or a few of the variables of a large system.

Another difficulty for the symplectic approach arises when the underlying fixed stepsize method which is most suited to the structure of the problem is a complicated construction, perhaps resulting from composition of a number of building blocks and coordinate transformations. In this case, the time-reversible approach is to be preferred, since its implementation is essentially independent of the details of how the fixed stepsize method is constructed; it is much easier to retain the explicit (or partially explicit) structure of the underlying fixed-stepsize scheme.

9.7 Exercises

1. *Sundman transformation.* Show that if a Sundman transformation is a function of the energy of a Hamiltonian system, the rescaled system is also Hamiltonian. Is this likely to give good results? Why or why not?

2. *Scaling invariance.* Show that (9.25) expresses a scaling invariance for the Kepler problem. How is the energy changed under this scaling of the variables?

3. *Poincaré transformation.* Show that the symplectic Euler method can be made explicit for the Poincaré transformation $g(\boldsymbol{q})$.

4. *Constrained dynamics and adaptivity.* Develop an adaptive method for simulating a constrained Hamiltonian system such as the spherical pendulum (unit length, unit mass, and unit gravitational constant) subject to soft collisions (inverse power repulsion) with a fixed body at position $\boldsymbol{q}_0$. The equations of motion can be written as

$$\dot{\boldsymbol{q}} = \boldsymbol{M}^{-1}\boldsymbol{p},$$
$$\dot{\boldsymbol{p}} = -\nabla_q V(\boldsymbol{q}) - \boldsymbol{q}\lambda,$$
$$\|\boldsymbol{q}\|^2 = 1,$$

where the potential

$$V(\boldsymbol{q}) = z + \phi(|\boldsymbol{q} - \boldsymbol{q}_0|), \qquad \phi(r) = r^{-\alpha},$$

is the sum of gravitational potential and the distance-dependent interaction potential between the pendulum bob and fixed body. The Sundman transformation should be chosen to reduce the stepsize in the vicinity of collisions between the bob and the fixed body. Test your method and report on its relative efficiency as a function of the power α in the repulsive wall and the choice of Sundman transformation function.

5. *Switching.* Discuss the construction of a time-reversible integrator to switch order of accuracy. As an example, consider the pair of methods consisting of the Störmer–Verlet method and the fourth order Yoshida method given in Chapter 6. Design a method which automatically changes the effective order of from 2 to 4 in a certain given region R of phase space.

10

Highly oscillatory problems

Hamiltonian systems often exhibit dynamical phenomena covering a vast range of different time scales. In this chapter, we will discuss systems with two well separated time scales. More specifically, we consider systems for which the fast motion is essentially oscillatory. Such systems can arise from very different applications such as celestial or molecular dynamics and they might manifest themselves in very different types of Hamiltonian equations. Hence, the discussion in this chapter is necessarily limited to special cases. However, the basic principles and ideas have a much wider range of applicability.

A standard integrator, whether symplectic or not, will, in general, have to use a stepsize that resolves the oscillations in the fast system and, hence, one might be forced to use very small timesteps in comparison to the slow dynamics which is of primary interest. However, in special cases, one might be able to individually exactly solve the fast oscillatory and the slow system. Following the idea of splitting methods, this suggests to compose these two exact propagators and to apply a stepsize that is large with respect to the period of the fast oscillations. Such a method is then called a *large timestep* (LTS) method. Often the fast oscillations cannot be integrated analytically. A natural idea for the construction of an LTS method is then to assign different timesteps to different parts of the system. This approach is called *multiple timestepping* (MTS) and can often even be implemented such that the overall timestepping procedure still generates a symplectic map.[1] We will explain the basic idea of symplectic LTS/MTS methods in Section 10.1.

While the idea of LTS/MTS methods is very appealing, they have severe limitations which are linked to numerically induced resonances [20, 67] and a break-down of backward error analysis. The instability problem can be partially overcome by combining LTS/MTS methods with the idea of *averaging*. Averaging over a highly oscillatory degree of freedom is a very powerful tool in classical

[1] Symplectic/reversible MTS methods were first discussed in the context of molecular dynamics by GRUBMÜLLER, HELLER, WINDEMUTH, AND SCHULTEN [75] and TUCKERMAN, BERNE, AND MARTYNA [195].

mechanics and a short introduction to the idea of averaging and the concept of an *adiabatic invariant* will be provided in Section 10.2. In Sections 10.3 and 10.4, we will describe two numerical methods that utilize averaging to eliminate or weaken resonance-induced instabilities in MTS methods. More specifically, the mollified impulse (MOLLY) method of García-Archilla, Sanz-Serna, and Skeel [67] was the first method to improve the stability properties of a symplectic MTS method by replacing the slow potential energy terms by an averaged (mollified) contribution. We will explain the basic idea of mollified MTS (MOLLY) in Section 10.4. MOLLY was subsequently extended to a wider class of averaging procedures by Hairer, Hochbruck, and Lubich (see the monograph [80]). A somewhat different approach, called *reversible averaging* (RA), was suggested by Leimkuhler and Reich [112]. Reversible averaging will be discussed in Section 10.3.

10.1 Large timestep methods

In this section we will investigate large timestep (LTS) methods for Hamiltonian systems with slow and highly oscillatory degrees of freedom. We will in particular develop the idea of multiple timestepping (MTS) methods and investigate their numerical behavior for simple model problems. We will observe that MTS methods work well except for instabilities near certain choices of the stepsize Δt. These instabilities are caused by *numerical resonances* which can already be observed for LTS methods applied to a single oscillatory degree of freedom and which are caused by a break-down of backward error analysis as discussed in Section 10.5.

10.1.1 A single oscillatory degree of freedom

Let us start with a numerical example.

Example 1 Consider a fast harmonic oscillator subject to a perturbation, for example

$$\dot{q} = \omega p, \qquad \dot{p} = -\omega q - q.$$

For $\omega \gg 1$, the solutions are highly oscillatory and the eigenvalues of the solution operator

$$\boldsymbol{W}(t) = \mathrm{e}^{t\boldsymbol{A}}, \qquad \boldsymbol{A} = \begin{bmatrix} 0 & \omega \\ -\omega - 1 & 0 \end{bmatrix},$$

are on the unit circle for all $\omega \geq 0$, i.e. the eigenvalues have modulus equal to one.

From a numerical point of view, one could be tempted to split the equations of motion into its highly oscillatory contribution

$$\dot{q} = \omega p, \qquad \dot{p} = -\omega q, \tag{10.1}$$

and the perturbation

$$\dot{q} = 0, \qquad \dot{p} = -q. \tag{10.2}$$

If we denote the associated matrix-valued solution operators by $\boldsymbol{W}_1(t)$ and $\boldsymbol{W}_2(t)$, respectively, a second-order numerical propagator is obtained, for example via the matrix product

$$\boldsymbol{M}_{\Delta t} = \boldsymbol{W}_2(\Delta t/2)\,\boldsymbol{W}_1(\Delta t)\,\boldsymbol{W}_2(\Delta t/2). \tag{10.3}$$

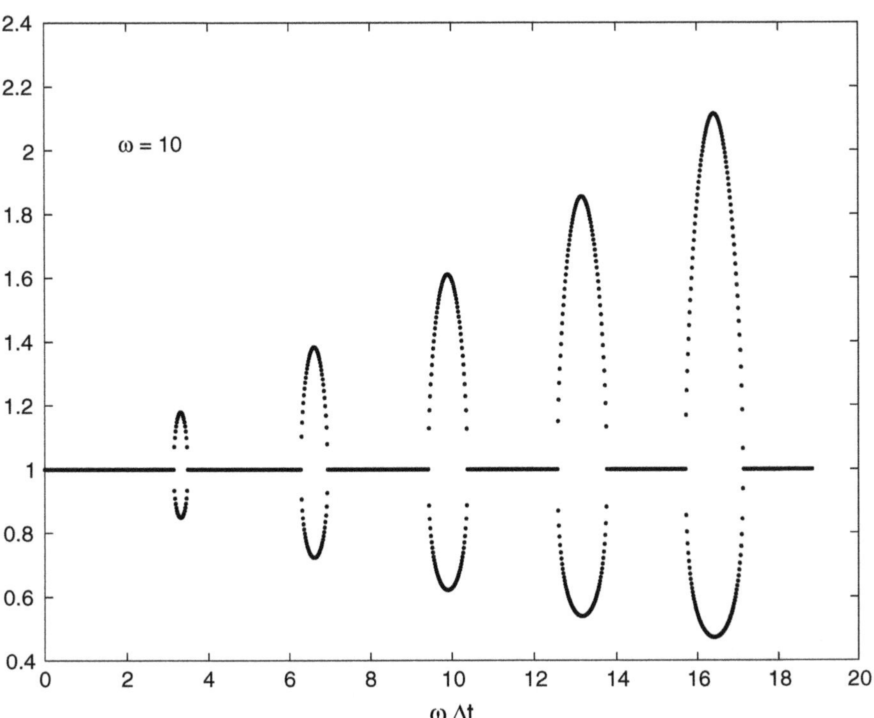

Figure 10.1 Modulus of eigenvalues of numerical propagator $\boldsymbol{M}_{\Delta t}$ as a function of $\omega\Delta t$.

We compute the modulus of the eigenvalues of $\boldsymbol{M}_{\Delta t}$ for $\omega = 10$ as a function of the stepsize $\Delta t \leq 0.2$. Regions of instabilities can be clearly seen in Fig. 10.1 for $\omega\Delta t \approx k\pi$, $k = 1, 2, 3$.

A similar behavior would be observed for the LTS splitting method (10.3) applied to a nonlinearly perturbed harmonic oscillator

$$\dot{q} = \omega p, \qquad \dot{p} = -\omega q - g'(q).$$

□

We now give a heuristic reasoning for these numerically observed instabilities. We first introduce action-angle variables (J, ϕ) via $q = \sqrt{2J}\cos\phi$ and $p =$

$-\sqrt{2J}\sin\phi$. Next we rewrite the given Hamiltonian $H = \omega(p^2+q^2)/2 + g(q)$ in the form

$$H(\phi, J) = \omega J + H_s(\phi, J), \tag{10.4}$$

where $H_s = g(\sqrt{2J}\cos\phi)$ and, hence, $H(\phi, J)$ is 2π-periodic in the angle ϕ. Both $H_f = \omega J$ and H_s can be solved exactly and we define the LTS method $\boldsymbol{\Psi}_{\Delta t}$ as a simple second-order composition of the associated flow maps, for example

$$\boldsymbol{\Psi}_{\Delta t} = \boldsymbol{\Phi}_{\Delta t/2, H_s} \circ \boldsymbol{\Phi}_{\Delta t, H_f} \circ \boldsymbol{\Phi}_{\Delta t/2, H_s}. \tag{10.5}$$

According to a result by KUKSIN AND PÖSCHEL [100] (see also MOAN [135]), the symplectic method $\boldsymbol{\Psi}_{\Delta t}$ can be written as the *exact* time-one-flow map of an extended Hamiltonian system $\tilde{H}_{\Delta t}(\phi, J, s, E)$, where $\tilde{H}_{\Delta t}$ is 2π-periodic in the parameter s. We may assume (see the Exercises) that the variable s evolves according to the differential equation $\dot{s} = -\pi$. The variable E is the "action" variable conjugate to s. The embedding of $\boldsymbol{\Psi}_{\Delta t}$ into such an exact time-one-flow map can be chosen such that $\tilde{H}$ is a real-analytic function of the general form

$$\tilde{H}_{\Delta t}(\phi, J, s, E) = \Omega_{\Delta t} J - \pi E + \Delta t f_{\Delta t}(\phi, J, s), \tag{10.6}$$

where $f_{\Delta t}$ is an appropriate function and $\Omega_{\Delta t} \in [-\pi, \pi)$ is defined by

$$\Omega_{\Delta t} = [\omega \Delta t + \pi]_{\mathrm{mod}\, 2\pi} - \pi.$$

The definition of $\Omega_{\Delta t}$ is motivated by the fact that

$$\boldsymbol{\Phi}_{\Delta t, H_f} = \boldsymbol{\Phi}_{\Delta t, \hat{H}_f},$$

with $\hat{H}_f = \Omega_{\Delta t} J/\Delta t = \hat{\omega} J$. Hence the numerical method (10.5) cannot distinguish between ω and $\hat{\omega} = \Omega_{\Delta t}/\Delta t$.

For $\Delta t\omega \to 0$, we obtain $\Omega_{\Delta t} = \Delta t \omega$ and one can average over the (fast) variable s in $\tilde{H}_{\Delta t}$ and that leads to the modified Hamiltonian of standard backward error analysis [143, 16]. However, for the LTS methods considered in this chapter, $\omega\Delta t$ and hence $\Omega_{\Delta t}$ are not necessarily small. In fact, $|\Omega_{\Delta t}|$ is only bounded by π. To understand the possible implications let us investigate the canonical equations of motion

$$\dot{\phi} = \Omega_{\Delta t} + \Delta t \nabla_J f_{\Delta t}(J, \phi, s), \qquad \dot{J} = -\Delta t \nabla_\phi f_{\Delta t}(J, \phi, s),$$
$$\dot{s} = -\pi, \qquad \dot{E} = -\Delta t \nabla_s f_{\Delta t}(\phi, J, s).$$

The associated flow map $\boldsymbol{\Phi}_{\tau, \tilde{H}}$ completely characterizes the behavior of the numerical method $\boldsymbol{\Psi}_{\Delta t}$. In particular, the numerical method is called effectively stable[2] if $J(\tau)$ and $E(\tau)$ remain bounded over long time-intervals $|\tau| \leq n \cdot \Delta t$,

[2]Effective stability is weaker than strong stability. However, boundedness of J and E over time intervals that are, for example, of order $\mathcal{O}(\mathrm{e}^{c/\Delta t})$ is clearly "almost" as strong as boundedness for all times.

$n \gg 1$. We only state at this point that the *averaging principle* (see, for example, [8]) implies that $J(\tau)$ and $E(\tau)$ are approximately constant provided

(i) the two natural frequencies $\Omega_1 = -\pi$ and $\Omega_2 = \Omega_{\Delta t}$ of the system are non-resonant[3] and

(ii) there is separation of time scales, i.e. the two frequencies Ω_1 and Ω_2 are much larger than $|\dot{J}|$ and $|\dot{E}|$.

Condition (ii) is clearly not satisfied for

$$\omega \Delta t = 2k\pi, \qquad k = 1, 2, 3, \ldots,$$

since this implies

$$\Omega_2 = \Omega_{\Delta t} = 0.$$

On the other hand, for

$$\omega \Delta t = k\pi, \qquad k = 1, 3, 5, \ldots,$$

we obtain

$$\Omega_2 = \Omega_{\Delta t} = -\pi = \Omega_1,$$

and Condition (i) is violated. These heuristic arguments provide some indication why numerical instabilities are observed for $\omega \Delta t$ equal to integer multiples of π. See the Exercises for more background material.

10.1.2 Slow-fast systems

We now extend the idea of LTS methods to a more general class of slow–fast Hamiltonian systems. This will lead us to the concept of multiple timestepping (MTS). Let us assume that we are given a Hamiltonian function H that can be split into two parts H_{f} and H_{s}; i.e. $H = H_{\mathrm{f}} + H_{\mathrm{s}}$. The assumption is that the equations of motion associated with H_{s} can be solved with a stepsize Δt_{s}, while the dynamics of H_{f} requires a stepsize $\Delta t_{\mathrm{f}} \ll \Delta t_{\mathrm{s}}$. A standard symplectic integrator will then require that a stepsize of $\Delta t \approx \Delta t_{\mathrm{f}}$ is used for the overall system with Hamiltonian H. Of course, one would like to make use of the fact that part of the system evolves on a slower time scale and to develop an LTS method.

[3] Two frequencies Ω_1 and Ω_2 are called resonant if there are two integers k_1 and k_2 such that $k_1\Omega_1 + k_2\Omega_2 = 0$. For linear LTS methods only the $k_1 = k_2 = 1$ resonance is of interest while for nonlinear systems higher-order resonances can play a role in the stability of an LTS method.

To illustrate the basic idea we consider a standard Newtonian system with Hamiltonian

$$H_s = \frac{1}{2}\boldsymbol{p}^T \boldsymbol{M}^{-1}\boldsymbol{p} + V(\boldsymbol{q})$$

and assume that it is coupled to a rapidly oscillating degree of freedom with position x, momentum p_x and Hamiltonian

$$H_f = \frac{1}{2\varepsilon}\left[p_x^2 + V_f(x, \boldsymbol{q})\right],$$

where $\varepsilon > 0$ is a small parameter and V_f is an appropriate potential energy function. An example of such a fast system is provided by a stiff harmonic oscillator whose position x vibrates about an equilibrium position $L > 0$ that depends on the slow coordinates $\boldsymbol{q}$; i.e. $L = L(\boldsymbol{q})$. The associated Hamiltonian is

$$H_f = \frac{1}{2\varepsilon}\left[p_x^2 + (x - L(\boldsymbol{q}))^2\right]. \tag{10.7}$$

Another example is provided by a stiff harmonic oscillator whose "spring" constant $K > 0$ depends on $\boldsymbol{q}$; i.e.

$$H_f = \frac{1}{2\varepsilon}\left[p_x^2 + K(\boldsymbol{q})x^2\right]. \tag{10.8}$$

In either case, we can write the total Hamiltonian H as

$$H = H_f + H_s$$

and obtain the equations of motion

$$\begin{aligned} \dot{\boldsymbol{p}} &= -\nabla_q V(\boldsymbol{q}) - \varepsilon^{-1}\nabla_q V_f(x, \boldsymbol{q}), & \dot{\boldsymbol{q}} &= \boldsymbol{M}^{-1}\boldsymbol{p}, \\ \dot{p}_x &= -\varepsilon^{-1}\nabla_x V_f(x, \boldsymbol{q}), & \dot{x} &= \varepsilon^{-1}p_x, \end{aligned}$$

which can be integrated by the Störmer–Verlet method. However, for stability reasons, the stepsize Δt has to be proportional to ε.

A simple idea to improve this situation is to integrate the equations of motion associated to the Hamiltonian H_f; i.e.

$$\begin{aligned} \dot{\boldsymbol{p}} &= -\varepsilon^{-1}\nabla_q V_f(x, \boldsymbol{q}), & \dot{\boldsymbol{q}} &= \boldsymbol{0}, \\ \dot{p}_x &= -\varepsilon^{-1}\nabla_x V_f(x, \boldsymbol{q}), & \dot{x} &= \varepsilon^{-1}p_x, \end{aligned}$$

either exactly or to apply the Störmer–Verlet method with a small stepsize $\delta t \sim \varepsilon$. We denote the associated propagator by $\boldsymbol{\Psi}_{\delta t, H_f}$. Then we apply $\boldsymbol{\Psi}_{\delta t, H_f}$ over N steps, where the integer N is chosen such that the slow system

$$\begin{aligned} \dot{\boldsymbol{p}} &= -\nabla_q V(\boldsymbol{q}), & \dot{\boldsymbol{q}} &= \boldsymbol{M}^{-1}\boldsymbol{p}, \\ \dot{p}_x &= 0, & \dot{x} &= 0, \end{aligned}$$

can be integrated by the Störmer–Verlet method with stepsize $\Delta t = \delta t \cdot N$. Let us denote this propagator by $\boldsymbol{\Psi}_{\Delta t, H_s}$. A symmetric one step method is now given by the composition

$$\boldsymbol{\Psi}_{\Delta t} = \boldsymbol{\Psi}_{\Delta t/2, H_s} \circ \left[\boldsymbol{\Psi}_{\delta t, H_f}\right]^N \circ \boldsymbol{\Psi}_{\Delta t/2, H_s}. \tag{10.9}$$

This is an example of a multiple timestepping (MTS) method. To summarize, the general idea of MTS is to split the given Hamiltonian into a slow and fast part, to integrate the fast and slow systems using a symplectic method with different timesteps, and to obtain an overall timestepping method by proper composition as in (10.9). The resulting one-step method is symplectic. However, the map $\boldsymbol{\Psi}_{\Delta t}$ is not necessarily close to the identity and standard backward error analysis does not apply. Hence it is not clear whether such a method will be stable and conserve energy.

10.1.3 Adiabatic invariants

To gain insight into the solution behavior of the slow–fast systems introduced in the previous subsection, let us discuss the limiting case $\varepsilon \to 0$ for bounded Hamiltonian H. We view H_f as a time-dependent Hamiltonian

$$H_f(x, p_x, t) = \frac{1}{2\varepsilon} p_x^2 + \frac{1}{\varepsilon} V_f(x, \boldsymbol{q}(t))$$

in the fast degree of motion (x, p_x), where $\boldsymbol{q}(t)$ is a given slowly varying function. Because of the bounded energy assumption, we necessarily have $p_x(t) = \mathcal{O}(\varepsilon^{1/2})$ and $x(t) = \mathcal{O}(\varepsilon^{1/2})$. Note also that, because of the explicit time dependence, the equations

$$\dot{p}_x = -\varepsilon^{-1} \nabla_x V_f(x, t), \qquad \dot{x} = \varepsilon^{-1} p_x \tag{10.10}$$

do not preserve the Hamiltonian H_f. However, for ε small enough, the solutions of (10.10) form nearly closed curves, with approximate period $T = \mathcal{O}(\varepsilon)$, due to the time dependence of $V_f(x, t)$. Hence, over one period, we can formally "shadow" the exact trajectory by the curve obtained for the same initial data and the same equations but with t temporarily frozen to its initial value $t = \bar{t}$. Let us denote the region enclosed by this periodic curve by $\mathcal{A}(\bar{t}) \subset \mathbb{R}^2$. We can compute the area A of the region $\mathcal{A}$ for any value of $\bar{t} = t$; i.e.

$$A(t) = \int_{\mathcal{A}(t)} dp \wedge dq.$$

Note that the area necessarily satisfies $A(t) = \mathcal{O}(\varepsilon)$ for bounded energy $H_f = \mathcal{O}(\varepsilon^0)$.

The flow map of (10.10) is symplectic and hence area preserving. It can be shown using normal form theory [7, 8] that this implies that the ratio $A(t)/A(0)$ remains equal to one along the flow of (10.10) up to order $\mathcal{O}(\varepsilon)$ terms. These small perturbation terms are due to the fact that the rapidly rotating solutions do not exactly form closed loops.

The scaled quantity

$$J := \frac{A}{2\pi}$$

is equivalent to the action variable in the standard transformation from (x, p_x) to action-angle variables (J, ϕ). Hence the action J is a "near-invariant" of (10.10) for ε sufficiently small and is called an *adiabatic invariant* [7, 8]. See also Section 5.2.3 in Chapter 5.

Let us discuss the precise form of J for (10.7) and (10.8). Assuming a fixed length L, the Hamiltonian (10.7) describes harmonic oscillations of frequency $\omega = \varepsilon^{-1}$ and the solutions are given by

$$x(t) = R\sin(\phi(t)) + L, \qquad p_x(t) = R\cos(\phi(t)),$$

where $\phi'(t) = \varepsilon^{-1}$ and R is a constant of integration determined by the initial conditions. The area of the circle described by $(x(t), p_x(t))$ is given by $A = \pi R^2$. On the other hand, we have $H_f = R^2/(2\varepsilon)$ and, hence, the associated adiabatic invariant is given by

$$J = \varepsilon H_f = \frac{1}{2}\left[p_x^2 + (x - L(\boldsymbol{q})^2)\right].$$

In a similar manner, we obtain

$$J = \frac{\varepsilon}{\sqrt{K(\boldsymbol{q})}} H_f = \frac{1}{2\sqrt{K(\boldsymbol{q})}}\left[p_x^2 + K(\boldsymbol{q})x^2\right]$$

as the adiabatic invariant for the Hamiltonian (10.8). Indeed, for fixed $K = K(\boldsymbol{q})$, solutions of (10.8) are of the form

$$x(t) = \frac{R}{K^{1/4}}\sin(\phi(t)), \qquad p_x(t) = RK^{1/4}\cos(\phi(t)),$$

where $\phi'(t) = \omega = \varepsilon^{-1}\sqrt{K}$ and R is again arbitrary. The area of the ellipse followed by $(x(t), p_x(t))$ is given by $A = \pi R^2$ and the energy by $H_f = K^{1/2}R^2/(2\varepsilon)$. Hence, we obtain J as given above. In both cases, the equality

$$H_f = \omega J \tag{10.11}$$

holds, where ω is the frequency of the rapidly rotating motion.

10.1.4 The effect of numerical resonances

Let us now return to the LTS methods. Composition methods have been considered before in this book. However, the method (10.5) is different from those composition methods since the map $\boldsymbol{\Psi}_{\Delta t}$ is *not* close to the identity. Hence, although (10.5) generates a symplectic map, long time energy conservation cannot be concluded from backward error analysis. Sharp increases in energy are indeed observed for particular values of Δt and these are associated with numerically induced resonances as briefly discussed in Section 10.1.1 for single frequency systems. To explore this issue in more detail in the context of slow–fast systems, we discuss two simple model problems.

A linear model problem

Consider a linear system with Hamiltonian

$$H = \frac{1}{2}\left[p^2 + q^2\right] + \frac{1}{2\varepsilon}\left[p_x^2 + (x-q)^2\right]. \tag{10.12}$$

The associated equations of motion are

$$\begin{bmatrix} \dot q \\ \dot p \\ \dot x \\ \dot p_x \end{bmatrix} = \begin{bmatrix} 0 & 1 & 0 & 0 \\ -1-\varepsilon^{-1} & 0 & \varepsilon^{-1} & 0 \\ 0 & 0 & 0 & \varepsilon^{-1} \\ \varepsilon^{-1} & 0 & -\varepsilon^{-1} & 0 \end{bmatrix} \begin{bmatrix} q \\ p \\ x \\ p_x \end{bmatrix}.$$

These equations can, of course, be solved analytically. This is also true for the fast and slow subsystems and we can implement an LTS method in the form

$$\boldsymbol{M}_{\Delta t} = e^{\Delta t/2\boldsymbol{B}} \cdot e^{\Delta t\boldsymbol{A}} \cdot e^{\Delta t/2\boldsymbol{B}}, \tag{10.13}$$

where

$$\boldsymbol{A} = \begin{bmatrix} 0 & 0 & 0 & 0 \\ -\varepsilon^{-1} & 0 & \varepsilon^{-1} & 0 \\ 0 & 0 & 0 & \varepsilon^{-1} \\ \varepsilon^{-1} & 0 & -\varepsilon^{-1} & 0 \end{bmatrix}, \qquad \boldsymbol{B} = \begin{bmatrix} 0 & 1 & 0 & 0 \\ -1 & 0 & 0 & 0 \\ 0 & 0 & 0 & 0 \\ 0 & 0 & 0 & 0 \end{bmatrix}.$$

The eigenvalues of the exact solution operator

$$\boldsymbol{W}_{\Delta t} = e^{\Delta t(\boldsymbol{A}+\boldsymbol{B})}$$

are all on the unit circle. For the numerical approximation $\boldsymbol{M}_{\Delta t}$ to be stable, we have to require that all its eigenvalues be on the unit circle as well. To check this property, we define

$$\Lambda(\Delta t) = \max_{i=1,\ldots,4} |\lambda_i(\Delta t)|,$$

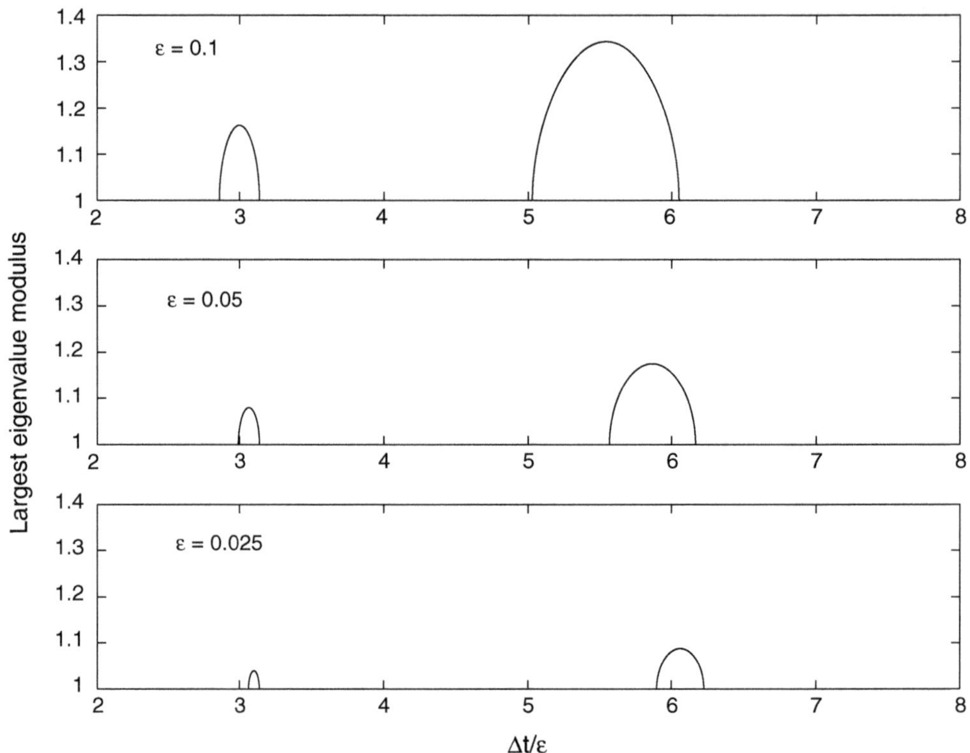

Figure 10.2 Stability of LTS method as a function of $\Delta t/\varepsilon$.

where $\lambda_i(\Delta t)$, $i = 1, \ldots, 4$, are the eigenvalues of $\boldsymbol{M}_{\Delta t}$. The linear stability boundary for the Störmer–Verlet method is $\omega \Delta t \leq 2$, where ω is the highest frequency in the system. Here we have $\omega \approx \varepsilon^{-1}$ and it can be seen from Fig. 10.2 that the LTS method becomes unstable for the first time at about $\Delta t \omega \approx \pi$. Of course, if one excludes the domains of instability by a proper choice of Δt, then a much larger timestep could be used in principle. One can also see from Fig. 10.2 that the domains of instability shrink for smaller values of ε. The precise motion of all four eigenvalues λ_i as a function of $\Delta t/\varepsilon$, $\varepsilon = 1/40$, can be found in Fig. 10.3. Note that the eigenvalues leave the unit circle near fast–fast or fast–slow eigenvalue crossings (resonances) which implies an exponential instability of the method.

A nonlinear model problem

We take a Hamiltonian of type (10.8) with $K(q) = 1 + \alpha q^2$, $\alpha = 0.1$; i.e.

$$H = \frac{1}{2}\left[p^2 + q^2\right] + \frac{1}{2\varepsilon}\left[p_x^2 + (1 + \alpha q^2)x^2\right]. \qquad (10.14)$$

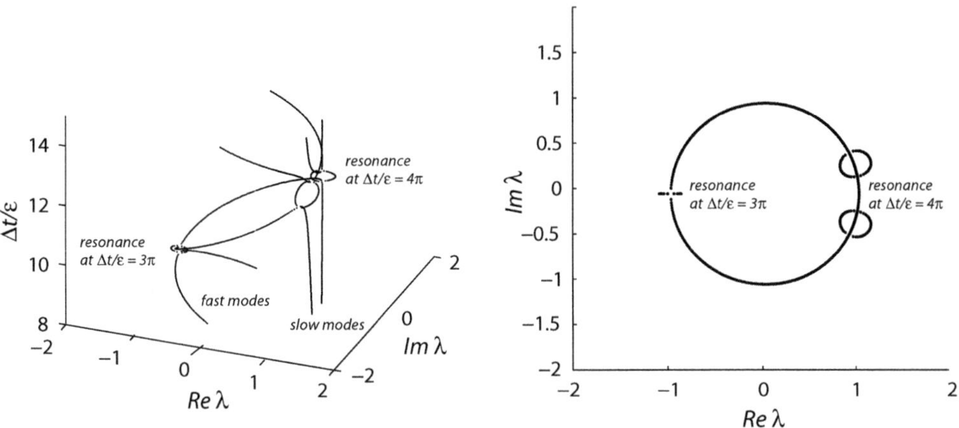

Figure 10.3 Eigenvalues of LTS propagator as a function of $\Delta t/\varepsilon$, $\varepsilon = 1/40$.

The equations of motion for the fast subsystem

$$\dot{x} = \varepsilon^{-1} p_x, \quad \dot{p}_x = -\varepsilon^{-1}(1+\alpha q^2)x, \quad \dot{q} = 0, \quad \dot{p} = -\frac{\alpha}{\varepsilon} x^2 q$$

can be solved analytically. The frequency of the fast motion is

$$\omega = \varepsilon^{-1}\sqrt{1+\alpha q^2}, \tag{10.15}$$

and the solution in (x, p_x) (for frozen q) is given by

$$x(t) = x(0)\cos(\omega t) + \frac{p_x(0)}{\sqrt{1+\alpha q^2}}\sin(\omega t),$$
$$p_x(t) = p_x(0)\cos(\omega t) - x(0)\sqrt{1+\alpha q^2}\sin(\omega t).$$

The update for the momentum p becomes

$$p(t) = p(0) - \frac{\alpha q(0)}{\sqrt{1+\alpha q(0)^2}}\left[\frac{x(0)^2}{2}(\omega t + \cos(\omega t)\sin(\omega t)) + \right.$$
$$\left. + \frac{p_x(0)^2}{2(1+\alpha q(0)^2)}(\omega t - \cos(\omega t)\sin(\omega t)) + \frac{x(0)p_x(0)}{\sqrt{1+\alpha q(0)^2}}\sin^2(\omega t)\right].$$

We denote the associated flow map by $\boldsymbol{\Phi}_{\Delta t, H_f}$. The timestepping method is given by

$$\boldsymbol{\Psi}_{\Delta t} = \boldsymbol{\Phi}_{\Delta t/2, H_s} \circ \boldsymbol{\Phi}_{\Delta t, H_f} \circ \boldsymbol{\Phi}_{\Delta t/2, H_s},$$

where $\boldsymbol{\Phi}_{t, H_s}$ is the solution operator for the (slow) harmonic oscillator

$$\dot{q} = p, \qquad \dot{p} = -q.$$

We set $\varepsilon = 0.025$, $q(0) = x(0) = 0$, $p(0) = 1$, $p_x(0) = 2\varepsilon^{1/2}$, and integrated the equations of motion over a time interval of $t \in [0, 200]$ for stepsizes Δt in the range of $\varepsilon^{-1}\Delta t \in [1, 10]$. In Fig. 10.4, we plot

$$\Delta E_{\text{max}} = \max_{t\in[0,200]} \frac{|H(t) - H(0)|}{H(0)}, \quad \Delta J_{\text{max}} = \max_{t\in[0,200]} \frac{|J(t) - J(0)|}{J(0)}$$

as a function of the scaled stepsize $\Delta t/\varepsilon$. Significantly increased errors in energy and adiabatic invariant are clearly visible in the vicinity of $\Delta t/\varepsilon = k\pi$, $k = 1, 2, 3$. However, since we observe exactly the same energy behavior for a simulation over the longer time interval $t \in [0, 800]$, the increased errors are not linked to an instability of the method. This is in contrast to the linear test problem (10.12), where exponential instabilities occurred near $\Delta t/\varepsilon = k\pi$, $k = 1, 2, \ldots$. The somewhat favorable behavior of nonlinear over linear problems is observed quite often in practice and can be traced back to the fact that the frequency (10.15) of the fast system is not constant but varies in time as a function of $q(t)$.

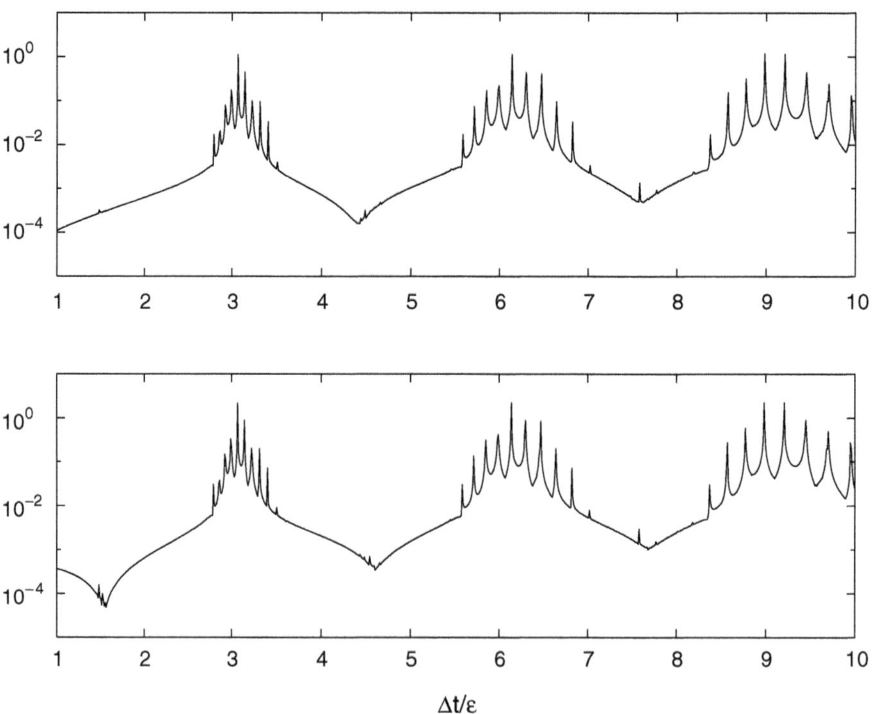

Figure 10.4 Maximum relative errors in total energy and adiabatic invariant as a function of $\Delta t/\varepsilon$, $\varepsilon = 1/40$, for LTS method.

10.2 Averaging and reduced equations

We have seen in Section 10.1.3 that the highly oscillatory motion in a single fast degree of freedom gives rise to the existence of an adiabatic invariant J. We will now use this adiabatic invariant to discuss the effect of the fast motion in (x, p_x) on the slow degrees of freedom $(\boldsymbol{q}, \boldsymbol{p})$. The relevant equations of motion in the variable $(\boldsymbol{q}, \boldsymbol{p})$ are

$$\dot{\boldsymbol{q}} = \boldsymbol{M}^{-1}\boldsymbol{p}, \qquad \dot{\boldsymbol{p}} = -\nabla_{\boldsymbol{q}} V(\boldsymbol{q}) - \varepsilon^{-1}\nabla_{\boldsymbol{q}} V_{\mathrm{f}}(x(t), \boldsymbol{q}), \tag{10.16}$$

where $x(t)$ is now assumed to be a given function of time. In fact, we can deduce from the previous discussions that $x(t)$ has to be of the form

$$x(t) \approx \sqrt{\frac{2J}{\varepsilon\omega}}\sin(\phi(t)) + b(\boldsymbol{q}),$$

with $\phi'(t) = \omega$ and $\omega = \omega(\boldsymbol{q}(t))$ is the instantaneous frequency of the fast oscillations. For example, consider the Hamiltonian (10.7), then $\omega = \varepsilon^{-1}$ and $b(\boldsymbol{q}) = L(\boldsymbol{q})$. On the other hand, we obtain $\omega = \varepsilon^{-1}\sqrt{K(\boldsymbol{q})}$ and $b(\boldsymbol{q}) = 0$ for the system (10.8). Recall that the action J is an adiabatic invariant. In accordance with this, we may set $J = \mathrm{const}$.

Let us now sketch the basic idea of averaging. See [197, 7, 8] for more details. Assume we are given a second-order time-dependent differential equation

$$\ddot{y} = f(y, \phi(t)),$$

where $\phi(t)$ is a given function such that $\phi'(t) = \omega(t) \gg 1$ and $f(y, \tau)$ is 2π-periodic in τ. One can think of these equations as a mechanical system driven by a highly oscillatory time-dependent excitation of instantaneous frequency ω. Let us assume that $\omega(t) \geq \varepsilon^{-1}$ and $\omega(t)$ is slowly varying in time, then the motion in the variable y is characterized by the averaged equations

$$\ddot{y} = \bar{f}(y), \qquad \bar{f}(y) = \frac{1}{2\pi}\int_0^{2\pi} f(y, \tau)\, d\tau,$$

over time intervals of order one up to an approximation error of $\mathcal{O}(\varepsilon)$ [197, 7, 8].

Let us apply this result to the system (10.16). We assume that $\omega(\boldsymbol{q}) \geq \varepsilon^{-1}$ and obtain the averaged equations

$$\dot{\boldsymbol{q}} = \boldsymbol{M}^{-1}\boldsymbol{p}, \qquad \dot{\boldsymbol{p}} = -\nabla_{\boldsymbol{q}} V(\boldsymbol{q}) + \bar{\boldsymbol{F}}_{\mathrm{f}}(\boldsymbol{q}), \tag{10.17}$$

where

$$\bar{\boldsymbol{F}}_{\mathrm{f}} = -\frac{1}{2\pi}\int_0^{2\pi} \varepsilon^{-1}\nabla_{\boldsymbol{q}} V_{\mathrm{f}}\left(\sqrt{\frac{2J}{\omega}}\sin(\tau) + b(\boldsymbol{q}), \boldsymbol{q}\right) d\tau.$$

Once the averaged force $\bar{\boldsymbol{F}}_{\rm f}$ is known, the equations (10.17) form a set of *reduced equations* in the slow variable $(\boldsymbol{q}, \boldsymbol{p})$.

Let us explicitly compute $\bar{\boldsymbol{F}}_{\rm f}$ for our two examples. We find that

$$\frac{1}{2\pi}\int_0^{2\pi} (x(\tau) - L)d\tau = 0,$$

and

$$\frac{1}{2\pi}\int_0^{2\pi} x(\tau)^2 d\tau = \frac{1}{2\pi}\int_0^{2\pi} \frac{2J}{K^{1/2}} \sin^2(\tau) d\tau = \frac{J}{K^{1/2}},$$

respectively. Hence we obtain $\bar{\boldsymbol{F}}_{\rm f} = \boldsymbol{0}$ for (10.7) and

$$\bar{\boldsymbol{F}}_{\rm f} = -\frac{J}{2\varepsilon\sqrt{K(\boldsymbol{q})}}\nabla_q K(\boldsymbol{q}) = -\varepsilon^{-1} J \nabla_q \sqrt{K(\boldsymbol{q})},$$

for (10.8), respectively. Furthermore, the averaged equations (10.17) can be written in canonical form with Hamiltonian

$$\bar{H}(\boldsymbol{q}, \boldsymbol{p}) = H_{\rm s}(\boldsymbol{q}, \boldsymbol{p}) + \omega(\boldsymbol{q})J,$$

and $\omega = \varepsilon^{-1}$ or $\omega = \varepsilon^{-1}\sqrt{K(\boldsymbol{q})}$, respectively. The value of the adiabatic invariant J is taken to be constant and is determined from the initial values of $(\boldsymbol{q}, x, p_x)$. We mention that $\bar{H}$ can, formally, be obtained from $H = H_{\rm s} + H_{\rm f}$, the identity (10.11) and the adiabatic invariance of J.

With this we complete the short discussion on the concept of averaging and the idea of reduced equations. See, for example, [165, 18, 28, 163] for further results on mechanical systems with a single fast degree of freedom.

Let us compare the full dynamics to the reduced dynamics for the Hamiltonian (10.14). The reduced Hamiltonian is given by

$$\bar{H} = \frac{1}{2}(p^2 + q^2) + \sqrt{1 + \alpha q^2}J.$$

We take the same initial conditions and parameter values as used in Section 10.1.4. In Fig. 10.5, we plot the solution curves in the slow (q, p) variable for the unreduced and reduced equations. Note that the equations obtained by simply dropping the high-frequency part altogether, i.e. $\dot{q} = p$, $\dot{p} = -q$, would have completely circular solutions.

The averaging approach and the complete elimination of the highly oscillatory degree of freedom, as just outlined, is difficult to generalize to systems with more than one fast degree of freedom. This is because of possible resonant interactions between the fast modes which lead to non-constant action variables [190, 27, 160]. However, the idea to average the fast forces can still be utilized in numerical methods and we will devote the rest of the chapter to two such numerical averaging approaches.

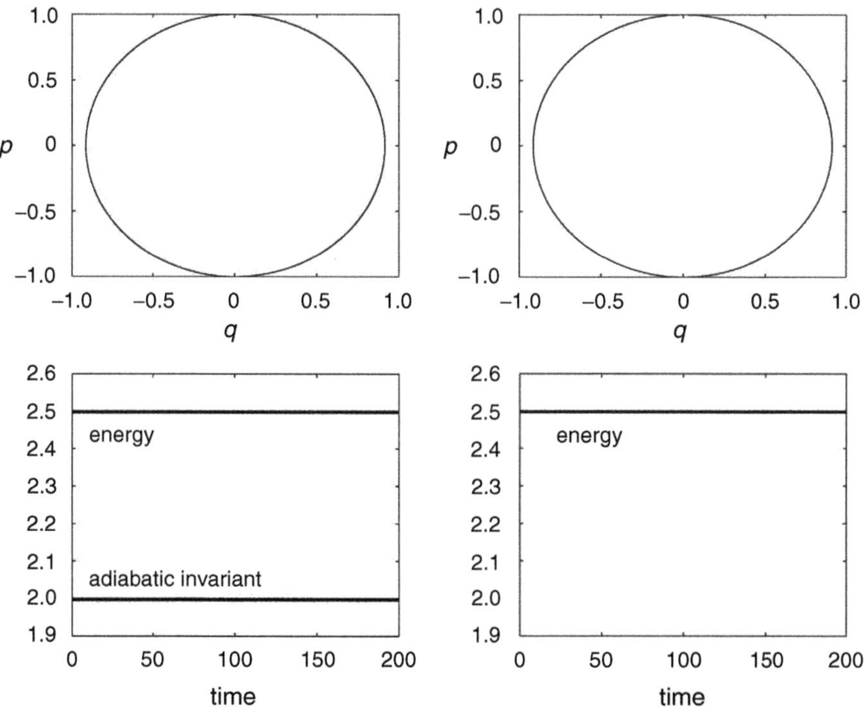

Figure 10.5 Comparison of slow q-component of the solutions to the unreduced and averaged equations.

10.3 The reversible averaging (RA) method

Let us now develop a numerical method that uses the idea of averaging to allow for larger timesteps. The main motivation is to avoid the loss of accuracy or, even worse, the instabilities near resonances that we observed for standard LTS methods. We wish to emphasize that the idea of averaging in the context of LTS methods and symplectic integration was first proposed by García-Archilla, Sanz-Serna, and Skeel [67]. The associated method MOLLY will be discussed in detail in the following section. In this section, we will focus on the *reversible averaging* (RA) method of Leimkuhler and Reich [112].

Let us assume that a numerical approximation $\boldsymbol{z}^n = (\boldsymbol{q}^n, \boldsymbol{p}^n, x^n, p_x^n)$ is given at time level t_n. We first describe a procedure to numerically compute a time-averaged force $\bar{\boldsymbol{F}}_{\mathrm{f}}$ at $t = t_n$. The idea is to freeze the value of $\boldsymbol{q}(t)$ at $\boldsymbol{q}^n$ and to compute the solution $(x(t), p_x(t))$ to the associated fast equations of motion

$$\dot{x} = \varepsilon^{-1} p_x, \qquad \dot{p}_x = \varepsilon^{-1} \nabla_x V_{\mathrm{f}}(x, \boldsymbol{q}^n),$$

over the time interval $t \in [t_{n-1}, t_{n+1}]$ with initial ("middle") conditions $x(t_n) = x^n$ and $p_x(t_n) = p_x^n$. We determine a time-averaged force along this solution

$$\bar{\boldsymbol{F}}_{\mathrm{f}}^n = -\frac{1}{2\Delta t}\int_{-\Delta t}^{\Delta t} \varepsilon^{-1}\nabla_{\boldsymbol{q}} V_{\mathrm{f}}(x(t_n+\tau), \boldsymbol{q}^n)\, d\tau. \tag{10.18}$$

The reversible averaging (RA) method is now defined by the following sequence of steps

THE REVERSIBLE AVERAGING (RA) METHOD

1. Compute the time-averaged force $\bar{\boldsymbol{F}}_{\mathrm{f}}^n$ using (10.18) and apply the Störmer–Verlet method to compute the new position $\boldsymbol{q}^{n+1}$ at t_{n+1}

$$\boldsymbol{p}^{n+1/2} = \boldsymbol{p}^n + \frac{\Delta t}{2}\left[\bar{\boldsymbol{F}}_{\mathrm{f}}^n - \nabla_{\boldsymbol{q}} V(\boldsymbol{q}^n)\right],$$
$$\boldsymbol{q}^{n+1} = \boldsymbol{q}^n + \Delta t \boldsymbol{M}^{-1}\boldsymbol{p}^{n+1/2}.$$

2. Compute the solution $(x(t), p_x(t))$ to the fast equations of motion

$$\dot{x} = \varepsilon^{-1} p_x, \qquad \dot{p}_x = \varepsilon^{-1}\nabla_x V_{\mathrm{f}}(x, \bar{\boldsymbol{q}}(t)),$$

over $t \in [t_n, t_{n+1}]$ with $\bar{\boldsymbol{q}}(t)$ varying along the linear path

$$\bar{\boldsymbol{q}}(t) = \frac{\boldsymbol{q}^{n+1} + \boldsymbol{q}^n}{2} + (t - t_{n+1/2})\boldsymbol{M}^{-1}\boldsymbol{p}^{n+1/2},$$

and initial conditions equal to $x(t_n) = x^n$, $p_x(t_n) = p_x^n$. Next set

$$x^{n+1} = x(t_{n+1}), \qquad p_x^{n+1} = p_x(t_{n+1}).$$

3. Compute the time-averaged force $\bar{\boldsymbol{F}}^{n+1}$ and update the momentum

$$\boldsymbol{p}^{n+1} = \boldsymbol{p}^{n+1/2} + \frac{\Delta t}{2}\left[\bar{\boldsymbol{F}}_{\mathrm{f}}^{n+1} - \nabla_{\boldsymbol{q}} V(\boldsymbol{q}^{n+1})\right].$$

If the force acting on the slow particles is linear in x, then the average to be computed for the slow momentum update simplifies to

$$\bar{\boldsymbol{F}}_{\mathrm{f}}^n = -\varepsilon\nabla_{\boldsymbol{q}} V_{\mathrm{f}}(\bar{x}^n, \boldsymbol{q}^n),$$

where

$$\bar{x}^n = \frac{1}{2\Delta t}\int_{-\Delta t}^{\Delta t} x(t_n + \tau)\, d\tau.$$

The same simplification may be used whenever $|x^n - \bar{x}^n|$ is sufficiently small.

The RA method is a symmetric one-step method. This implies that the method respects the time-reversal symmetry: $\boldsymbol{S} : (\boldsymbol{q}, \boldsymbol{p}, x, p_x) \to (\boldsymbol{q}, -\boldsymbol{p}, x, -p_x)$, since it is easy to verify that

$$\boldsymbol{S}\boldsymbol{\Phi}_{\Delta t}(z) = \boldsymbol{\Phi}_{-\Delta t}(\boldsymbol{S}z).$$

However, the RA method is, in general, not symplectic.

10.3.1 Numerical experiments

Let us explore the properties of the RA method by numerical investigation of three model problems.

A linear test problem

Let us apply the RA method to the Hamiltonian (10.12). Since the system is linear, one can explicitly construct the associated linear timestep map $\boldsymbol{M}_{\Delta t}$.

We first consider the motion of the fast system

$$\dot{x} = \varepsilon^{-1} p_x, \qquad \dot{p}_x = -\varepsilon^{-1}(x - q), \tag{10.19}$$

for $q = q^n$ fixed. We introduce the new variable $\xi = x - q$ and obtain the solution

$$\xi(t) = \xi(0)\cos(\omega t) + p_x(0)\sin(\omega t),$$

$\omega = \varepsilon^{-1}$. The averaged $\bar{\xi}^n$ is given by

$$\bar{\xi}^n = \frac{1}{2\Delta t}\int_{-\Delta t}^{\Delta t} \xi(t)dt = \xi^n \frac{1}{\omega \Delta t}\sin(\omega \Delta t), \tag{10.20}$$

where $\xi^n = x^n - q^n$. Hence the update of the slow variables becomes

$$p^{n+1/2} = p^n - \frac{\Delta t}{2}\left[q^n + C\frac{q^n - x^n}{\varepsilon}\right],$$
$$q^{n+1} = q^n + \Delta t p^{n+1/2},$$

where

$$C = \frac{\sin(\omega\Delta t)}{\omega\Delta t}.$$

Next we have to find the propagator for the linear system (10.19) along $q(t) = q^n + tp^{n+1/2}$. We again use the variable $\xi = x - q$ and obtain the linear system

$$\dot{\xi} = \omega(p_x - \varepsilon p^{n+1/2}), \qquad \dot{p}_x = -\omega\xi.$$

The solutions of this system are given by

$$\begin{aligned} \xi(t) &= \xi(0)\cos(\omega t) + (p_x(0) - \varepsilon p^{n+1/2})\sin(\omega t), \\ p_x(t) &= \varepsilon p^{n+1/2} + (p_x(0) - \varepsilon p^{n+1/2})\cos(\omega t) - \xi(0)\sin(\omega t), \end{aligned}$$

and we obtain the propagator

$$x^{n+1} = q^{n+1} + (x^n - q^n)\cos(\omega\Delta t) + (p_x^n - \varepsilon p^{n+1/2})\sin(\omega\Delta t) \quad , \tag{10.21}$$

$$p_x^{n+1} = \varepsilon p^{n+1/2} + (p_x^n - \varepsilon p^{n+1/2})\cos(\omega\Delta t) - (x^n - q^n)\sin(\omega\Delta t). \tag{10.22}$$

We finally update the momentum by

$$p^{n+1} = p^{n+1/2} - \frac{\Delta t}{2}\left[q^{n+1} + C\frac{q^{n+1} - x^{n+1}}{\varepsilon}\right].$$

It is found that the associated propagator is stable for all values of Δt. The precise motion of all four eigenvalues can be found in Fig. 10.6. Note that the eigenvalues stay on the unit circle.

We wish to point out that the proper choice of the average in (10.20) is crucial. For example, replacing (10.20) by

$$\bar{\xi}^n = \frac{1}{\Delta t}\int_{-\Delta t/2}^{\Delta t/2} \xi(t)dt = \xi^n \frac{2}{\omega\Delta t}\sin(\omega\Delta t/2),$$

leads to resonance instabilities similar to those observed in Fig. 10.2.

The linear system (10.19) could also be integrated along a constant $q = q^{n+1/2} = q^n + \Delta t/2p^n$. In this case, the linear propagator (10.21)–(10.22) simplifies to

$$\begin{aligned} x^{n+1} &= q^{n+1/2} + (x^n - q^{n+1/2})\cos(\omega\Delta t) + p_x^n \sin(\omega\Delta t), \\ p_x^{n+1} &= p_x^n \cos(\omega\Delta t) - (x^n - q^{n+1/2})\sin(\omega\Delta t). \end{aligned}$$

However, this modification degrades the performances of the RA method. See the numerical experiments in the following subsection.

A nonlinear model problem

Let us come back to the nonlinear model problem (10.14) of Section 10.1.4. The momentum update for the RA method becomes

$$p^{n+1/2} = p^{n-1/2} - \Delta t q^n - \frac{\alpha q^n}{2\sqrt{1+\alpha(q^n)^2}} \left[\left((x^n)^2 + \frac{(p_x^n)^2}{1+\alpha(q^n)^2} \right) \omega \Delta t \right.$$
$$\left. + \left((x^n)^2 - \frac{(p_x^n)^2}{1+\alpha(q^n)^2} \right) \cos(\omega \Delta t) \sin(\omega \Delta t) \right],$$

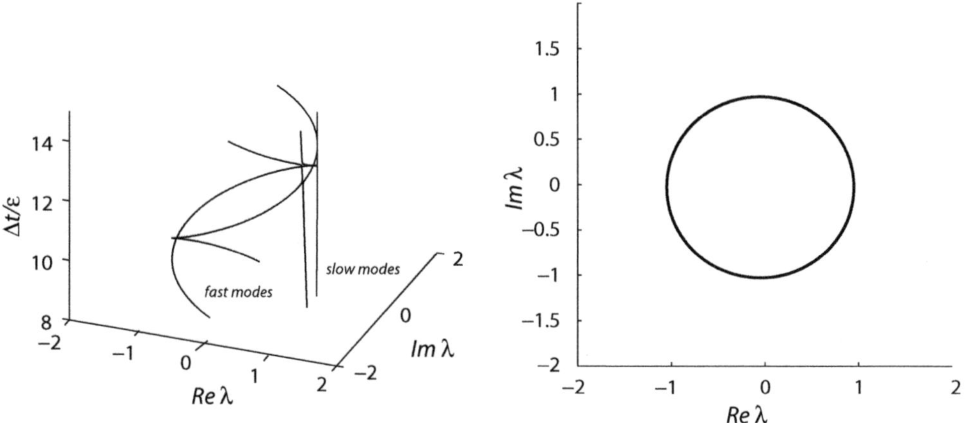

Figure 10.6 Eigenvalues of RA propagator as a function of $\Delta t/\varepsilon$, $\varepsilon = 1/40$.

where $\omega = \varepsilon^{-1}\sqrt{1+\alpha q^2}$. The fast system is propagated by integrating

$$\dot{q} = p^{n+1/2}, \quad \dot{p} = 0, \quad \dot{x} = \varepsilon^{-1} p_x \quad \dot{p}_x = -\varepsilon^{-1}(1+\alpha q^2)x,$$

using Störmer–Verlet with a small stepsize $\delta t \approx 0.1\varepsilon$ subject to the constraint that $\Delta t/\delta t$ is an integer. We call this the RA-1 method. For comparison, we also implemented the following version of the RA method. The momentum update in p is kept the same as for RA-1. But the fast system is propagated via

$$q^{n+1/2} = q^n + \frac{\Delta t}{2} p^{n+1/2},$$

followed by the exact solution of

$$\dot{x} = \varepsilon^{-1} p_x \quad \dot{p}_x = -\varepsilon^{-1}(1+\alpha(q^{n+1/2})^2)x,$$

and

$$q^{n+1} = q^{n+1/2} + \frac{\Delta t}{2} p^{n+1/2}.$$

We call this method RA-0. We performed the same numerical experiment as described in Section 10.1.2 for the MTS method. The results can be found in Fig. 10.7. While RA-0 is cheaper to implement than RA-1, its performance does not match that of RA-1. The largest relative error in energy over the whole range of stepsizes $\Delta t/\varepsilon \in [1, 10]$ is $\Delta E_{\max} \approx 0.0271$ for RA-1 and $\Delta E_{\max} \approx 1.2124$ for RA-0. RA-1 also clearly outperforms the MTS method of Section 10.2. Again, we emphasize that the increased errors in energy are not linked to instabilities but are to be attributed to a reduced accuracy of the method. The nearly level error in the adiabatic invariant for the RA-1 method is due to a constant resolution of the fast oscillations by choosing $\delta t \approx 0.1\varepsilon$ independently of Δt.

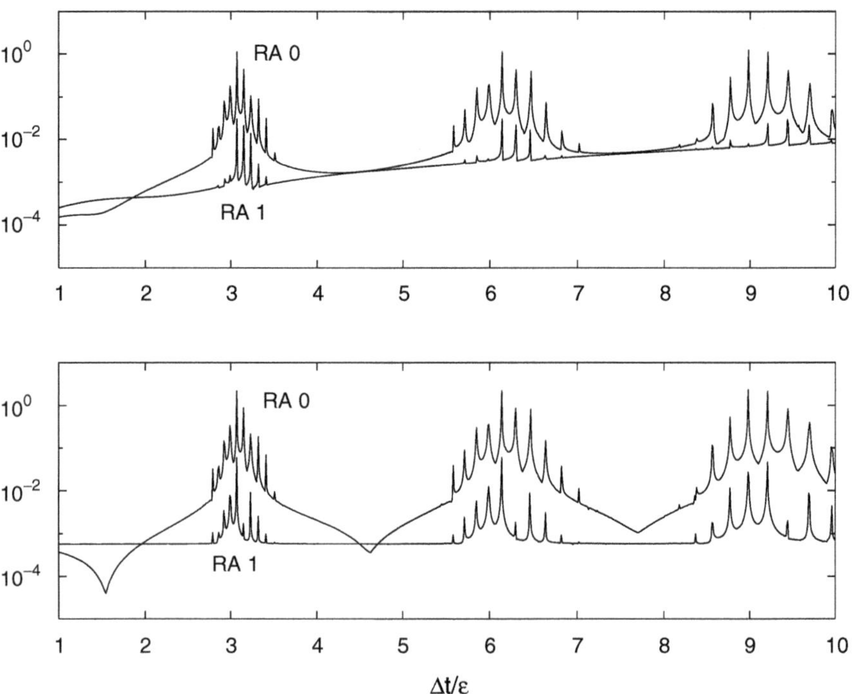

Figure 10.7 Maximum relative errors in total energy and adiabatic invariant as a function of $\Delta t/\varepsilon$, $\varepsilon = 1/40$, for RA methods.

10.4 The mollified impulse (MOLLY) method

The mollified impulse method (MOLLY) [67] has been designed for a slightly different class of Hamiltonian systems than those considered so far. Let us therefore assume that we are given a Hamiltonian

$$H = \frac{1}{2}\boldsymbol{p}^T \boldsymbol{M}^{-1}\boldsymbol{p} + V_{\mathrm{f}}(\boldsymbol{q}) + V_{\mathrm{s}}(\boldsymbol{q}),$$

such that the reduced Hamiltonian

$$H_{\mathrm{f}} = \frac{1}{2}\boldsymbol{p}^T \boldsymbol{M}^{-1}\boldsymbol{p} + V_{\mathrm{f}}(\boldsymbol{q})$$

well captures the motion of the overall system over short time intervals. Note that we do not have an associated splitting into slow and fast variables as assumed so far.

An MTS method can be implemented in the following way. We integrate H_{f} by the Störmer–Verlet method using a small timestep δt. Denote the associated timestepping map by $\boldsymbol{\Psi}_{\delta t, H_{\mathrm{f}}}$. Now consider the symplectic composition method

$$\boldsymbol{\Psi}_{\Delta t} := \boldsymbol{\Phi}_{\Delta t/2, V_{\mathrm{s}}} \circ \left[\boldsymbol{\Psi}_{\delta t, H_{\mathrm{f}}}\right]^N \circ \boldsymbol{\Phi}_{\Delta t/2, V_{\mathrm{s}}}, \tag{10.23}$$

$\Delta t = \delta t \cdot N$. If we apply this MTS method to a linear test problem, then resonance instabilities are, in general, observed (see Biesiadecki and Skeel [20] and García-Archilla, Sanz-Serna, and Skeel [67]).

The basic idea to eliminate or weaken some of these resonances is to replace the "slow" potential energy V_{s} by an averaged potential energy. In the mollified MTS (MOLLY) method, as suggested by García-Archilla, Sanz-Serna, and Skeel [67], this is achieved in the following way. Let $\boldsymbol{q}(t)$ denote the solution of H_{f} with initial conditions $\boldsymbol{q}(0) = \boldsymbol{q}$ and $\boldsymbol{p}(0) = \boldsymbol{0}$. Then one can compute an averaged $\bar{\boldsymbol{q}}$, for example the long average [67]

$$\bar{\boldsymbol{q}} = \frac{1}{2\Delta t}\int_{-\Delta t}^{\Delta t} \boldsymbol{q}(t)dt.$$

The relation between the initial $\boldsymbol{q}$ and the averaged $\bar{\boldsymbol{q}}$ can be viewed as a map

$$\bar{\boldsymbol{q}} = \mathcal{A}(\boldsymbol{q}).$$

This map is applied to the potential energy function V_{s} to yield a new (filtered) potential energy

$$\bar{V}_{\mathrm{s}}(\boldsymbol{q}) = V(\mathcal{A}(\boldsymbol{q})).$$

MOLLY is now characterized by the modified MTS method

$$\boldsymbol{\Psi}_{\Delta t} := \boldsymbol{\Phi}_{\Delta t/2, \bar{V}_{\mathrm{s}}} \circ \left[\boldsymbol{\Psi}_{\delta t, H_{\mathrm{r}}}\right]^N \circ \boldsymbol{\Phi}_{\Delta t/2, \bar{V}_{\mathrm{s}}}. \tag{10.24}$$

Note that $\boldsymbol{\Phi}_{\Delta t/2, \bar{V}_{\mathrm{s}}}$ denotes the exact time-$\Delta t/2$-flow map associated to the smoothed potential energy function $\bar{V}_{\mathrm{s}}$. Hence the overall timestepping generates a symplectic map.

An important limitation of MOLLY should be pointed out. Since averaging and differentiation do *not*, in general, commute, the smoothed potential energy can only be used when the difference between $\boldsymbol{q}$ and $\mathcal{A}(\boldsymbol{q})$ is small or the system is (nearly) linear.

Let us come back to the linear test example (10.12). We split the Hamiltonian into

$$H_{\rm f} = \frac{1}{2}\left[p^2 + \varepsilon^{-1}(p_x^2 + (x-q)^2)\right]$$

and

$$V_{\rm s}(x,q) = \frac{1}{2}q^2.$$

The propagator for the fast $H_{\rm f}$-system is $e^{\Delta t \boldsymbol{A}}$ with

$$\boldsymbol{A} = \begin{bmatrix} 0 & 1 & 0 & 0 \\ -\varepsilon^{-1} & 0 & \varepsilon^{-1} & 0 \\ 0 & 0 & 0 & \varepsilon^{-1} \\ \varepsilon^{-1} & 0 & -\varepsilon^{-1} & 0 \end{bmatrix},$$

and $\boldsymbol{z} = (q,p,x,p_x)^T$. The next step is to construct a mollifier $\mathcal{A}$. To do so, it is convenient to introduce transformed coordinates $q_1 = q$ and $q_2 = x - q$. The corresponding canonical momenta are given by $p = p_1 - p_2$ and $p_x = p_2$. Thus the transformed Hamiltonian $\tilde{H}_{\rm f}$ is

$$\tilde{H}_{\rm f} = \frac{1}{2}\left[(p_1 - p_2)^2 + \varepsilon^{-1}(p_2^2 + q_2^2)\right],$$

with associated equations of motion

$$\dot{p}_1 = 0, \quad \dot{q}_1 = p_1 - p_2, \quad \dot{p}_2 = -\varepsilon^{-1}q_2, \quad \dot{q}_2 = p_2 - p_1 + \varepsilon^{-1}p_2.$$

We solve these equations subject to zero initial conditions in p_1 and p_2, i.e. $p_1(0) = p_2(0) = 0$. We immediately conclude that $p_1(t) = 0$ and the equations of motion simplify to

$$\dot{q}_1 = -p_2, \quad \dot{p}_2 = -\varepsilon^{-1}q_2, \quad \dot{q}_2 = (1+\varepsilon^{-1})p_2.$$

The frequency of the fast oscillations in (q_2, p_2) is

$$\omega = \sqrt{\varepsilon^{-1}(1+\varepsilon^{-1})},$$

and solutions are given explicitly by

$$q_2(t) = q_2(0)\cos(\omega t), \qquad p_2(t) = -\frac{q_2(0)}{\varepsilon\omega}\sin(\omega t).$$

Hence, we obtain

$$q_1(t) = q_1(0) - \frac{q_2(0)}{1+\varepsilon^{-1}}\left(\cos(\omega t) - 1\right)$$

and the averaged value

$$\bar{q}_1 = \frac{1}{2\Delta t}\int_{-\Delta t}^{\Delta t} q_1(t)dt = q_1(0) - \frac{q_2(0)}{1+\varepsilon^{-1}}\left(\frac{\sin(\omega\Delta t)}{\Delta t\omega} - 1\right).$$

If we translate this result back to the original variables, then the mollifier $\mathcal{A}$ is given by

$$\bar{q} = \mathcal{A}(x, q) = q - \frac{x-q}{1+\varepsilon^{-1}}\left(\frac{\sin(\omega\Delta t)}{\omega\Delta t} - 1\right)$$

and the mollified potential energy V_s by

$$\bar{V}_s = \frac{1}{2}\bar{q}^2 = \frac{1}{2}q^2 + \frac{C^2}{2}(x-q)^2 + C(q-x)q,$$

where

$$C = \frac{1}{1+\varepsilon^{-1}}\left(\frac{\sin(\omega\Delta t)}{\omega\Delta t} - 1\right).$$

The propagator for the mollified slow system is $e^{\Delta t\bar{\boldsymbol{B}}}$ with

$$\bar{\boldsymbol{B}} = \begin{bmatrix} 0 & 0 & 0 & 0 \\ -(1+C)^2 & 0 & C+C^2 & 0 \\ 0 & 0 & 0 & 0 \\ C+C^2 & 0 & -C^2 & 0 \end{bmatrix}.$$

Fig. 10.8 reveals that the mollified LTS method is vastly more stable than the original LTS method (10.13). However, tiny resonance instabilities are still present near multiples of 2π.

See problem 4 in the Exercises for another example of a mollified MTS method.

10.5 Multiple frequency systems

So far we have exclusively dealt with systems that possess a single fast degree of freedom. This is, of course, not the situation typically encountered in applications such as molecular dynamics or numerical weather prediction [47]. The LTS methods described in Sections 10.3 and 10.4 can still be applied to such more general highly oscillatory systems but they encounter additional difficulties. Take the case of two fast degrees of freedom with frequencies ω_1 and ω_2. Following the

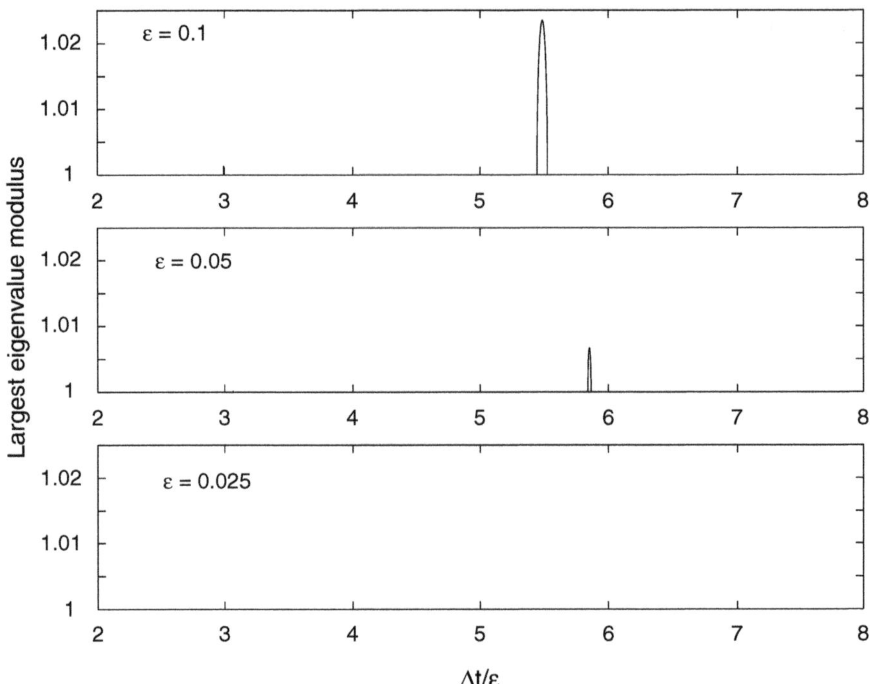

Figure 10.8 Stability of MOLLY as a function of $\Delta t/\varepsilon$.

discussion in Section 10.1.1, a standard LTS methods leads to the investigation of the two associated frequencies $\Omega_{1,\Delta t}$ and $\Omega_{2,\Delta t}$ which are both in the range $\mathcal{I} = [-\pi, \pi)$. While ω_1 and ω_2 will in general be different and non-resonant, we can now always find a stepsize Δt such that $\Omega_{1,\Delta t} = \Omega_{2,\Delta t}$. Hence, for that particular Δt, the two modes are in one-to-one resonance which, typically, manifests itself with a numerical instability of the associated integration scheme. Furthermore, these instabilities are not easily mollified because they can occur for any value of $\Omega_{1,\Delta t} = \Omega_{2,\Delta t}$ in the interval $\mathcal{I}$. See also [67].

The only known stable MTS method for general Hamiltonian systems of type

$$H = \frac{1}{2}\boldsymbol{p}^T \boldsymbol{M}^{-1}\boldsymbol{p} + V(\boldsymbol{q}) + \frac{1}{2\varepsilon^2}\boldsymbol{g}(\boldsymbol{q})^T \boldsymbol{g}(\boldsymbol{q}) \tag{10.25}$$

has been proposed by Izaguirre, Reich and Skeel [93] and makes use of the holonomic constraint $\boldsymbol{g}(\boldsymbol{q}) = \boldsymbol{0}$ in the definition of the mollifier $\mathcal{A}$. See the Exercises.

10.6 Exercises

1. *Non-autonomous backward error analysis.* Following a result by Kuksin and Pöschel [100] (see also Moan [135]), we can imbed a symplectic method into the time-one-flow map of a non-autonomous Hamiltonian system with Hamiltonian of the form

$$\tilde{H}_{\Delta t} = 2\pi e + \Delta t H(q, p) + \Delta t g_{\Delta t}(q, p, \tau),$$

where $g_{\Delta t}$ is an appropriate function which is 2π-periodic in τ. The symplectic structure of the extended equations in (q, p, τ, e) is

$$\omega = dp \wedge dq + de \wedge d\tau.$$

a. Perform a canonical change of variables from (e, τ) to $E = -2e$ and $s = -\tau/2$. Verify that the new equations of motion can be written in the form

$$\begin{aligned} \dot{q} &= +\Delta t \nabla_p [H(q, p) + h_{\Delta t}(q, p, s)], \\ \dot{p} &= -\Delta t \nabla_q [H(q, p) + h_{\Delta t}(q, p, s)], \\ \dot{s} &= -\pi, \\ \dot{E} &= -\Delta t \nabla_s h_{\Delta t}(q, p, s), \end{aligned}$$

and find $h_{\Delta t}$ in terms of $g_{\Delta t}$.

b. Use the result from (a) to "justify" the modified Hamiltonian (10.6).

2. *Resonance instabilities for single-degree-of-freedom systems.* Consider a one degree of freedom linear system

$$\dot{q} = \omega p, \qquad \dot{p} = -\omega q - q,$$

which we write as

$$\dot{\boldsymbol{z}} = \omega \boldsymbol{J} \boldsymbol{z} + \boldsymbol{B} \boldsymbol{z},$$

with $\boldsymbol{z} = (q, p)^T$. An LTS method can be defined by

$$\boldsymbol{M}_{\Delta t} = \boldsymbol{W}_2(\Delta t/2) \boldsymbol{W}_1(\omega \Delta t) \boldsymbol{W}_2(\Delta t/2),$$

where

$$\boldsymbol{W}_1(\omega \Delta t) = \mathbf{e}^{\omega \Delta t \boldsymbol{J}}, \qquad \boldsymbol{W}_2(\Delta t/2) = \mathbf{e}^{\Delta t/2 \boldsymbol{B}}.$$

a. According to standard stability results for symplectic matrices (see [7]), we know that $\boldsymbol{M}_{\Delta t}$ is stable provided

(i) Δt is sufficiently small,

(ii) the eigenvalues of $\boldsymbol{W}_1(\omega\Delta t)$ are on the unit circle and simple.

What values of $\tau = \omega\Delta t$ give rise to double eigenvalues of $\boldsymbol{W}_1(\tau)$?

b. Assume that Δt has been chosen such that

$$|\omega\Delta t - 2\pi| \ll \Delta t. \tag{10.26}$$

Show that $\boldsymbol{M}_{\Delta t}$ is equivalent to

$$\boldsymbol{M}_{\Delta t} = \boldsymbol{W}_2(\Delta t/2)\boldsymbol{W}_1(\omega\Delta t - 2\pi)\boldsymbol{W}_2(\Delta t/2).$$

Determine the parameter $\hat{\omega}$, $|\hat{\omega}| \leq \pi/\Delta t$, in the equation

$$\dot{z} = \hat{\omega}\boldsymbol{J}z + \boldsymbol{B}z, \tag{10.27}$$

such that the associated matrix exponential,

$$\tilde{\boldsymbol{W}}(\Delta t) = \mathbf{e}^{\Delta t(\hat{\omega}\boldsymbol{J}+\boldsymbol{B})},$$

satisfies

$$\boldsymbol{M}_{\Delta t} = \tilde{\boldsymbol{W}}(\Delta t) + \mathcal{O}(\Delta t^2),$$

as $\Delta t \to 0$ and $\omega \to \infty$ subject to (10.26).

Show that $\hat{\omega}$ can become negative for certain values of Δt. What implications can we deduce for the solution behavior of the modified equation (10.27)? Compare your findings with the results from Fig. 10.1.

3. *Resonance instabilities for single-degree-of-freedom systems.* If one applies the LTS method (10.5) to the nonlinear oscillator

$$\dot{q} = \omega p, \qquad \dot{p} = -\omega q - g'(q), \tag{10.28}$$

$g'(q)$ an odd function in q, numerical instabilities can be observed for $\omega\Delta t \approx k\pi$, k a positive integer. We have seen in Section 10.1.1 that the LTS method (10.5) leads to a modified Hamiltonian $\tilde{H}$ with an effective frequency $\hat{\omega} \approx 0$ for stepsizes Δt such that $\omega\Delta t \approx 2k\pi$. A similar statement can, in fact, be derived for the case $\omega\Delta t \approx k\pi$. We write the LTS method (10.5) applied to (10.28) in the abstract form

$$z^{n+1} = \boldsymbol{\Psi}_{\Delta t}(z^n), \qquad z^n = (q^n, p^n)^T.$$

Next we perform a change of variables

$$\hat{z}^n = (-1)^n z^n$$

and formally obtain a numerical method

$$\hat{z}^{n+1} = \hat{\boldsymbol{\Psi}}_{\Delta t}(\hat{z}^n),$$

in terms of the new sequence of variables $\{\hat{z}^n\}$.

a. Derive an explicit expression for $\hat{\boldsymbol{\Psi}}_{\Delta t}$ assuming that $g'(-q) = -g'(q)$.

b. Show that $\hat{\boldsymbol{\Psi}}_{\Delta t}$ is equivalent to the LTS method (10.5) applied to the nonlinear oscillator

$$\dot{q} = \left(\omega + \frac{\pi}{\Delta t}\right) p, \qquad \dot{p} = -\left(\omega + \frac{\pi}{\Delta t}\right) q - g'(q).$$

c. Following the arguments of Section 10.1.1, discuss the numerical behavior of $\hat{\boldsymbol{\Psi}}_{\Delta t}$ for $\omega\Delta t \approx k\pi$, k an odd integer. See also the previous problem.

4. *Mollified LTS methods.* We develop a mollified version of the LTS method (10.5) applicable to the nonlinear oscillator (10.28). The basic idea is to find a mollified slow Hamiltonian $\bar{H}_s$ such that

$$\boldsymbol{\Phi}_{\Delta t/2, \bar{H}_s} = \mathbf{id},$$

for $\omega\Delta t = k\pi$, k a positive integer. This implies that the numerical method reduces to

$$\boldsymbol{\Psi}_{\Delta t} = \boldsymbol{\Phi}_{\Delta t, H_f} = \mathbf{e}^{\omega \Delta t \boldsymbol{J}},$$

for resonant stepsizes Δt.

a. Consider the matrix

$$\boldsymbol{C}(\tau) = \frac{1}{2\tau} \boldsymbol{J}^{-1} \left(\mathbf{e}^{2\tau \boldsymbol{J}} - \boldsymbol{I}_2\right).$$

Show that (i) $\boldsymbol{C}(\tau) \to \boldsymbol{I}_2$ as $\tau \to 0$ and (ii) $\boldsymbol{C}(\tau) = \mathbf{0}$ for $\tau = k\pi$. Remark: Condition (i) is necessary for convergence of the method as $\tau \to 0$.

b. Given an arbitrary Hamiltonian $H_s(\boldsymbol{z})$, we define the associated mollified Hamiltonian

$$\bar{H}_s(\boldsymbol{z}; \tau) = H_s(\boldsymbol{C}(\tau)\boldsymbol{z}).$$

Find $\bar{H}_s = \bar{V}_s$ for $H_s = V_s = q^2/2$.

c. Using the result from (b), implement the mollified LTS method

$$\boldsymbol{\Psi}_{\Delta t} = \boldsymbol{\Phi}_{\Delta t/2,\bar{H}_s} \circ \boldsymbol{\Phi}_{\Delta t,H_f} \circ \boldsymbol{\Phi}_{\Delta t/2,\bar{H}_s},$$

for $H_s = q^2/2$. How does the method behave near $\omega\Delta t = k\pi$ as compared with the non-mollified method?

5. *Many-degrees-of-freedom systems.* Additional difficulties arise when considering systems with two fast degrees of freedom, for example

$$H = \frac{\omega_1}{2}\left(p_1^2 + q_1^2\right) + \frac{\omega_2}{2}\left(p_1^2 + q_2^2\right) + V_s(q_1, q_2).$$

For simplicity, we set $\omega_1 = \omega$ and $\omega_2 = 4\omega$ and concentrate again on linear systems.

An LTS method of type (10.5) will produce a symplectic matrix $\boldsymbol{M}_{\Delta t}$ which can be thought of as a small perturbation of the symplectic matrix

$$\boldsymbol{W}(\omega\Delta t) = \mathbf{e}^{\omega\Delta t \boldsymbol{A}}, \qquad \boldsymbol{A} = \begin{bmatrix} 0 & 1 & 0 & 0 \\ -1 & 0 & 0 & 0 \\ 0 & 0 & 0 & 2 \\ 0 & 0 & -2 & 0 \end{bmatrix}.$$

Hence we have to make use of the perturbation theory of symplectic matrices. A good introduction to this topic and the associated theory due to KREJN (1950) can be found in [7]. Below we give a brief discussion on the implications for LTS methods.

a. For which values of $\tau = \omega\Delta t \neq 0$ does $\boldsymbol{W}(\tau)$ have non-simple (repeated) eigenvalues on the unit circle? (Hint: sufficient condition is $\Delta t(\omega_1 + \omega_2) = \pi k$. Is this condition also necessary?) For future reference, let λ denote the double eigenvalue with $\operatorname{Im}\lambda > 0$. (Recall that the complex conjugate of λ is also an eigenvalue.)

b. Take one particular value of τ found under (a) and call it τ^*. Determine the two-dimensional plane spanned by the complex-valued eigenvectors $\boldsymbol{v}$ of $\boldsymbol{W}(\tau^*)$ corresponding to the double eigenvalue λ. Denote this plane by $\Pi_\lambda \subset \mathbb{C}^2$.

c. The matrix $\boldsymbol{W}(\tau)$ introduces a quadratic form

$$\kappa(\boldsymbol{x}, \boldsymbol{y}; \tau) = \boldsymbol{x}^T \boldsymbol{J}\boldsymbol{W}(\tau)\,\boldsymbol{y}, \qquad \boldsymbol{J} = \begin{bmatrix} 0 & 1 & 0 & 0 \\ -1 & 0 & 0 & 0 \\ 0 & 0 & 0 & 1 \\ 0 & 0 & -1 & 0 \end{bmatrix}.$$

Find an explicit expression of $\kappa(\boldsymbol{x}, \boldsymbol{x}; \tau^*)$ for your choice of $\tau = \tau^*$.

d. A standard result in symplectic perturbation theory [7] implies that the symplectic matrix $\boldsymbol{W}(\tau)$ is stable under small (symplectic) perturbations, if its eigenvalues are simple or, in case there is a non-simple eigenvalue λ, if the quadratic form $\kappa(\boldsymbol{x}, \boldsymbol{x}; \tau)$ has the same sign for all real vectors $\boldsymbol{x}$ in Π_λ.

Is $\boldsymbol{W}(\tau^*)$ stable for your choice of $\tau = \tau^*$? (Hint: The matrix is unstable under perturbations for all resonances satisfying $\Delta t(\omega_1 + \omega_2) = 2\pi k$, k a positive integer [67].)

6. *Averaged equations and adiabatic invariant.* Consider a particle of unit mass with coordinate $\boldsymbol{q} = (q_1, q_2)^T \in \mathbb{R}^2$ subject to the potential

$$V(\boldsymbol{q}) = \frac{1}{2}q_1^2 + \frac{1}{2\varepsilon^2}(\|\boldsymbol{q}\| - L(\phi))^2,$$

where the equilibrium length L is given by $L(\phi) = 1 + 0.2\sin^2\phi$ and $-\pi < \phi \leq \pi$ is the angle of $\boldsymbol{q} = (q_1, q_2)^T$ with the q_1-coordinate axis. We set $\varepsilon = 0.01$.

a. Reformulate the problem in polar coordinates (r, ϕ). Clearly, for $\varepsilon \to 0$, the motion in the radius $r(t)$ is highly oscillatory about its equilibrium value $L(\phi)$. Find the associated adiabatic invariant J, the fast frequency ω, and the reduced/averaged Hamiltonian.

b. Implement a symplectic method for the unreduced Hamiltonian system and compare the results with a simulation of the constrained formulation

$$H_{\text{constr}}(\boldsymbol{q}, \boldsymbol{p}) = \frac{1}{2}\|\boldsymbol{p}\|^2 + \frac{1}{2}q_1^2 + \lambda(\|\boldsymbol{q}\|^2 - L(\phi)^2),$$

subject to $\|\boldsymbol{q}\|^2 = L(\phi)^2$. Choose your initial conditions such that either $J = 0$ or $J = \varepsilon$.

Explain your numerical findings.

7. *The MTS method equilibrium.* We derive the LTS method Equilibrium [93] for a Hamiltonian of the form (10.25). Given a set of coordinates $\boldsymbol{q}$, we first define the "projection" map $\bar{\boldsymbol{q}} = \mathcal{A}(\boldsymbol{q})$ as the solution to the nonlinear system

$$\bar{\boldsymbol{q}} = \boldsymbol{q} + \boldsymbol{G}(\boldsymbol{q})^T\boldsymbol{\lambda},$$
$$\boldsymbol{0} = \boldsymbol{g}(\bar{\boldsymbol{q}}).$$

The map $\mathcal{A}(\boldsymbol{q})$ gives rise to the mollified Hamiltonian

$$H_{\text{molly}} = \frac{1}{2}\boldsymbol{p}^T \boldsymbol{M}^{-1}\boldsymbol{p} + V(\mathcal{A}(\boldsymbol{q})) + \frac{1}{2\varepsilon^2}\boldsymbol{g}(\boldsymbol{q})^T \boldsymbol{g}(\boldsymbol{q}). \tag{10.29}$$

a. Find the equations of motion corresponding to the mollified Hamiltonian (10.29).

b. Describe an implementation of the associated MTS method (10.24).

11

Molecular dynamics

Over the past several decades, molecular simulation has become increasingly important for chemists, physicists, bio-scientists, and engineers, and plays a role in applications such as rational drug design and the development of new types of materials. While many levels of detail can be incorporated if desired, in most cases work is performed with a simplified atomic model, consisting of a large number of mass points interacting in various types of forces, i.e. an N-body problem.

There are essentially two principal types of simulation methodology in common use. In *Monte-Carlo (MC) methods*, random steps are taken in order to achieve a rapid sampling of the most likely states of the molecule. In *molecular dynamics (MD)*, the idea is to construct approximate trajectories for the N-body problem and to use these to gain an understanding of how the molecule evolves in time, for example in response to a stimulus, during a transition between states, or as a means for calculating averages. It should be stressed that only MD and not MC methods allow the theoretical possibility of obtaining time-dependent quantities from simulation, while both schemes can in principle be used for the same statistical-mechanical calculations. Increasingly, one finds that MD and MC schemes are combined in various ways to seek improved efficiency. In this chapter we will focus only on (pure) MD methods, and in particular on the geometric integration issues associated to computing MD trajectories. For a more complete perspective on molecular simulation, the reader is referred to a text on the subject such as that of SCHLICK [174], ALLEN AND TILDESLEY [4], RAPPAPORT [152], or FRENKEL AND SMIT [66].

Because of the very rapid oscillatory motions that are characteristic of molecular systems, very small timesteps of about a femtosecond (10^{-15}sec) are typically used in molecular dynamics. This means that in order to get estimates for parameters of interest from simulation, it is necessary to perform runs involving enormous numbers of timesteps. Even a nanosecond is a very short time relative to physical phenomena, and that already represents around one million timesteps in a typical simulation. At the same time there are severe size limitations due to the complexity of the force evaluations that must be performed at each timestep.

The type of forces involved in the model can make a huge difference. If the atoms have charge, then a very long-ranged Coulombic potential will be needed. Important advances have been made in the design of efficient algorithms for these types of non-bonded interactions, but the costs are still prohibitive: at the time of this writing, systems of this type of dimension up to only several hundred thousand particles can be simulated, typically for times of at the most a few nanoseconds. The physical dimensions of such systems are only a few tens of angstroms at most. This is the typical situation for biomolecular models, which invariably include charge. Other types of molecular systems arising in materials science and condensed matter physics may not have any long-ranged forces; in such cases, the complexity of the force calculation is greatly reduced and systems of up to even a billion atoms can be simulated.

The first systems to be studied using direct simulation were purely collisional "hard sphere" systems [3]. Work on hard body models continues to the present [200, 181, 183, 88]. We will restrict ourselves to models involving a smooth potential energy function, especially those used most commonly for molecular liquids, as introduced by RAHMAN AND STILLINGER [151] and VERLET [198].

A simple gas such as argon can be described easily using a Lennard–Jones potential

$$\varphi(r) = 4\epsilon\left(\left(\frac{\sigma}{r}\right)^{12} - 2\left(\frac{\sigma}{r}\right)^{6}\right), \tag{11.1}$$

where r is atomic separation, ϵ the well depth, and σ is the equilibrium atomic separation.

To define a molecular system, we arrange N particles in a cubic box with edges of length L, with "periodic boundary conditions" imposed so that at any instant a particle with coordinates (x, y, z) inside the computational domain interacts with an infinite number of periodic images obtained by adding or subtracting multiples of L from each coordinate of each atom. When only short-range forces are involved, the potentials may be cut off so that particles in the computational domain do not interact with their own image. Ultimately, we are left with a system of Newton's equations

$$\boldsymbol{M}\ddot{\boldsymbol{q}} = \boldsymbol{F}(\boldsymbol{q}) := -\nabla_q V(\boldsymbol{q}),$$

where $\boldsymbol{M}$ is the mass matrix. We typically simulate the system from some given positions and velocities $\boldsymbol{q}(t_0) = \boldsymbol{q}_0$, $\dot{\boldsymbol{q}}(t_0) = \dot{\boldsymbol{q}}_0$.

Positions and velocities in a molecular system represent sample points from a certain ensemble of states. Because arbitrary initial data usually correspond to an unrepresentative state for the ensemble of interest, a crucial stage in molecular simulation is the so-called "equilibration" or "thermalization" of the system, usually performed by evolving the system for some time from provided data (typically a few thousand timesteps suffice). At the end of this initial phase, the positions and velocities are taken as the starting points for a sampling trajectory.

For simple homogeneous liquids, one sometimes begins by placing the atoms on a regular "crystal lattice" and introducing random velocities of appropriate magnitude. In the case of more complicated systems such as biomolecules, the approximate structure of the system must be encoded in the position data, for example based on a computation using *nuclear magnetic resonance* (NMR) imaging. Again, randomized initial velocities will likely be used at the start of the equilibration phase. It is sometimes necessary for the simulator to correct certain very nonphysical initial positions and/or velocities in order to be able to get a successful initial run.

There are several key computational challenges to be addressed when computing trajectories of a Lennard–Jones fluid. First, the greater part of the computational effort involves the calculation of the $\frac{1}{2}N(N-1)$ interaction forces and potentials between pairs of atoms, even though the force acting between distant pairs is very weak. Verlet resolved this difficulty by imposing *cutoffs* based on distance according to which the potential is replaced by zero for atomic separations greater than a certain cutoff value r_c (typically 2.5σ–3.3σ). A system for keeping track of close pairs, now referred to as the *Verlet table*, was introduced. All distances are computed from time to time, and those pairs within distance r_M, $r_M > r_c$, are recorded in the table. Force evaluations are limited to entries in the table. The "skin" $r_M - r_c$ must be chosen carefully in accordance with the temperature of simulation, typical particle speed, and frequency of updating of the Verlet table so that no particle pairs can move into the cutoff range between table updates. Other techniques are available for tracking local interactions, such as retaining linked lists of neighbor atoms for each given atom, and adjusting these lists during simulation.

In case long-ranged potentials (especially the r^{-1} Coulomb potential) are involved, we must include even very distant images in the force summation. If the interaction between particles is uniform, in a box of side L, then the potential energy of the system is

$$V(\boldsymbol{q}) = \sum_{\boldsymbol{k}} \sum_{i<j}^{N} \varphi(||\boldsymbol{q}_i - \boldsymbol{q}_j - L\boldsymbol{k}||),$$

where the sum is taken over pairs of atoms (i, j) and the repeating images defined by the multi-index $\boldsymbol{k}$ with three integer components. Various techniques are used to simplify the computation of this infinite sum. In particular the method of *fast Ewald summation* computes the Coulomb potential at a point due to a periodic lattice of point charges by dividing the computation into sums over near and far terms through introduction of a smooth "screening potential" of a particular type. The sum of the screened terms converges rapidly. The complementary summation converges very rapidly if calculated instead in the Fourier domain. A scaling parameter in the screening potential allows for balancing the work done in each part of the calculation for maximal efficiency. An alternative approach, the so-called *fast multipole method*, is in some ways more flexible than Ewald summation since it does not depend on having an infinite periodic sum, although there are complex efficiency trade-offs between the two schemes, particularly in the context of parallel computing. Both methods are employed in MD simulations. Other methods of treating long-ranged solvent forces include implicit solvent models based on a simplified continuum description. While the development and assessment of force calculation techniques is an important area of research, it does not directly influence the time integration issues (as long as a sufficiently accurate approximation is used).

The next issue is the choice of integrator. In Chapter 4, we introduced the Störmer–Verlet method, a popular explicit second-order symplectic scheme and applied it to a "planar molecular system"; in fact this was one of the first used for continuous potential molecular dynamics simulation. Verlet observed in his 1967 paper, "small irregularities in the total energy" but commented that "the error is of no consequence." What Verlet and those who followed him observed was the lack of any systematic or *secular* drift when the Störmer–Verlet method was properly implemented. We have seen similar behavior in our simulation of the planar system in Chapter 4. Given the symplecticness of the method, and our understanding of the properties of symplectic discretizations (see especially Chapter 5), it is perhaps not surprising that the Störmer–Verlet method achieves good conservation of energy.[1] The reason that backward error analysis seems to be accurate for molecular models is probably related to the presence of the Lennard–Jones potential (or other short-ranged repulsive component of the energy); through repulsion, very close approaches of atoms are ruled out. As we shall see below, the presence of a perturbed energy function is also of dramatic importance for validating computed statistical mechanics from MD trajectories.

11.1 From liquids to biopolymers

During the 1970s and 1980s, work continued on MD simulations of water [151] and other molecular liquids with internal degrees of freedom. Today molecular dynamics methods are being applied routinely to simulate large flexible biomolecules.

[1]Note: careless implementations of cut-offs, boundary conditions and other simplifications of the force calculation can easily introduce systematic drifts, as symplecticness may be broken if the force does not remain an exact gradient.

The simple model of an MD pair potential requires a number of changes for more complicated systems. For example, the interaction between atoms of polyatomic molecules typically requires both Lennard–Jones and Coulombic terms between all atoms, not just simplified potentials between molecular centres. The potential energy function then must maintain internal geometries by including bonds (lengths, angles, and dihedral angles). This results in a more complicated molecular energy function of the form

$$V = V_{\mathrm{l.j.}} + V_{\mathrm{C}} + V_{\mathrm{l.b.}} + V_{\mathrm{a.b.}} + V_{\mathrm{d.b.}} + V_{\mathrm{i.d.}}, \tag{11.2}$$

where $V_{\mathrm{L.J.}}$ and V_{C} represent Lennard–Jones and Coulombic nonbonded potentials, $V_{\mathrm{l.b.}}$, $V_{\mathrm{a.b.}}$, $V_{\mathrm{d.b.}}$ correspond to length bonds, angle bonds, and dihedral angle bending, respectively. The latter terms consist of sums over various pairs, triples and quadruples of spatially localized groups of atoms. Table 11.1 describes the typical functional form of each term. All coefficients κ_b, κ_θ, κ_ϕ, and the equilibrium values $\bar{r}_{ij}$, $\bar{\theta}_{ijk}$, $\bar{\phi}_{ijkl}$ are derived from empirical values; these coefficients may be different for each pair, triple, quadruple of atoms admitting the corresponding type of bond. (Another type of term, a so-called *improper dihedral* term, may also be present.)

Some of the most popular parameterizations for potentials are those provided by the software packages CHARMM, AMBER, NAMD and GROMACS.[2]

Because molecular simulation is concerned with very small regions of space and small time intervals, it is useful and numerically sensible to use an appropriate system of units. In practice, typical MD codes perform an internal conversion to a standard system such as the AKMA convention detailed in the following table:

AKMA Units

length	angstrom
energy	kilocalorie/mole
mass	atomic mass unit (amu)
charge	electron charge e
time	20.455 ps
k_B	1.987191×10^{-3}
coulomb constant $1/(4\pi\epsilon_0)$	3.320716×10^{2}

In typical MD codes, unit conversions are performed internally.

[2]Because of the great volume of parameters involved in a model for even a modest molecule, it is essential to have access to a suitable software package. GROMACS and NAMD are freely distributed. Both CHARMM and AMBER are available for purchase; for academic users, the cost is relatively very modest. For CHARMM pricing and availability see the website at http://yuri.harvard.edu/. For AMBER pricing and availability, see the website at http://www.amber.ucsf.edu/amber/amber.html. For GROMACS, see http://www.gromacs.org/. For NAMD see http://www.ks.uiuc.edu/Research/namd/.

Table 11.1 Typical components of the molecular dynamics energy function.

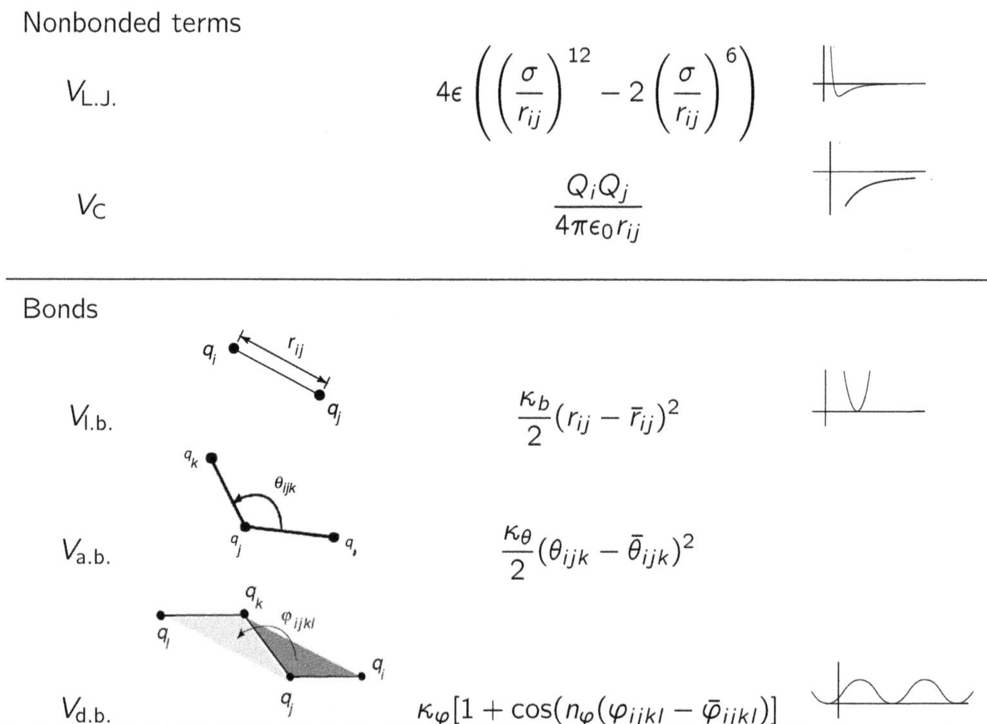

Nonbonded terms	
$V_{\text{L.J.}}$	$4\epsilon\left(\left(\frac{\sigma}{r_{ij}}\right)^{12} - 2\left(\frac{\sigma}{r_{ij}}\right)^{6}\right)$
V_{C}	$\frac{Q_i Q_j}{4\pi\epsilon_0 r_{ij}}$
Bonds	
$V_{\text{l.b.}}$	$\frac{\kappa_b}{2}(r_{ij} - \bar{r}_{ij})^2$
$V_{\text{a.b.}}$	$\frac{\kappa_\theta}{2}(\theta_{ijk} - \bar{\theta}_{ijk})^2$
$V_{\text{d.b.}}$	$\kappa_\varphi[1 + \cos(n_\varphi(\varphi_{ijkl} - \bar{\varphi}_{ijkl}))]$

The terms in the potential energy include interactions on a broad range of spatial scales (including bonds with equilibrium length 1 Å$=10^{-10}$ m, as well as very long-ranged Coulombic terms). Likewise the variation of timescales is vast, ranging from rapid oscillations of the CH bond (around 10^{-14}s) to conformational changes such as the rotation or folding process of a large molecular component that may take milliseconds or even seconds. We have already learned in Chapter 2 that the numerical stability of an integrator will be limited by the fastest harmonic components (i.e., the bond stretch), which means that in molecular dynamics stepsize is limited to around a femtosecond. Much work is therefore concentrated on developing advanced multi-scale timestepping schemes (such as those we have encountered in Chapter 10), and on rapid evaluation of non-bonded forces.

The MD potential is highly nonlinear. When many terms are present, the potential surface will be very rough. If we think of dynamics in terms of the analogy of a ball rolling on a potential energy surface, then it is easy to see how the dynamics may be trapped for long intervals in highly localized regions. We have mentioned that MD is often used as a sampling device. We typically assume that MD models are *ergodic*, implying that all but a few special trajectories will visit the entire allowable phase space. The critical practical question for MD simulation is how long does it take for this to happen; on a highly corrugated landscape sampling may be achieved very slowly, and algorithmic devices are needed to accelerate the process. Another related problem sometimes treated using dynamics is *minimization* of the potential energy; this problem is also very challenging, since there are likely to be many local minima.

Example 1 One of the most popular small models used in biomolecular simulation is the alanine dipeptide (N-Acetylalanyl-N'-Methylamide), which serves as a common test example for demonstrating enhancements in simulation methodology. The all-atom model contains 22 atoms (carbon, hydrogen, nitrogen and oxygen) as described by Figure 11.1. The labels in 11.1 are those used in describing the molecule and its interactions for input to a molecular dynamics code. These labels are defined in a "PDB" (protein databank) file which gives representative atomic positions, atomic types, and linking topology of the molecule. The force field parameters are then provided automatically by the MD software. There may be slight or even substantial differences in force parameters depending on which software package is used, so care is needed in comparing numerical simulation results.

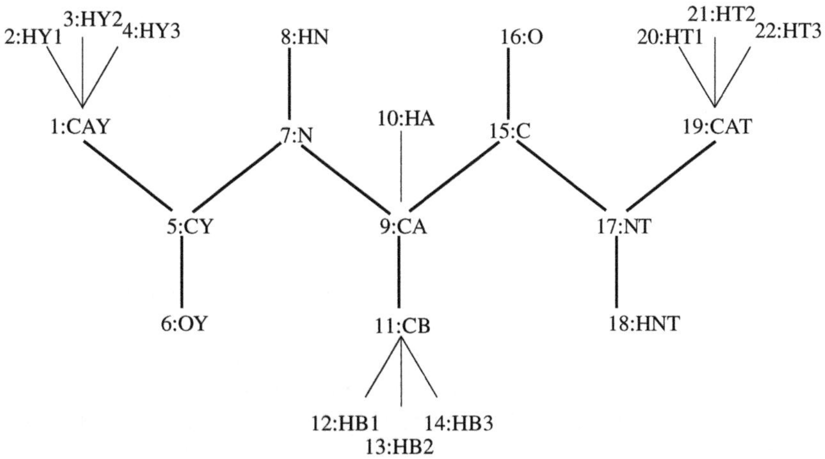

Figure 11.1 Diagram showing atom numbers and labels for the alanine dipeptide used in a PDB file and for CHARMM simulation.

The Alanine Dipeptide exhibits many of the features characteristic of biological molecules. While the dynamics is complicated, it turns out that the study of the energy surface can be simplified by reference to two central dihedral angles (ψ and ϕ corresponding to the successive atom quadruples 5-7-9-15 and 7-9-15-17 in Fig. 11.1, respectively). For each pair of dihedral values, a global

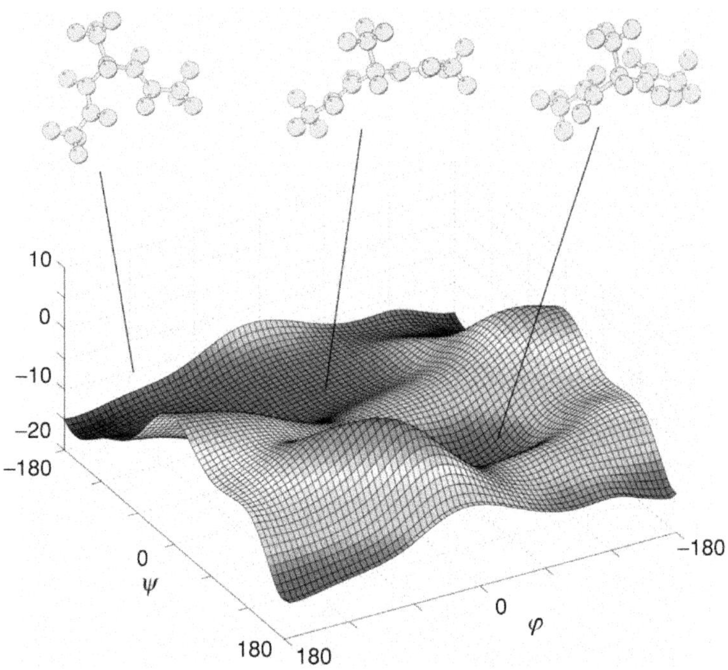

Figure 11.2 Reduced energy surface for the alanine dipeptide, see text for description.

constrained minimization is performed over all degrees of freedom. The resulting "reduced energies" are plotted in the lower diagram in Fig. 11.2, with three states marked by the molecular conformations corresponding to the three stable local minima.

Together with the dipeptide molecule (the "solute"), one has to introduce as part of the modeling process a collection of surrounding water molecules (the "solvent"). Typically for the alanine dipeptide, this might require 500 additional water molecules. The whole system is then typically treated using periodic boundary conditions and Ewald summation for the long-ranged forces. The result is a model Hamiltonian of considerable detail with which simulations can be performed.

As an illustration, we simulate the all-atom model using CHARMM. We first "equilibrated the system" to a temperature of 300K by solving with a thermostat (see later in this chapter for a discussion of thermostatting). The thermostat was then turned off and constant energy simulation was performed. On the left in Fig. 1, we illustrate a short-term (2500 timestep) simulation with 1 femtosecond timesteps by plotting the trajectory of the central dihedral angles. At the energy level used, the constant energy trajectory will visit both of the two local minima near the initial point (the two left-most conformations shown in Fig. 11.2). This is illustrated by plotting snapshots every 100 timesteps of a longer (1M timestep) simulation (Fig. 1, right).

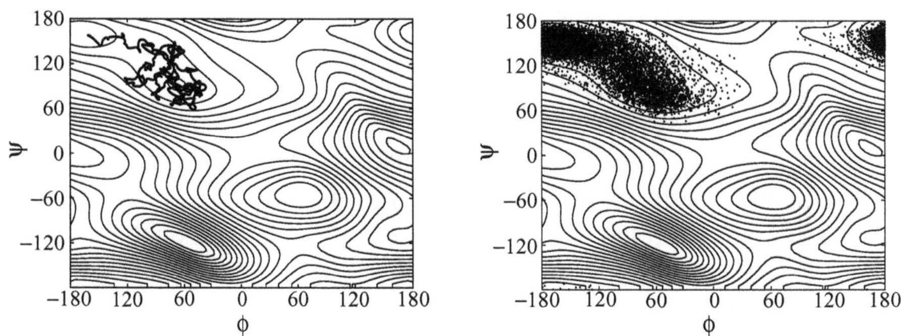

Figure 11.3 Short (left) and medium-length (right) trajectories for the alanine dipeptide. See text.

The relative smoothness and simplicity of the reduced energy surface for the alanine dipeptide suggests that, if the "right" variables in a molecular system can be identified (the so-called *essential degrees of freedom*), then the study of molecular conformations might be made much more tractable. However, one should keep in mind that identifying essential degrees of freedom is not always so easy a task and that most biomolecules will have many more atoms than the alanine dipeptide and many more conformational states, and even the reduced energy landscape will be highly complicated. Even when, as here, the system reduces elegantly in the essential degrees of freedom, one must be cautious about the interpretations that are made regarding the "dynamics" on this surface, since no simple formula relates the slow timescale to the rapid timescales of the original system. □

11.1.1 Constraints

We have already discussed in Chapter 7 the use of constraints to remove the fastest oscillatory terms due to the presence of stiff springs with rest length. Since a bond stretch is nothing other than a stiff spring with rest length, this device is commonly used in molecular simulations. In this way, the energy function will be slightly smoothed, but it has often been argued in the molecular dynamics literature that the resulting simplification typically does not alter the dynamics appreciably on the timescales used in current simulation practice. However some caution must be exercised with regard to the thermodynamic equilibrium behavior as has been first pointed out by FIXMAN [61].

See the Exercises for an example of using SHAKE in the molecular modeling context.

11.2 Statistical mechanics from MD trajectories

Given the small size of any atom, it is not obvious that simulating an atomic system consisting of several thousands or millions of particles can provide useful information about even the smallest realistic system. Only when the principles of statistical mechanics are brought into play can we extrapolate from the tiny simulated system to that of a much larger aggregate. Statistical mechanics begins with the identification of an appropriate density or ensemble of states with respect to which all averages are computed. The key premise is that any given state is representative of a continuum of similar states, and that the probability of finding the system in any given region R of phase space at any arbitrary instant can be viewed as proportional to an integral of the provided density over R.

We will assume that any individual motion in the system is governed by a Hamiltonian H. In general, we expect the density to be invariant under the evolution of the system, so that ρ is a steady state of the Liouville equation, i.e.

$$\frac{d\rho}{dt} = \{\rho, H\} = \nabla_{\mathbf{q}}\rho \cdot \nabla_{\mathbf{p}}H - \nabla_{\mathbf{p}}\rho \cdot \nabla_{\mathbf{q}}H.$$

One way, and in the generic case the only way, to guarantee this is to demand that ρ be a function of H.

In general several parameters must be chosen to define the characteristics of the system. One approach is to fix the number N of atoms and assume that these atoms comprise an isolated system which evolves in a fixed volume in space. While evidently unrealistic from a physical point of view, this finite model provides the simplest framework for simulation on a computer. (The restriction to a finite volume is achieved most readily by the use of periodic boundary conditions as mentioned previously, which introduces a modification of the potential energy function.) Because the system is isolated, its evolution is completely defined by the system Hamiltonian and the energy is a constant of motion. We still must define a suitable density. Since the density is to be a function of energy and the energy is a constant of motion, it seems that a natural choice for the density in these circumstances is

$$\rho_{\text{microcanonical}} = \delta[H - E],$$

which, when normalized, implies that the relative likelihood of finding the system in a given bounded measurable subset of the energy surface $H = E$ is proportional to the corresponding surface area. Because this approach relies on the assumptions of constant particle number N, volume V, and energy E we typically refer to it as the NVE ensemble. It is also commonly termed the *microcanonical* ensemble.

A more realistic assumption than constant energy is that our finite system is continually exchanging energy with a larger bath of atoms at a fixed temperature.

The *instantaneous temperature* of a molecular system with N atoms is defined by

$$T_{\text{inst}} := \frac{1}{3k_B N} \sum_{i=1}^{N} m_i \|\mathbf{v}_i\|^2. \tag{11.3}$$

The *canonical* ensemble assumes the density

$$\rho_{\text{canonical}} = \exp(-\frac{1}{k_B T} H) =: \exp(-\beta H).$$

Here k_B is Boltzmann's constant, T represents the fixed temperature of the bath, and $\beta = 1/k_B T$. Again, the number of particles N and the volume of the system V are fixed, but the energy is now allowed to fluctuate. (The precise justification for the exponential function is due to Gibbs and is based on a simplified kinetic theory.) Whereas simulating the dynamics of a system in the microcanonical ensemble requires nothing more than numerical integration of the Hamiltonian system, the NVT ensemble demands some sort of stochastic or dynamic mechanism to model the effect of the bath. We will return to this issue momentarily. Note that just because the temperature of the environment is assumed to be fixed, it does not imply that T_{inst} is constant. Rather, this quantity, which is a function of the N variables of the system, will fluctuate along trajectories, although its long-term average will tend to T.

Some other examples of useful ensembles include the NPT (constant pressure, temperature and particle number) ensemble and the *grand canonical ensemble* (in which both energy and particle number are allowed to fluctuate). It is also possible to work with a more general density. In some cases, generalized ensembles are introduced based on physical arguments; in other instances they may be used as modeling devices to increase the sampling rate in phase space.

In defining statistical mechanics as we have done here, we make several assumptions about the Hamiltonians involved, in particular that the energy is bounded below and that the total integral of the density function is bounded. An important assumption we commonly make about a physical model is that averages taken along trajectories reproduce the appropriate ensemble average. This is what is meant by *ergodicity*: that a typical trajectory will eventually visit all states with $\rho > 0$. If we treat a system in the NVE ensemble, then we can infer that the energy is constant, but we cannot necessarily infer that all possible states on the constant energy surface will be reachable from any particular state. For small systems, it is common to find periodic orbits or quasi-periodic tori motion that restrict a typical trajectory to a small portion of phase space; more generally, the presence of strong potentials such as those due to harmonic bonds may introduce barriers which similarly confine trajectories. If ergodicity is assumed, however, then we can use trajectories as sampling devices, simply averaging a given quantity along the computed dynamics to obtain ensemble averages.

11.2.1 Ensemble computations

As we have stated, the typical use made of molecular simulation is to compute averages with respect to the underlying density. Examples of such computations include finding the internal energy, spatial extent, or the time constants associated with local relaxation of a perturbation. We will denote the average of some function $A(\boldsymbol{q}, \boldsymbol{p})$ with respect to the NVE ensemble for a given Hamiltonian H by $\langle A \rangle^H_{\mathrm{NVE}}$, and use similar notation for other averages. Under the ergodicity assumption, a macroscopic quantity, which for real systems could be observed or measured, can be thought of as a long-time average $\bar{A}(\tau)$, $\tau \gg 1$, of some (instantaneous) function $A(t) = A(\boldsymbol{q}(t), \boldsymbol{p}(t))$ which depends on the collective position and velocity at time t. The temporal average is defined as

$$\langle A \rangle := \lim_{\mathcal{T}\to\infty} \bar{A}(\tau) = \lim_{\mathcal{T}\to\infty} \frac{1}{\mathcal{T}} \int_0^{\mathcal{T}} A(\boldsymbol{q}(t), \boldsymbol{p}(t)) \, dt.$$

Ergodicity in the NVE ensemble now becomes equivalent to the statement that $\langle A \rangle = \langle A \rangle^H_{\mathrm{NVE}}$.

In practice, trajectories are computed at a large finite number of discrete times $\tau_1, \ldots, \tau_N$, in which case the integral is replaced by a (finite) discrete sum.

Some macroscopic quantities of interest are collected in Table 11.2. In this table, the Boltzman constant is denoted k_B, and instantaneous temperature T is proportional to kinetic energy and is defined as in (11.3). Specific heat at constant volume measures the rate of change of temperature due to a change in energy. The velocity autocorrelation function measures how the velocities at time t are related to velocities at a later time $t + \tau$. The pair correlation function, or radial distribution function for a system of particles with volume V, gives the number of particles $n(r)$ situated at a distance between r and $r + \Delta r$ from another particle. The mean square displacement $R(\tau)$ measures average atomic fluctuations over time windows of length τ. With $t = 0$, the value of τ at which R ceases to change significantly can be understood as the time required for a simulated system to achieve equilibrium. The diffusion coefficient D is proportional to the slope of $R(\tau)$ over long times via the Einstein relation given in Table 11.2. More details for computing these quantities can be found in ALLEN AND TILDESLEY [4], RAPPAPORT [152], and FRENKEL AND SMIT [66].

Table 11.2 Some computable quantities

Specific heat at constant volume

$$C_V = \left[\frac{2}{3N} - \frac{4}{9} \frac{\langle (T - \langle T \rangle)^2 \rangle}{\langle T \rangle^2} \right]^{-1} k_B$$

Velocity autocorrelation function

$$Z(\tau) = \left\langle \frac{1}{N} \mathbf{v}(.) \cdot \mathbf{v}(. + \tau) \right\rangle$$

Pair correlation function (radial distribution function)

$$g(r, \Delta r) = \left\langle \frac{V}{N} \frac{n(r)}{(4\pi r^2 \Delta r)} \right\rangle$$

Mean square displacement after time τ

$$R(\tau) = \left\langle \frac{1}{N} \sum_{i=1}^{N} (r_i(. + \tau) - r_i(.))^2 \right\rangle$$

Diffusion coefficient, D,

$$2\tau D = \frac{1}{3} R(\tau)$$

11.3 Dynamical formulations for the NVT, NPT and other ensembles

Some modification of the dynamical system (or the introduction of a stochastic perturbation) is necessary to make the dynamics sample the other ensembles, for example the canonical one. There are many ways to do this, but the most popular techniques are *Nosé* dynamics and *Langevin dynamics*. In Langevin dynamics, the combination of a damping force and a stochastic term maintains the system at a given temperature. (The stochastic component models the interactions with a temperature bath; the damping simulates the transfer of energy to the bath.)

The dynamics take the form

$$\boldsymbol{M}\ddot{\boldsymbol{q}} = -\nabla_q V(\boldsymbol{q}) + \boldsymbol{R}(t) - \gamma\dot{\boldsymbol{q}},$$

where the dissipation rate γ and the vector $\boldsymbol{R}(t)$ of white noise processes are linked via the fluctuation–dissipation relation. Because of the presence of damping and the introduction of a random forcing, the dynamics are not any longer Hamiltonian. As the geometric integration of stochastic differential equations is beyond the scope of the current text (and indeed still at a preliminary stage of development compared with the deterministic theory), we consider rather alternatives which compute trajectories in the canonical ensemble from a modified continuous dynamics.

NOSE [146] proposed to augment the phase space by an additional variable s and its canonical momentum π_s, and to work with the extended Hamiltonian

$$\mathcal{H}_{\text{Nosé}} = H(\boldsymbol{q}, \tilde{\boldsymbol{p}}/s) + \frac{\pi_s^2}{2Q} + g\beta^{-1}\ln s. \tag{11.4}$$

Here $\beta = 1/k_B T$ and $g = N_f + 1$, where N_f is the number of degrees of freedom of the real system (typically $N_f = 3N$, where N is the number of atoms, but this may be reduced if constraints are present). The constant Q represents an artificial "mass" associated with s. One should note that $\tilde{\boldsymbol{p}}$ is the canonical momenta associated with the position variable, $\boldsymbol{q}$. The tilde is used to distinguish it from the from the real momenta given by $\boldsymbol{p} = \tilde{\boldsymbol{p}}/s$.

To understand how Nosé's extended Hamiltonian yields canonical averages, we consider the following sequence of calculations.

1. From elementary calculus, we know that for any a, the following holds

$$e^{-a} = \int_0^\infty \delta[a + \ln s]\, ds.$$

   Since here and below, all integration is over the entire real line, we will suppress the limits and simply write

$$e^{-a} = \int \delta[a + \ln s]\, ds.$$

2. A very slight generalization results in

$$e^{-\beta H(\boldsymbol{q},\boldsymbol{p})} = \beta \int \delta[H(\boldsymbol{q}, \boldsymbol{p}) + \beta^{-1}\ln s]\, ds.$$

3. We can also include a power of s outside the δ-function, as in

$$e^{-\beta H(\boldsymbol{q},\boldsymbol{p})} = \frac{\beta}{g} \int \delta[H(\boldsymbol{q}, \boldsymbol{p}) + g\beta^{-1}\ln s]\, s^{g-1}\, ds.$$

4. Note that for a scalar function $h(\boldsymbol{x})$ of an N_f-vector $\boldsymbol{x}$, we have, via change of variables,

$$\int\int \ldots \int h(\tilde{x}/s)\, d\tilde{x}_1 d\tilde{x}_2 \ldots d\tilde{x}_{N_f} = \int\int \ldots \int s^{N_f} h(\boldsymbol{x})\, dx_1 dx_2 \ldots dx_{N_f}.$$

Again, the integration is over all real space. We introduce another shorthand at this point, writing, simply,

$$\int h(\tilde{\boldsymbol{x}}/s)\, d^{N_f}\tilde{\boldsymbol{x}} = \int s^{N_f} h(\boldsymbol{x})\, d^{N_f}\boldsymbol{x}.$$

As a consequence, we may write:

$$\int A(\boldsymbol{q}, \boldsymbol{p})\, e^{-\beta H(\boldsymbol{q},\boldsymbol{p})}\, d^{N_f}\boldsymbol{p} = \frac{\beta}{g} \int\int A(\boldsymbol{q}, \tilde{\boldsymbol{p}}/s)\, \delta[H(\boldsymbol{q}, \tilde{\boldsymbol{p}}/s) + g\beta^{-1} \ln s]\, d^{N_f}\tilde{\boldsymbol{p}}\, ds,$$

where, as before, $g = N_f + 1$.

5. Finally, we note that

$$\int\int A(\boldsymbol{q}, \tilde{\boldsymbol{p}}/s)\, \delta[H(\boldsymbol{q}, \tilde{\boldsymbol{p}}/s) + g\beta^{-1} \ln s]\, d^{N_f}\tilde{\boldsymbol{p}}\, ds = C \int\int\int A(\boldsymbol{q}, \tilde{\boldsymbol{p}}/s)\, \delta[\mathcal{H}_{\text{Nosé}} - E]\, d^{N_f}\tilde{\boldsymbol{p}}\, d\pi_s\, ds,$$

where C is a constant, and E is the pseudo-energy level. It follows that the NVE ensemble average for the Hamiltonian (11.4) is equivalent to the canonical NVT ensemble for the original system with Hamiltonian H in the following sense:

$$\langle A(\boldsymbol{q}, \tilde{\boldsymbol{p}}/s)\rangle_{\text{NVE}}^{\mathcal{H}_{\text{Nosé}}} = \langle A(\boldsymbol{q}, \boldsymbol{p})\rangle_{\text{NVT}}^{H}.$$

The conclusion is that trajectories of the Nosé extended Hamiltonian generate canonically distributed averages, given assumptions of equal *a priori* probabilities and ergodicity, hence Nosé dynamics can be used for canonical sampling. The Nosé formulation is the starting point for numerical schemes for realizing the canonical ensemble from trajectories, but it is not yet clear precisely how the extended Hamiltonian dynamics are related to the those of the original system. We take up this issue in the next subsections.

11.3.1 Coordinate transformations: the separated form

Consider first a canonical Hamiltonian of the form

$$H = a(\boldsymbol{q})\frac{\boldsymbol{p}^T \boldsymbol{M}^{-1}\boldsymbol{p}}{2} + V(\boldsymbol{q}), \tag{11.5}$$

where a is a positive scalar-valued function of position $\boldsymbol{q}$, V is a potential and $\boldsymbol{M}$ is a mass matrix. Introducing a Poincaré type time transformation, we can reduce (11.5) to mechanical form

$$\hat{H} = \frac{H - E}{a(\boldsymbol{q})} = \frac{\boldsymbol{p}^T \boldsymbol{M}^{-1}\boldsymbol{p}}{2} + \frac{V(\boldsymbol{q}) - E}{a(\boldsymbol{q})},$$

where E is a constant representing the value of the energy integral along a given trajectory. The problem of separating the variables in the Hamiltonian therefore means finding a change of variables to the form (11.5).

While we cannot flatten out just any metric using such simple coordinate and time transformations, it is possible to do this neatly for Nosé dynamics. First replace s by its logarithm, also modifying the momentum to make the transformation canonical:

$$\theta = \ln s, \qquad \pi_s = \exp(-\theta)\pi_\theta.$$

We then follow with a rescaling of time by $e^{2\theta}$, so that the Hamiltonian becomes

$$\hat{H} = \frac{\boldsymbol{p}^T \boldsymbol{M}^{-1}\boldsymbol{p}}{2} + \frac{\pi_\theta^2}{2Q} + e^{2\theta}(V(\boldsymbol{q}) + g\beta^{-1}\theta - E).$$

Because the transformation is canonical, it is also area preserving, and it can be shown that the averages of functions of $\boldsymbol{q}$ and $\boldsymbol{p}/s$ are not disturbed.

The elementary form of the separated Hamiltonian can simplify the interpretation of Nosé dynamics. With g/β replaced by an arbitrary positive parameter γ, we can define an "isothermal potential" by

$$\tilde{V}(\boldsymbol{q}, \theta) = e^{2\theta}(\gamma\theta + V(\boldsymbol{q}) - E).$$

This has stationary points $(\bar{\boldsymbol{q}}, \bar{\theta})$, where $\bar{\boldsymbol{q}}$ is a stationary point of V and $\bar{\theta} = (E - V(\bar{\boldsymbol{q}}))/\gamma - 1/2$. The stability is determined from the original system. A stable center in the original mechanical system is thus mapped to a corresponding stable center in the constant temperature dynamics.

Due to the shift of energy in the Poincaré transformation, all motion takes place on the zero-energy surface of the Nosé Hamiltonian. Since the kinetic energy is everywhere non-negative, the entire range of negative isothermal potential energy will be explored. For a fixed value of θ, the new potential is a shift and scaling of the original potential. For positive θ, the motion in a bounded potential is uniformly bounded, while in principle the dynamics may make very long

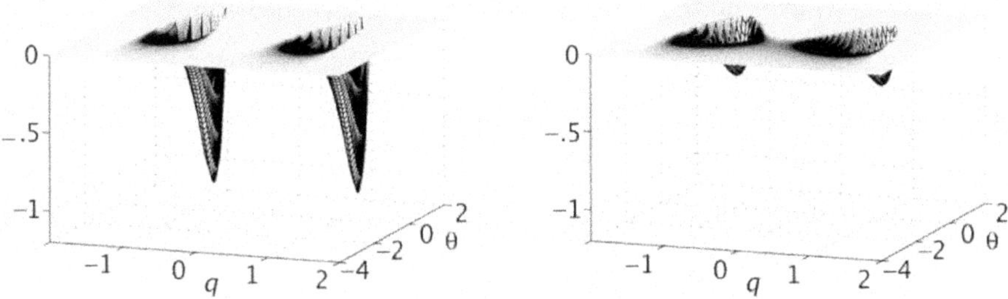

Figure 11.4 "Isothermal" potential surface corresponding to the double well potential at small (left) and large (right) values of r.

excursions in the negative θ direction. With $\theta = 0$ we have $\tilde{V}(\boldsymbol{q}, 0) = V(\boldsymbol{q}) - E$, and

$$\lim_{\theta \to +\infty} \tilde{V}(\boldsymbol{q}, \theta) = +\infty, \qquad \lim_{\theta \to -\infty} \tilde{V}(\boldsymbol{q}, \theta) = 0.$$

The modified potential has a unique global minimum in θ, so the modified potential can be viewed as a "valley" with sides rising in θ.

Let us illustrate this with the example of a double well potential

$$V(q) = q^2(q-1)^2.$$

In the constant energy dynamics, low energy orbits are automatically confined to one or the other of the two energy wells. The isothermal system does not exhibit such a separation, but the transition between wells becomes rare at low temperature. See Fig. 11.4.

11.3.2 Time-reparameterization and the Nosé–Hoover method

The Nosé Hamiltonian generates configurations from the canonical distribution, but it also introduces an unnatural scaling of the intrinsic time. This introduces some practical computational difficulties, since the configurations are not available at equally spaced points in real time. A similar problem is encountered when the separated formulation is used as the basis for computation, as described above. An alternative real-variable reformulation of the equations of motion was proposed by NOSE [146] which incorporates a time transformation. Simplifications to the real-variable system were suggested by HOOVER [87] resulting in the traditional treatment of the Nosé Hamiltonian called the Nosé–Hoover method. Let us briefly review the properties of the Nosé–Hoover system of equations, and discuss some of the common numerical methods.

The Nosé Hamiltonian generates the equations of motion

$$\begin{aligned}
\frac{d\boldsymbol{q}_i}{d\tau} &= \frac{\tilde{\boldsymbol{p}}_i}{m_i s^2},\\
\frac{ds}{d\tau} &= \frac{\pi_s}{Q},\\
\frac{d\tilde{\boldsymbol{p}}_i}{d\tau} &= -\nabla_{\boldsymbol{q}_i} V(\boldsymbol{q}),\\
\frac{d\pi_s}{d\tau} &= \sum_i \frac{\|\tilde{\boldsymbol{p}}_i\|^2}{m_i s^3} - \frac{\beta^{-1} g}{s}.
\end{aligned}$$

The following change of variables can be applied

$$\boldsymbol{p}_i = \frac{\tilde{\boldsymbol{p}}_i}{s} \qquad \pi = \pi_s/s. \tag{11.6}$$

This is followed by a time-transformation, $\frac{d\tau}{dt} = s$ yielding the non-Hamiltonian system in terms of the "real" variables

$$\begin{aligned}
\dot{\boldsymbol{q}}_i &= \frac{\boldsymbol{p}_i}{m_i},\\
\dot{\boldsymbol{p}}_i &= -\nabla_{\boldsymbol{q}_i} V(\boldsymbol{q}) - \boldsymbol{p}_i \frac{s\,\pi}{Q},\\
\dot{s} &= \frac{s^2\,\pi}{Q},\\
\dot{\pi} &= \frac{1}{s}\left(\sum_i \frac{\|\boldsymbol{p}\|_i^2}{m_i} - \beta^{-1} g\right) - \frac{s\,\pi^2}{Q}.
\end{aligned}$$

Introducing $\xi = s\,\pi/Q$ and replacing $\ln s$ by η, one not only eliminates the variable π, but also decouples the variable s from the system; this yields the *Nosé–Hoover* formulation:

$$\dot{\boldsymbol{q}}_i = \frac{\boldsymbol{p}_i}{m_i}, \tag{11.7}$$

$$\dot{\boldsymbol{p}}_i = -\nabla_{\boldsymbol{q}_i} V(\boldsymbol{q}) - \boldsymbol{p}_i\,\xi, \tag{11.8}$$

$$\dot{\eta} = \xi, \tag{11.9}$$

$$\dot{\xi} = \frac{1}{Q}\left(\sum_i \frac{\boldsymbol{p}_i^2}{m_i} - \beta^{-1} g\right). \tag{11.10}$$

Canonical sampling in Nosé–Hoover can be shown to require $g = N_f$ (the number of degrees of freedom of the real system). This reduction in the degrees of freedom is needed to recover configurations at the correct temperature. Although this system is not Hamiltonian, it does have a conserved quantity, the *total extended energy*

$$E_{ext} = \sum_i \frac{\|\boldsymbol{p}\|_i^2}{2\, m_i} + V(\boldsymbol{q}) + \frac{Q\,\xi^2}{2} + g\beta^{-1}\eta. \tag{11.11}$$

This system is time reversible, and it is advisable to solve the equations of motion with a reversible integrator. A number of schemes are available, for example: [66, 126, 94].

11.4 A symplectic approach: the Nosé–Poincaré method

We introduce two different kinds of error when calculating an ensemble average using a molecular dynamics sampling technique. First, every timestepping scheme introduces some error at each timestep, and these accumulate in a complicated way. Second, the calculation of stochastic quantities from simulation is generally a slow-converging process, so very large numbers of timesteps will often be needed to deliver good estimates. Indeed, in systems with a large number of local minima, we may need to greatly restrict the space over which our trajectories sample in order to achieve any sort of useful estimates at all.

In the most general setting, there is also not much we can say about the effect of the numerical error on the ensemble calculation, other than that if we use an off-the-shelf integrator, we expect the error to grow fairly rapidly with time. Methods which do not stay near the energy surface very soon lead to unrealistic averages. On the other hand, is it enough for the integrator to preserve energy approximately in order for us to trust the averages? The answer is of course, no: in the simplest case, we could think of using any arbitrary method projected on to the energy surface. We could apply this scheme in conjunction with any arbitrary method to obtain perfect energy conservation, but it would not generally be effective. For example if the method introduced dissipation, causing trajectories to spiral towards equilibria, we would expect to see similar artifacts in the projected method's dynamics.

If we use a symplectic method, we usually expect good energy conservation, but something else comes into play. There is also an associated perturbed statistical mechanics – a consequence of the Hamiltonian structure of the modified equations! Thus a symplectic scheme not only stays near the modified energy surface $\tilde{H}_{\Delta t}$, it allows us the possibility of computing averages in a modified microcanonical ensemble, for example $\delta\left[\tilde{H}_{\Delta t} - E\right]$. Does this idea carry over to Nosé dynamics? If we solve the Nosé system at constant energy using a symplectic method, then what we are actually able to compute (assuming ergodicity) is a sampling with respect to a modified microcanonical ensemble of the form

$$\rho_{\text{NVE}}^{\tilde{\mathcal{H}}_{\Delta t}} = \delta\left[\mathcal{H}_{\text{Nosé}} + \Delta t^2 \Delta\mathcal{H}_2 + \ldots - E\right].$$

In order to move things forward we must show how to complete each of the steps described in the proof of Nosé dynamics of integrating out the variables s and π_s on which the above density depends, resulting in a modified canonical distribution. This introduces some new challenges, since all transformations involved must be global, and it is difficult to say anything about the perturbative expansion far from the trajectory of interest. However if we are willing to assume that trajectories of our system stay uniformly bounded for all time, then it is possible to invoke the implicit function theorem, which allows us to compute the integral over s as in the proof of Nosé's theorem, and using this we can show that the effective combination of Nosé dynamics and symplectic discretization is real variable sampling of a modified canonical ensemble of the form $\exp(-\beta\tilde{H}_{\Delta t})$. This viewpoint is addressed in [25]. This perspective offers concrete motivation for using a symplectic method, since for bounded systems, at least, the symplectic property is crucial for establishing effective sampling.

Note that the separated formulation makes possible a very simple symplectic discretization of Nosé dynamics based on Störmer–Verlet, see [110]. However, although the method of implementation is trivial, the computation of autocorrelation functions typically requires data equally spaced in time, and since the separated form works in an artificial timescale, this requires some interpolation. It simplifies things to work with a scheme that is formulated in real time.

In this section we will outline a procedure for scaling time while preserving the Hamiltonian structure. The method used here was first given in [26]. We have already seen in the separable formulation (and earlier in the context of variable stepsize) that a time transformation can be implemented as a rescaling of the Hamiltonian of the system, shifted to zero

$$\tilde{\mathcal{H}} = f(q,p)(H - E), \tag{11.12}$$

where $f > 0$ is a "time scaling" function, and the constant E is the initial value of H along a trajectory of interest. With $H = E$, the dynamics of the transformed system will be equivalent to those of the original system, up to a transformation of time $dt/dt' = f$.

In order to correct the Nosé timescale, the appropriate Poincaré transformation is $dt/dt' = s$, applied to a slightly modified version of the Nosé extended Hamiltonian in (11.4)

$$\tilde{\mathcal{H}} = \left(\sum_i \frac{\|\tilde{\boldsymbol{p}}_i\|^2}{2m_i s^2} + V(\boldsymbol{q}) + \frac{\pi_s^2}{2Q} + g\beta^{-1}\ln s - E\right)s. \tag{11.13}$$

The modification comes in that we must take the constant $g = N_f$ (as opposed to $\tilde{g} = N_f + 1$), as in Nosé–Hoover. The constant E is chosen to be the initial value of the Nosé Hamiltonian, $\mathcal{H}_{\text{Nosé}}$. It is not difficult to show that this transformed Hamiltonian (11.13), which we call Nosé–Poincaré, generates configurations from the canonical distribution in the variables $\boldsymbol{q}$ and $\tilde{\boldsymbol{p}}/s$, see [26] for details.

The disadvantage of the general Poincaré transformation in (11.12) is that it mixes the variables so that a straightforward explicit symplectic treatment of the extended Hamiltonian is not, in general, possible. However, the fact that the scaling function f depends only on s means that the discrete equations can often be decoupled, and we can easily formulate semi-explicit symplectic methods.

Returning to the Nosé–Poincaré Hamiltonian, $\tilde{\mathcal{H}}$, we write the equations of motion

$$\frac{d}{dt}\boldsymbol{q}_i = \frac{\tilde{\boldsymbol{p}}_i}{m_i\, s},$$
$$\frac{d}{dt}\tilde{\boldsymbol{p}}_i = -s\nabla_{\boldsymbol{q}_i} V(\boldsymbol{q}),$$
$$\frac{ds}{dt} = s\frac{\pi_s}{Q},$$
$$\frac{d\pi_s}{dt} = \sum_i \frac{\|\tilde{\boldsymbol{p}}_i\|^2}{m_i\, s^2} - g\beta^{-1} - \Delta\mathcal{H}\left(\boldsymbol{q}, \tilde{\boldsymbol{p}}, s, \pi_s\right),$$

where

$$\Delta\mathcal{H}\left(\boldsymbol{q}, \tilde{\boldsymbol{p}}, s, \pi_s\right) = \sum_i \frac{\|\tilde{\boldsymbol{p}}_i\|^2}{2\, m_i\, s^2} + V(\boldsymbol{q}) + \frac{\pi_s^2}{2Q} + g\beta^{-1}\ln s - E. \tag{11.14}$$

The value of E is chosen such that $\Delta\mathcal{H}\left(\boldsymbol{q}(0), \tilde{\boldsymbol{p}}(0), s(0), \pi_s(0)\right) = 0$. We apply to this the generalized leapfrog algorithm (see Chapter 4). Since we are treating a time-reversible Hamiltonian system, the resulting method is symplectic and time reversible

$$\tilde{\boldsymbol{p}}_i^{n+1/2} = \tilde{\boldsymbol{p}}_i^n - \frac{\Delta t}{2} s^n \nabla_{\boldsymbol{q}_i} V(\boldsymbol{q}^n), \tag{11.15}$$

$$\pi_s^{n+1/2} = \pi_s^n + \frac{\Delta t}{2}\left(\sum_i \frac{1}{m_i}\left(\frac{\|\tilde{\boldsymbol{p}}_i^{n+1/2}\|}{s^n}\right)^2 - g\beta^{-1}\right)$$
$$-\frac{\Delta t}{2}\Delta\mathcal{H}\left(\boldsymbol{q}^n, \tilde{\boldsymbol{p}}^{n+1/2}, s^n, \pi_s^{n+1/2}\right), \tag{11.16}$$

$$s^{n+1} = s^n + \frac{\Delta t}{2}\left(s^{n+1} + s^n\right)\frac{\pi_s^{n+1/2}}{Q}, \tag{11.17}$$

$$\boldsymbol{q}_i^{n+1} = \boldsymbol{q}_i^n + \frac{\Delta t}{2}\left(\frac{1}{s^{n+1}} + \frac{1}{s^n}\right)\frac{\tilde{\boldsymbol{p}}_i^{n+1/2}}{m_i}, \tag{11.18}$$

$$\pi_s^{n+1} = \pi_s^{n+1/2} + \frac{\Delta t}{2}\left(\sum_i \frac{1}{m_i}\left(\frac{\|\tilde{\boldsymbol{p}}_i^{n+1/2}\|}{s^{n+1}}\right)^2 - g\beta^{-1}\right)$$
$$-\frac{\Delta t}{2}\Delta\mathcal{H}\left(\boldsymbol{q}^{n+1}, \tilde{\boldsymbol{p}}^{n+1/2}, s^{n+1}, \pi_s^{n+1/2}\right), \quad (11.19)$$
$$\tilde{\boldsymbol{p}}_i^{n+1} = \tilde{\boldsymbol{p}}_i^{n+1/2} - \frac{\Delta t}{2} s^{n+1}\nabla_{\boldsymbol{q}_i} V(\boldsymbol{q}^{n+1}). \quad (11.20)$$

Solving (11.16) requires us to find the roots of a quadratic equation for $\pi_s^{n+1/2}$

$$\frac{\Delta t}{4Q}\left(\pi_s^{n+1/2}\right)^2 + \pi_s^{n+1/2} + \delta = 0, \quad (11.21)$$

where

$$\delta = \frac{\Delta t}{2}\left(g\beta^{-1}\left(1 + \ln s^n\right) - \sum_i \frac{\left(\tilde{\boldsymbol{p}}_i^{n+1/2}\right)^2}{2m_i\left(s^n\right)^2} + V(\boldsymbol{q}^n) - E\right) - \pi_s^n.$$

To avoid subtractive cancellation, we use the following quadratic formula to solve (11.21)

$$\pi_s^{n+1/2} = \frac{-2\delta}{1 + \sqrt{1 - \delta\Delta t/Q}}. \quad (11.22)$$

The remaining steps of the method are explicit.

11.4.1 Generalized baths

In [102], an alternative to Nosé dynamics is proposed based on a more powerful family of extended Hamiltonians. By examination of the proof that Nosé dynamics recovers the canonical ensemble, we can observe that also the following Hamiltonian would have this property

$$\mathcal{H}_{GN} = H(\boldsymbol{q}, \tilde{\boldsymbol{p}}/s) + g\beta^{-1}\ln s + G(\pi_s, \sigma_1, \sigma_2, \ldots, \sigma_k, \pi_1, \pi_2, \ldots, \pi_k), \quad (11.23)$$

Here new configuration variables $\sigma_1, \sigma_2, \ldots, \sigma_k$ and the corresponding momenta $\pi_1, \pi_2, \ldots, \pi_k$ represent augmenting variables designed to (i) increase the rate of convergence to the canonical ensemble, or (ii) enable the bath to resonate with several components of the Hamiltonian system. It is possible to show that, under certain assumptions, the constant energy trajectories of the generalized bath Hamiltonian sample the canonical ensemble. The precise form of the term G will be important both for the dynamical behavior and for the issues of numerical integrator design. However, there are some limitations on the choice of G, as illustrated in the following examples.

Example 2 Nosé–Hoover chains are often used to enhance ergodicity. The idea is to introduce additional thermostatting variables which successively "thermostat the thermostats." It is natural to look for a Hamiltonian formulation of Nosé–Hoover chains, such as

$$\mathcal{H}_{\text{chain}} = H(\boldsymbol{q}, \tilde{\boldsymbol{p}}/s) + \frac{\pi_s^2}{2s_1^2 Q} + \frac{\pi_1^2}{2s_2^2 Q_1} + \ldots + \frac{\pi_{r-1}^2}{2s_r^2 Q_{r-1}} + \frac{\pi_r^2}{2Q_r}$$
$$+g\beta^{-1}\ln s + g_1\beta^{-1}\ln s_1 + \ldots g_r\beta^{-1}\ln s_r, \tag{11.24}$$

which is clearly in the form (11.23).

Take the case $r = 1$, in which case the microcanonical partition function is

$$\mathcal{P} = \int \delta\left[\mathcal{H}_{\text{chain}} - E\right] d\tilde{V}\, ds\, ds_1\, d\pi_s\, d\pi_1$$
$$= \int \delta\left[H(\boldsymbol{q}, \tilde{\boldsymbol{p}}/s) + \frac{\pi_s^2}{2s_1^2 Q} + \frac{\pi_1^2}{2Q_1} + g\beta^{-1}\ln s + g_1\beta^{-1}\ln s_1 - E\right] d\tilde{V}\, ds\, ds_1\, d\pi_s\, d\pi_1,$$

where $d\tilde{V} = d^{N_f}\boldsymbol{q}\, d^{N_f}\tilde{\boldsymbol{p}}$ is the volume form in phase space. We attempt now to integrate out with respect to s_1. Set $\pi_s/s_1 = \bar{\pi}$, so $d\pi = s_1 d\bar{\pi}$ and

$$\mathcal{P} = \int \delta\left[H(\boldsymbol{q}, \tilde{\boldsymbol{p}}/s) + \frac{\bar{\pi}^2}{2Q_0} + \frac{\pi_1^2}{2Q_1} + g\beta^{-1}\ln s + g_1\beta^{-1}\ln s_1 - E\right] s_1\, d\tilde{V}\, ds\, ds_1\, d\bar{\pi}\, d\pi_1.$$

This yields, for $g_1 = 2$

$$\mathcal{P} = \int \exp\left(-\beta\left[H(\boldsymbol{q}, \tilde{\boldsymbol{p}}/s) + \frac{\bar{\pi}^2}{2Q} + \frac{\pi_1^2}{2Q_1} + g\beta^{-1}\ln s - E\right]\right) d\tilde{V}\, ds\, d\bar{\pi}\, d\pi_1,$$

but now following this by $\tilde{\boldsymbol{p}}/s = \boldsymbol{p}$, we obtain

$$\mathcal{P} = \int \exp\left(-\beta\left[H(\boldsymbol{q}, \boldsymbol{p}) + \frac{\bar{\pi}^2}{2Q} + \frac{\pi_1^2}{2Q_1} + g\beta^{-1}\ln s - E\right]\right) s^{N_f}\, dV\, ds\, d\bar{\pi}\, d\pi_1,$$

$dV = d^{N_f}\boldsymbol{q}\, d^{N_f}\boldsymbol{p}$, if we then choose $g = N_f$, it appears in the exponential that $\exp(-g\ln s) = s^{-N_f}$, but we must still perform the integration with respect to s which is unbounded! □

In order to correct this problem there are several approaches one can take, including modification of the chain Hamiltonian in order to bound the s-integral. On the other hand we can also try completely different kinds of couplings, such as the following from [102]:

Example 3

$$\mathcal{H}_{\text{vertex}} = \frac{\tilde{\boldsymbol{p}}^T \boldsymbol{M}^{-1}\tilde{\boldsymbol{p}}}{2s^2} + V(\boldsymbol{q}) + g\beta^{-1}\ln s + \frac{(1 + \sum \sigma_i^2)\pi_s^2}{2Q_s} + \sum_i \frac{\sigma_i^2}{2} + \sum_{i=1,m} \frac{\pi_i^2}{2Q_i}. \tag{11.25}$$

The σ_i are coupled to the original variables through π_s (a common vertex). To this generalized Hamiltonian we add a time transformation of Poincaré type in the same manner as for Nosé dynamics, resulting in

$$\tilde{\mathcal{H}}_{\text{vertex}}=s\left[\frac{\tilde{\boldsymbol{p}}^T\boldsymbol{M}^{-1}\tilde{\boldsymbol{p}}}{2s^2}+V(\boldsymbol{q})+g\beta^{-1}\ln s+\frac{(1+\sum\sigma_i^2)\pi_s^2}{2Q_s}+\sum_i\frac{\sigma_i^2}{2}+\sum_{i=1,m}\frac{\pi_i^2}{2Q_i}-E\right],$$

$E=\mathcal{H}_{\text{vortex}}(0)$. In this case a splitting method can be generated from the composition of the flows on

$$H_1=-s\left[\frac{(1+\sum\sigma_i^2)\pi_s^2}{2Q_s}+\frac{1}{2}\sum_i\sigma_i^2+g\beta^{-1}\ln s\right], \tag{11.26}$$

$$H_2=s\left[\frac{\tilde{\boldsymbol{p}}^T\boldsymbol{M}^{-1}\tilde{\boldsymbol{p}}}{2s^2}+\sum\frac{\pi_i^2}{2Q_i}\right], \tag{11.27}$$

$$H_3=s[V(\boldsymbol{q})-E]. \tag{11.28}$$

This splitting is very similar to one proposed by Nosé for the original Nosé–Poincaré method [147]. The flows for H_2 and H_3 are easily computed. For H_1 we propose the use of a symplectic method such as the generalized Leapfrog scheme. □

Example 4 A well-known system with reasonably good ergodicity properties is the three-ball billiard system with three hard-sphere particles in a square box. While we could use a hard-sphere bath [88], this would introduce some complications. Instead, we suggest to use a simplified bath consisting of three "soft" spheres (with pair potentials of the form r^{-p}) constrained to a small region via a harmonic tether to the origin. We can use this system as a generalized bath. Define

$$\mathcal{H}_{\text{billiard}}=\mathcal{H}_{\text{Nosé}}(\tilde{\boldsymbol{p}},\boldsymbol{q},s,\pi_s)+\mathcal{H}_{\text{bath}}(\{\boldsymbol{\sigma}_i,\boldsymbol{\pi}_i\})+\mathcal{H}_{\text{int}}(\pi_s,\{\boldsymbol{\sigma}_i,\boldsymbol{\pi}_i\}),$$

where

$$\mathcal{H}_{\text{bath}}=\sum_{i=1}^{3}\frac{\|\boldsymbol{\pi}_i\|^2}{2Q_i}+\sum_{i=1}^{3}\frac{\|\boldsymbol{\sigma}_i\|^2}{2}+\sum_i\sum_{j>i}\|\boldsymbol{\sigma}_i-\boldsymbol{\sigma}_j\|^{-12}, \tag{11.29}$$

and

$$\mathcal{H}_{\text{int}}=\left(\sum_{i=1}^{3}\|\boldsymbol{\sigma}_i\|^2\right)\frac{\pi_s^2}{2Q_s}. \tag{11.30}$$

The bath positions and momenta $\boldsymbol{\sigma}_i$ and $\boldsymbol{\pi}_i$ are vectors in $\mathbb{R}^3$. The Nosé–Poincaré equations for this system can be integrated using a Hamiltonian splitting like that used in the vertex coupling. This is left as an exercise.

Experiments with both of these generalized baths are discussed in [102] and, at the time of this writing, research is on-going into design of optimal baths. □

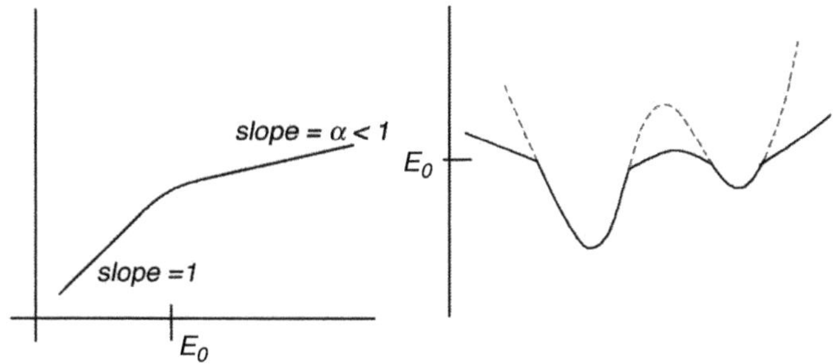

Figure 11.5 A simple energy rescaling function (left) and effect on energy of a double well model (right).

The development of flexible é-like schemes with good ergodicity properties is a topic of current. The interested reader is referred to the articles by LEIMKUHLER and SWEET [113a, b].

11.4.2 Simulation in other ensembles

It is interesting to consider alternatives to computation in the microcanonical or canonical ensembles. One example is the constant temperature and pressure (NPT) ensemble. Schemes for NPT calculations can be derived which are very similar to the methods discussed above for NVT simulation. For example, a variant of Nosé–Poincaré has been adapted for NPT simulation [182].

The development of non-standard distributions is often seen as a way to accelerate the configurational sampling of complex molecular systems having highly corrugated potential landscapes, for which the time scale of standard molecular-dynamics simulations is insufficient for convergence of statistical averages. This is especially true of macromolecules, biomolecules, and amorphous materials. Over past decades, numerous schemes have been proposed for enhancing sampling efficiency, usually based on the systematic deformation of the potential (or total) energy surface to accelerate barrier crossings, either by lowering the barriers or raising the potential valleys. From a statistical mechanical perspective, such energetic modifications induce a corresponding alteration of the phase-space distribution by enhancing the statistical weight of configurations which otherwise would be rarely visited. If we have detailed knowledge of the modified sampling distribution, it is possible to "reweight" the computed trajectories to obtain averages in the original ensemble.

One of the simplest smoothing consists of introducing an energy scaling at high energy. We consider a function $\eta(E)$, as depicted in Fig. 11.5, and define the "smoothed energy" by

$$H' = \eta(H)H.$$

Canonical sampling with respect to this energy implies an ensemble density of the form

$$\rho_{\text{canonical}}(H') = \exp(-\beta H') = \exp(-\beta\eta(H)H) = \rho'(H),$$

which can be viewed as a noncanonical sampling with respect to the original energy. Now suppose that we have computed a sampling trajectory for this ensemble using Nosé dynamics, then

$$\lim \frac{1}{T}\int_0^T f(\boldsymbol{q}(t), \tilde{\boldsymbol{p}}(t)/s(t))\, dt \propto \int f(\boldsymbol{q}, \boldsymbol{p})\, \rho'(H)\, dV.$$

Of course this is not what we want! What we would like is to be able to recover the correct sampling with respect to the canonical ensemble. However, if we define

$$\Delta H = H - H',$$

then

$$\lim \frac{1}{T}\int_0^T \exp[-\beta\Delta H(\boldsymbol{q}(t), \tilde{\boldsymbol{p}}(t)/s(t))]\, f(\boldsymbol{q}(t),$$
$$\tilde{\boldsymbol{p}}(t)/s(t))\, dt \propto \int f(\boldsymbol{q}, \boldsymbol{p})\rho_{\text{can}}(H)\, dV.$$

This is what we mean by ensemble "reweighting."

Alternatively, if the density is given, expressed as a function of the phase variables, one can typically derive both extended Hamiltonians and numerical methods from a reformulation of the problem to one couched in the canonical ensemble. If $\rho(\boldsymbol{q}, \boldsymbol{p})$ is the given (positive) ensemble density, we solve

$$\rho = \exp(-\beta\tilde{H})$$

for $\tilde{H}$, where β is essentially a free parameter. Averages with respect to Hamiltonian H in the ρ ensemble are now recovered directly by computing canonical averages of the system with Hamiltonian $\tilde{H}$. However, in this work, care is needed to develop sensible geometric integrators. The Nosé Hamiltonian for Generalized Density Dynamics (GDD) is

$$H^{\rho}_{\text{Nosé}} = -\frac{1}{\beta}\ln\rho(\boldsymbol{q}, \tilde{\boldsymbol{p}}/s) + \frac{\pi^2}{2Q} + g\beta^{-1}\ln s. \tag{11.31}$$

After Poincaré transformation, this results in the equations of motion

$$\begin{aligned}
\dot{\boldsymbol{q}} &= -\frac{\beta^{-1}}{\rho(\boldsymbol{q}, \tilde{\boldsymbol{p}}/s)} \nabla_{\tilde{\boldsymbol{p}}/s} \rho(\tilde{\boldsymbol{p}}/s, \boldsymbol{q}), \\
\dot{\tilde{\boldsymbol{p}}} &= \frac{\beta^{-1}}{\rho(\boldsymbol{q}, \tilde{\boldsymbol{p}}/s)} \nabla_{\tilde{q}} \rho(\boldsymbol{q}, \tilde{\boldsymbol{p}}/s), \\
\dot{s} &= \frac{s\pi_s}{Q}, \\
\dot{\pi}_s &= \frac{\beta^{-1}\tilde{p}}{s\rho(\boldsymbol{q}, \tilde{\boldsymbol{p}}/s)} \nabla_{\tilde{p}} \rho(\boldsymbol{q}, \tilde{\boldsymbol{p}}/s) - g\beta^{-1} - \Delta H^{\rho}_{\text{Nosé}} \, .
\end{aligned}$$

A separable Hamiltonian H will not result in a separable $\tilde{H}$, and the result is that a straightforward treatment based on applying Nosép–Poincaré to simulate $\tilde{H}$ would have to be fully implicit. In [14] this difficulty was analyzed in some special cases, for example where $\rho = \rho_A(\boldsymbol{q})\rho_B(\boldsymbol{p})$ and for $\rho = \rho(H)$. In both of these cases, it turns out to be possible to derive semi-explicit second-order symplectic schemes which require only a single force evaluation per timestep.

11.5 Exercises

1. *Smooth potential cutoff.* Often, one replaces the Lennard–Jones potential by a *cut-off* version whose derivative has compact support. This should be done so that the potential remains at least smooth (continuously differentiable), for example

$$\phi_{\text{LJ}}(r) \approx \hat{\phi}_{\text{LJ}}(r) = \begin{cases} \phi_{\text{LJ}}(r), & r < R_1 \\ P(r), & R_1 \le r < R_2, \\ 0, & r \ge R_2 \end{cases}$$

   where $P(r)$ is an interpolating polynomial (Fig. 11.6). A C^m-approximation (meaning that the approximation can be differentiated m times) can be obtained by introducing a higher-order Hermite interpolant $P(r) = A_0 + A_1 r + \ldots + A_{2m-1} r^{2m-1}$, and determining the $2m$ coefficients to satisfy the $2m$ conditions $P^{(k)}(R_1) = \phi^{(k)}(R_1)$, $P^{(k)}(R_2) = 0$, $k = 0, 1, 2, \ldots, m-1$. Find a C^2-cutoff of the Lennard–Jones potential.

2. *SHAKE discretization of Methane.* Each molecule of Methane has five atoms (a carbon and four hydrogens) (see Fig. 11.7), arranged at the vertices and center of a regular tetrahedron. Let us order the atoms as follows: the central

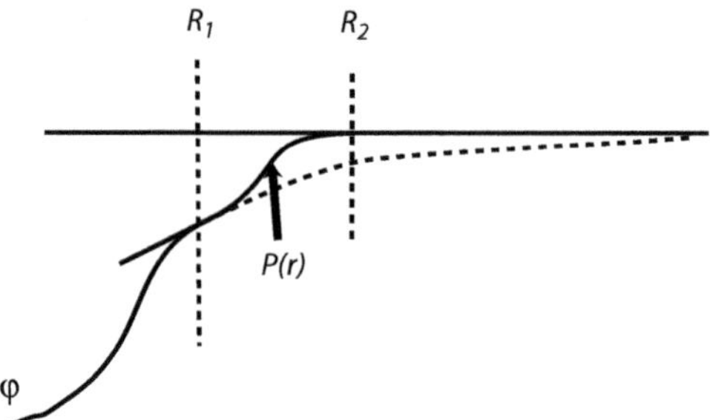

Figure 11.6 Smooth potential cutoff by Hermite interpolation.

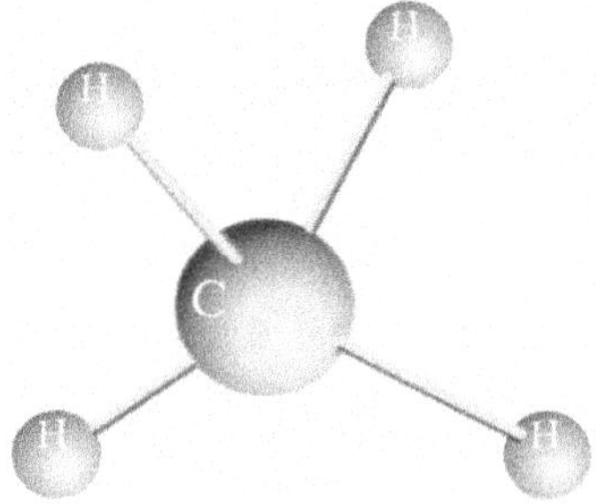

Figure 11.7 Methane.

carbon of the first methane molecule, first hydrogen of first methane, second hydrogen of first methane, . . . , fourth hydrogen of first methane, central carbon of second methane, first hydrogen of second methane, etc., indexing them from 1. Just considering the equations for the first methane we have constrained equations of the form

$$\begin{aligned}
\dot{\boldsymbol{q}}_1 &= m_{\mathrm{C}}^{-1}\boldsymbol{p}_1, \\
\dot{\boldsymbol{q}}_k &= m_{\mathrm{H}}^{-1}\boldsymbol{p}_k, \qquad\qquad k = 2,3,4,5, \\
\dot{\boldsymbol{p}}_1 &= \boldsymbol{F}_{\mathrm{C}} - \sum_{k=1}^{4}\lambda_k(\boldsymbol{q}_1 - \boldsymbol{q}_{k+1}), \\
\dot{\boldsymbol{p}}_k &= \boldsymbol{F}_{\mathrm{H}[k-1]} - \lambda_{k-1}(\boldsymbol{q}_k - \boldsymbol{q}_1), \qquad\qquad k = 2,3,4,5, \\
\frac{1}{2}\|\boldsymbol{q}_k - \boldsymbol{q}_1\|^2 &= \frac{1}{2}L_{\mathrm{CH}}^2.
\end{aligned}$$

Here m_C, m_H represent the masses of the carbon and hydrogen atoms, respectively, $\boldsymbol{F}_{\mathrm{C}}$ and $\boldsymbol{F}_{\mathrm{H}[k]}$, $k = 1,2,3,4$ are the forces acting on the carbon and each hydrogen atom due both to interactions with the other atoms of the first methane as well as to intermolecular interactions, and L_{CH} is the natural length of this bond. The shape of the molecule is maintained by angle potentials between the hydrogens and the central carbon atom. After application of SHAKE and simplification, we find that we are left with a relatively simple system of four quadratic equations in $\lambda_1, \ldots, \lambda_4$. Since the equations look the same for each Methane, the SHAKE equations will only involve the Lagrange multipliers of the local group, so in a system of s methanes, we will face only s decoupled four-dimensional nonlinear subsystems, each of which can be solved by an iterative method.

Determine the coupled quadratic equations that must be solved at each step of SHAKE discretization.

3. *Generalized Nosé dynamics.* State and prove conditions under which canonical sampling can be recovered from (11.23).

4. *Generalized Nosé dynamics.* Write a discretization scheme for the soft-particle ("billiard") bath model in Example 4.

12

Hamiltonian PDEs

Many physical processes of interest not only evolve continuously in time but also possess a continuous spatial structure and, hence, can be described by partial differential equations (PDEs). Furthermore, many fundamental laws of physics, such as quantum mechanics, electrodynamics, ideal continuum mechanics, can be formulated within an extension of the Hamiltonian framework discussed so far to PDEs. In this chapter we focus on two particular examples of such Hamiltonian PDEs and discuss a number of numerical discretization techniques. The reader should, however, keep in mind that the solution behavior of PDEs is much more complex than that of ODEs and that the choice of an appropriate discretization will depend very much on the anticipated type of solutions. The techniques described in this chapter are very much restricted to *smooth* solutions such as solitons [53, 201] and balanced geophysical flows [169]. This excludes, in particular, the consideration of shocks [201]. A general introduction to numerical methods for PDEs can be found, for example, in [140].

12.1 Examples of Hamiltonian PDEs

12.1.1 The nonlinear wave equation

Let us consider the nonlinear wave equation

$$u_{tt} = \partial_x \sigma'(u_x) - f'(u), \qquad u = u(x,t), \tag{12.1}$$

where σ and f are smooth functions. If $\sigma(u_x) = u_x^2/2$, then the semi-linear wave equation

$$u_{tt} = u_{xx} - f'(u)$$

is obtained. Other choices for $\sigma(u_x)$ lead to idealized one-dimensional models for fluids and materials.

Throughout this chapter, solutions $u = u(x,t)$ of (12.1) are assumed to be smooth in the independent variables x and t and we impose periodic boundary conditions $u(x,t) = u(x+L,t)$, $L > 0$.

We introduce the total energy $\mathcal{E}[u]$ by

$$\mathcal{E}[u] = \int_0^L \left[\frac{1}{2}u_t^2 + \sigma(u_x) + f(u)\right] dx,$$

and observe that, using integration by parts,

$$\begin{aligned}\frac{d}{dt}\mathcal{E}[u] &= \int_0^L \left[u_t u_{tt} + \sigma'(u_x)u_{xt} + f'(u)u_t\right] dx \\ &= \int_0^L u_t \left[u_{tt} - \partial_x\sigma'(u_x) + f'(u)\right] dx.\end{aligned}$$

However, the term in brackets is equal to zero along solutions of (12.1) and, hence, the total energy $\mathcal{E}[u]$ is conserved.

Let us denote the space of smooth and L-periodic functions in x by $S = C^\infty[0, L]$. Our assumption then is that $u(., t) \in S$ for $t \geq 0$ or in short hand $u(t) \in S$. This smoothness assumption explicitly excludes the consideration of shock-type solutions [201].

Furthermore, upon rewriting (12.1) as

$$\begin{aligned}u_t &= v, \\ v_t &= \partial_x\sigma'(u_x) - f'(u),\end{aligned}$$

the wave equation can, formally, be viewed as a Hamiltonian system with phase space $(u, v)^T \in S \times S$, symplectic form,

$$\bar{\omega} = \int_0^L du \wedge dv\, dx, \tag{12.2}$$

and Hamiltonian functional

$$\mathcal{H}[u, v] = \int_0^L \left[\frac{1}{2}v^2 + \sigma(u_x) + f(u)\right] dx. \tag{12.3}$$

The Hamiltonian equations of motion are derived in the following way. First, the gradient ∇_z of classical mechanics is replaced by the "variational" gradient $\boldsymbol{\delta}_z = (\delta_u, \delta_v)^T$, $\mathbf{z} = (u, v)^T \in S \times S$. The variational derivative $\delta_u\mathcal{G}$ of a functional $\mathcal{G}[u]$ is defined by

$$\int_0^L (\delta_u\mathcal{G}[u]\,\delta u)\, dx = \lim_{\varepsilon\to 0} \frac{\mathcal{G}[u + \varepsilon\delta u] - \mathcal{G}[u]}{\varepsilon},$$

for any $\delta u \in S$. Let us demonstrate this for $\delta_u\mathcal{H}[u, v]$ which is equivalent to the variational derivative of $\mathcal{G}[u] = \int_0^L [\sigma(u_x) + f(u)]\,dx$

$$\int_0^L (\delta_u \mathcal{G}[u]\,\delta u)\,dx = \lim_{\varepsilon\to 0}\frac{1}{\varepsilon}\left\{\int_0^L [\sigma([u+\varepsilon\delta u]_x) + f(u+\varepsilon\delta u)]\,dx - \int_0^L [\sigma(u_x)+f(u)]\,dx\right\}$$
$$= \int_0^L [\sigma'(u_x)(\delta u)_x + f'(u)\delta u]\,dx$$
$$= \int_0^L [-\partial_x\sigma'(u_x)\delta u + f'(u)\delta u]\,dx$$
$$= \int_0^L [-\partial_x\sigma'(u_x) + f'(u)]\,\delta u\,dx.$$

Comparison of the left- and right-hand side yields

$$\delta_u\mathcal{H}[u,v] = \delta_u\mathcal{G}[u] = -\partial_x\sigma'(u_x) + f'(u).$$

One also obtains $\delta_v\mathcal{H}[u,v] = v$.

Next we rewrite the symplectic form (12.2) as

$$\bar{\omega} = \frac{1}{2}\int_0^L (\boldsymbol{J}_2^{-1}d\boldsymbol{z})\wedge d\boldsymbol{z}\,dx,$$

with the (local) structure matrix

$$\boldsymbol{J}_2 = \begin{bmatrix} 0 & +1 \\ -1 & 0 \end{bmatrix},$$

and $d\boldsymbol{z} = (du, dv)^T$. Then the wave equation (12.1) becomes equivalent to an abstract Hamiltonian system

$$\boldsymbol{z}_t = \boldsymbol{J}_2\boldsymbol{\delta}_z\mathcal{H}[\boldsymbol{z}]. \tag{12.4}$$

We finally note that the nonappearance of the independent variable x in the functions f and σ implies another conserved functional for the PDE (12.1), namely the total momentum

$$\mathcal{M}[u,v] = \int_0^L v u_x\,dx.$$

Indeed

$$\frac{d}{dt}\mathcal{M} = \int_0^L (v_t u_x + v u_{xt})\,dx$$
$$= \int_0^L (u_x[\partial_x\sigma'(u_x) - f'(u)] + v v_x)\,dx$$

$$
\begin{aligned}
&= \int_0^L [-\sigma(u_x) - f(u) + v^2/2]_x \, dx \\
&= \Big[-\sigma(u_x) - f(u) + v^2/2\Big]_{x=0}^{x=L} \\
&= 0.
\end{aligned}
$$

12.1.2 Soliton solutions

Waves are one of the most important features of fluid dynamics [201]. Particular types of waves are those that travel at a constant speed $c \neq 0$ without changing their shape. These waves are called *traveling waves* or *solitons* [201, 53]. Mathematically a soliton is described by a (smooth) function ϕ such that

$$
u(x, t) = \phi(x - ct).
$$

Let us introduce the new variable $\xi = x - ct$, then $u_x = \phi_\xi$, $u_t = -c\phi_\xi$, etc. Hence, assuming a solitary solution, the wave equation (12.1) gives rise to a second-order ODE,

$$
c^2 \phi_{\xi\xi} = \partial_\xi \sigma'(\phi_\xi) - f'(\phi), \tag{12.5}
$$

in the independent variable ξ. Equation (12.5) is a Euler–Lagrange equation and to obtain the corresponding Hamiltonian formulation we introduce the new dependent variable (conjugate momentum) $\psi = c^2\phi_\xi - \sigma'(\phi_\xi)$. Let us assume that this relation is invertible, i.e., there is a function $g(\psi)$ such that $\phi_\xi = g'(\psi)$. Hence one can rewrite the second-order ODE (12.5) as a conservative system

$$
\psi_\xi = -f'(\phi), \qquad \phi_\xi = g'(\psi), \tag{12.6}
$$

with Hamiltonian

$$
H = g(\psi) + f(\phi).
$$

A solution $\phi(\xi)$ gives rise to a soliton solution if the boundary conditions $\phi_\xi(\pm\infty) = 0$ are satisfied, i.e., $\phi(\xi)$ approaches some constant value as $\xi \to \pm\infty$. In particular, let (ϕ_i, ψ_i) denote the equilibrium solutions of (12.6), then any *homoclinic* or *heteroclinic* solution of (12.6) gives rise to a (not necessarily stable) soliton solution of the nonlinear wave equation (12.1).[1]

[1]A homoclinic solution is a solution connecting an equilibrium point (ϕ_i, ψ_i) with itself, i.e., $\lim_{\xi\to\pm\infty} \phi(\xi) = \phi_i$, and a heteroclinic solution is a solution connecting two different equilibrium points (ϕ_i, ψ_i), (ϕ_j, ψ_j), i.e., $\lim_{\xi\to+\infty} \phi(\xi) = \phi_i$, $\lim_{\xi\to-\infty} \phi(\xi) = \phi_j$.

Example 1 Let us consider the sine-Gordon equation

$$u_{tt} = u_{xx} - \sin(u).$$

Traveling wave solutions $u(x,t) = \phi(x - ct)$ must satisfy the second-order ODE

$$c^2\phi_{\xi\xi} = \phi_{\xi\xi} - \sin(\phi).$$

We introduce the momentum $\psi = c^2\phi_\xi - \phi_\xi$. For $c \neq \pm 1$, this relation can be inverted and we obtain $\phi_\xi = \psi/(c^2-1)$ and $g(\psi) = \frac{1}{2}\psi^2/(c^2-1)$. Hence the Hamiltonian equations are

$$\psi_\xi = -\sin(\phi), \qquad \phi_\xi = \frac{1}{c^2-1}\psi$$

with Hamiltonian

$$H = \frac{\psi^2}{2(c^2-1)} + (1 - \cos\phi).$$

These are the equations of motion for a nonlinear pendulum with mass $m = c^2 - 1$. Hence we first consider the condition $|c| > 1$. The nonlinear pendulum possesses heteroclinic solutions connecting pairs of hyperbolic equilibria $(\phi,\psi) = (k\pi, 0)$ and $(\phi,\psi) = ((k+2)\pi, 0)$ for $k = 1, \pm 3, \pm 5, \ldots$. These heteroclinic connections are easily found as the contour lines of constant energy $H = 0$. Unfortunately, the associated soliton solutions are all unstable [201, 53]. For $|c| < 1$, the equilibrium points $(k\pi, 0)$, $((k+2)\pi, 0)$, $k = 0, \pm 2, \pm 4, \ldots$, become hyperbolic and give rise to stable soliton solutions. An explicit soliton solution with wave speed c, $|c| < 1$, is given by

$$u(x,t) = 4 \arctan \exp\left(\frac{x - ct}{\sqrt{1-c^2}}\right).$$

Because of their special shape, these solutions are called kink solitons. □

12.1.3 The two-dimensional rotating shallow-water equations

Large-scale geophysical flows in the atmosphere and ocean are essentially incompressible and often stratified into nearly two-dimensional layers. Furthermore, the effect of the earth's rotation significantly affects large-scale patterns away from the equator, for example in mid-latitude or near the poles. See ANDREWS [6] and SALMON [169] for an introduction to geophysical fluid dynamics and MORRISON [139] for further details on the Hamiltonian formalism of geophysical fluid dynamics.

A simple one-layer model system is provided by the two-dimensional rotating shallow-water equations (SWEs) [139, 169]

$$u_t + uu_x + vu_y = +fv - c_0^2 h_x, \tag{12.7}$$

$$v_t + uv_x + vv_y = -fu - c_0^2 h_y, \tag{12.8}$$

$$h_t + uh_x + vh_y = -h(u_x + v_y), \tag{12.9}$$

where $\boldsymbol{u} = (u, v)^T \in \mathbb{R}^2$ is the horizontal velocity field, $c_0 = \sqrt{gH}$, g is the gravitational constant, H is the mean layer depth of the fluid, $h > 0$ is the normalized layer depth with mean value scaled equal to one, and $f > 0$ is twice the angular velocity of the rotating fluid. For simplicity, we will consider the SWEs over a double periodic domain of size $L \times L$ and keep $f = f_0$ constant.

We next introduce the material time derivative of a function $w(x, y, t)$

$$\frac{Dw}{Dt} = w_t + uw_x + vw_y = w_t + \boldsymbol{u} \cdot \nabla_{\boldsymbol{x}} w,$$

and rewrite the SWEs (12.7)–(12.9) in the form

$$\frac{D\boldsymbol{u}}{Dt} = f_0 \boldsymbol{J}_2 \boldsymbol{u} - c_0^2 \nabla_{\boldsymbol{x}} \eta, \tag{12.10}$$

$$\frac{Dh}{Dt} = -h \nabla_{\boldsymbol{x}} \cdot \boldsymbol{u}, \tag{12.11}$$

where $\boldsymbol{x} = (x, y)^T$, $h = 1 + \eta$, and

$$\boldsymbol{J}_2 = \begin{bmatrix} 0 & +1 \\ -1 & 0 \end{bmatrix}.$$

Given a function $w(\boldsymbol{x}, t)$, the material time derivative characterizes the change of w along motion of a fluid particle $\boldsymbol{X}(t) = (X(t), Y(t))^T \in \mathbb{R}^2$, which is passively advected under the velocity field $\boldsymbol{u}$, i.e.

$$\frac{D\boldsymbol{X}}{Dt} = \boldsymbol{u}. \tag{12.12}$$

As an example consider absolute vorticity

$$\zeta = v_x - u_y + f_0 = \nabla_{\boldsymbol{x}} \times \boldsymbol{u} + f_0.$$

Using

$$\begin{aligned} \nabla_{\boldsymbol{x}} \times \frac{D\boldsymbol{u}}{Dt} &= \frac{\partial}{\partial x}\frac{D}{Dt} v - \frac{\partial}{\partial y}\frac{D}{Dt} u \\ &= (v_t + uv_x + vv_y)_x - (u_t + uu_x + vu_y)_y \\ &= (v_x - u_y)_t + u(v_x - u_y)_x + v(v_x - u_y)_y + (v_x - u_y)(u_x + v_y) \\ &= \frac{D}{Dt}\zeta + (\zeta - f_0)\nabla_{\boldsymbol{x}} \cdot \boldsymbol{u}, \end{aligned}$$

it is easy to conclude from (12.10) that absolute vorticity satisfies the continuity equation

$$\frac{D\zeta}{Dt} = -\zeta \nabla_{\boldsymbol{x}} \cdot \boldsymbol{u}. \tag{12.13}$$

The ratio of absolute vorticity ζ and layer depth h is called *potential vorticity* (PV) $q = \zeta/h$ [169]. The PV field q is materially conserved since, using (12.11) and (12.13)

$$\frac{Dq}{Dt} = \frac{1}{h}\left\{\frac{D\zeta}{Dt} - q\frac{Dh}{Dt}\right\} = 0.$$

Let us now assume that, for the scales and motion of interest, we find that the velocity field $\boldsymbol{u}$ satisfies

$$\frac{1}{f_0}\frac{D\boldsymbol{u}}{Dt} \approx \mathbf{0}.$$

Then the momentum equation (12.10) reduces to

$$\mathbf{0} \approx J_2\boldsymbol{u} - \frac{c_0^2}{f_0}\nabla_x\eta. \tag{12.14}$$

The right-hand side can be solved for the velocity and one obtains the *geostrophic wind* approximation

$$\boldsymbol{u}^{\rm g} = -\frac{c_0^2}{f_0}J_2\nabla_x\eta. \tag{12.15}$$

We next assume that potential vorticity q is materially advected along the geostrophic wind $\boldsymbol{u}^{\rm g}$, i.e.

$$q_t + \boldsymbol{u}^{\rm g}\cdot\nabla_x q = 0. \tag{12.16}$$

Potential vorticity q, with $\boldsymbol{u}$ replaced by $\boldsymbol{u}^{\rm g}$, can be expressed as

$$q = \frac{\nabla_x \times \boldsymbol{u}^{\rm g} + f_0}{1+\eta} = f_0\frac{L_R^2\nabla_x^2\eta + 1}{1+\eta},$$

where we used $h = 1 + \eta$ and introduced the Rossby deformation radius $L_R = c_0/f_0$. The final step in our approximation is provided by the assumption

$$\frac{L_R^2\nabla_x^2\eta + 1}{1+\eta} \approx L_R^2\nabla_x^2\eta + 1 - \eta,$$

which leads to

$$q/f_0 - 1 \approx L_R^2\nabla_x^2\eta - \eta.$$

This approximation gives rise to the (linear) *PV inversion* relation

$$\eta^{\rm g} = -\left(1 - L_R^2\nabla_x^2\right)^{-1}(q/f_0 - 1). \tag{12.17}$$

Equation (12.15) with η replaced by η^g together with (12.16) provide a closed set of equations called the *quasi-geostrophic approximation* [169]. The quasi-geostrophic approximation is a valid approximation to the full SWEs for a layer depth h evolving on a typical horizontal length-scale $L \geq L_R$ and for velocities $\boldsymbol{u}$ with a typical scale U such that the associated Rossby number satisfies $\mathrm{Ro} = U/(f_0 L) \ll 1$.[2] The ratio of L_R to L is called the Burger number: $\mathrm{Bu} = (L_R/L)^2$.

Instead of solving the reduced quasi-geostrophic equations, one is often interested in numerically approximating the full SWEs (12.10)–(12.11) subject to initial data satisfying (12.14). We will describe a numerical method in Section 12.2.2. The method is based on a Lagrangian formulation[3] of fluid dynamics, which we outline next.

The Lagrangian description of fluid dynamics is a particle-based formulation with the continuum of particles being advected according to (12.12). Hence, the positions of all fluid particles $\boldsymbol{X} = (X, Y)^T$ are given as a time-dependent transformation from a label space $\mathbb{A} \subset \mathbb{R}^2$ to position space $\mathbb{X} \subset \mathbb{R}^2$

$$\boldsymbol{X} = \boldsymbol{X}(\boldsymbol{a}, t), \quad \boldsymbol{a} = (a, b)^T \in \mathbb{A}, \quad \boldsymbol{X} = (X, Y)^T \in \mathbb{X}^2.$$

The labels are fixed for each particle and a natural choice is provided by the particle's initial conditions, i.e. $\boldsymbol{a} = \boldsymbol{X}(\boldsymbol{a}, 0)$, which we assume from now on.

The fluid layer depth h is defined as a function of the determinant of the 2×2 Jacobian matrix

$$\boldsymbol{X}_a = \frac{\partial(X, Y)}{\partial(a, b)}$$

through the relation

$$h(\boldsymbol{X}, t)\,|\boldsymbol{X}_a| = h_o(\boldsymbol{a}), \tag{12.18}$$

where, for the specific labels defined above, $h_o(\boldsymbol{a}) = h(\boldsymbol{a}, 0)$ is the initial layer-depth at $t = 0$. Differentiation of (12.18) with respect to time and using (12.12) yields an expression that is equivalent to the continuity equation (12.11). Hence (12.18) and (12.11) are essentially equivalent statements.

Consider the integral identity

$$h(\boldsymbol{x}, t) = \int h(\boldsymbol{X}, t)\,\delta(\boldsymbol{x} - \boldsymbol{X})\,dX\,dY$$

[2]More precisely, we also have to request that $|\eta| \ll 1$ [169].

[3]The Lagrangian formulation of fluid dynamics is not to be confused with the Lagrangian variational principle. The Lagrangian formulation of fluid dynamics, as opposed to the Eulerian formulation, is based on fluid particles and their velocities as dependent variables, while the Lagrangian variational principle is complementary to the Hamiltonian formalism and provides a way to derive conservative equations of motion.

defining the layer-depth h at time t and Eulerian position $\mathbf{x}$. Here δ denotes the Dirac delta function. Using $dX\,dY = |\mathbf{X_a}|\,da\,db$ and (12.18), we can pull this integral back to label space, arriving at the relation

$$h(\mathbf{x},t) = \int h_o(\mathbf{a})\,\delta(\mathbf{x} - \mathbf{X}(\mathbf{a},t))\,da\,db, \tag{12.19}$$

which can be taken as the definition of the layer depth in a Lagrangian description of fluid mechanics.

The SWEs (12.10)–(12.11) are now reformulated to

$$\frac{\partial \mathbf{u}}{\partial t} = f_0 J_2 \mathbf{u}^\perp - c_0^2 \nabla_{\mathbf{X}} h, \tag{12.20}$$

$$\frac{\partial \mathbf{X}}{\partial t} = \mathbf{u}, \tag{12.21}$$

where h is defined by (12.18) or (12.19). Note that the material time derivative was replaced by the partial derivative with respect to time. This reflects the fact that the material time derivative becomes a partial derivative in a Lagrangian formulation of fluid dynamics, where time t and labels $\mathbf{a} \in \mathbb{A}$ are now the independent variables. Next we introduce the canonical momenta

$$\mathbf{p} = h_o\,\mathbf{u},$$

and the equations (12.20)–(12.21) become canonical with Hamiltonian

$$\mathcal{H} = \frac{1}{2}\int \frac{\mathbf{p}\cdot\mathbf{p}}{h_o}\,dadb + \frac{c_0^2}{2}\int h_o h\,da\,db$$

and symplectic two-form

$$\bar{\omega} := \int \left(\frac{h_o f_0}{2} d\mathbf{X} \wedge J_2 d\mathbf{X} + d\mathbf{p} \wedge d\mathbf{X}\right) da\,db. \tag{12.22}$$

12.1.4 Noncanonical Hamiltonian wave equations

We have already encountered in Chapter 8 the rigid body as an example of a noncanonical Hamiltonian system of the general form

$$\frac{d}{dt}z = J(z)\nabla_z H(z).$$

For PDEs, this generalizes to

$$u_t = \mathcal{J}(u)\delta_u \mathcal{H}.$$

Here $\mathcal{J}(u)$ is a linear (in general, differential) operator, called the Poisson operator, that has to satisfy certain properties similar to those for the matrix $\mathbf{J}(\mathbf{z})$ [139].

A well-known example is provided by the Korteweg–de Vries (KdV) equation [53]

$$u_t + uu_x + u_{xxx} = 0. \tag{12.23}$$

Here the operator $\mathcal{J}$ is equal to

$$\mathcal{J} = -\frac{\partial}{\partial x},$$

and the Hamiltonian functional is given by

$$\mathcal{H} = \int \left[\frac{1}{6}u^3 - \frac{1}{2}u_x^2\right] dx.$$

We can also introduce a noncanonical Poisson bracket

$$\{\mathcal{F}, \mathcal{G}\} = -\int (\delta_u \mathcal{F})\partial_x(\delta_u \mathcal{G})\, dx,$$

and the skew-symmetry of $\{\mathcal{F}, \mathcal{G}\}$ follows upon integration by parts. Similar to the rigid body, the Poisson operator $\mathcal{J}$ is not invertible and this gives rise to the Casimir function

$$C[u] = \int u\, dx.$$

Indeed, it is easy to verify that

$$\frac{d}{dt}C = \{C, \mathcal{H}\} = 0,$$

along solutions of the KdV equation.

Inviscid fluid dynamics leads to Eulerian equations of motion that are also noncanonical. However, compared with KdV, the situation is made more complicated by the fact that the Poisson operator $\mathcal{J}$ is now no longer constant and independent of the dynamical variables. In fact, one can draw an analogy between two-dimensional incompressible fluid dynamics and an infinite-dimensional version of rigid body dynamics. See the review article by Morrison [139] for further details.

12.2 Symplectic discretizations

The basic idea of symplectic discretization methods for Hamiltonian PDEs consists of two steps:

(i) A spatial truncation that reduces the PDE to a system of Hamiltonian ODEs.

(ii) Timestepping of the finite-dimensional Hamiltonian ODE using an appropriate symplectic method.

The crucial new step is the construction of a finite-dimensional ODE model that retains the Hamiltonian character of the given PDE. The most popular approach is based on the introduction of a spatial grid over which the equations of motion can be truncated. Another approach, particularly well suited for Lagrangian fluid dynamics, reduces the PDE to a set of moving particles interacting through an appropriate potential energy function. Both approaches will be described below.

Finally we give a note of warning. Certain noncanonical Hamiltonian PDEs resist a spatial truncation to a finite-dimensional Hamiltonian system. This is true in particular for the Eulerian formulation of inviscid fluid dynamics. The only significant exception is provided by incompressible fluids on a plane with double periodic boundary conditions. See Zeitlin [208] and McLachlan [128] for a numerical implementation.

12.2.1 Grid-based methods

Consider the nonlinear wave equation (12.1). The first step towards a numerical algorithm is to introduce N grid points $x_i = i\,\Delta x$, $\Delta x = L/N$, $i = 1, \ldots, N$, and to approximate functions $u \in S$ by vectors $\boldsymbol{u} = (u_1, u_2, \ldots, u_N)^T \in \mathbb{R}^N$ with $u(x_i) \approx u_i$. We define $u_{i+N} = u_i$, reflecting the fact that periodic boundary conditions are imposed. The new state space is $\boldsymbol{z} = \{\boldsymbol{z}_i\} \in \mathbb{R}^{2N}$, $\boldsymbol{z}_i = (u_i, v_i)^T \in \mathbb{R}^2$. The symplectic form (12.2) is naturally truncated to

$$\bar{\omega}_N = \sum_{i=1}^{N} du_i \wedge dv_i\,\Delta x = \frac{1}{2}\sum_{i=1}^{N} d\boldsymbol{z}_i \wedge \boldsymbol{J}_2^{-1} d\boldsymbol{z}_i\,\Delta x,$$

and the Hamiltonian functional (12.3) is approximated by the sum

$$H = \sum_{i=1}^{N}\left[\frac{1}{2}v_i^2 + \sigma\left(\frac{u_i - u_{i-1}}{\Delta x}\right) + f(u_i)\right]\Delta x.$$

Hence, we obtain the system of Hamiltonian ODEs

$$\frac{d}{dt}\boldsymbol{z}_i = \frac{1}{\Delta x} J_2 \nabla_{\boldsymbol{z}_i} H(\boldsymbol{z}), \qquad i = 1, \ldots, N.$$

Note that

$$\lim_{\Delta x \to 0} \frac{1}{\Delta x}\nabla_{\boldsymbol{u}} H \to \delta_u \mathcal{H}[u, v] \quad \text{and} \quad \lim_{\Delta x \to 0} \frac{1}{\Delta x}\nabla_{\boldsymbol{v}} H \to \delta_v \mathcal{H}[u, v],$$

and the spatially discrete equations formally converge to the PDE limit (12.4).

For the specific Hamiltonian, as given above, the finite-dimensional truncation becomes

$$\frac{d}{dt}v_i = \frac{\sigma'(w_{i+1}) - \sigma'(w_i)}{\Delta x} - f'(u_i), \tag{12.24}$$

$$\frac{d}{dt}u_i = v_i, \tag{12.25}$$

$i = 1, \dots, N$, with

$$w_{i+1} = \frac{u_{i+1} - u_i}{\Delta x}, \qquad w_i = \frac{u_i - u_{i-1}}{\Delta x}.$$

The equations of motion can be integrated in time using any canonical method such as a symplectic Euler method, for example,

$$v_i^{n+1} = v_i^n + \Delta t \left[\frac{\sigma'(w_{i+1}^n) - \sigma'(w_i^n)}{\Delta x} - f'(u_i^n) \right],$$
$$u_i^{n+1} = u_i^n + \Delta t v_i^{n+1},$$

where

$$w_i^n = \frac{u_i^n - u_{i-1}^n}{\Delta x}, \qquad w_{i+1}^n = \frac{u_{i+1}^n - u_i^n}{\Delta x}.$$

See McLachlan [129] for further details on this classical approach to the numerical solution of Hamiltonian PDEs.

This might appear to be the end of the story. However, the interpretation of the wave equation (12.1) as an infinite-dimensional Hamiltonian system has masked some of the interesting local features of the PDE. For example, let us have another look at the Hamiltonian functional $\mathcal{H}[u, v]$. We can write this functional as

$$\mathcal{H}[u, v] = \int_0^L E(u, v)\, dx, \qquad E(u, v) = \frac{1}{2}v^2 + \sigma(u_x) + f(u).$$

The function E is called the *energy density*. Let us compute the time derivative of E

$$\begin{aligned} E_t &= vv_t + \sigma'(u_x)u_{xt} + f'(u)u_t \\ &= v\left[\partial_x\sigma'(u_x) - f'(u)\right] + \sigma'(u_x)v_x + f'(u)v \\ &= [v\sigma'(u_x)]_x. \end{aligned}$$

We have obtained what is called an *energy conservation law*

$$E_t + F_x = 0, \tag{12.26}$$

where $F = -v\sigma'(u_x)$ is called the *energy flux*. Under periodic boundary conditions, the conservation law (12.26) immediately implies conservation of total energy since $\int_0^L F_x = [F]_{x=0}^{x=L} = 0$. But the energy conservation law (12.26) is valid independently of any boundary conditions. Hence it is more fundamental than conservation of total energy.

Let us repeat the above calculation for the spatially truncated system (12.24)–(12.25). We define the discrete energy density

$$E_i = \frac{1}{2}v_i^2 + \sigma(w_i) + f(u_i),$$

and find:

$$\begin{aligned}\frac{d}{dt}E_i &= v_i\dot{v}_i + \sigma'(w_i)\frac{\dot{u}_i - \dot{u}_{i-1}}{\Delta x} + f'(u_i)\dot{u}_i \\ &= v_i\frac{\sigma'(w_{i+1}) - \sigma'(w_i)}{\Delta x} + \sigma'(w_i)\frac{v_i - v_{i-1}}{\Delta x} \\ &= \frac{v_i\sigma'(w_{i+1}) - \sigma'(w_i)v_{i-1}}{\Delta x} \\ &= -\frac{F_{i+1/2} - F_{i-1/2}}{\Delta x},\end{aligned}$$

where

$$F_{i+1/2} = -v_i\sigma'(w_{i+1}), \quad F_{i-1/2} = -v_{i-1}\sigma'(w_i).$$

Hence we have obtained a semi-discrete energy conservation law

$$\frac{d}{dt}E_i + \frac{F_{i+1/2} - F_{i-1/2}}{\Delta x} = 0,$$

and $F_{i\pm1/2}$ are approximations to the energy flux $F(u(x))$ at $x = i\Delta x \pm \Delta x/2$. Again this local energy conservation law is more fundamental than conservation of total energy H. Applying a symplectic integration method in time, we can now monitor the residual,

$$R_i^{n+1/2} := \frac{E_i^{n+1} - E_i^n}{\Delta t} + \frac{F_{i+1/2}^{n+1/2} - F_{i-1/2}^{n+1/2}}{\Delta x}, \tag{12.27}$$

of a fully discretized local energy conservation law with

$$E_i^n = \frac{1}{2}(v_i^n)^2 + \sigma(w_i^n) + f(u_i^n),$$

and

$$F_{i+1/2}^{n+1/2} = -\frac{1}{2}\left[v_i^n\sigma'(w_{i+1}^n) + v_i^{n+1}\sigma'(w_{i+1}^{n+1})\right],$$

etc. Similar to the non-conservation of total energy under symplectic time integration, we cannot, in general, expect the residual $R_i^{n+1/2}$ to be zero. But a formal backward error analysis has been developed by MOORE AND REICH [137, 138] to explain the remarkable global and local energy conservation of symplectic PDE discretizations observed, for example, in [162].

One can also derive methods that exactly conserve energy. See, for example, [96] and [115]. Such methods will, in general, not be symplectic.

Example 2 Let us discuss the sine-Gordon equation

$$u_{tt} = u_{xx} - \sin(u).$$

The energy density is

$$E = \frac{1}{2}((u_t)^2 + (u_x)^2) + (1 - \cos u),$$

and the energy flux is

$$F = -u_t u_x.$$

The spatial discretization (12.24)–(12.25) followed by a symplectic Euler discretization in time yields

$$\frac{v_i^{n+1} - v_i^n}{\Delta t} = \frac{w_{i+1}^n - w_i^n}{\Delta x} - \sin(u_i^n), \qquad \frac{u_i^{n+1} - u_i^n}{\Delta t} = v_i^{n+1}, \qquad \frac{u_{i+1}^{n+1} - u_i^{n+1}}{\Delta x} = w_{i+1}^{n+1}. \tag{12.28}$$

The semi-discretized energy conservation law is

$$\frac{d}{dt}\left[\frac{1}{2}(v_i^2 + w_i^2) + (1 - \cos(u_i))\right] + \left[\frac{-v_i w_{i+1} + v_{i-1} w_i}{\Delta x}\right] = 0. \tag{12.29}$$

Note that, upon eliminating w_i^n and v_i^n, the method (12.28) is equivalent to the classical centered leap-frog scheme

$$\frac{u_i^{n+1} - 2u_i^n + u_i^{n-1}}{\Delta t^2} = \frac{u_{i+1}^n - 2u_i^n + u_{i-1}^n}{\Delta x^2} - \sin(u_i^n).$$

□

We will come back to the local aspects of Hamiltonian PDEs and their numerical counterparts in Section 12.3. Numerical results will be presented in Section 12.3.4.

Let us briefly discuss a spatial discretization for the KdV equation (12.23). The Hamiltonian is easily discretized to

$$H = \sum_{i=1}^{N}\left[\frac{1}{6}u_i^3 - \frac{1}{2}\left(\frac{u_i - u_{i-1}}{\Delta x}\right)^2\right]\Delta x.$$

The spatial truncation of the Poisson operator $\mathcal{J}$ is potentially more challenging. However, for KdV, any skew symmetric approximation of the differential operator $\mathcal{J} = \partial_x$ is sufficient and will lead to a finite-dimensional Hamiltonian system. This Hamiltonian ODE can be integrated in time by a symplectic integrator. Symplectic methods based on splitting the Hamiltonian into integrable problems are available. See, for example, QUISPEL AND MCLACHLAN [132] and ASCHER AND MCLACHLAN [9].

12.2.2 Particle-based methods

Standard grid-based methods are, in general, not applicable to Lagrangian fluid dynamics. This is due to the ill-conditioning of the map from an initial grid (labels) $\{\boldsymbol{X}_{ij}(0) = \boldsymbol{a}_{ij}\}$, $\boldsymbol{a}_{ij} = (i\Delta x, j\Delta y)$ to the advected particle positions $\{\boldsymbol{X}_{ij}(t)\}$. Since this map determines the layer-depth approximation in a standard mesh-based Lagrangian method via a discrete approximation of (12.18), the quality of the simulation results is usually rather poor and instabilities are observed.

On the other hand, general grid-based methods are very easy to implement for an Eulerian formulation of fluid dynamics. See DURRAN [55] for an overview of such methods. However, none of these methods respects the Hamiltonian nature of the inviscid equations of motion. This is due to the already mentioned difficulty of finding a spatial truncation of the underlying noncanonical formulation of Eulerian fluid dynamics.

All these problems disappear if we give up the grid and work with the Lagrangian instead of the reduced Eulerian formulation of inviscid fluid dynamics. The resulting so-called *mesh-free methods* are based on an approximation of the layer-depth h via the identity (12.19). The most well-known mesh-free method for Lagrangian fluid dynamics simulations is the *Smoothed Particle Hydrodynamics* (SPH) method of LUCY [120] and GINGOLD AND MONAGHAN [71]. Many different variants of the basic SPH method have been proposed over the years. The first application of SPH to the shallow-water equations is due to SALMON [168]. We follow here the general framework of FRANK AND REICH [64].

Any spatial discretization will lead to a finite spatial resolution. For grid-based methods that resolution is directly related to the mesh-size Δx. For a mesh-free method, we have to instead introduce a smoothing or filter length $\alpha > 0$. Any fluid motion below that length scale will not be properly resolved. Hence we may replace the SWEs (12.20)–(12.21) by the 'regularized/smoothed' formulation

$$\frac{\partial}{\partial t}\boldsymbol{u} = f_0 J_2 \boldsymbol{u} - c_0^2 \nabla_{\boldsymbol{X}}(\mathcal{A} * h), \tag{12.30}$$

$$\frac{\partial}{\partial t}\boldsymbol{X} = \boldsymbol{u}, \tag{12.31}$$

where the convolution $\mathcal{A} * h$ is defined in terms of a smooth kernel function $\psi(\boldsymbol{x}, \boldsymbol{y}) \geq 0$ satisfying

- $\psi(\boldsymbol{x}, \boldsymbol{y}) = \psi(\boldsymbol{y}, \boldsymbol{x})$ (symmetry),
- $\int \psi(\boldsymbol{x}, \boldsymbol{y})\, dx\, dy = 1$ (conservation of mass),
- $\psi(\boldsymbol{x}, \boldsymbol{y}) = \phi(\|\boldsymbol{x} - \boldsymbol{y}\|)$ (radial symmetry).

More explicitly, we have

$$
\begin{aligned}
(\mathcal{A} * h)(\boldsymbol{x}, t) &= \int \left(\psi(\boldsymbol{x}, \bar{\boldsymbol{x}}) \int h_o(\boldsymbol{a})\, \delta(\bar{\boldsymbol{x}} - \boldsymbol{X}(\boldsymbol{a}, t))\, da\, db \right) d\bar{x}\, d\bar{y} \\
&= \int h_o(\boldsymbol{a}) \left(\int \psi(\boldsymbol{x}, \bar{\boldsymbol{x}})\, \delta(\bar{\boldsymbol{x}} - \boldsymbol{X}(\boldsymbol{a}, t))\, d\bar{x}\, d\bar{y} \right) da\, db \\
&= \int h_o(\boldsymbol{a})\, \psi(\boldsymbol{x}, \boldsymbol{X}(\boldsymbol{a}, t))\, da\, db.
\end{aligned}
$$

A kernel often used in the SPH method is the Gaussian

$$
\psi(\boldsymbol{x}, \boldsymbol{y}; \alpha) = \frac{1}{\pi^{3/2}\alpha} \mathrm{e}^{-\|\boldsymbol{x} - \boldsymbol{y}\|^2/\alpha^2},
$$

where $\alpha > 0$ is the smoothing length scale.

To set up the numerical method, we introduce a mesh in label space with equally spaced grid points $\{\boldsymbol{a}_{ij}\}$ and mesh-size $\Delta a = \Delta b$. The grid points are enumerated by integers $k = 1, \ldots, N$, which serve as discrete labels, so the map $s : (i, j) \to k$ is one-to-one. The particle positions at time t are denoted by $\boldsymbol{X}_k(t)$ and initially $\boldsymbol{X}_k(0) = \boldsymbol{a}_{ij}$. Each particle has a "mass" $m_k = h_o(\boldsymbol{a}_{ij})$ and a velocity $\boldsymbol{u}_k(t) \in \mathbb{R}^2$. The layer depth at $\boldsymbol{X}_k(t)$ is then approximated by

$$
h_k(t) = \sum_l m_l\, \psi(\boldsymbol{X}_k(t), \boldsymbol{X}_l(t))\, \Delta a \Delta b.
$$

Similarly, any integral of the form

$$
I = \int h_o(\boldsymbol{a})\, w(\boldsymbol{X}(\boldsymbol{a}))\, da\, db
$$

is approximated by

$$
I \approx \sum_k m_k w_k\, \Delta a \Delta b, \qquad w_k = w(\boldsymbol{X}_k).
$$

From here on we can follow exactly the same approach as outlined for the symplectic discretization of grid-based methods. The Hamiltonian functional

$$
\mathcal{H} = \frac{1}{2} \int \frac{\mathbf{p} \cdot \mathbf{p}}{h_o}\, da\, db + \frac{c_0^2}{2} \int h_o\, (\mathcal{A} * h)\, da\, db
$$

is approximated by the Hamiltonian

$$H = \frac{1}{2}\left(\sum_{k=1}^{N} \frac{1}{m_k}\|\boldsymbol{p}_k\|^2 + c_0^2 \sum_{k,l=1}^{N} m_k m_l\, \psi(\boldsymbol{X}_k, \boldsymbol{X}_l)\, \Delta a \Delta b\right) \Delta a \Delta b,$$

$\boldsymbol{p}_k = m_k \boldsymbol{u}_k$, and (12.22) is replaced by the symplectic form

$$\bar{\omega}_N := \sum_{k=1}^{N} \left(\frac{m_k f_0}{2} d\boldsymbol{X}_k \wedge \boldsymbol{J}_2 d\boldsymbol{X}_k + d\boldsymbol{p}_k \wedge d\boldsymbol{X}_k\right) \Delta a \Delta b.$$

The resulting equations of motion are

$$\frac{d}{dt}\boldsymbol{u}_k = f_0 \boldsymbol{J}_2 \boldsymbol{u}_k - c_0^2 \sum_{l=1}^{N} \bar{m}_l \nabla_{\boldsymbol{X}_k} \psi(\boldsymbol{X}_k, \boldsymbol{X}_l),$$

$$\frac{d}{dt}\boldsymbol{X}_k = \boldsymbol{u}_k,$$

$k = 1, \ldots, N$, where we have made use of the symmetry of $\psi(\boldsymbol{x}, \boldsymbol{y})$ as well as $\boldsymbol{p}_k/m_k = \boldsymbol{u}_k = \frac{d}{dt}\boldsymbol{X}_k$, and defined

$$\bar{m}_k = m_k \, \Delta a \Delta b.$$

Note that the mesh $\{\boldsymbol{a}_{ij}\}$ is only needed at time $t = 0$ to give each particle $\boldsymbol{X}_k$ an initial position and velocity and to allocate a mass m_k.

The behavior of the particle method depends crucially on a proper choice of the kernel ψ. The classical SPH method uses a radially symmetric kernel of the form $\psi(\boldsymbol{x}, \boldsymbol{y}) = \phi(\|\boldsymbol{x} - \boldsymbol{y}\|)$, with $\phi(r)$, for example, a Gaussian or a spline function.

In [63], FRANK, GOTTWALD, AND REICH dropped the radial symmetry condition and adopted the following somewhat different strategy called the *Hamiltonian Particle-Mesh* (HPM) method. Following the standard *particle-mesh* procedure [21, 85], one defines a computational grid $\{\boldsymbol{x}_{mn}\}$, $\boldsymbol{x}_{mn} = (m\Delta x, n\Delta y)$, with, for simplicity, $\Delta x = \Delta y$. The choice $\Delta x = 4\Delta a$ seems to work well in practice. This implies that, initially, there are 16 particles per computational grid cell. We introduce the interpolation function

$$\phi(\boldsymbol{x} - \boldsymbol{X}_k) := \frac{1}{\Delta x \,\Delta y}\rho\left(\frac{|x - X_k|}{\Delta x}\right)\rho\left(\frac{|y - Y_k|}{\Delta y}\right),$$

where $\rho(r)$ is given by the cubic spline

$$\rho(r) = \begin{cases} \frac{2}{3} - r^2 + \frac{1}{2}r^3, \; r \le 1 \\ \frac{1}{6}(2-r)^3, & 1 < r \le 1 \cdot \\ 0, & r > 2 \end{cases}$$

We define the scaling factor

$$\gamma = \int \phi(\mathbf{x})\, dx\, dy$$

and approximate the layer-depth h^{mn} at a grid point $\mathbf{x}_{mn}$ by

$$h^{mn}(t) = \sum_k \frac{m_k}{\gamma}\, \phi_{mn}(\mathbf{X}_k(t))\, \Delta a \Delta b, \qquad \phi_{mn}(\mathbf{X}) := \phi(\mathbf{x}_{mn} - \mathbf{X}).$$

Let $\mathbf{S} = \{S_{pq}^{mn}\}$ denote the representation of a spatial averaging operator S over the given grid $\{\mathbf{x}_{mn}\}$.[4] Since the cubic splines form a partition of unity on the grid, i.e.

$$\sum_{mn} \phi_{mn}(\mathbf{x}) = 1,$$

we can define a continuous approximation of a smoothed/averaged layer-depth in space by means of

$$(\mathcal{A} * h)\,(\mathbf{x}, t) = \sum_{pq,mn} \phi_{pq}(\mathbf{x})\, S_{mn}^{pq}\, h^{mn}(t),$$

and the kernel function ψ, as used in the HPM method, is finally given by

$$\psi(\mathbf{x}, \mathbf{y}) = \frac{1}{\gamma} \sum_{pq,mn} \phi_{mn}(\mathbf{x})\, S_{pq}^{mn} \phi_{pq}(\mathbf{y})\,. \tag{12.32}$$

Note that

$$\int \psi(\mathbf{x}, \bar{\mathbf{x}})\, dx\, dy = \sum_{mn,pq} S_{pq}^{mn} \phi_{pq}(\bar{\mathbf{x}}) = \sum_{pq} \phi_{pq}(\bar{\mathbf{x}}) = 1,$$

and the constructed basis function is symmetric and satisfies conservation of mass.

It is found that the inverse of the modified Helmholtz operator $H = 1 - \alpha^2 \nabla_x^2$ computed over the grid $\{\mathbf{x}_{mn}\}$ using FFT provides a good choice for the matrix operator $\mathbf{S}$. The smoothing length α is typically chosen between $\alpha = 2\Delta x$ and $\alpha = 4\Delta x$. We assume that α is shorter than the Rossby deformation radius L_R. Hence the matrix operator $\mathbf{S}$ will filter out high-frequency waves but will leave the balanced geostrophic layer-depth (12.17) essentially unaffected.

Note that the mesh has only been introduced to efficiently compute the kernel function (12.32).

The geometric conservation properties of the particle method, including conservation of circulation and potential vorticity, have been discussed in [64] and [32].

[4] The approximation $\mathbf{S}$ should be symmetric and $\sum_{pq} S_{pq}^{mn} = 1$.

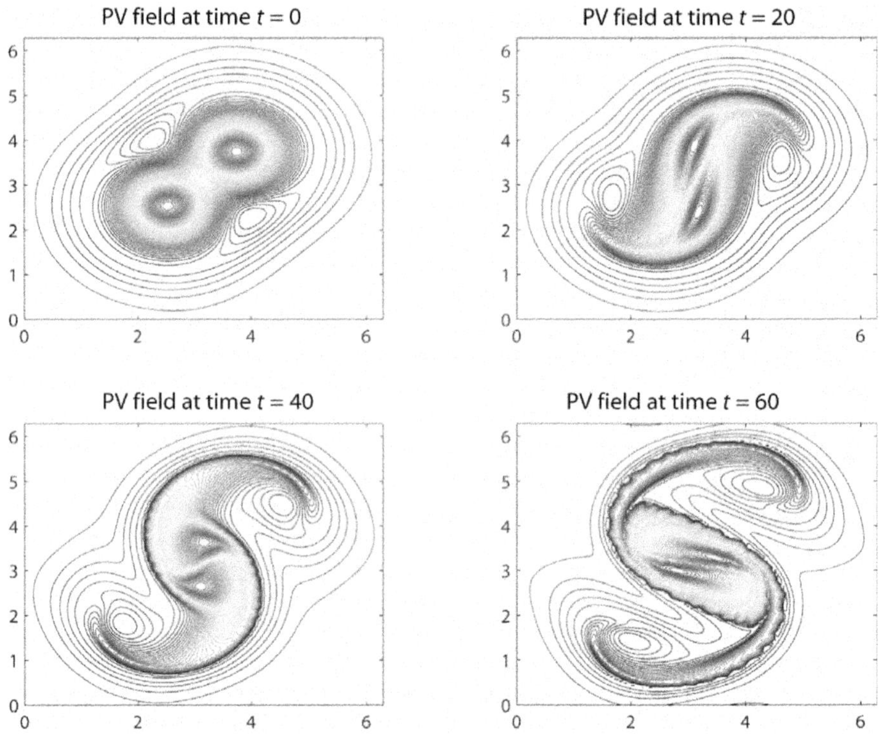

Figure 12.1 Time evolution of PV field under merger of two vortices.

As a simple demonstration we consider the merger of two vortices with parameters $f_0 = c_0 = L_R = 1$. The initial velocity field is set equal to $\boldsymbol{u}^{\mathrm{g}}$ with an initial layer-depth h such that the Rossby number satisfies $\mathrm{Ro} \approx 0.1$ and $L \approx L_R$, i.e. $\mathrm{Bu} \approx 1$. This setting complies with quasi-geostrophic theory. We show snapshots of the time evolution of the PV field q over 60 time units in Fig. 12.1. In Fig. 12.2 we demonstrate the excellent conservation of total energy and maintenance of geostrophic balance by monitoring the ratio of the L_2-norm of the ageostrophic velocity $\boldsymbol{u}^{\mathrm{ag}} := \boldsymbol{u} - \boldsymbol{u}^{\mathrm{g}}$ to the L_2-norm of the geostrophic wind approximation $\boldsymbol{u}^{\mathrm{g}}$

$$R(t) := \frac{\|\boldsymbol{u}^{\mathrm{ag}}(t)\|_2}{\|\boldsymbol{u}^{\mathrm{g}}(t)\|_2}.$$

Note that $R(t) = 0$ at $t = 0$. The simulation was run with $N = 262144$ particles, $\Delta x = 2\pi/128$ and a smoothing length $\alpha = 2\Delta x$.

One can apply the Hamiltonian particle-mesh method to the shallow-water equations on the sphere which provides a simple model for global atmospheric circulation [65]. Furthermore, the Hamiltonian particle-mesh method can be extended to fully three-dimensional flows. The basic approach and its relation to

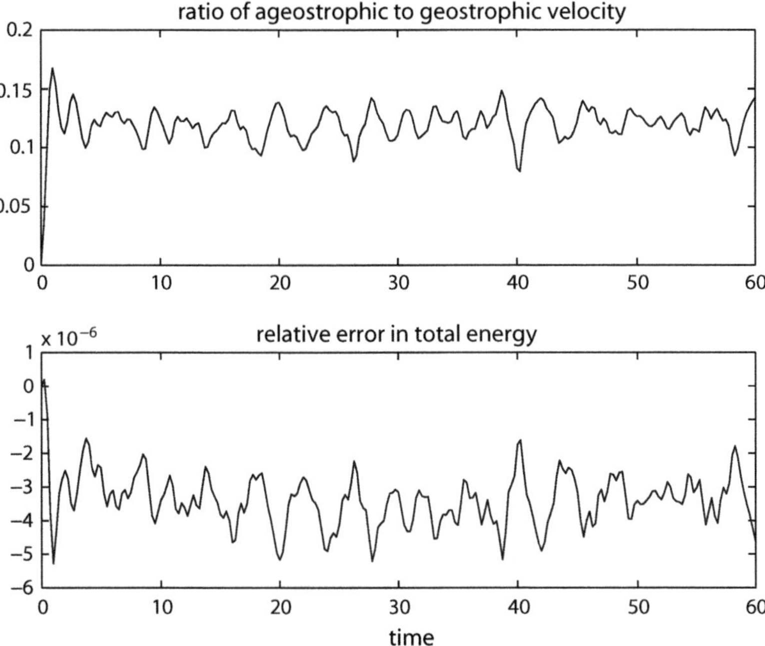

Figure 12.2 Ratio $R(t)$ of ageostrophic to geostrophic velocity contributions and relative error in energy.

geostrophic and hydrostatic balance have been outlined by COTTER AND REICH [47].

12.3 Multi-symplectic PDEs

When we derived the Hamiltonian formulation of the wave equation (12.1), the PDE (12.1) was rewritten as a system of equations with only first-order derivatives in time. Let us now rewrite (12.1) as a system of equations containing only first-order derivatives in space *and* time. We obtain, for example,

$$\begin{aligned} -v_t - p_x &= f'(u), \\ u_t &= v, \\ u_x &= w, \\ 0 &= p + \sigma'(w). \end{aligned}$$

As first noted by BRIDGES [31, 30], this system is of the general from

$$\boldsymbol{K} z_t + \boldsymbol{L} z_x = \nabla_z S(z), \tag{12.33}$$

where $z \in \mathbb{R}^d$ is the state variable, $\boldsymbol{K}, \boldsymbol{L} \in \mathbb{R}^{d \times d}$ are two (constant) skew-symmetric matrices, $S : \mathbb{R}^d \to \mathbb{R}$ is a smooth function, and ∇_z is the standard gradient in $\mathbb{R}^d$.

Example 3 In case of the nonlinear wave equation (12.1), we can take $z = (u, v, p, w)^T \in \mathbb{R}^4$

$$S(z) = \frac{1}{2}v^2 + wp + \sigma(w) + f(u),$$

and

$$\boldsymbol{K} = \begin{bmatrix} 0 & -1 & 0 & 0 \\ 1 & 0 & 0 & 0 \\ 0 & 0 & 0 & 0 \\ 0 & 0 & 0 & 0 \end{bmatrix}, \qquad \boldsymbol{L} = \begin{bmatrix} 0 & 0 & -1 & 0 \\ 0 & 0 & 0 & 0 \\ 1 & 0 & 0 & 0 \\ 0 & 0 & 0 & 0 \end{bmatrix}.$$

□

According to BRIDGES [31, 30], a PDE of the form (12.33) is now called a *multi-symplectic* PDE for the following reason. With each of the two skew-symmetric matrices we identify a pre-symplectic form;[5] i.e.

$$\omega = \frac{1}{2}dz \wedge \boldsymbol{K}dz, \qquad \kappa = \frac{1}{2}dz \wedge \boldsymbol{L}dz.$$

These pre-symplectic forms satisfy a *conservation law of symplecticness*

$$\omega_t + \kappa_x = 0. \tag{12.34}$$

Indeed, the variational equation

$$\boldsymbol{K}dz_t + \boldsymbol{L}dz_x = \boldsymbol{A}(x,t)dz, \qquad \boldsymbol{A}(x,t) = S_{zz}(z(x,t)),$$

implies

$$dz \wedge \boldsymbol{K}dz_t + dz \wedge \boldsymbol{L}dz_x = dz \wedge \boldsymbol{A}(x,t)dz,$$

and, since $dz \wedge \boldsymbol{A}(x,t)dz = \boldsymbol{0}$

$$\frac{1}{2}\partial_t(dz \wedge \boldsymbol{K}dz) + \frac{1}{2}\partial_x(dz \wedge \boldsymbol{L}dz) = 0.$$

Note that

$$\frac{d}{dt}\bar{\omega} = \frac{d}{dt}\int_0^L \omega\, dx = \int_0^L \omega_t\, dx = -\left[\kappa\right]_{x=0}^{x=L} = 0,$$

[5]The skew-symmetric matrices $\boldsymbol{K}$ and $\boldsymbol{L}$ are, in general, singular and, hence, do not define a symplectic structure on $\mathbb{R}^d$. This situation is somewhat similar to the Lie–Poisson formulation of rigid bodies.

and conservation of the total symplectic form (12.2) is a consequence of the conservation law of symplecticness (12.34).

The Lagrangian formulation (12.20)–(12.21) of the shallow-water equations can also be rewritten as a multi-symplectic PDE of the form

$$\boldsymbol{K}\boldsymbol{z}_t + \boldsymbol{L}_1\boldsymbol{z}_a + \boldsymbol{L}_2\boldsymbol{z}_b = \nabla_z S(\boldsymbol{z}),$$

where $\boldsymbol{K}$, $\boldsymbol{L}_1$, $\boldsymbol{L}_2$ are again constant skew-symmetric matrices associated with the independent variables time t and labels $\boldsymbol{a} = (a, b)^T$. The details can be found in [32]. As shown above for the nonlinear wave equation, the associated multi-symplectic conservation law is of the form

$$\partial_t \omega + \partial_a \kappa_1 + \partial_b \kappa_2 = 0, \tag{12.35}$$

where

$$\omega = \frac{1}{2} d\boldsymbol{z} \wedge \boldsymbol{K} d\boldsymbol{z} = \frac{h_o f_0}{2} d\boldsymbol{X} \wedge J_2 d\boldsymbol{X} + d\boldsymbol{p} \wedge d\boldsymbol{X}$$

is the density of the symplectic form (12.22).

12.3.1 Conservation laws

Every multi-symplectic formulation (12.33) implies conservation laws of energy, momentum, and symplecticness [31, 30]. To derive those explicitly, let us introduce a decomposition of the two skew-symmetric matrices $\boldsymbol{K}$ and $\boldsymbol{L}$ such that

$$\boldsymbol{K} = \boldsymbol{K}_+ + \boldsymbol{K}_- \quad \text{and} \quad \boldsymbol{L} = \boldsymbol{L}_+ + \boldsymbol{L}_-,$$

with

$$\boldsymbol{K}_+^T = -\boldsymbol{K}_- \quad \text{and} \quad \boldsymbol{L}_+^T = -\boldsymbol{L}_-.$$

A decomposition of this form immediately implies that

$$d\boldsymbol{z} \wedge \boldsymbol{K}_+ d\boldsymbol{z} = d\boldsymbol{z} \wedge \boldsymbol{K}_- d\boldsymbol{z} \quad \text{and} \quad d\boldsymbol{z} \wedge \boldsymbol{L}_+ d\boldsymbol{z} = d\boldsymbol{z} \wedge \boldsymbol{L}_- d\boldsymbol{z},$$

hence, the conservation law (12.34) holds with

$$\omega = d\boldsymbol{z} \wedge \boldsymbol{K}_+ d\boldsymbol{z} \quad \text{and} \quad \kappa = d\boldsymbol{z} \wedge \boldsymbol{L}_+ d\boldsymbol{z}.$$

An energy conservation law (12.26) is obtained by taking the inner product of (12.33) and $\boldsymbol{z}_t$. Since $\langle \boldsymbol{z}_t, \boldsymbol{K}\boldsymbol{z}_t \rangle = 0$, we obtain

$$\langle \boldsymbol{z}_t, \boldsymbol{L}\boldsymbol{z}_x \rangle = \langle \boldsymbol{z}_t, \nabla_z S(\boldsymbol{z}) \rangle = \partial_t S(\boldsymbol{z}).$$

Note that

$$\begin{aligned}\langle \boldsymbol{z}_t, \boldsymbol{L}\boldsymbol{z}_x\rangle &= \langle \boldsymbol{z}_t, \boldsymbol{L}_+\boldsymbol{z}_x\rangle + \langle \boldsymbol{z}_t, \boldsymbol{L}_-\boldsymbol{z}_x\rangle \\ &= \langle \boldsymbol{z}_t, \boldsymbol{L}_+\boldsymbol{z}_x\rangle - \langle \boldsymbol{z}_x, \boldsymbol{L}_+\boldsymbol{z}_t\rangle \\ &= \partial_x\langle \boldsymbol{z}_t, \boldsymbol{L}_+\boldsymbol{z}\rangle - \partial_t\langle \boldsymbol{z}_x, \boldsymbol{L}_+\boldsymbol{z}\rangle,\end{aligned}$$

and the energy conservation law (12.26) is satisfied by

$$E = S(\boldsymbol{z}) + \langle \boldsymbol{z}_x, \boldsymbol{L}_+\boldsymbol{z}\rangle \quad \text{and} \quad F = -\langle \boldsymbol{z}_t, \boldsymbol{L}_+\boldsymbol{z}\rangle. \tag{12.36}$$

Similarly, taking the inner product with $\boldsymbol{z}_x$, the momentum conservation law

$$\partial_t M + \partial_x I = 0, \tag{12.37}$$

is satisfied by

$$M = -\langle \boldsymbol{z}_x, \boldsymbol{K}_+\boldsymbol{z}\rangle \quad \text{and} \quad I = S(\boldsymbol{z}) + \langle \boldsymbol{z}_t, \boldsymbol{K}_+\boldsymbol{z}\rangle.$$

Example 4 In case of the multi-symplectic formulation of the nonlinear wave equation (12.1), we can take a decomposition of the matrices $\boldsymbol{K}$ and $\boldsymbol{L}$ defined by

$$\boldsymbol{K}_+ = \begin{bmatrix} 0 & -1 & 0 & 0 \\ 0 & 0 & 0 & 0 \\ 0 & 0 & 0 & 0 \\ 0 & 0 & 0 & 0 \end{bmatrix} \quad \text{and} \quad \boldsymbol{L}_+ = \begin{bmatrix} 0 & 0 & -1 & 0 \\ 0 & 0 & 0 & 0 \\ 0 & 0 & 0 & 0 \\ 0 & 0 & 0 & 0 \end{bmatrix}.$$

The conservation law of symplecticness becomes

$$\partial_t[du \wedge dv] + \partial_x[du \wedge dp] = 0.$$

It is also easy to show that (12.26) is satisfied with

$$E = \frac{1}{2}v^2 + \sigma(w) + f(u) = \frac{1}{2}u_t^2 + \sigma(u_x) + f(u) \quad \text{and} \quad F = vp = -u_t\sigma'(u_x).$$

Similarly, the momentum conservation law (12.37) is satisfied for

$$M = wv = u_x u_t$$

and

$$I = \sigma(w) + pw - \frac{1}{2}v^2 + f(u) = \sigma(u_x) - \sigma'(u_x)u_x - \frac{1}{2}u_t^2 + f(u).$$

□

An additional conservation law is obtained from the restriction of the multi-symplectic conservation law (12.34) to the space of solutions $\boldsymbol{z}(x,t)$ via

$$d\boldsymbol{z} = \boldsymbol{z}_t\, dt + \boldsymbol{z}_x\, dx.$$

Hence

$$\omega_{|z} = dz \wedge \boldsymbol{K}dz = 2\langle z_t, \boldsymbol{K}z_x\rangle dt \wedge dx, \quad \kappa_{|z} = dz \wedge \boldsymbol{L}dz = 2\langle z_t, \boldsymbol{L}z_x\rangle dt \wedge dx,$$

and (12.34) reduces to

$$\partial_t \langle z_t, \boldsymbol{K}z_x\rangle + \partial_x \langle z_t, \boldsymbol{L}z_x\rangle = 0.$$

If one applies a similar restriction

$$dz = z_a\, da + z_b\, db$$

to the multi-symplectic conservation law (12.35) of the shallow-water equations, then one finds that

$$\omega_{|z} = q, \qquad \kappa^1_{|z} = \kappa^2_{|z} = 0,$$

and the restricted multi-symplectic conservation law is equivalent to conservation of potential vorticity, i.e. $q_t = 0$ in the Lagrangian setting [32].

12.3.2 Traveling waves and dispersion

Traveling wave or soliton solutions are very easy to characterize once a multi-symplectic formulation has been found. We make the *ansatz*

$$z(x,t) = \boldsymbol{\phi}(x - ct) = \boldsymbol{\phi}(\xi).$$

Then the equation (12.33) becomes

$$-c\boldsymbol{K}\boldsymbol{\phi}_\xi + \boldsymbol{L}\boldsymbol{\phi}_\xi = \nabla_{\boldsymbol{\phi}} S(\boldsymbol{\phi}).$$

If we assume that the skew-symmetric matrix $\boldsymbol{L} - c\boldsymbol{K}$ is invertible, then we obtain the Hamiltonian ODE

$$\frac{d}{d\xi}\boldsymbol{\phi} = J_c \nabla_{\boldsymbol{\phi}} S(\boldsymbol{\phi}), \qquad J_c = [\boldsymbol{L} - c\boldsymbol{K}]^{-1},$$

in the state variable $\boldsymbol{\phi} \in \mathbb{R}^n$ and with Hamiltonian S. It should be noted that the symplectic structure matrix $\boldsymbol{J}_c$ is a superposition of the temporal and spatial pre-symplectic structures ω and κ, respectively.

As an example, let us consider the KdV equation (12.23). We introduce the velocity potential ψ with $\psi_x = u$ and obtain

$$\psi_{xt} + \frac{1}{2}\partial_x(\psi_x)^2 + \psi_{xxxx} = 0.$$

It should be kept in mind that ψ is only uniquely determined up to a constant. To remove this ambiguity we apply the normalization $\int \psi dx = 0$. A multi-symplectic formulation of type (12.33) is provided by

$$u_t + p_x = 0, \tag{12.38}$$
$$-\psi_t - 2w_x = -p + u^2, \tag{12.39}$$
$$2u_x = 2w, \tag{12.40}$$
$$-\psi_x = -u, \tag{12.41}$$

$\boldsymbol{z} = (\psi, u, w, p)^T$ and

$$S = \frac{1}{3}u^3 + w^2 - pu.$$

If we look for soliton solutions with wave speed c, then the associated structure matrix $\boldsymbol{J}_c$ is given by

$$\boldsymbol{J}_c^{-1} = \begin{bmatrix} 0 & -c & 0 & 1 \\ c & 0 & -2 & 0 \\ 0 & 2 & 0 & 0 \\ -1 & 0 & 0 & 0 \end{bmatrix}.$$

The equilibrium solutions are given by $\nabla_z S(\boldsymbol{z}) = \boldsymbol{0}$ and, hence, $u = w = p = 0$. Since the value of ψ does not enter the equations, we set it to $\psi = 0$ in accordance with our normalization condition $\int \psi dx = 0$. Hence there is only one equilibrium and a soliton solution must correspond to a homoclinic connection. Indeed, such a homoclinic orbit exists for any $c > 0$ and the soliton solutions with wave speed $c > 0$ are of the form

$$u(x,t) = 3c \,\mathrm{sech}^2\left(\frac{1}{2}\sqrt{c}(x - ct)\right).$$

As in the analysis of Section 5.1.1, for linear PDEs we can make the (harmonic) wave *ansatz*

$$\boldsymbol{z}(x,t) = \boldsymbol{\phi}(x - ct) = \boldsymbol{a}\, \mathrm{e}^{j(kx - \omega t)},$$

where $\boldsymbol{a} \in \mathbb{C}^d$ is a constant vector, $k \geq 0$ is the wave number, ω is the frequency[6] of the wave with wave speed $c = \omega/k$, and $j = \sqrt{-1}$. Upon substituting this into a linear multi-symplectic PDE

$$\boldsymbol{K}\boldsymbol{z}_t + \boldsymbol{L}\boldsymbol{z}_x = \boldsymbol{A}\boldsymbol{z},$$

one obtains the linear system of equations

$$[j\omega \boldsymbol{K} - jk\boldsymbol{L} + \boldsymbol{A}]\,\boldsymbol{a} = \boldsymbol{0}.$$

[6]One should not confuse the frequency ω with the two form ω defined earlier.

For a nontrivial solution $\boldsymbol{a} \neq \boldsymbol{0}$, we must require that

$$\det\left[j\omega \boldsymbol{K} - jk\boldsymbol{L} + \boldsymbol{A}\right] = 0. \tag{12.42}$$

The resulting polynomial expression in ω and k is called the *dispersion relation* [201]. Consider, for example, the linear KdV equation

$$u_t + u_x + u_{xxx} = 0,$$

which possesses a multi-symplectic structure identical to the nonlinear KdV equation (12.23), but with the Hamiltonian S replaced by

$$S = u^2 + w^2 - pu.$$

The associated linear system of equations is

$$\begin{bmatrix} 0 & j\omega & 0 & -jk \\ -j\omega & 2 & 2jk & -1 \\ 0 & -2jk & 2 & 0 \\ jk & -1 & 0 & 0 \end{bmatrix} \begin{bmatrix} a_\psi \\ a_u \\ a_w \\ a_p \end{bmatrix} = \boldsymbol{0},$$

which gives rise to the dispersion relation

$$\det \begin{bmatrix} 0 & j\omega & 0 & -jk \\ -j\omega & 2 & 2jk & -1 \\ 0 & -2jk & 2 & 0 \\ jk & -1 & 0 & 0 \end{bmatrix} = 4\omega k - 4k^2 + 4k^4 = 0.$$

The more common form of this relation is $\omega = k - k^3$.

12.3.3 Multi-symplectic integrators

In this section, we discuss numerical schemes that preserve a discrete version of the conservation law of symplecticness. Such schemes are called *multi-symplectic*. The idea is to apply symplectic discretization methods in space *and* in time. This is in contrast to the numerical approach outlined in Section 12.2 where only the time discretization was explicitly treated by a symplectic method. In other words, multi-symplectic methods treat space and time on equal footing.

Here and throughout the remainder of the chapter, we use the notation $\boldsymbol{z}_i^n$ to denote a numerical approximation of $\boldsymbol{z}(x_i, t_n)$, where x_i makes reference to a particular point on the spatial mesh and t_n refers to a specific point in time. We also define $\Delta x = x_{i+1} - x_i$ and $\Delta t = t_{n+1} - t_n$ and assume that both quantities are constant throughout the mesh. Then using both forward and backward differences, we define discrete approximations to $\boldsymbol{z}_x$ by

$$\partial_x^+ \boldsymbol{z}_i^n := \frac{\boldsymbol{z}_{i+1}^n - \boldsymbol{z}_i^n}{\Delta x} \quad \text{and} \quad \partial_x^- \boldsymbol{z}_i^n := \frac{\boldsymbol{z}_i^n - \boldsymbol{z}_{i-1}^n}{\Delta x},$$

and discrete approximations to $\mathbf{z}_t$ by

$$\partial_t^+ \mathbf{z}_i^n := \frac{\mathbf{z}_i^{n+1} - \mathbf{z}_i^n}{\Delta t} \quad \text{and} \quad \partial_t^- \mathbf{z}_i^n := \frac{\mathbf{z}_i^n - \mathbf{z}_i^{n-1}}{\Delta t}.$$

We have seen that the multi-symplectic formulation (12.33) provides a pre-symplectic form for each of the independent variables x and t. BRIDGES AND REICH [33] suggested to apply a symplectic discretization to each of the independent variables. We now give two specific examples for this approach.

Euler box scheme

Given a Hamiltonian ODE, written as

$$\boldsymbol{K}\mathbf{z}_t = \nabla_z H(\mathbf{z}), \quad \boldsymbol{K} = \boldsymbol{J}^{-1},$$

consider the scheme

$$\boldsymbol{K}_+ \partial_t^+ \mathbf{z}^n + \boldsymbol{K}_- \partial_t^- \mathbf{z}^n = \nabla_z H(\mathbf{z}^n), \tag{12.43}$$

where $\boldsymbol{K} = \boldsymbol{K}_+ + \boldsymbol{K}_-$ is an appropriate decomposition in the sense discussed in Section 12.3.1.

As suggested by MOORE AND REICH [137], the discretization (12.43) can also be applied to the spatial part of a multi-symplectic PDE (12.33). This yields the semi-discretization

$$\boldsymbol{K}\partial_t \mathbf{z}_i + \boldsymbol{L}_+ \partial_x^+ \mathbf{z}_i + \boldsymbol{L}_- \partial_x^- \mathbf{z}_i = \nabla_z S(\mathbf{z}_i). \tag{12.44}$$

We next discretize in time to obtain the fully discrete equations

$$\boldsymbol{K}_+ \partial_t^+ \mathbf{z}_i^n + \boldsymbol{K}_- \partial_t^- \mathbf{z}_i^n + \boldsymbol{L}_+ \partial_x^+ \mathbf{z}_i^n + \boldsymbol{L}_- \partial_x^- \mathbf{z}_i^n = \nabla_z S(\mathbf{z}_i^n).$$

In order to show that this discretization satisfies a discrete conservation law of symplecticness, consider the discrete variational equation

$$\boldsymbol{K}_+ \partial_t^+ d\mathbf{z}_i^n + \boldsymbol{K}_- \partial_t^- d\mathbf{z}_i^n + \boldsymbol{L}_+ \partial_x^+ d\mathbf{z}_i^n + \boldsymbol{L}_- \partial_x^- d\mathbf{z}_i^n = S_{zz}(\mathbf{z}_i^n) d\mathbf{z}_i^n.$$

Now take the wedge product of this equation with $d\mathbf{z}_i^n$, and notice that we have

$$d\mathbf{z}_i^n \wedge S_{zz}(\mathbf{z}_i^n) d\mathbf{z}_i^n = 0,$$

because $S_{zz}(\mathbf{z}_i^n)$ is a symmetric matrix. Then, for the terms containing $\partial_t^\pm$, we obtain

$$d\mathbf{z}_i^n \wedge \boldsymbol{K}_+ \partial_t^+ d\mathbf{z}_i^n + d\mathbf{z}_i^n \wedge \boldsymbol{K}_- \partial_t^- d\mathbf{z}_i^n = \partial_t^+ \left(d\mathbf{z}_i^{n-1} \wedge \boldsymbol{K}_+ d\mathbf{z}_i^n \right).$$

Doing the same for the terms containing $\partial_x^{\pm}$, we derive a discrete conservation law

$$\partial_t^+ \omega_i^n + \partial_x^+ \kappa_i^n = 0, \tag{12.45}$$

for

$$\omega_i^n = d\mathbf{z}_i^{n-1} \wedge \mathbf{K}_+ d\mathbf{z}_i^n \quad \text{and} \quad \kappa_i^n = d\mathbf{z}_{i-1}^n \wedge \mathbf{L}_+ d\mathbf{z}_i^n .$$

Example 5 The sine-Gordon equation can be rewritten as a multi-symplectic PDE

$$-v_t - p_x = \sin(u), \quad u_t = v, \quad u_x = -p.$$

Spatial discretization by a method (12.43) yields

$$-\frac{d}{dt}v_i - \frac{p_{i+1} - p_i}{\Delta x} = \sin(u_i), \quad \frac{d}{dt}u_i = v_i, \quad \frac{u_i - u_{i-1}}{\Delta x} = -p_i,$$

or, equivalently

$$-\frac{d}{dt}v_i + \frac{u_{i+1} - 2u_i + u_{i-1}}{\Delta x} = \sin(u_i), \quad \frac{d}{dt}u_i = v_i.$$

The same spatial truncation is obtained from the truncated Hamiltonian

$$H = \sum_i \left[\frac{1}{2}(v_i)^2 + \frac{1}{2}\left(\frac{u_{i+1} - u_i}{\Delta x} \right)^2 + (1 - \cos u_i) \right] \Delta x$$

and the Hamiltonian equations of motion

$$\frac{d}{dt}v_i = -\Delta x^{-1} \nabla_{u_i} H, \quad \frac{d}{dt}u_i = \Delta x^{-1} \nabla_{v_i} H.$$

Another application of the discretization (12.43) in time yields the method (12.28). □

Let us now go back to the spatially discretized PDE (12.44). Taking the inner product with $\partial_t \mathbf{z}_i$ yields

$$\langle \partial_t \mathbf{z}_i, \mathbf{L}_+ \partial_x^+ \mathbf{z}_i \rangle + \langle \partial_t \mathbf{z}_i, \mathbf{L}_- \partial_x^- \mathbf{z}_i \rangle = \partial_t S(\mathbf{z}_i),$$

because $\langle \partial_t \mathbf{z}_i, \nabla_z S(\mathbf{z}_i) \rangle = \partial_t S(\mathbf{z}_i)$. Next we note that

$$\begin{aligned} \partial_t E_i &= \partial_t S(\mathbf{z}_i) + \partial_t \langle \partial_x^- \mathbf{z}_i, \mathbf{L}_+ \mathbf{z}_i \rangle \\ &= \langle \partial_t \mathbf{z}_i, \mathbf{L}_+ \partial_x^+ \mathbf{z}_i \rangle + \langle \partial_t \mathbf{z}_i, \mathbf{L}_- \partial_x^- \mathbf{z}_i \rangle + \partial_t \langle \partial_x^- \mathbf{z}_i, \mathbf{L}_+ \mathbf{z}_i \rangle \\ &= \langle \partial_t \mathbf{z}_i, \mathbf{L}_+ \partial_x^+ \mathbf{z}_i \rangle + \langle \partial_t \partial_x^- \mathbf{z}_i, \mathbf{L}_+ \mathbf{z}_i \rangle \\ &= \partial_x^+ \langle \partial_t \mathbf{z}_{i-1}, \mathbf{L}_+ \mathbf{z}_i \rangle . \end{aligned}$$

Hence, the spatially discrete energy conservation law

$$\partial_t E_i + \partial_x^- F_{i+1/2} = 0 \tag{12.46}$$

is obtained for the discrete energy density

$$E_i = S(\boldsymbol{z}_i) + \langle \partial_x^- \boldsymbol{z}_i, \boldsymbol{L}_+ \boldsymbol{z}_i \rangle,$$

and the energy flux

$$F_{i+1/2} = -\langle \partial_t \boldsymbol{z}_i, \boldsymbol{L}_+ \boldsymbol{z}_{i+1} \rangle.$$

Now consider the momentum conservation law (12.37). A semi-discretized version of this law can be obtained in the same manner as the spatially discretized version of the energy conservation law by simply exchanging the roles of space and time. Hence, we formally discretized (12.33) in time using (12.43) and take the inner product with $\partial_x \boldsymbol{z}^n$ to get the semi-discretized momentum conservation law

$$\partial_t^+ M^{n-1/2} + \partial_x I^n = 0, \tag{12.47}$$

where

$$M^{n+1/2} = -\langle \partial_x \boldsymbol{z}^n, \boldsymbol{K}_+ \boldsymbol{z}^{n+1} \rangle \quad \text{and} \quad I^n = S(\boldsymbol{z}^n) + \langle \partial_t^- \boldsymbol{z}^n, \boldsymbol{K}_+ \boldsymbol{z}^n \rangle.$$

See [137] for further details on the energy-momentum conservation of the fully discretized equations using modified equation analysis.

Example 6 The semi-discretized energy conservation law for the Euler box scheme applied to the sine-Gordon equation is given by (12.29). Let us also derive the semi-discretized momentum conservation law. We formally introduce a symplectic Euler time-discretization of the multi-symplectic sine-Gordon equation

$$-\partial_t^+ v^n - p_x^n = \sin(u^n), \quad \partial_t^- u^n = v^n, \quad u_x^n = -p^n.$$

Taking the appropriate inner product and rearranging terms we obtain

$$-u_x^n \partial_t^+ v^n + v_x^{n+1} \partial_t^+ u^n = \left(1 - \cos(u^n) + (v^{n+1})^2/2 - (p^n)^2/2\right)_x.$$

The left-hand side is equivalent to

$$-u_x^n \partial_t^+ v^n + v_x^{n+1} \partial_t^+ u^n = -\partial_t^+ (u_x^n v^n) + \left(v^{n+1} \partial_t^+ u^n\right)_x = -\partial_t^+ (u_x^n v^n) + \partial_x (v^{n+1})^2.$$

Hence the semi-discretized momentum conservation law is given by

$$\partial_t^+ (p^n v^n) - \partial_x \left(1 - \cos(u^n) - (v^{n+1})^2/2 - (p^n)^2/2\right) = 0. \tag{12.48}$$

□

Preissman box scheme

The Preissman box scheme can be viewed as the space-time version of the implicit midpoint scheme for ODEs and is widely used in hydraulics [1]. The scheme, applied to a multi-symplectic PDE, is given by

$$\boldsymbol{K} \partial_t^+ \boldsymbol{z}_{i+1/2}^n + \boldsymbol{L} \partial_x^+ \boldsymbol{z}_i^{n+1/2} = \nabla_z S(\boldsymbol{z}_{i+1/2}^{n+1/2}), \tag{12.49}$$

where we use the notation

$$z_i^{n+1/2} = \frac{1}{2}\left(z_i^n + z_i^{n+1}\right), \qquad z_{i+1/2}^n = \frac{1}{2}\left(z_i^n + z_{i+1}^n\right),$$

and

$$z_{i+1/2}^{n+1/2} = \frac{1}{4}\left(z_i^n + z_i^{n+1} + z_{i+1}^n + z_{i+1}^{n+1}\right).$$

Example 7 The implicit midpoint method in space applied to the sine-Gordon equation results in

$$\begin{aligned} -\frac{d}{dt}\frac{v_{i+1}+v_i}{2} - \frac{p_{i+1}-p_i}{\Delta x} &= \sin\left(\frac{u_{i+1}+u_i}{2}\right), \\ \frac{d}{dt}\frac{u_{i+1}+u_i}{2} &= \frac{v_{i+1}+v_i}{2}, \\ \frac{u_{i+1}-u_i}{\Delta x} &= -\frac{p_{i+1}+p_i}{2}. \end{aligned}$$

Note that this discretization is different from a midpoint discretization of the Hamiltonian

$$H = \sum_i \left[\frac{1}{2}\left(\frac{v_{i+1}+v_i}{2}\right)^2 + \frac{1}{2}\left(\frac{u_{i+1}-u_i}{\Delta x}\right)^2 + \left(1 - \cos\left(\frac{u_{i+1}+u_i}{2}\right)\right)\right]\Delta x$$

and associated Hamiltonian equations of motion

$$\frac{d}{dt}v_i = -\Delta x^{-1}\nabla_{u_i} H, \qquad \frac{d}{dt}u_i = \Delta x^{-1}\nabla_{v_i} H.$$

□

It was shown by BRIDGES AND REICH [33] that the scheme (12.49) satisfies a discrete conservation law of symplecticness (12.45) with

$$\omega_i^n = \frac{1}{2} dz_{i+1/2}^n \wedge \boldsymbol{K} dz_{i+1/2}^n \quad \text{and} \quad \kappa_i^n = \frac{1}{2} dz_i^{n+1/2} \wedge \boldsymbol{L} dz_i^{n+1/2}.$$

Similar to the Euler box scheme, a semi-discretized energy conservation law of type (12.46) can be derived for the spatially discretized system

$$\boldsymbol{K}\partial_t z_{i+1/2} + \boldsymbol{L}\partial_x^+ z_i = \nabla_z S(z_{i+1/2}). \tag{12.50}$$

In particular, taking the inner product with $\partial_t z_{i+1/2}$, we obtain

$$\langle \partial_t z_{i+1/2}, \boldsymbol{L}\partial_x^+ z_i \rangle = \partial_t S(z_{i+1/2}),$$

which, after a few formal manipulations, leads to the spatially discrete energy conservation law

$$\partial_t E_{i+1/2} + \partial_x^+ F_i = 0,$$

with

$$E_{i+1/2} = S(\boldsymbol{z}_{i+1/2}) + \langle \partial_x^+ \boldsymbol{z}_i, \boldsymbol{L}_+ \boldsymbol{z}_{i+1/2} \rangle,$$

and

$$F_i = -\langle \partial_t \boldsymbol{z}_i, \boldsymbol{L}_+ \boldsymbol{z}_i \rangle.$$

Upon applying the implicit midpoint method to (12.50) in time, the Preissman box scheme is obtained and one can consider the fully discretized energy conservation law

$$\partial_t^+ E_{i+1/2}^n + \partial_x^+ F_i^{n+1/2} = \mathrm{R}_{i+1/2}^{n+1/2}.$$

The residual $\mathrm{R}_{i+1/2}^{n+1/2}$ will be non-zero, in general. However, for a linear multi-symplectic PDE

$$\boldsymbol{K} \boldsymbol{z}_t + \boldsymbol{L} \boldsymbol{z}_x = \boldsymbol{A} \boldsymbol{z}, \tag{12.51}$$

$\boldsymbol{A}$ a (constant) symmetric matrix, the residual is identically equal to zero and

$$\partial_t^+ E_{i+1/2}^n + \partial_x^+ F_i^{n+1/2} = 0,$$

along numerical solutions $\{\boldsymbol{z}_i^n\}$, where

$$E_{i+1/2}^n = \langle \boldsymbol{z}_{i+1/2}^n, \boldsymbol{A} \boldsymbol{z}_{i+1/2}^n \rangle + \langle \partial_x^+ \boldsymbol{z}_i^n, \boldsymbol{L}_+ \boldsymbol{z}_{i+1/2}^n \rangle,$$

and

$$F_i^{n+1/2} = -\langle \partial_t^+ \boldsymbol{z}_i^n, \boldsymbol{L}_+ \boldsymbol{z}_i^{n+1/2} \rangle.$$

A semi-discretized momentum conservation law is derived in a similar manner by switching the roles of space and time. Thus the semi-discrete conservation law

$$\partial_t^+ M^n + \partial_x I^{n+1/2} = 0$$

is obtained, where

$$M^n = -\langle \partial_x \boldsymbol{z}^n, \boldsymbol{K}_+ \boldsymbol{z}^n \rangle,$$

and

$$I^{n+1/2} = S(\boldsymbol{z}^{n+1/2}) + \langle \partial_t^+ \boldsymbol{z}^n, \boldsymbol{K}_+ \boldsymbol{z}^{n+1/2} \rangle.$$

Again, the Preissman box scheme applied to a linear PDE (12.51) yields a fully discretized momentum conservation law

$$\partial_t^+ M_{i+1/2}^n + \partial_x^+ I_i^{n+1/2} = 0,$$

that is exactly satisfied by the numerical solutions $\{\boldsymbol{z}_i^n\}$.

See [33, 138] for more details on the energy and momentum conservation properties of the Preissman box scheme. The Preissman box scheme has been successfully applied to the KdV equation and the nonlinear Schrödinger equation (see, for example [209, 92]). A generalization to higher-order Gauss–Legendre RK methods has been given by REICH in [162].

Example 8 A Preissman box scheme discretization of the multi-symplectic formulation (12.38)–(12.41) of the KdV equation results in

$$\begin{aligned}
\partial_t^+ u_{i+1/2}^n + \partial_x^+ p_i^{n+1/2} &= 0,\\
-\partial_t^+ \phi_{i+1/2}^n - 2\partial_x^+ w_i^{n+1/2} &= -p_{i+1/2}^{n+1/2} + (u_{i+1/2}^{n+1/2})^2,\\
2\partial_x^+ u_i^{n+1/2} &= 2w_{i+1/2}^{n+1/2},\\
-\partial_x^+ \phi_i^{n+1/2} &= -u_{i+1/2}^{n+1/2}.
\end{aligned}$$

We assume that initially $\partial_x^+ u_i^0 = w_{i+1/2}^0$ and $\partial_x^+ \phi_i^0 = u_{i+1/2}^0$. Upon collecting all approximations u_i^n at time-level t_n into a vector $\boldsymbol{u}^n = \{u_i^n\}$ etc. and upon introducing two appropriate matrices $\boldsymbol{A}$ and $\boldsymbol{D}$, we can rewrite the scheme as

$$\begin{aligned}
\partial_t^+ \boldsymbol{A}\boldsymbol{u}^n + \boldsymbol{D}\boldsymbol{p}^{n+1/2} &= \boldsymbol{0},\\
-\partial_t^+ \boldsymbol{A}\boldsymbol{\phi}^n - 2\boldsymbol{D}\boldsymbol{w}^{n+1/2} &= -\boldsymbol{A}\boldsymbol{p}^{n+1/2} + (\boldsymbol{A}\boldsymbol{u}^{n+1/2}) * (\boldsymbol{A}\boldsymbol{u}^{n+1/2}),\\
\boldsymbol{D}\boldsymbol{u}^n &= \boldsymbol{A}\boldsymbol{w}^n,\\
\boldsymbol{D}\boldsymbol{\phi}^n &= \boldsymbol{A}\boldsymbol{u}^n,
\end{aligned}$$

where $\boldsymbol{c} = \boldsymbol{a} * \boldsymbol{b}$ denotes the component-wise vector product. Note that the averaging operator $\boldsymbol{A}$ and the differentiation operator $\boldsymbol{D}$ commute. We also assume that $\boldsymbol{A}$ is invertible. Hence we can eliminate $\boldsymbol{\phi}^n$, $\boldsymbol{p}^n$ and $\boldsymbol{w}^n$ from the system and obtain the timestepping scheme

$$\frac{\boldsymbol{A}\boldsymbol{u}^{n+1} - \boldsymbol{A}\boldsymbol{u}^n}{\Delta t} + \frac{1}{2}\boldsymbol{A}^{-1}\boldsymbol{D}\left[(\boldsymbol{A}\boldsymbol{u}^{n+1/2}) * (\boldsymbol{A}\boldsymbol{u}^{n+1/2})\right] + (\boldsymbol{A}^{-1}\boldsymbol{D})^3\boldsymbol{A}\boldsymbol{u}^{n+1/2} = \boldsymbol{0}.$$

Finally, introduce the variable $\bar{\boldsymbol{u}}^n = \boldsymbol{A}\boldsymbol{u}^n$, then

$$\frac{\bar{\boldsymbol{u}}^{n+1} - \bar{\boldsymbol{u}}^n}{\Delta t} + \frac{1}{2}\boldsymbol{A}^{-1}\boldsymbol{D}\left[(\bar{\boldsymbol{u}}^{n+1/2}) * (\bar{\boldsymbol{u}}^{n+1/2})\right] + (\boldsymbol{A}^{-1}\boldsymbol{D})^3\bar{\boldsymbol{u}}^{n+1/2} = \boldsymbol{0}.$$

See [161] for a similar approach to the shallow-water Boussinesq equation. □

Discrete variational methods

The nonlinear wave equation (12.1) can also be derived from a Lagrangian variational principle. Consider the Lagrangian functional

$$\mathcal{L}[u] = \int_{t=0}^{T}\int_{x=0}^{L} \mathrm{L}(u_t, u_x, u)\, dxdt,$$

with the Lagrangian density

$$\mathrm{L}(u_t, u_x, u) = \frac{1}{2}u_t^2 - \sigma(u_x) - f(u).$$

Then, using the Lagrangian variational principle, we derive the Euler–Lagrange equation

$$\partial_t \mathrm{L}_{u_t} + \partial_x \mathrm{L}_{u_x} - \mathrm{L}_u = 0. \tag{12.52}$$

For the particular Lagrangian density given above, the Euler–Lagrange equation reduces to the nonlinear wave equation (12.1).

The concept of multi-symplectic integration was first proposed in the context of the Euler–Lagrange equation (12.52) by MARSDEN, PATRICK, AND SHKOLLER [123] as a generalization of the discrete variational principle to first-order field theories.

Introduce, for example, the discrete density

$$\mathrm{L}_i^n = \mathrm{L}(\partial_t^+ u_i^n, \partial_x^+ u_i^n, u_i^n)$$

and the associated discrete Lagrangian functional

$$\mathcal{L}[u_i^n] = \sum_i \sum_n \mathrm{L}_i^n \, \Delta x \Delta t.$$

Minimization of this functional along sequences $\{u_i^n\}$ yields the discrete Euler–Lagrange equations

$$\partial_t^- \mathrm{L}_{\partial_t^+ u_i^n} + \partial_x^- \mathrm{L}_{\partial_x^+ u_i^n} - \mathrm{L}_{u_i^n} = 0.$$

See MARSDEN, PATRICK, AND SHKOLLER [123] for details on the underlying discrete multi-symplectic form formula and a proper differential geometric treatment.

For the nonlinear wave equation (12.1) and the above discrete Lagrangian density, the discrete variational principle yields the discretization

$$\frac{u_i^{n+1} - 2u_i^n + u_i^{n-1}}{\Delta t^2} = \frac{\sigma'\left(\frac{u_{i+1}^n - u_i^n}{\Delta x}\right) - \sigma'\left(\frac{u_i^n - u_{i-1}^n}{\Delta x}\right)}{\Delta x} - f'(u_i^n),$$

which is equivalent to Euler box scheme discretization applied to the nonlinear wave equation. In fact, the Euler–Lagrange equation (12.52) can be rewritten as a multi-symplectic PDE

$$\begin{aligned}
-q_t - p_x &= L_u, \\
u_t &= v, \\
u_x &= w, \\
0 &= q + L_v, \\
0 &= p + L_w,
\end{aligned}$$

and, hence, the variational and the multi-symplectic approach are complementary for first-order field theories, i.e. for Lagrangian densities of the form $\mathrm{L}(u_t, u_x, u) = \mathrm{L}(v, w, u)$.

The discrete variational approach can be generalized to second-order field theories such as the KdV equation. See, for example, [209].

12.3.4 Numerical dispersion and soliton solutions

We have seen in Section 12.3.2 that linear multi-symplectic PDEs can be discussed in terms of their dispersion relation. Following the idea of von Neumann stability analysis [140], a numerical dispersion relation can be defined by making the *ansatz*

$$\boldsymbol{z}_i^n = \boldsymbol{a}\,\mathrm{e}^{j(Kx_i - \Omega t_n)},$$

where $x_i = i\,\Delta x$, $t_n = n\,\Delta t$, and $j = \sqrt{-1}$. It is easy to derive that, for example,

$$\begin{aligned}\partial_x^+ \boldsymbol{z}_{i+1/2}^n &= \frac{\mathrm{e}^{-j\Omega\Delta t/2} - \mathrm{e}^{j\Omega\Delta t/2}}{\Delta t}\,\frac{\mathrm{e}^{jK\Delta x/2} + \mathrm{e}^{-jK\Delta x/2}}{2}\,\bar{\boldsymbol{z}}_{i+1/2}^{n+1/2} \\ &= -\frac{2j}{\Delta t}\sin(\Omega\Delta t/2)\cos(K\Delta x/2)\,\bar{\boldsymbol{z}}_{i+1/2}^{n+1/2},\end{aligned}$$

where we used the "exact" midpoint approximation

$$\bar{\boldsymbol{z}}_{i+1/2}^{n+1/2} := \boldsymbol{a}\,\mathrm{e}^{j(Kx_{i+1/2} - \Omega t_{n+1/2})},$$

$t_{n+1/2} = (n+1/2)\,\Delta t$, $x_{i+1/2} = (i+1/2)\,\Delta x$, which should not be confused with the approximation $\boldsymbol{z}_{i+1/2}^{n+1/2}$ used in the numerical scheme, i.e.

$$\boldsymbol{z}_{i+1/2}^{n+1/2} = \cos(\Omega\Delta t/2)\cos(K\Delta x/2)\,\bar{\boldsymbol{z}}_{i+1/2}^{n+1/2}.$$

If one substitutes these formulas and the corresponding expression for ∂_x^+ into the Preissman box scheme, then the following linear system appears

$$\left[\frac{2j}{\Delta t}\tan(\Omega\Delta t/2)\boldsymbol{K} - \frac{2j}{\Delta x}\tan(K\Delta x/2)\boldsymbol{L} + \boldsymbol{A}\right]\boldsymbol{a} = \boldsymbol{0}.$$

This equation leads to the *numerical dispersion relation* via

$$\det\left[\frac{2j}{\Delta t}\tan(\Omega\Delta t/2)\boldsymbol{K} - \frac{2j}{\Delta x}\tan(K\Delta x/2)\boldsymbol{L} + \boldsymbol{A}\right] = 0.$$

Upon defining

$$\omega = \frac{2}{\Delta t}\tan(\Omega\Delta t/2) \tag{12.53}$$

and

$$k = \frac{2}{\Delta x} \tan(K\Delta x/2), \tag{12.54}$$

the numerical dispersion relation turns into the analytic relation (12.42) for k and ω.

The important point is that only wave numbers K with

$$-\pi < K\Delta x \leq \pi$$

and frequencies Ω with

$$-\pi < \Omega\Delta t \leq \pi$$

are distinguishable in terms of the associated numerical solution z_i^n. Hence (12.53)–(12.54) can be uniquely solved for (K, Ω) in terms of $\Delta t\omega$ and $\Delta x k$, respectively. In other words, the numerical discretization does not introduce any spurious modes. This is a desirable property of the multi-symplectic Preissman box scheme [9, 136].

One should, however, keep in mind that multi-symplectic methods can be subject to resonance instabilities very much like those discussed for large time step (LTS) methods in Chapter 10. The concept of a Krein signature has been generalized to multi-symplectic PDEs by BRIDGES in [30] and has been used to classify instabilities. The same concept could be used to study numerical-induced resonance instabilities similar to what has been outlined in Chapter 10.

We finally conduct a numerical experiment for the sine-Gordon equation over a periodic domain $x \in (-L/2, L/2]$, $L = 60$, and initial data

$$u_o(x) = 4\tan^{-1}\left(e^{(x-L/4)/\sqrt{1-c^2}}\right) + 4\tan^{-1}\left(e^{(-x-L/4)/\sqrt{1-c^2}}\right),$$

and

$$v_o(x) = \frac{-4c}{\sqrt{1-c^2}}\left[\frac{e^{(x-L/4)/\sqrt{1-c^2}}}{1+e^{2(x-L/4)/\sqrt{1-c^2}}} - \frac{e^{(-x-L/4)/\sqrt{1-c^2}}}{1+e^{2(-x-L/4)/\sqrt{1-c^2}}}\right],$$

$c = 0.5$. The solution consists of a pair of kink solitons moving with speed $c = 0.5$ in opposite directions which, due to the periodicity of the domain, meet at $x = 0$ and $x = L/2$.

We use the Euler box scheme and set the spatial mesh-size to $\Delta x = 0.0187$ and the timestep to $\Delta t = 0.01$. In Figs. 12.3 and 12.4, we plot the numerical errors in the local energy and momentum conservation laws under a soliton collision near $t = 100$ and $x = 0$. These errors are defined in terms of fully discretized formulations of the semi-discrete conservation laws (12.29) and (12.48),

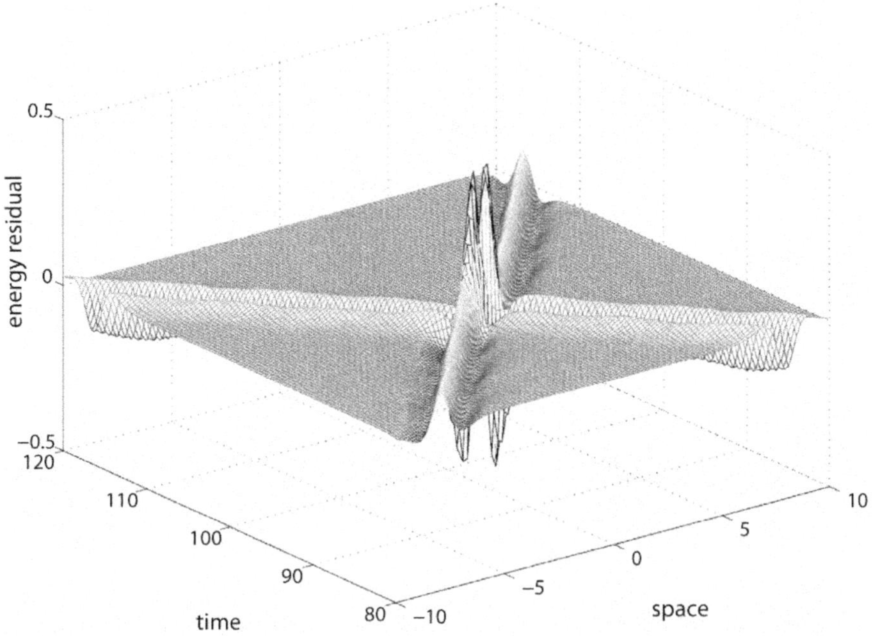

Figure 12.3 Numerical residual in the energy conservation law.

for example we use formula (12.27) for the energy conservation law. One can clearly see that the errors are concentrated along the moving fronts of the two solitons. The error in energy increases locally whenever the two solitons collide. Interestingly enough this is not the case for the momentum conservation law. This indicates that the larger energy error near collisions is due to numerical timestepping errors in the "pendulum part"

$$u_{tt} = -\sin(u).$$

12.3.5 Summary

At present the range of applicability and the advantages of classical symplectic methods, as described in Section 12.2, and multi-symplectic methods is an open question. A careful comparison of several methods has been carried out by Ascher and McLachlan [9] for the KdV equation. One should keep in mind that a large number of "classical" symplectic methods are in fact multi-symplectic. Furthermore, whatever method is chosen, it should correctly reflect the important physical quantities on a local discretization level. This is clearly the point of view taken with the multi-symplectic approach with regard to energy and momentum

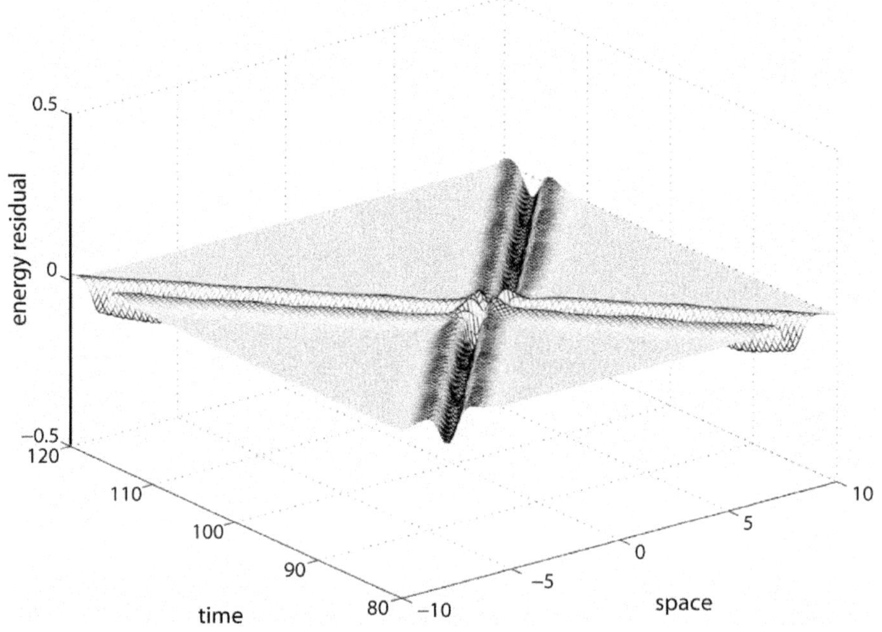

Figure 12.4 Numerical residual in the momentum conservation law.

conservation. Other applications might lead to other relevant quantities such as potential vorticity (PV) and hence to other preferable discretization methods. In some cases, such as turbulent fluid flows, one also has to address the problem of numerically unresolved motion on sub-grid length scales. These issues are clearly beyond this exposition. But they would lead into the area of stochastic sub-grid modeling and averaging.

12.4 Exercises

1. *Symplectic Euler scheme.* Apply the scheme (12.43) to the canonical Hamiltonia

$$\frac{d}{dt}\boldsymbol{p} = -\nabla_{\boldsymbol{q}} H(\boldsymbol{q},\boldsymbol{p}), \qquad \frac{d}{dt}\boldsymbol{q} = +\nabla_{\boldsymbol{p}} H(\boldsymbol{q},\boldsymbol{p}),$$

with $\boldsymbol{z} = (\boldsymbol{q},\boldsymbol{p})^T$, and

$$K_+ \boldsymbol{z} = \begin{pmatrix} \boldsymbol{q} \\ \boldsymbol{0} \end{pmatrix}.$$

Show that the scheme preserves the symplectic form

$$d\boldsymbol{q}^{n+1} \wedge d\boldsymbol{p}^{n} = d\boldsymbol{q}^{n} \wedge d\boldsymbol{p}^{n-1}.$$

How can the scheme be related to a standard symplectic Euler method?

2. *Nonlinear Schrödinger equation, Preissman box scheme.* Consider complex-valued functions $\psi(x,t)$ which satisfy the nonlinear Schrödinger equation

$$j\,\psi_t = \psi_{xx} + |\psi|^2\psi,$$

over the spatial domain $x \in [0, L)$ subject to periodic boundary conditions. We introduce the real part $a(x,t)$ and the imaginary part $b(x,t)$ of the wave function, i.e.

$$\psi(x,t) = a(x,t) + j\,b(x,t),$$

and rewrite the nonlinear Schrödinger equation as

$$-b_t = a_{xx} + (a^2+b^2)a, \tag{12.55}$$

$$a_t = b_{xx} + (a^2+b^2)b. \tag{12.56}$$

a. Write (12.55)–(12.56) in multi-symplectic form.

b. Apply the Preissman box scheme to the multi-symplectic formulation derived under (a).

c. The norm of the wave function is a conserved quantity for the nonlinear Schrödinger equation, i.e.

$$\frac{d}{dt}\int_0^L (a^2+b^2)\,dx = 0.$$

A discrete analog of this conservation law is

$$\sum_i \left[(a_{i+1/2}^n)^2 + (b_{i+1/2}^n)^2\right] = \sum_i \left[(a_{i+1/2}^{n+1})^2 + (b_{i+1/2}^{n+1})^2\right].$$

Does this identity hold for the multi-symplectic scheme derived under (b)?

3. *Sine-Gordon equation, energy-momentum conservation.* Implement the Euler box scheme for the sine-Gordon equation over a periodic domain $x \in (-L/2, L/2]$, $L = 60$, and initial data

$$u_o(x) = 4\tan^{-1}\left(e^{(x-L/4)/\sqrt{1-c^2}}\right) + 4\tan^{-1}\left(e^{(-x-L/4)/\sqrt{1-c^2}}\right)$$

and

$$v_o(x) = \frac{-4c}{\sqrt{1-c^2}} \left[\frac{e^{(x-L/4)/\sqrt{1-c^2}}}{1+e^{2(x-L/4)/\sqrt{1-c^2}}} - \frac{e^{(-x-L/4)/\sqrt{1-c^2}}}{1+e^{2(-x-L/4)/\sqrt{1-c^2}}} \right],$$

with wave speed $c = 0.5$.

a. Compute the numerical solution over a time-interval $[0, 200]$ using a stepsize of $\Delta t = 0.01$ and a spatial mesh-size of $\Delta x = L/3200$.

b. Implement formula (12.27) into your scheme to monitor the residual in the energy conservation law. You should reproduce the results from Fig. 12.3.

c. Find the analog of (12.27) for the momentum conservation law. Implement the formula into your scheme and monitor the residual in the discrete momentum conservation law.

4. *Dispersion relation.* Consider the linear advection equation

$$u_t + u_x = 0.$$

We introduce the velocity potential $u = \phi_x$ and write the advection equation in the multi-symplectic form

$$u_t + w_x = 0, \quad -\phi_t = -w + 2u, \quad -\phi_x = -u. \tag{12.57}$$

a. Find the linear dispersion relation for the formulation (12.57).

b. Apply the Preissman box scheme to the multi-symplectic formulation (12.57). Discuss the associated numerical dispersion relations. Verify the result from Fig. 12.5.

c. Compare the results from (b) with a direct application of the box scheme to the linear advection equation.

d. The advection equation is an example of a noncanonical Hamiltonian system with Poisson operator $\mathcal{J} = -\partial_x$ and Hamiltonian $\mathcal{H} = \int u^2/2dx$. Does the direct application of the box scheme to the linear wave equation lead to a "classical" symplectic method?

5. *KdV and Schrödinger equation, mixed multi-symplectic discretization.* Besides the two box schemes discussed in this chapter so far, another useful class of multi-symplectic methods can be obtained by applying the symplectic Euler scheme (12.43) in space and the implicit midpoint method in time.

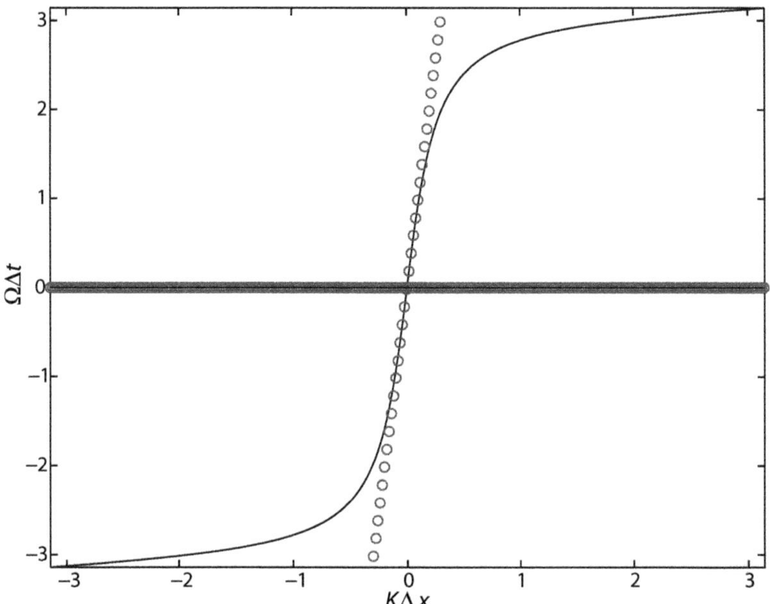

Figure 12.5 Numerical dispersion relation for Preissman box scheme applied to (12.57) and a Courant number $c = \Delta t/\Delta x = 10$. We also plot the exact dispersion relation ('o') for comparison.

a. What scheme do you obtain when applying this discretization to the multi-symplectic KdV formulation (12.38)–(12.41)? Show that, in terms of the variable u, the scheme is equivalent to

$$\partial_t^+ u_{i+1/2}^n + \frac{1}{2}\partial_x^+ (u_i^{n+1/2})^2 + \partial_x^+ \partial_x^- \partial_x^+ u_i^{n+1/2} = 0.$$

b. Discuss the dispersion relation of the linear KdV scheme

$$\partial_t^+ u_{i+1/2}^n + \partial_x^+ u_i^{n+1/2} + \partial_x^+ \partial_x^- \partial_x^+ u_i^{n+1/2} = 0.$$

c. Apply the same space-time discretization to the nonlinear Schrödinger equations (12.55)–(12.56). Compare the scheme with the Preissman box scheme discretization of the same equations in terms of conservation properties and computational complexity.

6. *Modified equation analysis.* Application of the Euler box scheme to a multi-symplectic PDE results in the discrete formula

$$\boldsymbol{K}_+ \partial_t^+ \boldsymbol{z}_i^n + \boldsymbol{K}_- \partial_t^- \boldsymbol{z}_i^n + \boldsymbol{L}_+ \partial_x^+ \boldsymbol{z}_i^n + \boldsymbol{L}_- \partial_x^- \boldsymbol{z}_i^n = \nabla_z S(\boldsymbol{z}_i^n).$$

Performing standard Taylor expansions, we find that

$$\partial_t^+ \boldsymbol{z}(x_i, t_n) = \boldsymbol{z}_t(x_i, t_n) + \frac{\Delta t}{2} \boldsymbol{z}_{tt}(x_i, t_n) + \mathcal{O}(\Delta t^2)$$

and similar terms for the other discrete derivative approximations.

a. Let us drop all terms of order $\mathcal{O}(\Delta t^2, \Delta x^2)$ and higher in the Taylor expansions for the discrete derivatives. Verify that, following standard local error analysis, the Euler box scheme becomes formally equivalent to the modified PDE

$$\boldsymbol{K} z_t + \frac{\Delta t}{2} \boldsymbol{A} z_{tt} + \boldsymbol{L} z_x + \frac{\Delta x}{2} \boldsymbol{B} z_{xx} = \nabla_z S(z), \tag{12.58}$$

where

$$\boldsymbol{A} = \boldsymbol{K}_+ - \boldsymbol{K}_-, \qquad \boldsymbol{B} = \boldsymbol{L}_+ - \boldsymbol{L}_-$$

are both symmetric matrices.

b. Write (12.58) as a multi-symplectic PDE over some enlarged phase space.

c. Use the multi-symplectic formulation, derived under (b), to find modified energy and momentum conservation laws. Note that these conservation laws can also be obtained by direct manipulation of the modified equation (12.58).

d. Show that (12.58) is the Euler–Lagrange equation arising from a Lagrangian functional $\mathcal{L}$ with density $L(z, z_t, z_x, z_{tt}, z_{xx})$. Find the specific form of L.

e. The modified conservation laws, derived under (c), are satisfied by the Euler box scheme to second-order accuracy. What needs to be done to derive modified conservation laws that are satisfied to third-order by the Euler box scheme?

See [137, 138] for more details.

References

[1] M.B. Abbott and D.R. Basco. *Computational Fluid Dynamics*. Harlow, Essex: Longman Scientific & Technical, 1989.

[2] R. Abraham and J.E. Marsden. *Foundations of Mechanics*. Reading, MA: Benjamin Cummings Pub. Co. 2nd edition, 1978.

[3] B.J. Alder and T.E. Wainwright. Hard-sphere molecular dynamics. *J. Chem. Phys.*, **27**:1208–1209, 1957.

[4] M.P. Allen and D.J. Tildesley. *Computer Simulation of Liquids*. Oxford: Clarendon Press, 1987.

[5] H.C. Andersen. Rattle: a "velocity" version of the Shake algorithm for molecular dynamics calculations. *J. Comput. Phys.*, **52**:24–34, 1983.

[6] D.G. Andrews. *An Introduction to Atmospheric Physics*. Cambridge: Cambridge University Press, 2000.

[7] V.I. Arnold. *Mathematical Methods of Classical Mechanics*. New York: Springer-Verlag, 2nd edition, 1989.

[8] V.I. Arnold, V.V. Kozlov, and A.I. Neishtadt. *Mathematical Aspects of Classical and Celestial Mechanics*. Berlin Heidelberg: Springer-Verlag, 2nd edition, 1993.

[9] U. Ascher and R.I. McLachlan. Multisymplectic box schemes and the Korteweg-de Vries equation. *Appl. Numer. Math.*, **48**:255–269, 2004.

[10] U.M. Ascher, H. Chin, and S. Reich. Stablization of DAEs and invariant manifolds. *Numer. Math.*, **67**:131–149, 1994.

[11] K. Atkinson. *Elementary Numerical Analysis*. New York: John Wiley & Sons, 1985.

[12] S.P. Auerbach and A. Friedman. Long-time behavior of numerically computed orbits: small and intermediate timestep analysis of one-dimensional systems. *J. Comput. Phys.*, **93**:189–223, 1991.

[13] E. Barth, K. Kuczera, B. Leimkuhler, and R.D. Skeel. Algorithms for constrained molecular dynamics. *J. Comput. Chem.*, **16**:1192–1209, 1995.

[14] E. Barth, B. Laird, and B. Leimkuhler. Generating generalized distributions from dynamical simulation. *J. Chem. Phys.*, **118**:5759–5768, 2003.

[15] E. Barth and B. Leimkuhler. Symplectic methods for conservative multibody systems. In J.E. Mardsen, G.W. Patrick, and W.F. Shadwick, editors, *Integration Algorithms and Classical Mechanics*, volume 10 of *Fields Inst. Comm.*, pages 25–44. Amer. Math. Soc., 1996.

[16] G. Benettin and A. Giorgilli. On the Hamiltonian interpolation of near to the identity symplectic mappings with application to symplectic integration algorithms. *J. Stat. Phys.*, **74**:1117–1143, 1994.

[17] G. Benettin, A. M. Cherubini, and F. Fassò. A changing charts symplectic algorithm for rigid bodies and other dynamical systems on manifolds. *SIAM J. Sci. Comput.*, **23**:1189–1203, 2001.

[18] G. Benettin, L. Galgani, and A. Giorgilli. Realization of holonomic constraints and freezing of high frequency degrees of freedom in the light of classical perturbation theory. Part I. *Comm. Math. Phys.*, **113**:87–103, 1987.

[19] F. Benvenuto, G. Casati, and D. Shepelyansky. Dynamical localization: Hydrogen atom in magnetic and microwave fields. *Phys. Rev. A*, **55**:1732–1735, 1997.

[20] J.J. Biesiadecki and R.D. Skeel. Dangers of multiple-time-step methods. *J. Comput. Phys.*, **109**:318–328, 1993.

[21] C.K. Birdsall and A.B. Langdon. *Plasma Physics via Computer Simulations*. New York: McGraw-Hill, 1981.

[22] S. Blanes. High order numerical integrators for differential equations using composition and processing of low order methods. *Appl. Numer. Math.*, **37**:289–306, 2001.

[23] S. Blanes and P.C. Moan. Practical symplectic partitioned Runge–Kutta methods and Runge–Kutta–Nyström methods. *J. Comput. Appl. Math*, **142**:313–330, 2002.

[24] S. Bond and B. Leimkuhler. Time transformations for reversible variable stepsize integration. *Numer. Algorithms*, **19**:55–71, 1998.

[25] S.D. Bond. *Numerical methods for extended Hamiltonian systems with applications in statistical mechanics*. Ph.D. thesis, Department of Mathematics, University of Kansas, Lawrence, KS 66045, 2000.

[26] S.D. Bond, B.J. Leimkuhler, and B.B. Laird. The Nosé–Poincaré method for constant temperature molecular dynamics. *J. Comput. Phys.*, **151**:114–134, 1999.

[27] F.A. Bornemann. *Homogenization in Time of Singularly Perturbed Conservative Mechanical Systems*, volume 1687 of *Lecture Notes in Mathematics*. Berlin Heidelberg: Springer-Verlag, 1998.

[28] F.A. Bornemann and Ch. Schütte. Homogenization of Hamiltonian systems with a strong constraining potential. *Physica D*, **102**:57–77, 1997.

[29] R.W. Brankin, I. Gladwell, and L.F. Shampine. RKSUITE: a suite of Runge–Kutta codes for the initial value problem of ODEs. Technical Report Softreport 92-S1, Department of Mathematics, Southern Methodist University, Dallas, TX, 1992.

[30] T.J. Bridges. A geometric formulation of the conservation of wave action and its implications for signature and the classification of instabilities. *Proc. R. Soc. Lond. A*, **453**(A):1365–1395, 1997.

[31] T.J. Bridges. Multi-symplectic structures and wave propagation. *Math. Proc. Cambridge*, **121**:147–190, 1997.

[32] T.J. Bridges, P. Hydon, and S. Reich. Vorticity and symplecticity in Lagrangian fluid dynamics. Technical Report, Department of Mathematics, University of Surrey, 2002.

[33] T.J. Bridges and S. Reich. Multi-symplectic integrators: numerical schemes for Hamiltonian PDEs that conserve symplecticity. *Phys. Lett. A*, **284**:184–193, 2001.

[34] B. Brooks, R. Bruccoleri, B. Olafson, D. States, S. Swaminathan, and M. Karplus. CHARMm: a program for macromolecular energy, minimization, and dynamics calculations. *J. Comput. Chem.*, **4**:187–217, 1983.

[35] R.C. Buck. *Advanced Calculus.* New York: McGraw-Hill, 3rd edition, 1978.

[36] C. Budd, B. Leimkuhler, and M. Piggott. Scaling invariance and adaptivity. *Appl. Numer. Math.*, **39**:261–288, 2001.

[37] R. Burden and J. Faires. *Numerical Analysis.* Brooks Cole Publishing, 7th edition, 2001.

[38] S.R. Buss. Accurate and efficient simulation of rigid body rotations. *J. Comput. Phys.*, **164**:377–406, 2000.

[39] J.C. Butcher. *Numerical Methods for Ordinary Differential Equations.* New York: John Wiley & Sons, 2003.

[40] M.P. Calvo, A. Murua, and J.M. Sanz-Serna. Modified equations for ODEs. *Contemp. Math.*, **172**:63–74, 1994.

[41] M.P. Calvo and A. Portillo. Are higher order equistage initializers better than standard starting algorithms? *J. Comput. Appl. Math*, **169**:333–344, 2004.

[42] M.P. Calvo and J.M. Sanz-Serna. Variable steps for symplectic integrators. In *Dundee Conference on Numerial Analysis 1991*, volume 260 of *Pitman Res. Notes Math. Ser.*, pages 34–48. Pitman, London, 1992.

[43] P.J. Channel. Symplectic integration algorithms. Technical Report AT-6:ATN 83–9, Los Alamos National Laboratory, 1983.

[44] P.J. Channel and J.S. Scovel. Integrators for Lie–Poisson dynamical systems. *Physica D*, **50**:80–88, 1991.

[45] S. Cirilli, E. Hairer, and B. Leimkuhler. Asymptotic error analysis of the Adaptive Verlet method. *BIT*, **39**:25–33, 1999.

[46] G.J. Cooper. Stability of Runge–Kutta methods for trajectory problems. *IMA J. Numer. Anal.*, **7**:1–13, 1987.

[47] C.J. Cotter and S. Reich. Adiabatic invariance and applications to molecular dynamics and numerical weather prediction. *BIT*, to appear.

[48] G. Dahlquist and A. Björk. *Numerical Methods.* Englewood Cliffs, NJ: Prentice-Hall, 1974.

[49] R. de Vogelaere. Methods of integration which preserve the contact transformation property of the Hamiltonian equations. Technical Report 4, Dept. Math. Univ. of Notre Dame, 1956.

[50] P.A.M. Dirac. Lectures on Quantum Mechanics. Technical Report 3, Belfer Graduate School Monographs, Yeshiva University, 1964.

[51] A.J. Dragt and J.M. Finn. Lie series and invariant functions for analytic symplectic maps. *J. Math. Phys.*, **17**:2215–2227, 1976.

[52] A.J. Dragt, F. Neri, G. Rangarajan, D.R. Douglas, L.M. Healy, and R.D. Ryne. Lie algebraic treatment of linear and nonlinear beam dynamics. *Ann. Rev. Nucl. Part. Sci.*, **38**:455–496, 1988.

[53] P.G. Drazin. *Solitons*, volume 85 of *London Mathematical Society Lecture Note Series*. Cambridge: Cambridge University Press, 1983.

[54] A. Dullweber, B. Leimkuhler, and R.I. McLachlan. Split-Hamiltonian methods for rigid body molecular dynamics. *J. Chem. Phys.*, **107**:5840–5851, 1997.

[55] D.R. Durran. *Numerical Methods for Wave Equations in Geophysical Fluid Dynamics*. Berlin: Springer-Verlag, 1998.

[56] F. Fassò. Comparison of splitting algorithms for the rigid body. *J. Comput. Phys.*, **189**:527–538, 2003.

[57] K. Feng. On difference schemes and symplectic geometry. In K. Feng, editor, *Proceedings of the 1984 Beijing Symposium on Differential Geometry and Differential Equations*, pages 42–58, Beijing: Science Press, 1985.

[58] K. Feng. Canonical difference schemes for Hamiltonian canonical differential equations. In *International Workshop on Applied Differential Equations (Beijing, 1985)*, pages 59–73, Singapore: World Sci. Publishing, 1986.

[59] K. Feng. Formal power series and numerical algorithms for dynamical systems. In T. Chan and Chong-Ci Shi, editors, *Proceedings of International Conference on Scientific Computation (Hangzhao, China)*, volume 1 of *Series on Appl. Math.*, pages 28–35, 1991.

[60] E. Fermi, J. Pasta, and S. Ulam. Studies of nonlinear problems-I. Technical Report LA-1940, Los Alamos National Laboratory, Los Alamos, NM, 1955.

[61] M. Fixman. Classical statistical mechanics of constraints: a theorem and applications to polymers. *Proc. Nat. Acad. Sci.*, **71**:3050–3053, 1974.

[62] E. Forest and R.D. Ruth. Fourth order symplectic integration. *Physica D*, **43**:105–117, 1990.

[63] J. Frank, G. Gottwald, and S. Reich. The Hamiltonian particle-mesh method. In M. Griebel and M.A. Schweitzer, editors, *Meshfree Methods for Partial Differential Equations*, volume 26 of *Lect. Notes Comput. Sci. Eng.*, pages 131–142, Berlin Heidelberg: Springer-Verlag, 2002.

[64] J. Frank and S. Reich. Conservation properties of Smoothed Particle Hydrodynamics applied to the shallow-water equations. *BIT*, **43**:40–54, 2003.

[65] J. Frank and S. Reich. The Hamiltonian particle-mesh method for the spherical shallow water equations. *Atmos. Sci. Lett.*, **5**:89–95, 2004.

[66] D. Frenkel and B. Smit. *Understanding Molecular Simulation: From Algorithms to Applications*. New York: Academic Press, 2nd edition, 2002.

[67] B. García-Archilla, J.M. Sanz-Serna, and R.D. Skeel. The mollified impulse method for oscillatory differential equations. *SIAM J. Sci. Comput.*, **20**:930–963, 1998.

[68] Z. Ge. Equivariant symplectic difference schemes and generating functions. *Physica D*, **49**:376–386, 1991.

[69] Z. Ge and J. Marsden. Lie–Poisson integrators and Lie–Poisson Hamiltonian–Jacobi theory. *Phys. Lett. A*, **133**:134–139, 1988.

[70] C.W. Gear. *Numerical Initial Value Problems in Ordinary Differential Equations*. Englewood Cliffs, NJ: Prentice-Hall, 1971.

[71] R.A. Gingold and J.J. Monaghan. Smoothed Particle Hydrodynamics: theory and application to non-spherical stars. *Mon. Not. R. Astr. Soc.*, **181**:375–389, 1977.

[72] B. Gladman, M. Duncan, and J. Candy. Symplectic integrators for long-term integration in celestial mechanics. *Celest. Mech.*, **52**:221–240, 1991.

[73] H. Goldstein, C. Pole, and J. Safko. *Classical Mechanics*. San Francisco, CA: Addison Wesley, 3nd edition, 2002.

[74] O. Gonzalez, D.J. Higham, and A.M. Stuart. Qualitative properties of modified equations. *IMA J. Numer. Anal.*, **19**:169–190, 1999.

[75] H. Grubmüller, H. Heller, A. Windemuth, and K. Schulten. Generalized Verlet algorithm for efficient molecular dynamics simulations with long-range interactions. *Mol. Sim.*, **6**:121–142, 1991.

[76] J. Guckenheimer and P. Holmes. *Nonlinear Oscillations, Dynamical Systems and Bifurcations of Vector Fields*. New York: Springer-Verlag, 1983.

[77] E. Hairer. Backward analysis of numerical integrators and symplectic methods. *Annals of Numer. Math.*, **1**:107–132, 1994.

[78] E. Hairer. Variable time step integration with symplectic methods. *Appl. Numer. Math.*, **25**:219–227, 1997.

[79] E. Hairer and Ch. Lubich. The life-span of backward error analysis for numerical integrators. *Numer. Math.*, **76**:441–462, 1997.

[80] E. Hairer, Ch. Lubich, and G. Wanner. *Geometric Numerical Integration*. Berlin Heidelberg: Springer-Verlag, 2002.

[81] E. Hairer, Ch. Lubich, and G. Wanner. Geometric numerical integration illustrated by the Störmer–Verlet method. *Acta Numerica*, **12**:399–450, 2003.

[82] E. Hairer, S.P. Nørsett, and G. Wanner. *Solving Ordinary Differential Equations I, Nonstiff Problems*. Berlin Heidelberg: Springer-Verlag, 2nd edition, 1993.

[83] E. Hairer and D. Stoffer. Reversible long-term integration with variable step sizes. *SIAM J. Sci. Comput.*, **18**:257–269, 1997.

[84] E. Hairer and G. Wanner. *Solving Ordinary Differential Equations II, Stiff and Differential-Algebraic Problems*. Berlin Heidelberg: Springer-Verlag, 2nd edition, 1996.

[85] R.W. Hockney and J.W. Eastwood. *Computer Simulations Using Particles*. Bristol and Philadelphia: Institute of Physics Publisher, 1988.

[86] T. Holder, B. Leimkuhler, and S. Reich. Explicit, time-reversible and variable stepsize integration. *Appl. Numer. Math.*, **39**:367–377, 2001.

[87] W.G. Hoover. Canonical dynamics: Equilibrium phase-space distributions. *Phys. Rev. A*, **31**:1695–1697, 1985.

[88] Y.A. Houndonougbo, B.B. Laird, and B.J. Leimkuhler. Molecular dynamics algorithms for mixed hard-core/continuous potentials. *Mol. Phys.*, **98**:309–316, 1999.

[89] W. Huang and B. Leimkuhler. The adaptive Verlet method. *SIAM J. Sci. Comput.*, **18**:239–256, 1997.

[90] P. Hut, J. Makino, and S. McMillan. Building a better leapfrog. *Astrophys. J. Lett.*, **443**:L93–L96, 1995.

[91] A. Iserles. *A First Course in the Numerical Analysis of Differential Equations.* Cambridge: Cambridge University Press, 1996.

[92] A.L. Islas, D.A. Karpeev, and C.M. Schober. Geometric integrators for the nonlinear Schrödinger equation. *J. Comput. Phys.*, **173**:116–148, 2001.

[93] J.A. Izaguirre, S. Reich, and R.D. Skeel. Longer time steps for molecular dynamics. *J. Chem. Phys.*, **110**:9853–9864, 1999.

[94] S. Jang and G.A. Madden. Simple reversible molecular dynamics algorithms for Nosé–Hoover chain dynamics. *J. Chem. Phys.*, **107**:9514–9526, 1997.

[95] L.O. Jay. Runge–Kutta type methods for index three differential-algebraic equations with applications to Hamiltonian systems. Ph.D. thesis, Department of Mathematics, University of Geneva, Switzerland, 1994.

[96] S. Jiménez. Derivation of the discrete conservation laws for a family of finite difference schemes. *Appl. Math. Comput.*, **64**:13–45, 1994.

[97] W. Kahan and R.C. Li. Composition constants for raising the order of unconventional schemes for ordinary differential equations. *Math. Comp.*, **66**:1089–1099, 1997.

[98] A. Kol, B. Laird, and B. Leimkuhler. A symplectic method for rigid-body molecular simulation. *J. Chem. Phys.*, **107**:2580–2588, 1997.

[99] F. Krogh. Issues in the design of a multistep code. *Annals of Numer. Math.*, **1**:423–437, 1994.

[100] S. Kuksin and J. Pöschel. On the inclusion of analytic symplectic maps in analytic Hamiltonian flows and its applications. In S. Kuksin, V. Lazutkin, and J. Pöschel, editors, *Seminar on Dynamical Systems (St. Petersburg, 1991)*, volume 12 of *Progr. Nonlinear Differential Equations Appl.*, pages 96–116, Basel: Birkhäuser Verlag, 1994.

[101] A. Kværnø and B. Leimkuhler. A time-reversible, regularized, switching integrator for the n-body problem. *SIAM J. Sci. Comput.*, **22**:1016–1035, 2001.

[102] B. Laird and B. Leimkuhler. Generalized dynamical thermostatting technique. *Phys. Rev. E*, **68**:art. no. 016704, 2003.

[103] P. Lancaster. *Theory of Matrices*. New York: Academic Press, 1969.

[104] C. Lanczos. *The Variational Principles of Mechanics*. Toronto: University of Toronto Press, 1949.

[105] L.D. Landau and E.M. Lifshitz. *Course in Theoretical Physics, Vol. 1: Mechanics*. Oxford: Pergamon Press, 1976.

[106] F.M. Lasagni. Canonical Runge–Kutta methods. *Z. Angew. Math. Phys.*, **39**:952–953, 1988.

[107] F.M. Lasagni. Integration methods for Hamiltonian differential equations. Unpublished manuscript, 1990.

[108] E. Lee, A. Brunello, and D. Farrelly. Coherent states in a Rydberg atom: classical mechanics. *Phys. Rev. A*, **55**:2203–2221, 1997.

[109] B. Leimkuhler. Reversible adaptive regularization: perturbed Kepler motion and classical atomic trajectories. *Philos. T. Roy. Soc. A*, **357**:1101–1133, 1999.

[110] B. Leimkuhler. A separated form of Nosé dynamics for constant temperature and pressure simulation. *Comput. Phys. Comm.*, **148**:206–213, 2002.

[111] B. Leimkuhler and S. Reich. Symplectic integration of constrained Hamiltonian systems. *Math. Comp.*, **63**:589–606, 1994.

[112] B. Leimkuhler and S. Reich. A reversible averaging integrator for multiple time-scale dynamics. *J. Comput. Phys.*, **171**:95–114, 2001.

[113] B. Leimkuhler and R.D. Skeel. Symplectic numerical integrators in constrained Hamiltonian systems. *J. Comput. Phys.*, **112**:117–125, 1994.

[113a] B. Leimkuhler and C. Sweet. The canonical ensemble via symplectic integrators using é and é-Poincaré chains. *J. Chem. Phys.* **121**:106–116, 2004a.

[113b] B. Leimkuhler and C. Sweet. A Hamiltonian formulation for recursive multiple thermostats in a common timescale. *SIAM J. Appl. Dyn. Syst.*, in press.

[114] D. Lewis and J.C. Simo. Conserving algorithms for the n-dimensional rigid body. In J.E. Mardsen, G.W. Patrick, and W.F. Shadwick, editors, *Integration Algorithms and Classical Mechanics*, volume 10 of *Fields Inst. Comm.*, pages 121–140. Amer. Math. Soc., 1996.

[115] S. Lie and L. Vu-Quoc. Finite difference calculus invariant structure of a class of algorithms for the nonlinear Klein–Gordon equation. *SIAM J. Numer. Anal.*, **32**:1839–1875, 1995.

[116] L. Lochak and A.I. Neishtadt. Estimates of stability time for nearly integrable systems with quasiconvex Hamiltonian. *Chaos*, **2**:495–499, 1992.

[117] M. López-Marcos, J.M. Sanz-Serna, and R.D. Skeel. Cheap enhancement of symplectic integrators. In D.F. Griffiths and G.A. Watson, editors, *Proceedings, 1995 Dundee Conference on Numerical Analysis*, pages 107–122, Harlow, Essex: Longman Scientific & Technical, 1996.

[118] M. López-Marcos, J.M. Sanz-Serna, and R.D. Skeel. Explicit symplectic integrators using Hessian-vector products. *SIAM J. Sci. Comput.*, **18**:223–238, 1997.

[119] E.N. Lorenz. Deterministic nonperiodic flow. *J. Atmos. Sci.*, **20**:130–141, 1963.

[120] L.B. Lucy. A numerical approach to the testing of the fission hypothesis. *Astron. J.*, **82**:1013–1024, 1977.

[121] J.B. Marion and S.T. Thornton. *Classical Dynamics of Particles and Systems.* Ft. Worth, TX: Saunders, 1995.

[122] J.E. Marsden. *Lectures on Mechanics.* Cambridge: Cambridge University Press, 1992.

[123] J.E. Marsden, G.P. Patrick, and S. Shkoller. Multi-symplectic geometry, variational integrators, and nonlinear PDEs. *Comm. Math. Phys.*, **199**:351–395, 1999.

[124] J.E. Marsden and T. Ratiu. *Mechanics and Symmetry.* New York: Springer-Verlag, 2nd edition, 1998.

[125] J.E. Marsden and M. West. Disrete mechanics and variational integrators. *Acta Numerica*, **10**:357–514, 2001.

[126] G.J. Martyna, M.E. Tuckerman, D.J. Tobias, and M.L. Klein. Explicit reversible integrators for extended systems dynamics. *Mol. Phys.*, **87**:1117–1157, 1996.

[127] R. McLachlan and A. Zanna. The discrete Moser–Veselov algorithm for the free rigid body, revisited. Technical Report 255, Informatics, University of Bergen, Bergen, Norway, 2003.

[128] R.I. McLachlan. Explicit Lie-Poisson integration and the Euler equations. *Phys. Rev. Lett.*, **71**:3043–3046, 1993.

[129] R.I. McLachlan. Symplectic integration of Hamiltonian wave equations. *Numer. Math.*, **66**:465–492, 1994.

[130] R.I. McLachlan. On the numerical integration of ordinary differential equations by symmetric composition methods. *SIAM J. Sci. Comput.*, **16**:151–168, 1995.

[131] R.I. McLachlan. More on symplectic correctors. In J.E. Mardsen, G.W. Patrick, and W.F. Shadwick, editors, *Integration Algorithms and Classical Mechanics*, volume 10 of *Fields Inst. Comm.*, pages 141–149. Amer. Math. Soc., 1996.

[132] R.I. McLachlan and G.R.W. Quispel. Splitting methods. *Acta Numerica*, **11**:341–434, 2002.

[133] R.I. McLachlan and C. Scovel. Equivariant constrained symplectic integration. *J. Nonlinear Sci.*, **5**:233–256, 1995.

[134] S. Mikkola and S. Aarseth. A time-transformed leapfrog scheme. *Celestial Mechanics and Dynamical Astronomy*, **84**:343–354, 2002.

[135] P.C. Moan. On the KAM and Nekhoroshev theorems for symplectic integrators and implications for error growth. *Nonlinearity*, **17**:67–83, 2003.

[136] B.E. Moore. *A Modified Equations Approach for Multi-Symplectic Integration Methods*. Ph.D. thesis, Department of Mathematics, University of Surrey, 2003.

[137] B.E. Moore and S. Reich. Backward error analysis for multi-symplectic integration methods. *Numer. Math.*, **95**:625–652, 2003.

[138] B.E. Moore and S. Reich. Multi-symplectic integration methods for Hamiltonian PDEs. *Future Gener. Comp. Sys.*, **19**:395–402, 2003.

[139] P.J. Morrison. Hamiltonian description of the ideal fluid. *Rev. Modern Phys.*, **70**:467–521, 1998.

[140] K.W. Morton and D.F. Mayers. *Numerical Solution of Partial Differential Equations.* Cambridge: Cambridge University Press, 1994.

[141] J. Moser and A.P. Veselov. Discrete version of some classical integrable systems and factorization of matrix polynomials. *Comm. Math. Phys.*, **139**:217–243, 1991.

[142] A. Murua and J.M. Sanz-Serna. Order conditions for numerical integrators obtained by composing simpler integrators. *Proc. R. Soc. Lond. A*, **357**(A):1079–1100, 1999.

[143] A.I. Neishtadt. The separation of motions in systems with rapidly rotating phase. *J. Appl. Math. Mech.*, **48**:133–139, 1984.

[144] N.N. Nekhoroshev. An exponential estimate of the time of stability of nearly integrable Hamiltonian systems. *Russ. Math. Surveys*, **32**:1–65, 1977.

[145] F. Neri. Lie algebras and canonical integration. Technical Report, Department of Physics, University of Maryland, 1987.

[146] S. Nosé. A molecular dynamics method for simulations in the canonical ensemble. *Mol. Phys.*, **52**:255–268, 1984.

[147] S. Nosé. An improved symplectic integrator for Nosé–Poincaré thermostat. *J. Phys. Soc. Jpn.*, **70**:75–77, 2001.

[148] D. Okunbor and R.D. Skeel. Explicit canonical methods for Hamiltonian systems. *Math. Comp.*, **59**:439–455, 1992.

[149] P.J. Olver. *Applications of Lie Groups to Differential Equations.* New York: Springer-Verlag, 1986.

[150] J. Ortega and W. Rheinboldt. *Iterative Solution of Nonlinear Equations in Several Variables.* New York: Academic Press, 1970.

[151] A. Rahman and F.H. Stillinger. Molecular dynamics study of liquid water. *J. Chem. Phys.*, **55**:3336–3359, 1971.

[152] D.C. Rappaport. *The Art of Molecular Dynamics Simulation*. New York: Cambridge University Press, 1995.

[153] S. Reich. Numerical integration of the generalized Euler equation. Technical Report TR 93-20, University of British Columbia, 1993.

[154] S. Reich. Momentum conserving symplectic integrators. *Physica D*, **76**:375–383, 1994.

[155] S. Reich. Symplectic integration of constrained Hamiltonian systems by composition methods. *SIAM J. Numer. Anal.*, **33**:475–491, 1996.

[156] S. Reich. Symplectic methods for conservative multibody systems. In J.E. Mardsen, G.W. Patrick, and W.F. Shadwick, editors, *Integration Algorithms and Classical Mechanics*, volume 10 of *Fields Inst. Com.*, pages 181–192. Amer. Math. Soc., 1996.

[157] S. Reich. On higher-order semi-explicit symplectic partitioned Runge–Kutta methods for constrained Hamiltonian systems. *Numer. Math.*, **76**:249–263, 1997.

[158] S. Reich. Backward error analysis for numerical integrators. *SIAM J. Numer. Anal.*, **36**:475–491, 1999.

[159] S. Reich. Conservation of adiabatic invariants under symplectic discretization. *Appl. Numer. Math.*, **29**:45–55, 1999.

[160] S. Reich. Multiple times-scales in classical and quantum-classical molecular dynamics. *J. Comput. Phys.*, **151**:49–73, 1999.

[161] S. Reich. Finite volume methods for multi-symplectic PDEs. *BIT*, **40**:559–582, 2000.

[162] S. Reich. Multi-symplectic Runge–Kutta collocation methods for Hamiltonian wave equations. *J. Comput. Phys.*, **157**:473–499, 2000.

[163] S. Reich. Smoothed Langevin dynamics of highly oscillatory systems. *Physica D*, **138**:210–224, 2000.

[164] E.J. Routh. *Dynamics of a System of Rigid Bodies, Elementary Part*. New York: Dover, 7th edition, 1960.

[165] H. Rubin and P. Ungar. Motion under a strong constraining force. *Comm. Pure and Appl. Math.*, **10**:65–87, 1957.

[166] R.D. Ruth. A canonical integration technique. *IEEE Trans. Nucl. Sci.*, **30**:2669–2671, 1983.

[167] J.P. Ryckaert, G. Ciccotti, and H.J.C. Berendsen. Numerical integration of the cartesian equations of motion of a system with constraints: molecular dynamics of n-alkanes. *J. Comput. Phys.*, **23**:327–341, 1977.

[168] R. Salmon. Practical use of Hamilton's principle. *J. Fluid Mech.*, **132**:431–444, 1983.

[169] R. Salmon. *Lectures on Geophysical Fluid Dynamics*. Oxford: Oxford University Press, 1999.

[170] J.M. Sanz-Serna. Runge–Kutta schemes for Hamiltonian systems. *BIT*, **28**:877–883, 1988.

[171] J.M. Sanz-Serna. Symplectic integrators for Hamiltonian problems: an overview. *Acta Numerica*, **1**:243–286, 1992.

[172] J.M. Sanz-Serna and M.P. Calvo. *Numerical Hamiltonian Problems*. London: Chapman & Hall, 1994.

[173] T. Sauer and J.A. York. Rigorous verification of trajectories for the computer simulation of dynamical systems. *Nonlinearity*, **4**:961–979, 1994.

[174] T. Schlick. *Molecular Modeling and Simulation*. New York: Springer-Verlag, 2002.

[175] M. Shimada and H. Yoshida. Long-term conservation of adiabatic invariants by using symplectic integrators. *Publ. Astron. Soc. Japan*, **48**:147–155, 1996.

[176] R.D. Skeel and C.W. Gear. Does variable step size ruin a symplectic integrator? *Physica D*, **60**:311–313, 1992.

[177] R.D. Skeel and D.J. Hardy. Practical construciton of modified Hamiltonians. *SIAM J. Sci. Comput.*, **23**:1172–1188, 2001.

[178] D. Stoffer. On reversible and canonical integration methods. Technical Report SAM-Report No. 88-05, ETH Zürich, 1988.

[179] D. Stoffer. Variable steps for reversible integration methods. *Computing*, **55**:1–22, 1995.

[180] G. Strang. On the construction and comparison of difference schemes. *SIAM J. Numer. Anal.*, **5**:506–517, 1968.

[181] R.M. Stratt, S.L. Holmgren, and D. Chandler. Constrained impulsive molecular dynamics. *Mol. Phys.*, **42**:1233–1243, 1981.

[182] J. Sturgeon and B. Laird. Symplectic algorithm for constant-pressure molecular dynamics using a Nosé–Poincaré thermostat. *J. Chem. Phys.*, **112**:3474–3482, 2000.

[183] S.-H. Suh, M.-Y. Teran, H.S. White, and H.T. Davis. Molecular dynamics study of the primative model of 1–3 electrolyte solutions. *Chem. Phys.*, **142**:203–211, 1990.

[184] G. Sun. Symplectic Partitioned Runge–Kutta methods. *J. Comput. Math.*, **11**:365–372, 1993.

[185] Y.B. Suris. The canonicity of mappings generated by Runge–Kutta type methods when integrating the systems $\ddot{x} = -\frac{\partial U}{\partial x}$. *USSR Comput. Maths. Math. Phys.*, **29**:138–144, 1989.

[186] Y.B. Suris. Hamiltonian methods of Runge–Kutta type and their variational interpretation. *Math. Model.*, **2**:78–87, 1990. (In Russian).

[187] G.J. Sussman and J. Wisdom. Chaotic evolution of the solar system. *Science*, **257**:56–62, 1992.

[188] M. Suzuki. Fractal decomposition of exponential operators with applications to many-body theories and Monte-Carlo simulations. *Phys. Lett. A*, **146**:319–323, 1990.

[189] M. Suzuki. General theory of higher-order decompositions of exponential operators and symplectic integrators. *Phys. Lett. A*, **165**:387–395, 1992.

[190] F. Takens. Motion under the influence of a strong constraining force. In Z. Nitecki and C. Robinson, editors, *Lecture Notes in Mathematics*, volume 819, pages 425–445, Berlin: Springer, 1980.

[191] Y.F. Tang. Formal energy of a symplectic scheme for Hamiltonian systems and its applications. *Computers Math. Applic.*, **27**:31–39, 1994.

[192] M. Toda, R. Kubo, and N. Saitô. *Statistical Physics, Volume I.* New York: Springer-Verlag, 2nd edition, 1992.

[193] J. Touma and J. Wisdom. Lie–Poisson integrators for rigid body dynamics in the solar system. *Astron. J.*, **107**:1189–1202, 1994.

[194] H.F. Trotter. On the product of semi-groups of operators. *Proc. Am. Math. Soc.*, **10**:545–551, 1959.

[195] M. Tuckerman, B.J. Berne, and G.J. Martyna. Reversible multiple time scale molecular dynamics. *J. Chem. Phys.*, **97**(3):1990–2001, 1992.

[196] V.S. Varadarajan. *Lie Groups, Lie Algebras and Their Representations*. Englewood Cliffs, NJ: Prentice-Hall, 1974.

[197] F. Verhulst. *Nonlinear Differential Equations and Dynamical Systems*. Berlin Heidelberg: Springer-Verlag, 1989.

[198] L. Verlet. Computer experiments on classical fluids. I. Thermodynamical properties of Lennard–Jones molecules. *Phys. Rev.*, **159**:98–103, 1967.

[199] R.F. Warming and B.J. Hyett. The modified equation approach to the stability and accuracy analysis of finite difference methods. *J. Comput. Phys.*, **14**:159–179, 1974.

[200] J.D. Weeks, D. Chandler, and H.C. Anderson. Role of repulsive forces in determining the equilibrium structure of simple liquids. *J. Chem. Phys.*, **54**:5237–5247, 1971.

[201] G.B. Whitham. *Linear and Nonlinear Waves*. New York: Wiley-Interscience, 1974.

[202] J.H. Wilkinson. Error analysis of floating point computation. *Numer. Math.*, **2**:319–340, 1960.

[203] J. Wisdom and M. Holman. Symplectic maps for the N-body problem. *Astron. J.*, **102**:1528–1538, 1991.

[204] J. Wisdom, M. Holman, and J. Touma. Symplectic correctors. In J.E. Mardsen, G.W. Patrick, and W.F. Shadwick, editors, *Integration Algorithms and Classical Mechanics*, volume 10 of *Fields Inst. Com.*, pages 217–244. Amer. Math. Soc., 1996.

[205] H. Yoshida. Construction of higher order symplectic integrators. *Phys. Lett. A*, **150**:262–268, 1990.

[206] H. Yoshida. Recent progress in the theory and application of symplectic integrators. *Celest. Mech. and Dyn. Astro.*, **56**:27–43, 1993.

[207] L.-S. Young. Large deviations in dynamical systems. *Trans. Amer. Math. Soc.*, **318**:525–543, 1990.

[208] V. Zeitlin. Finite-mode analogs of 2D ideal hydordynamcis: coadjoint orbits and local canonical structure. *Physica D*, **49**:353–362, 1991.

[209] P.-F. Zhao and M.-Z. Qin. Multisymplectic geometry and multisymplectic Preissman scheme for the KdV equation. *J. Phys. A*, **33**:3613–3626, 2000.

Index

For EU product safety concerns, contact us at Calle de José Abascal, 56–1°, 28003 Madrid, Spain or eugpsr@cambridge.org.

www.ingramcontent.com/pod-product-compliance
Ingram Content Group UK Ltd.
Pitfield, Milton Keynes, MK11 3LW, UK
UKHW020905110726
473066UK00017B/54

* 9 7 8 0 5 2 1 7 7 2 9 0 7 *